Lecture Notes in Artificial Intelligence   5914

Edited by R. Goebel, J. Siekmann, and W. Wahlster

Subseries of Lecture Notes in Computer Science

T0238572

Dimitris Karagiannis   Zhi Jin (Eds.)

Knowledge Science, Engineering and Management

Third International Conference, KSEM 2009
Vienna, Austria, November 25-27, 2009
Proceedings

 Springer

Series Editors

Randy Goebel, University of Alberta, Edmonton, Canada
Jörg Siekmann, University of Saarland, Saarbrücken, Germany
Wolfgang Wahlster, DFKI and University of Saarland, Saarbrücken, Germany

Volume Editors

Dimitris Karagiannis
University of Vienna, Faculty of Computer Science
Institute for Knowledge and Business Engineering
Brünner Straße 72, 1210 Vienna, Austria
E-mail: dk@dke.univie.ac.at

Zhi Jin
Peking University
School of Electronic Engineering and Computer Science
No. 5 Yiheyuan Road, Beijing 100871, China
E-mail: zhi.jin.deng@gmail.com

CR Subject Classification (1998): I.2.6, I.2, H.2.8, H.3-5, F.2.2, K.3

LNCS Sublibrary: SL 7 – Artificial Intelligence

ISSN 0302-9743

ISBN 978-3-642-10487-9 Springer Berlin Heidelberg New York

springer.com

© Springer-Verlag Berlin Heidelberg 2009

Typesetting: Camera-ready by author, data conversion by Scientific Publishing Services, Chennai, India
Printed on acid-free paper   SPIN: 12800339   06/3180   5 4 3 2 1 0

Preface

Following two successful events in Guilin, People's Republic of China (KSEM 2006) and in Melbourne, Australia (KSEM 2007) the third event in this conference series was held for the first time in Europe, namely, in Vienna, Austria. KSEM 2009 aimed to be a communication platform and meeting ground for research on knowledge science, engineering and management, attracting high-quality, state-of-the-art publications from all over the world. It offers an exceptional opportunity for presenting original work, technological advances, practical problems and concerns of the research community.

The importance of studying "knowledge" from different viewpoints such as science, engineering and management has been widely acknowledged. The accelerating pace of the "Internet age" challenges organizations to compress communication and innovation cycles to achieve a faster return on investment for knowledge. Thus, next-generation business solutions must be focused on supporting the creation of value by adding knowledge-rich components as an integral part to the work process. Therefore, an integrated approach is needed, which combines issues from a large array of knowledge fields such as science, engineering and management.

Based on the reviews by the members of the Program Committee and the additional reviewers, 42 papers were selected for this year's conference. Additionally, two discussion panels dealing with "Knowware: The Third Star after Hardware and Software" and "Required Knowledge for Delivering Services" took place under the auspices of the conference. The papers and the discussions covered a great variety of approaches of knowledge science, management and engineering, thus making KSEM a unique conference.

A large scientific community was involved in setting up KSEM 2009. We would like to express our warm thanks to everybody who contributed to making it a success. First of all, this includes all the authors who submitted a paper to the review process, the members of the Program Committee and the additional reviewers who made such an effort to select the best papers and to ensure a high-quality program. Our thanks go to the Organizing Committee at the University of Vienna and the university itself for providing an excellent environment for the conference. Last but not least, we would like to thank the General Conference Chairs, Ruqian Lu from the Chinese Academy of Sciences and A Min Tjoa from the Vienna University of Technology, for their support.

November 2009

Zhi Jin
Dimitris Karagiannis

Conference Organization

Program Chairs

Zhi Jin
Dimitris Karagiannis

Program Committee

Klaus-Dieter Althoff
Nathalie Aussenac-Gilles
Philippe Besnard
Cungen Cao
Key-Sun Choi
James Delgrande
Xiaotie Deng
Andreas Dengel
Kevin Desouza
Juan Manuel Dodero
Brian Donnellan
Joaquim Filipe
Aldo Gangemi
Ulrich Geske
Lluis Godo
Yoshinori Hara
Remko Helms
Melanie Hilario
Knut Hinkelmann
Achim Hoffmann
Zhisheng Huang
Anthony Hunter
Takayuki Ito
Manfred Jeusfeld
Byeong Ho Kang
Gabriele Kern-Isberner
John Kidd
Jérome Lang
Weiru Liu
James Lu
Ronald Maier
Vladimir Marik
Simone Marinai

Pierre Marquis
John-Jules Meyer
Michele Missikof
Takeshi Morita
John Mylopoulos
Patricia Ordonez de Pablos
Ewa Orlowska
Maurice Pagnucco
Sven-Volker Rehm
Peter Reimann
Ulrich Reimer
Ulrich Remus
Bodo Rieger
Gerold Riempp
Martin Schrder
Heiner Stuckenschmidt
Kaile Su
A Min Tjoa
Mirek Truszczynski
Eric Tsui
Abel Usoro
Kewen Wang
Hui Wang
Ju Wang
Zhongtuo Wang
Herbert Weber
Rosina Weber
Mary-Anne Williams
Robert Woitsch
Takahira Yamaguchi
Jia-Huai You
Qingtian Zeng
Mingyi Zhang

Shichao Zhang
Chunxia Zhang

Zhi-Hua Zhou
Meiyun Zuo

External Reviewers

Hidenao Abe
Lina Al-Jadir
Ralf Biedert
Kang Chen
Taolue Chen
Naoki Fukuta
Gunnar Grimnes
Christophe Gueret
Alexandre Hanft
Caiyan Jia

Guohua Liu
Kedian Mu
Jun Sun
Nils Urbach
Hui Wang
Li Xiong
Shichao Zhang
Ludger van Elst
Chonghai Wang

Table of Contents

Models in Knowledge Management

John Mylopoulos

University of Trento

Knowledge Management (KM) consists of a set of concepts, techniques and tools for creating, representing, distributing and evolving knowledge within an organizational context. This knowledge may be about the domain (e.g., cars and the car market for an auto maker), the organization itself (e.g., its current state, objectives, plans, finances), but also about insights and experiences of its members. This knowledge may be explicit in documents or other artifacts, implicit in members of the organization, or embedded in organisational processes or practices. KM is an established discipline since the early '90s [Nonaka94] and is taught in the fields of business administration, information systems, management, and library and information sciences.

Models are used in KM to create useful abstractions that can be disseminated, analyzed and managed. We review the history of such models in Information Systems and Computer Science. We then present on-going work that proposes new concepts for capturing strategic knowledge within an organization. These concepts are used through an extension of UML class diagrams to build models that represent organizational vision, objectives, strategies and tactics. This work is inspired by the Business Motivation Model ([BMM]), an OMG standard intended to capture governance knowledge.

The presentation is based on on-going research with Daniel Amyot (University of Ottawa), Daniele Barone and Lei Jiang (University of Toronto).

References

[BMM] The Business Motivation Model, OMG Release,
   http://www.omg.org/spec/BMM/1.0/
[Nonaka94] Nonaka, I.: A Dynamic Theory of Organizational Knowledge Creation. Organization Science 5(1), 14–37 (1994)

D. Karagiannis and Z. Jin (Eds.): KSEM 2009, LNAI 5914, p. 1, 2009.
© Springer-Verlag Berlin Heidelberg 2009

Two Applications of Computer-Aided Theorem Discovery and Verification

Fangzhen Lin

Department of Computer Science
Hong Kong University of Science and Technology
Clear Water Bay, Hong Kong

In this talk I present two applications of computer-aided theorem discovery and verification. One is discovering classes of finite games that have unique pure Nash equilibria payoffs. The other is on using computer to verify the base case of Arrow's impossibility theorem in social choice theory.

D. Karagiannis and Z. Jin (Eds.): KSEM 2009, LNAI 5914, p. 2, 2009.
© Springer-Verlag Berlin Heidelberg 2009

Knowware: The Third Star after Hardware and Software

David Bell[1], Ying Jiang[2], Ruqian Lu[3], Kaile Su[4], and Songmao Zhang[3]

[1] School of Electronics, Electrical Engineering and Computer Science,
Queen's University Belfast, UK
da.bell@queens-belfast.ac.uk
[2] Institute of Software, Chinese Academy of Sciences, China
jy@ios.ac.cn
[3] Academy of Mathematics and System Sciences, Chinese Academy of Sciences, China
rqlu@math.ac.cn, smzhang@math.ac.cn
[4] School of Electronics Engineering and Computer Science, Peking University, China
sukl@pku.edu.cn

Ruqian Lu first presented the notion of *knowware* in 2005 through his IEEE Intelligent Systems paper entitled "From hardware to software to knowware: IT's third liberation?". He further elaborated the notion in the book "Kowware, the third star after hardware and software", published in 2007 by Polimetrica. Lu defines knowware as an independent and commercialized knowledge module that is computer operable, but free of any built-in control mechanism (in particular, not bund to any software), meeting some industrial standards and embeddable in software and/or hardware. Lu proposes to separate knowledge from software, and that knowware, software and hardware should be considered as three equally important underpinnings of IT industry.

This panel is to discuss knowware and its related topics in knowledge-based research. Basic issues about knowware will be introduced by Lu, including the architecture, representation and functions of knowware, the principles of knowware engineering and its three life cycle models (furnace model, crystallization model and spiral model), and techniques of software/knowware co-engineering. Particularly, mixware refers to a software component whose knowledge is replaced by knowware, and an object- and component-oriented development schema of mixware is introduced, including the tower model and ladder model. Other panelists will address issues related to knowware and of broad knowledge-related formalisms and systems. Particularly we will focus on the need for handling inconsistencies between and within knowwares, and for handling individual and collective knowledge, the need for transductive reasoning as well as inductive and deductive inference, the definition of knowledge middleware and its relation to some hot topics in artificial intelligence research such as belief revision, co-algebra as the potential formal semantics for specifying software/Knowware co-engineering, ontology-related topics including ontology alignment techniques and their application in domains such as biomedicine. The goal of the panel is to raise the awareness of knowware among people from both academy and industry, and to invite views and insights, towards making knowware a reality in the future.

D. Karagiannis and Z. Jin (Eds.): KSEM 2009, LNAI 5914, p. 3, 2009.
© Springer-Verlag Berlin Heidelberg 2009

Required Knowledge for Delivering Services

Brian Donnellan[1], Diem Ho[2], John Mylopoulos[3], Stefan Schambron[4],
and Hans-Georg Fill[5]

[1] National University of Ireland Maynooth, Ireland
brian.donnellan@nuim.ie
[2] IBM, France
diem_ho@fr.ibm.com
[3] University of Trento, Italy
jm@disi.unitn.it
[4] Together Internet Services GmbH, Austria
[5] Faculty of Computer Science, University of Vienna
hans-georg.fill@univie.ac.at

Despite constant growth of service economy there exist different approaches which mainly provide different perspectives toward the definition of service and have proposed service realization methods, technologies, productivity, quality measurement and innovation in services merely as areas for further research. One of the most important factors which influence this fact is the multidisciplinary nature of this field of study and its wide variety. While in some Industries service is observed as a component of the product-service bundle, in other it can be seen as an autonomous unit. Regardless of the context in which service is being provided and the discipline through which the service is being studied, knowledge is of the essence for using and providing services. However, the need for a multi-perspective analysis of "required knowledge for realizing services" is still immense.

This Panel discussion aims to confluence different attitudes of service scientists, service managers and service engineers toward service and will benefit from knowledge as the convergence point for all existing approaches.

Following questions will be discussed in this panel:

- How can science, management and engineering co-contribute to this subject?
- What are the commonalities and differences among the view points of different disciplines?
- Which potential research directions may target and solve this need?
- Which technologies may be exploited to realize services?
- Which courses should be planned in a curriculum for service science, management and engineering?
- Which role does modeling play in service realization?

D. Karagiannis and Z. Jin (Eds.): KSEM 2009, LNAI 5914, p. 4, 2009.
© Springer-Verlag Berlin Heidelberg 2009

Mapping Relational Databases to the Semantic Web with Original Meaning

Dmitry V. Levshin

NICEVT (Science and Research Center on Computer Technology)
Moscow, Russia
`levshin@nicevt.ru`

Abstract. Most of nowadays Web content is stored in relational databases. It is important to develop ways for representation of this information in the Semantic Web to allow software agents to process it intelligently. The paper presents an approach to translation of data, schema and the most important constraints from relational databases into the Semantic Web without any extensions to the Semantic Web languages.

1 Introduction

The Semantic Web is aimed in allowing computer agents to process intelligently information from different sources. To use semantic technologies, one needs to represent information in standardized machine readable data formats. The World Wide Web Consortium has recommended several formats for representing data in the Semantic Web including RDF, RDF Schema and OWL. The more data is available in the Semantic Web, the more powerful these technologies become.

Wikipedia [1] and databases [2] are good candidates for populating the Semantic Web, because they contain great amounts of information in a structured form. Moreover, today the majority of Web content (so-called Deep Web) is stored in relational databases. Ability to publish it in the Semantic Web is important not only for development of the latter, but also allows search engines to return more relevant Deep Web search results [3]. Mappings databases to OWL ontologies are also applicable in integrating distributed data sources [4].

Many efforts (e.g. [4,5]) are concerned on translation of data and schema from databases into OWL and mostly disregard relations implicitly given by primary and foreign key constraints. Several reasons not to loose this information during mappings are presented in [6]: it can be useful if an exported RDF graph is imported in a relational database at another place, or if updates on a materialized RDF graph have to be performed such that key and foreign key properties have to be checked. It also can be used for semantic query optimizations.

However, semantics of OWL does not allow it to represent integrity constraints including relational keys and it leads to development of different extensions to OWL [7,8]. At the same time, computational and conceptual complexity of the existing formats is relatively high. As consequence, practical applications often use very poor ontologies [9], and there are proposals to simplify OWL for better performance and scalability [10].

D. Karagiannis and Z. Jin (Eds.): KSEM 2009, LNAI 5914, pp. 5–16, 2009.

Therefore, it seems natural to use for mappings existing languages without any extensions. Two formats are useful in this case: SWRL is suggested as rule language and SPARQL is recommended as query language for the Semantic Web.

The main contribution of the paper can be summarized as follows:

- *an approach to mapping databases to the Semantic Web with the most important constraints is proposed;*
- *it uses the full strength of the existing formats without any extensions.*

Important aspect of the approach is usage of SPARQL. It is also proposed in [11] to use RDF and SPARQL for representing constraints. In contrast to [11], the approach presented in the paper does not propose any extensions to SPARQL and performs mappings in more natural way due to using SWRL.

The paper is organized as follows. Section 2 overviews basic notions of relational model and the Semantic Web. Section 3 presents our approach to mapping of relational databases into ontologies and verifying database-style constraints. Section 4 discusses some reasoning aspects involved in the approach. Section 5 overviews related work before Section 6 concludes.

2   Preliminaries

This Section briefly recalls basic notions of relational databases that are considered in the paper and then introduces the Semantic Web languages.

2.1   Relational Terminology

A *relation schema* Hr is a set of ordered pairs $\langle A, T \rangle$, where A is a name of an *attribute*, and T is it's domain. A *tuple* t corresponding to the schema Hr is a set of ordered triples $\langle A, T, v \rangle$, one triple for each attribute in Hr, where v is a value of the attribute (written $t.A = v$). Relation body Br is a set of tuples corresponding to Hr. A relation *instance* (*relational table*) is a pair $\langle Hr, Br \rangle$.

A *potential key* over a relation instance is a subset K of it's attributes such that the following holds:

$$\forall t_1, t_2 \in Br : t_1 \neq t_2 \longrightarrow \neg(\forall A \in K : t_1.A = t_2.A) \ , \qquad (1)$$

and there is no any $K' \subset K$ for which (1) holds too. Any attribute from K can have NULL value. Although $(a = \text{NULL})$ in general is computed to be *unknown* for each value a, NULL values are considered to be equal in (1). A *primary key* PK is a potential key such that no any attribute from PK can be NULL.

A *foreign key* over a relation instance R_1 is a subset FK of its attributes such that a relation instance R_2 with the potential key FK exists and the following condition holds:

$$\forall t_1 \in Br_1 : \begin{array}{l} (\forall A \in FK : t_1.A \text{ IS NULLL}) \ \lor \\ (\exists t_2 \in Br_2 : \forall A \in FK : t_1.A = t_2.A) \end{array} \ . \qquad (2)$$

In addition to primary and foreign keys constraints, CHECK-constraints can be specified on particular attributes, tables or on the whole database. A database is consistent, if all specified constraints hold.

2.2   Semantic Web Language Stack

The Semantic Web has stack architecture assuming that upper layers (languages) are based on underlying ones and extend them with new features.

First, *Resource Description Framework* (RDF) uses triples of the form *'subject-property-object'* with URIs to represent data. RDF documents can be presented in graph notation and, thus, are also called RDF graphs.

Next, *RDF Schema* and *Web Ontology Language* extends RDF with the ability to express statements about domain of interest. OWL species, OWL Lite and OWL DL[1], are based on Description Logics (DLs) [12]. DLs allow definition of relations between binary *roles* and unary *concepts*. DLs differ one from each other in that which constructs and in which way are allowed for concept description. Usually these constructs include set operations (intersection, union, complement) and property restrictions.

$$Person \sqcap (\geq 1\, hasChild) \sqsubseteq Parent \qquad (3)$$

For instance, the inclusion axiom (3) states that all instances of class *Person*, who has at least one value for property *hasChild*, are also instances of class *Parent*. Although OWL allows usage of restrictions in concept descriptions, it should be clear that OWL restrictions are not the same as database-style constraints. While constraints specify conditions to be satisfied by stored data, restrictions are used to infer some new information from it. For instance, let an ontology consist of (3) and the following assertions:

$$Person(Peter) \qquad (4)$$
$$hasChild(Peter,\, Pablo) \qquad (5)$$

Treating (3) as constraint, one can decide that the ontology is inconsistent due to absence of the following assertion:

$$Parent(Peter)\ . \qquad (6)$$

On the contrary, OWL reasoners will use (3) to infer (6). This difference between semantics is studied in [7,8].

SWRL [13] extends OWL DL with the ability to specify Horn-like rules, where only unary and binary predicates are permitted. Although SWRL rules are similar to Datalog rules, SWRL applies OWL semantics. Therefore, negation-as-failure and constraints are not supported by SWRL. In the paper, rules are written in SWRL human readable syntax for the better understanding.

These languages are developed for knowledge representation allowing its intelligent processing by software agents. *SPARQL* query language was developed for querying data from the Semantic Web. Its basal notion is a triple pattern in which variables are denoted by the leading symbol '?' or '$'. For instance, the pattern {?p hasChild ?c} specifies that variables p and c should be bound to

---

[1] OWL Full is not considered in the paper, because it is undecidable and not supported by many popular tools.

resources connected by the *hasChild* property in the queried RDF graph. More complicated queries with graph patterns can be postulated using the SPARQL features, some of which have SQL analogues: patterns can be joined with '.', query results can be united with `UNION` feature, `OPTIONAL` is analogue of `OUTER JOIN`, and `FILTER` is used to restrict query results. For instance, the following query is used to find persons with their children, who have not consorts:

```
SELECT $x $y WHERE {
        ?x rdf:type Person. ?x hasChild ?y.
        OPTIONAL{ ?y hasConsort ?z}.  FILTER(!bound(?z))}
```

Note that **bound** built-in predicate is used to check, if the variable was bound in the query, and '!' operator is used to specify that this condition does not hold. Therefore, SPARQL is the only one from the developed Semantic Web languages implements negation-as-failure. More detailed description of standardized formats can be found in [14].

3   Mapping Approach

The Section first presents mapping of relations into OWL ontology, and then presents mappings for database-style constraints allowing consistency checking.

3.1   Relation Mapping

Relations are mapped in a very straightforward way close to value-based approach in [6], because it simplifies expressing constraints in SWRL.

Classes and functional data-valued properties are created for relations and their attributes, respectively. It is required that different relations (attributes) correspond to different classes (resp., properties). It can be done in the following way: class with URI `host/database/Rel` corresponds to relation *Rel*, and property `host/database/Rel/Att` – to its attribute *Att*. The following notation is used further: created class has the same name as base relation, and name of property is concatenation of relation's and attribute's names (`relAtt`).

A vocabulary (further denoted as `db:`) was created to represent database terminology considered in the paper. All created classes are defined to be subclasses of `db:Tuple` to distinguish them from any other OWL concepts. Assertions ⟨`relAtt rdfs:domain Rel`⟩ are added to the ontology to mark attributes of a relation. Domain of the attribute is pointed with `rdfs:range` property.

Each tuple *t* of a relation instance *Rel* is translated into an instance **t** of the class Rel (i.e. ⟨`t rdf:type Rel`⟩), where **t** is URI which might be obtained by attaching tuples counter value to Rel. If a tuple *t* has not-NULL value *v* for an attribute *Att*, ⟨`t relAtt v`⟩ statement is added. Because domain of a column is specified as `range` of the corresponding property, assertion of a value of incompatible type leads to inconsistency of the ontology. Therefore, DL reasoners can be used to verify types' compatibility[2].

---

[2] It is worth to note that data types support is not strong aspect of existing reasoners.

It is impossible to use `relAtt` properties for NULL values of attributes. Absence of any assertions about some attribute's value is treated as inconsistency in the presented approach. Therefore, NULL values are specified explicitly in the generated ontology: sub-class of the `Rel` class (denoted `RelAttNo`) is created for any attribute *Att*; assertion ⟨t rdf:type RelAttNo⟩ is added to the ontology for all tuples *t* such that (*t.Att* IS NULL).

3.2 Constraints Mapping

Development of a constraint mapping was directed by the following task. It is assumed that generated RDF graph can be updated, and checking its consistency with respect to given constraints must be as simple as possible. Therefore, constraints are represented as SWRL rules, and their verification is performed as answering on a fixed query using SPARQL engine coupled with reasoner. All constraints are satisfied in the RDF graph, if the query has empty answer.

Two kinds of constraints can be outlined which we call *positive* and *negative*. Positive constraints are violated, if incorrect data exists. Their violation condition can be specified in SWRL rule. To do so, `db:` vocabulary contains property `violates` to relate incorrect tuples with violated constraints.

A negative constraint is violated, if some expected data is absent. Verification of such constraints requires negation-as-failure. Since it is supported only by SPARQL (see Sect. 2.2), these constraints are handled in a more complicated way. Property `db:hasNC` is used to specify in the ontology that instances of some class must satisfy a negative constraint. Then SPARQL query is used to find all instances of this class which do not satisfy it, i.e. have no property `db:satisfies`. Domain of these three properties is `db:Tuple`, and range is `db:Constraint`. Mappings for both kinds of constraints are presented below.

Positive Constraints. Mapping for positive constraints is demonstrated on example of NOT NULL-constraints, potential and primary keys. The following query is used to find all positive constraints c violated by tuple x from table t:

```
Q1:  SELECT $c $t $x WHERE {
          ?t rdfs:subClassOf db:Tuple.
          ?x rdf:type ?t.
          ?x db:violates ?c}
```

Let NOT NULL-constraint be specified for an attribute *Att* of a relation instance *Rel*. Then its violation condition is specified using the following rule:

$$\text{RelAttNo(?t)} \rightarrow \text{db:violates(?t, relAttNN)} . \qquad (7)$$

Note that the mapping for NULL values makes this specification very simple. In the rule, `relAttNN` is an instance of `db:Constraint` tailored by *Att*.

Let $K = \{Att_1, \ldots, Att_n\}$ be a potential key on a relation instance *Rel*. The violation condition for a potential key can be obtained from (1):

$$\exists t_1, t_2 \in Hr : (t_1 \neq t_2) \land (\forall A \in K : t_1.A = t_2.A) . \qquad (8)$$

As it was mentioned in Sect. 2.1, NULL values are considered to be equal in this case. Thus, property `relAttEq` is created with `Rel` as its domain and range, and the following rules are added for any Att from K:

$$\text{RelAttNo(?t1) \& RelAttNo(?t2)} \rightarrow \text{relAttEq(?t1,?t2)} \qquad (9)$$

$$\text{relAtt(?t1,?v) \& relAtt(?t2,?v)} \rightarrow \text{relAttEq(?t1,?t2)} \qquad (10)$$

When this is done, (8) is obviously represented as the SWRL rule:

$$\begin{aligned} &\text{relAtt1Eq(?t,?s) \& ...\&}\\ &\text{relAttnEq(?t,?s) \& differentFrom(?t,?s)} \rightarrow \text{db:violates(?t,K)} \end{aligned} \qquad (11)$$

Although relational model and SQL standard handle NULL values in this manner, in some DBMSs (e.g. PostgreSQL) two NULL values are not considered equal in (8). In the latter case, the following rule can be used instead of (11) and (9-10):

$$\begin{aligned} &\text{relAtt1(?t1,?v1) \& relAtt1(?t2,?v1) \&}\\ &\hspace{5cm}\text{...\&}\\ &\text{relAttn(?t1,?vn) \& relAttn(?t2,?vn) \&}\\ &\hspace{1.5cm}\text{differentFrom(?t1,?t2)} \rightarrow \text{db:violates(?t1,K)} \end{aligned} \qquad (12)$$

This rule does not infer constraint violation for a tuple t, if ($t.Att_i$ IS NULL) for some Att_i from K, because individual t has no property `relAtti`.

A primary key PK can be considered as a potential key such that each attribute $Att \in PK$ is defined to be NOT NULL. Therefore, $(n+1)$ SWRL rules are used to specify it: n rules (one for each attribute from PK) are similar to (7) and the last one is similar to (12)[3].

Negative Constraints. The following query verifies negative constraints:

```
Q2:  SELECT $c $t $x WHERE {
     1:     ?x rdf:type ?t.    ?t rdfs:subClassOf db:Tuple.
     2:     ?p rdfs:domain ?t.    ?p rdfs:subPropertyOf db:hasNC.
     3:     ?p rdfs:range ?e.    ?c rdf:type ?e.
     4:     OPTIONAL { ?y db:satisfies ?c FILTER(sameTerm(?x,?y)) }
     5:     FILTER(!bound(?y) && !sameTerm(?p,db:hasNC))}
```

Lines 2-3 allow linking tables t with their positive constraints c. Lines 4-5 implement negation-as-failure to return only tuples x violating these constraints.

Let us consider how a negative constraint is mapped into OWL axioms and SWRL rules to allow one to verify it using the above query. A foreign key constraint is an example of such constraint: it is violated for some tuple without NULL values for attributes from FK, if the corresponding tuple in the referenced table does **not exist**. Therefore, the mapping approach is shown on this example.

Let a foreign key $FK = \{Att_1, \ldots, Att_n\}$ be defined for a relation instance Rel_1 and referring to Rel_2. First, a new individual FK is created, and a class CFK

---

[3] Object in heads of the rules is PK.

is defined as to be owl:oneOf(FK). Next, a property pFK is created with Rel1 as its domain and CFK as range and assigned to be sub-property of hasNC[4]. Then the following two rules should be created to specify (2):

$$Rel1Att1No(?t) \ \& \ \ldots \ \& \ Rel1AttnNo(?t) \ \rightarrow \ db:satisfies(?t,FK) \quad (13)$$

$$rel1Att1(?t,?v1) \ \& \ rel2Att1(?s,?v1)$$
$$\& \ \ldots \ \& \quad (14)$$
$$rel1Attn(?t,?vn) \ \& \ rel2Attn(?s,?vn) \ \rightarrow \ db:satisfies(?t,FK)$$

Rule (13) specifies that foreign-key constraint is satisfied for a tuple, if all its attributes from FK are NULLs. Rule (14) specifies that it is satisfied for a tuple without NULL values for attributes in FK, if corresponding tuple exists in Rel_2.

If relAttNp is sub-property of db:hasNC with range oneOf(relAttSt) and domain Rel, the following rules guarantee attribute Att is set in tuples from Rel:

$$RelAttNo(?t) \ \rightarrow \ db:satisfies(?t,relAttSt) \quad (15)$$

$$relAtt(?t,?v) \ \rightarrow \ db:satisfies(?t,relAttSt) \quad (16)$$

Unknown Source of Violation. Let consider *anti-key* constraint AK of the form $AntiKey(Rel, [Att_1, \ldots, Att_n])$. This constraint is satisfied, if potential key $PK = \{Att_1, \ldots, Att_n\}$ is violated in Rel (i.e. at least one pair of tuples with identical values for attributes Att_1 through Att_n exists). Note that when anti-key is violated for some relation instance, particular violation source cannot be determined. Although all tuples of this relation instance can be returned as violation sources with Q2, it would not be informative. Therefore, constraints like anti-key are considered separately in the paper.

For an anti-key AK, an individual AK, a class CAK and a property pAK are created in the same way as for FK; the only difference is that pAK is sub-property of hasAntiKey. Rule specifying satisfaction condition for AK is similar to violation condition (12) for PK, but has the differenet consequent:

$$relAtt1(?t,?v1) \ \& \ relAtt1(?s,?v1) \ \&$$
$$\ldots \&$$
$$relAttn(?t,?vn) \ \& \ relAttn(?s,?vn) \ \& \quad (17)$$
$$differentFrom(?t,?s) \hspace{3cm} \rightarrow \ db:satisfies(?t,AK)$$

To verify anti-keys, the following query is used, which mainly distinguishes from Q2 in that it does not return violations sources:

```
Q3: SELECT $c $t WHERE {
          ?t rdfs:subClassOf db:Tuple.
          ?p rdfs:domain ?t.   ?p rdfs:subPropertyOf db:hasAntiKey.
          ?p rdfs:range ?e.     ?c rdf:type ?e.
          OPTIONAL{ ?y rdf:type ?t.  ?y db:satisfies ?c  }.
          FILTER(!bound(?y) && !sameTerm(?p,db:hasAntiKey))}
```

---

[4] Three entities are created for one constraint, because OWL Lite and OWL DL require disjoint sets of individuals and classes and, therefore, it is impossible to express that *"c is constraint on t"* in one statement.

4   Discussion

This Section discusses some aspects which may remain unclear. Constraint verification can be performed with only one query uniting queries Q1, Q2 and Q3 presented above. The most important constraints (potential, primary and foreign keys and NOT NULL) can be mapped to SWRL rules automatically as it is described above. Rules for CHECK-constraints are added manually. This task is obvious for interval constraints and some other popular constraints, but is not solvable in general. The reason is that the Semantic Web languages do not support aggregation, and this is a topic of researches now [15]. Therefore, table constraints with aggregates are not expressible in any of these formats with exception to some particular cases, when number of aggregated entities is fixed. This decreases applicability of the mapping only slightly, because such constraints are used not very frequently and not supported by many commercial DBMSs.

4.1   OWL vs. SPARQL

It may be preferable to create violates-rules for CHECK-constraints. The reason is that their implementation is simpler and can be reduced to standard DL reasoning task of ontology consistency checking with the following axiom:

$$\exists violates.\top \sqsubseteq \bot \tag{18}$$

However, (18) has side effects which can be shown on the following ontology O_1:

$$Aircraft \sqsubseteq \exists locate.(Airport \sqcup Repair) \tag{19}$$

$$functionalProperty(locate) \tag{20}$$

$$RaceAir \sqsubseteq Aircraft \tag{21}$$

$$RaceAir(os606) \tag{22}$$

$$locate(os606, aprt) \tag{23}$$

The following rules are defined in O_1 to implement constraints for $RaceAir$:

$$\texttt{RaceAir(?a)\&locate(?a,?l)\&Repair(?l)} \rightarrow \texttt{db:violates(?a,R)} \tag{24}$$

$$\texttt{RaceAir(?a)\&locate(?a,?l)\&Airport(?l)} \rightarrow \texttt{db:satisfies(?a,A)} \tag{25}$$

Using (19)-(23), one can entail that $aprt$ is an instance of either $Airport$ or $Repair$. Rule (24) itself does not allow choosing one class among them, and the SPARQL query returns $os606$ as violating constraint A defined in (25). However, if (18) is used, it can be verified that $Repair(aprt)$ leads to inconsistency, and $Airport(aprt)$ does not. Thus, the axiom allows new fact to be entailed. The ontology is not inconsistent, and no constraints are violated in this case.

4.2   Safe Rules

Let ontology O_2 contain assertions (19)-(22) with the following ones:

$$hold(aprt, os606) \tag{26}$$

$$locate \equiv hold^- \tag{27}$$

RaceAir is constrained in O_2 to have *locate* property using the negative rule:

$$\text{locate}(?x,?y) \rightarrow \text{db:satisfies}(?x,CL) . \tag{28}$$

Although O_2 does not contain (23), it can use (26) and (27) to entail it. It seems natural for constraint CL not to be violated in this case. Let now ontology O_3 be obtained from O_2 by removing assertion (27). Instead of (23), it can be entailed from O_3 using (19) that the following holds for an unnamed individual α :

$$locate(os606, \alpha) . \tag{29}$$

In contrary to (23), this assertion does not give any explicit information about $os606$, stating only that it has some value for *locate*. Intention of constraint CL is to guarantee that we explicitly know this value. Therefore, the constraint must be violated in this case, and (28) must not be fired. These reasons lead to applying the following semantics to SWRL rules: *variables in rules can be instantiated only with values explicitly used in underlying ontology*. Note that this semantics is applied in safe rules [16] to solve problems with SWRL decidability [17].

4.3   How to Distinguish Individuals

Some constraints like potential or primary keys can be violated only by a pair of different individuals. Two individuals are considered to be different in the corresponding SWRL rules, if they are related via owl:differentFrom property. There are several ways to distinguish individuals built by tuples from a database. First, since attributes are mapped to functional properties, any two individuals with different values for some of these properties are also different. Next, the following rule is created for any attribute *Att* of a relation *Rel* to distinguish individuals t_1 and t_2 such that ($t_1.Att$ IS NULL) and ($t_2.Att$ NOT IS NULL):

$$\text{RelAttNo}(?t1) \text{ \& } \text{relAtt}(?t2,?v) \rightarrow \text{differentFrom}(?t1,?t2) . \tag{30}$$

Finally, individuals not distinguished in these ways must be explicitly related via owl:differentFrom property.

Another possible way to distinguish individuals is to apply Unique Name Assumption (UNA), i.e. individuals with different names are considered to be different. Although OWL does not apply UNA [18], it can be done with SWRL, if built-in atom URI2literal converting URI to string is added. In this case, UNA is implemented using a functional property db:hasName instantiated with the following rule:

$$\text{db:Tuple}(?t) \text{ \& } \text{URI2literal}(?t,?y) \rightarrow \text{db:hasName}(?t,?y) . \tag{31}$$

However, the author claims that UNA is not useful in Web environment. As noted in [9], one of the most useful features of the Semantic Web is owl:sameAs allowing better integration. Let us consider ontology O_{DB} created from a database. One way to achieve benefits from this mapping is to integrate O_{DB} with an expressively rich ontology O_{SEM} available in the Web. To do so, it may be required to state that individuals i_1 from O_{DB} and i_2 from O_{SEM} are the same. These individuals have different URIs, and (31) entails that they have different values of db:hasName. Because db:hasName is functional, the individuals will be entailed to be different. This contradiction leads to inconsistency of $O_{DB} \cup O_{SEM}$. It means that integration with other ontologies becomes impossible, if UNA is used. Therefore, it is not used in the presented approach.

5   Related Work

There are a plenty of approaches to mapping of relational databases to RDF, and some of them are surveyed by W3C RDB2RDF Incubator Group in [19]. The survey also presents different criterions for classifying mappings.

- Data mapping can be either static (RDF dump) or dynamic (query-driven). Since the approach presented in the paper is focused on constraints, it implements RDF dump. Although it can be enhanced to support dynamic instantiation, the current implementation, as [19] notes, has advantages if entailment rules are applied to the RDF repository to infer new facts.
- Approaches are based either on automatic creation of new ontology from database (e.g. [4,20]) or on manual mapping to an existing ontology [2].
- Also these approaches use different languages like [5].

All the approaches have their advantages and drawbacks, but they mostly disregard semantics of local databases (i.e. primary and foreign keys, constraints).

The recent works [7,8] consider these constraints and show inability of OWL to express them because of difference between the semantics discussed in Sect. 2. As consequence, they propose to extend OWL with features for support of relational constraints. OWL 2 [21] supports keys in OWL semantics. Different ways to extend OWL with integrity constraints support are reviewed in [18].

Although [13] mentions that SWRL can be straightforwardly extended to enable negation-as-failure and database-style constraints, [22] points that it is not possible because SWRL applies the same logic foundations as in OWL and proposes multi-stack architecture to overcome this drawback. Several proposals to extending DLs with rules have applied this architecture, but it does not meet common acceptance, because it leads to establishment of two Semantic Webs.

In [23], SWRL is used to represent primary keys for more accurate translation in Automapper. However, constraint verification is not considered in this work.

In [11], it is proposed to use RDF and SPARQL query language to check consistency of ontologies generated from relational databases. It supposes dynamic generation of SPARQL queries to implement semantics of database-style constraints. However, our approach uses the Semantic Web languages in more natural way: OWL and SWRL are used to implement the constraints' semantics, and static SPARQL query is used to check consistency. In this case SWRL rules can be used for semantic query optimization in the same way as it is shown in [11]. Our approach supports mapping and verification of so-called negative constraints like anti-keys, for which extension of SPARQL is proposed in [11].

In contrast to the above-mentioned approaches, the mapping presented in the paper does not extend any Semantic Web language to support constraints, what is important to establish and not to complicate the Semantic Web.

6   Conclusion

The most of nowadays information is stored in relational databases. Therefore, populating the Semantic Web with it is an important step for making future Web more intelligent. It leads to development of a plenty of proposals for mappings databases to ontologies. However, the most of them disregards constraints which can be useful in many ways including semantic query optimizations, database dumps, and supporting materialized RDF graphs. In the distant future, it may simplify migration from relational databases to RDF stores like [24].

As since the data representation formats separately do not allow constraints representation due to the differences in logical foundations, the paper presents the approach to database mapping which uses the full strength of the Semantic Web stack in a natural way: data is presented in RDF; OWL and SWRL are used to express constraints' satisfaction or validation conditions; and the static SPARQL query is used to verify these constraints. The approach allows one to handle the most important constraints (potential, primary, and foreign keys) as like as many CHECK-constraints (including NOT NULL) without any extensions to the Semantic Web languages. As proof of concept, it was implemented on Java using JDBC for automatic mapping from PostgreSQL databases.

It is shown that SPARQL better suits for constraints verification than OWL. Safe rules are shown to be applicable not only in solving the decidability problem, but in constraints representation too. Finally, the author claims that UNA is not really useful in the Web. The future work may be directed on extending the presented approach to express integrity constraints for arbitrary OWL concepts. Also it is worth to investigate ways for introducing aggregation into some Semantic Web language for more complete support of database-style constraints.

References

1. Wu, F., Weld, D.S.: Automatically refining the wikipedia infobox ontology. In: Proc. of the 17th International Conference on World Wide Web, Beijing, China, pp. 635–644 (2008)
2. Auer, S., Dietzold, S., Lehmann, J., Hellmann, S., Aumueller, D.: Triplify: lightweight linked data publication from relational databases. In: Proc. of the 18th International Conference on World Wide Web, Madrid, Spain, pp. 621–630 (2009)
3. Geller, J., Chun, S.A., An, Y.J.: Towards the semantic deep web. IEEE Computer 41(9), 95–97 (2008)
4. Suwanmanee, S., Benslimane, D., Champin, P.A., Thiran, P.: Wrapping and integrating heterogeneous relational data with OWL. In: Proc. of the 7th International Conference on Enterprise Information Systems (ICEIS 2005), pp. 11–18 (2005)
5. Bizer, C., Seaborne, A.: D2RQ - treating non-RDF databases as virtual RDF graphs. In: 3rd International Semantic Web Conference (ISWC 2004), Hiroshima, Japan (2004) (poster)

6. Lausen, G.: Relational databases in rdf: Keys and foreign keys. In: Christophides, V., Collard, M., Gutierrez, C. (eds.) SWDB-ODBIS 2007. LNCS, vol. 5005, pp. 43–56. Springer, Heidelberg (2008)

7. Calvanese, D., De Giacomo, G., Lembo, D., Lenzerini, M., Rosati, R.: Can OWL model football leagues? In: OWLED 2007. CEUR Workshop Proceedings, vol. 258 (2007)

8. Motik, B., Horrocks, I., Sattler, U.: Bridging the gap between OWL and relational databases. In: Proc. of the 16th International Conference on World Wide Web, Banff, Alberta, Canada, pp. 807–816 (2007)

9. Hendler, J.: The dark side of the semantic web. IEEE Intelligent Systems 22(1), 2–4 (2007)

10. Volz, R.: Change paths in reasoning! In: Proc. of the First International Workshop New forms of reasoning for the Semantic Web: scalable, tolerant and dynamic, co-located with ISWC 2007 and ASWC 2007 (2007)

11. Lausen, G., Meier, M., Schmidt, M.: SPARQLing constraints for RDF. In: EDBT 2008, 11th International Conference on Extending Database Technology, Nantes, France, pp. 499–509. ACM, New York (2008)

12. Baader, F., Calvanese, D., McGuinness, D., Nardi, D., Patel-Schneider, P.: The Description Logic Handbook: Theory, Implementation, and Applications. Cambridge University Press, Cambridge (2003)

13. Horrocks, I., Patel-Schneider, P.F., Boley, H., Tabet, S., Grosof, B., Dean, M.: SWRL: A semantic web rule language combining OWL and RuleML. W3C Member Submission (May 21 2004),
http://www.w3.org/Submission/2004/SUBM-SWRL-20040521/

14. Herman, I.: W3C semantic web activity (2001), http://www.w3.org/2001/sw/

15. Barbieri, D.F., Braga, D., Ceri, S., Valle, E.D., Grossniklaus, M.: C-SPARQL: SPARQL for continuous querying. In: Proc. of the 18th International Conference on World Wide Web, Madrid, Spain, pp. 1061–1062 (2009)

16. Patel-Schneider, P.: Safe rules for owl 1.1. In: Fourth International Workshop OWL: Experiences and Directions (OWLED 2008 DC), Washington, DC (2008)

17. Levy, A.Y., Rousset, M.C.: The limits on combining recursive horn rules with description logics. In: AAAI/IAAI, vol. 1, pp. 577–584 (1996)

18. Sirin, E., Smith, M., Wallace, E.: Opening, closing worlds - on integrity constraints. In: OWLED 2008. CEUR Workshop Proceedings, vol. 432 (2008)

19. Sahoo, S.S., Halb, W., Hellmann, S., Idehen, K., Thibodeau Jr., T., Auer, S., Sequeda, J., Ezzat, A.: A survey of current approaches for mapping of relational databases to rdf, January 8 (2009),
http://www.w3.org/2005/Incubator/rdb2rdf/RDB2RDF_SurveyReport.pdf

20. Ghawi, R., Cullot, N.: Database-to-ontology mapping generation for semantic interoperability. In: Third International Workshop on Database Interoperability (InterDB 2007), Vienna, Austria (2007)

21. Cuenca Grau, B., Horrocks, I., Motik, B., Parsia, B., Patel-Schneider, P., Sattler, U.: OWL 2: The next step for OWL. J. of Web Semantics 6(4), 309–322 (2008)

22. Kifer, M., de Bruijn, J., Boley, H., Fensel, D.: A realistic architecture for the semantic web. In: Adi, A., Stoutenburg, S., Tabet, S. (eds.) RuleML 2005. LNCS, vol. 3791, pp. 17–29. Springer, Heidelberg (2005)

23. Fisher, M., Dean, M., Joiner, G.: Use of owl and swrl for semantic relational database translation. In: Fourth International Workshop OWL: Experiences and Directions (OWLED 2008 DC), Washington, DC (2008)

24. Ramanujam, S., Gupta, A., Khan, L., Seida, S., Thuraisingham, B.: Relationalizing RDF stores for tools reusability. In: Proc. of the 18th International Conference on World Wide Web, Madrid, Spain, pp. 1059–1060 (2009)

Computing Knowledge-Based Semantic Similarity from the Web: An Application to the Biomedical Domain

David Sánchez, Montserrat Batet, and Aida Valls

Department of Computer Science and Mathematics
University Rovira i Virgili
Av. Països Catalans, 26. 43007, Tarragona
{david.sanchez,montserrat.batet,aida.valls}@urv.cat

Abstract. Computation of semantic similarity between concepts is a very common problem in many language related tasks and knowledge domains. In the biomedical field, several approaches have been developed to deal with this issue by exploiting the knowledge available in domain ontologies (SNOMED-CT) and specific, closed and reliable corpuses (clinical data). However, in recent years, the enormous growth of the Web has motivated researchers to start using it as the base corpus to assist semantic analysis of language. This paper proposes and evaluates the use of the Web as background corpus for measuring the similarity of biomedical concepts. Several classical similarity measures have been considered and tested, using a benchmark composed by biomedical terms and comparing the results against approaches in which specific clinical data were used. Results shows that the similarity values obtained from the Web are even more reliable than those obtained from specific clinical data, manifesting the suitability of the Web as an information corpus for the biomedical domain.

Keywords: Semantic similarity, ontologies, information content, Web, biomedicine, UMLS, SNOMED.

1 Introduction

The computation of the semantic similarity/distance between concepts has been a very active trend in computational linguistics. It gives a clue to quantify how words extracted from documents or textual descriptions are alike. Semantically, similarity is usually based on taxonomical relations between concepts. For example, *bronchitis* and *flu* are similar because both are disorders of the respiratory system.

From a domain independent point of view, the assessment of similarity has many direct applications such as, word-sense disambiguation [21], document categorization or clustering [7], word spelling correction [3], automatic language translation [7], ontology learning [23] or information retrieval [13].

In the biomedical domain, semantic similarity measures can improve the performance of Information Retrieval tasks, since they are able, for example, to map a user's specific search query (e.g. patient cohort identification) to multiple equivalent formulations [19]. Other authors have applied them to discover similar protein

D. Karagiannis and Z. Jin (Eds.): KSEM 2009, LNAI 5914, pp. 17–28, 2009.

sequences [15] or to the automatic indexing and retrieval of biomedical documents (e.g. in the PubMed digital library) [25].

In general, similarity assessment is based on the estimation of semantic evidence observed in one or several knowledge or information sources. So, background data or knowledge is needed in order to measure the degree of similarity between concepts.

Domain-independent approaches [22][14][11] typically rely on WordNet [9], which is a freely available lexical database that represents an ontology of more than 100,000 general English concepts and SemCor [16], a semantically tagged text repository consisting of 100 passages from the Brown Corpus. A domain-independent application is typically characterized by its high amount of ambiguous words, represented as polysemy and synonymy. In those cases, the more background knowledge is available (i.e. textual corpus, dictionaries, ontologies, etc.) and the more pre-processing of the corpus data (e.g. manual tagging, disambiguation, etc.), the better the estimation will be [11].

However, for specialized domains such as biomedicine, words are much less polysemic, thus, ambiguity is reduced and the conclusions extracted from the available data may be more accurate.

In the past, some classical similarity computation measures have been adapted to the biomedical domain, exploiting medical ontologies (such as UMLS or MESH) and clinical data sources in order to extract the semantic evidence in which they base the similarity assessment. The general motivation is that the lack of domain coverage of the resources included in classical domain-independent sources (such as the Brown Corpus or WordNet) makes them ineffective in domain specific tasks [19]. However, the use of a domain dependant corpus introduces some problems: *i)* the preparation of the input data for each domain in a format which can be exploited (i.e. data pre-processing or filtering is typically required), *ii)* data sparseness of scarce concepts when the corpus does not provide an enough semantic evidence to extract accurate conclusions, and *iii)* the availability of enough domain data (this is especially critical in the medical domain, due to the privacy of data). So, even though a domain-dependant approach may lead to more accurate results, the dependency on the domain knowledge and data availability hampers the real applicability of the similarity measures.

The situation has changed in the recent years with the enormous development of the Web. Nowadays, the Web is the biggest electronic repository available [2]. Web data, understood as individual observations in web resources may seem unreliable due to its uncontrolled publication and "dirty" form. However, taking the Web as a global source, it has been demonstrated that the amount and heterogeneity of the information available is so high that it can approximate the real distribution of information at a social scale [7]. In fact, some authors [7][24] have exploited web information distribution in order to evaluate word relatedness in an unsupervised fashion (i.e. no domain knowledge is employed). However, their performance is still far [10] from the supervised (ontology-driven) approaches considered in this paper.

Following these premises, our hypothesis is that the amount of information related to the biomedical domain (and in general to any concrete domain) contained in the Web is enough to obtain similarity values as robust (or even better) as those extracted from a reliable, pre-processed and domain specific corpus (i.e. clinical data, in the case of the biomedical domain). In order to prove it, we have studied the performance of classical ontology-based similarity measures applied to rank the similarity between concepts of the biomedical domain; in our experiments the Web is used as the source from which to perform the semantic assessment, in comparison to other approaches using reliable domain specific data.

This paper presents the analysis and results of this study. In section 2 we will present ontology-based similarity computation paradigms and the way in which they have been used in the past when dealing with biomedical concepts. In sections 3 and 4, we will study and adapt some measures to the Web environment (particularly in the way in which word statistics are computed). Then, in section 5, we will evaluate the Web-based measures using a standard benchmark composed by 30 medical terms whose similarity has been assessed by expert physicians of the Mayo Clinic [19] and compare the results against previous approaches evaluating the same measures but using only domain-specific data. The final section will present the conclusions of this study.

2 Ontology-Based Semantic Similarity

In the literature, there exist several semantic similarity computation paradigms using ontologies as background knowledge.

First, there are approaches which consider an ontology as a graph model in which semantic interrelations are modeled as links between concepts. They exploit this geometrical model, computing concept similarity as inter-link distance (also called path length) [26][20][12]. In the past, this idea has been applied to the MeSH semantic network, which consists of biomedical terms organized in a hierarchy [20]. Taking a similar approach, several authors [5][18] developed measures for finding path lengths in the UMLS hierarchy. The advantage of this kind of measures is that they only use a domain ontology as the background knowledge, so, no corpus with domain data is needed.

The main problem of those path-length-based approaches is that they heavily depend on the degree of completeness, homogeneity and coverage of the semantic links represented in the ontology [8]. Moreover, it is worth to note that the presence of a semantic link between two concepts gives an evidence of a relationship but not about the strength of their semantic distance (i.e. all individual links have the same length and, in consequence, represent uniform distances [1]). As this kind of measures does not use a background corpus to compute their similarity, we will not consider them in this study.

On the other hand, there exist other ontology-based similarity measures which combine the knowledge provided by an ontology and the Information Content (IC) of the concepts that are being compared. IC measures the amount of information provided by a given term from its probability of appearance in a corpus. Consequently, infrequently appearing words are considered more informative than common ones.

Based on this premise, Resnik [22] presented a seminal work in which the similarity between two terms is estimated as the amount of information they share in common. In a taxonomy, this information is represented by the Least Common Subsumer (LCS) of both terms. So, the computation of the IC of the LCS results in an estimation of the similarity of the subsumed terms. The more specific the subsumer is (higher IC), the more similar the subsumed terms are, as they share more information.

Several variations of this measure have been developed (as will be presented in section 3).

These measures have been evaluated by Pedersen et al. [19] in the biomedical domain by using SNOMED-CT as ontology and a source of clinical data as corpus. They were able to outperform path length-based ones in this domain specific environment [19].

In the next sections, we present different approaches for IC-based similarity computation and we study how they can be modified in order to use the Web as a corpus, instead of specific clinical data (which may introduce applicability limitations as will be discussed in section 4).

3 IC-Based Similarity Measures

The *Information content* (IC) of a concept is the inverse to its probability of occurrence. The IC calculation is based on the probability $p(a)$ of encountering a concept a in a given corpus, by applying eq 1. In this way, infrequent words obtain a higher IC than more common ones.

$$IC(a) = -\log p(a) \tag{1}$$

As mentioned above, Resnik [22] introduced the idea of computing the similarity between a pair of concepts (a and b) as the IC of their *Least Common Subsumer* (*LCS(a,b)*, i.e., the most concrete taxonomical ancestor common to a and b in a given ontology) (eq, 2). This gives an indication of the amount of information that concepts share in common. The more specific the subsummer is (higher IC), the more similar the terms are.

$$sim_{res}(a,b) = IC(LCS(a,b)) \tag{2}$$

The most commonly used extensions to Resnik measure are Lin [14] and Jiang & Conrath [11].

Lin's [14] similarity depends on the relation between the information content of the LCS of two concepts and the sum of the information content of the individual concepts (eq. 3).

$$sim_{lin}(a,b) = \frac{2 \times sim_{res}(a,b)}{(IC(a) + IC(b))} \tag{3}$$

Jiang & Conrath [11] subtract the information content of the LCS from the sum of the information content of the individual concepts (eq. 4).

$$dis_{jcn}(a,b) = (IC(a) + IC(b)) - 2 \times sim_{res}(a,b) \tag{4}$$

Note that this is a dissimilarity measure because the more different the terms are, the higher the difference from their IC to the IC of their LCS will be.

Original approaches use those measures relying on WordNet [9] as the ontology from where obtain the LCS of the terms and SemCor [16] as a general purpose pre-tagged corpus from which to compute concept probabilities from manually computed term appearances. However, in specific domains such as biomedicine, those general-purpose corpuses present a reduced coverage [4]. In addition, WordNet ontology has a limited coverage of biomedical terms. These two issues result in a poor performance of the described similarity measures when applying them to concrete domain concepts [19]. Consequently, as stated in the previous section, in this work they have been adapted to the biomedical domain by exploiting SNOMED-CT taxonomy instead of WordNet and the Mayo Clinic Corpus of Clinical Notes corpus instead of SemCor.

On one hand, SNOMED-CT[1] (*Systematized Nomenclature of Medicine, Clinical Terms*) is an ontological/terminological resource distributed as part of the UMLS and it is used for indexing electronic medical records, ICU monitoring, clinical decision support, medical research studies, clinical trials, computerized physician order entry, disease surveillance, image indexing and consumer health information services. It contains more than 311,000 medical concepts organized into 13 hierarchies conforming 1.36 million relationships, from which *is-a* relationships are exploited to extract the LCS between a pair of terms.

On the other hand, the Mayo Clinic Corpus consists of 1,000,000 clinical notes collected over the year 2003 which cover a variety of major medical specialties at the Mayo Clinic. Clinical notes have a number of specific characteristics that are not found in other types of discourse, such as news articles or even scientific medical articles found in MEDLINE. Clinical notes are generated in the process of treating a patient at a clinic and contain the record of the patient-physician encounter. The notes were transcribed by trained personnel and structured according to the reasons, history, diagnostic, medications and other administrative information. Patient's history, diagnostic and medication notes were collected as the domain-specific and pre-processed data corpus from which to assess the semantic similarity between different pairs of diseases (more details in the evaluation section) [19].

4   Computing IC from the Web

Using a domain-specific and reliable corpus like the Mayo Clinic repository to compute IC-based semantic similarity may lead to accurate results. However, the availability of those corpuses (i.e. the use of patient data should ensure privacy and anonymity) and their coverage with respect to the evaluated terms (i.e. what happens if the concepts compared are not considered in typical clinical histories) are the main problems which hamper the applicability of those domain-dependant approaches. In fact, data sparseness (i.e. the fact that not enough data is available for certain concepts to reflect an appropriate semantic evidence) is the main problem of those approximations [2].

On the other hand, the Web and, more particularly, web search engines are able to index almost any possible term. Considering its size and heterogeneity, it can be considered as a social-scale general purpose corpus. Its main advantages are its free and direct access and its wide coverage of any possible domain. In comparison with other general purpose repositories (such as the Brown Corpus or SemCor) which have shown a poor performance for domain-dependant problems [19], the Web's size is

---

[1] http://www.nlm.nih.gov/research/umls/Snomed/snomed_main.html

millions of orders of magnitude higher. In fact, the Web offers more than 1 trillion of accessible resources which are directly indexed by web search engines[2]. It has been demonstrated [2] the convenience of using such a wide corpus to improve the sample quality for statistical analysis.

In order to study the performance of the presented similarity measures when using the Web as a corpus with biomedical concepts, in this section, we adapt them by computing term occurrences from the Web instead of a reliable, closed and domain-specific repository of clinical data.

The main problem of computing term's Web occurrences is that the analysis of such an enormous repository for computing appearance frequencies is impracticable. However, the availability of massive Web Information Retrieval tools (general-purpose search engines like Google) can help in this purpose, because they provide the number of pages (hits) in which the searched terms occur. This possibility was exploited in [24], in which a modification of the Point-wise Mutual Information measure [6] approximating concept probabilities from web search engine hit counts is presented.

This idea was later developed in [7], in which it is claimed that the probabilities of Web search engine terms, conceived as the frequencies of page counts returned by the search engine divided by the number of indexed pages, approximate the relative frequencies of those searched terms as actually used in society. So, exploiting Web Information Retrieval (IR) tools and concept's usage at a social scale as an indication of its generality, one can estimate, in an unsupervised fashion, the concept probabilities from Web hit counts.

Even though web-based statistical analyses brought benefits to domain-independent unsupervised approaches (i.e. no background ontology is exploited) [24], their performance is still far from supervised measures [10] like those considered in this paper.

Taking those aspects into consideration, in the following, we will adapt the ontology-based similarity measures introduced in section 3 to use the Web as a corpus, by estimating concept's IC from web hit counts. The web-based IC computation is specified as follows.

$$IC\_IR(a) = -\log p_{web}(a) = -\log \frac{hits(a)}{total\_webs} \qquad (5)$$

Being, $p_{web}(a)$ the probability of appearance of string 'a' in a web resource. This probability is estimated from the Web hit counts returned by Web IR tool -hits- when querying the term 'a'. Total_webs is the total number of resources indexed by a web search engine (estimated as 1 trillion, as stated in section 3).

In this manner, IC-based measures presented in section 3 can be directly rewritten to compute concept probabilities from the Web by incorporating the Web-based IC computation (IC_IR).

Resnik measure can be rewritten as follows.

$$sim_{res}\_IR(a,b) = IC\_IR(LCS(a,b)) = -\log \frac{hits(LCS(a,b))}{total\_webs} \qquad (6)$$

---

[2] http://googleblog.blogspot.com/2008/07/we-knew-web-was-big.html

Lin measure can be rewritten as follows.

$$sim_{lin} \_ IR(a,b) = \frac{2 \times sim_{res} \_ IR(a,b)}{(IC\_IR(a) + IC\_IR(b))} =$$

$$= \frac{2 \times \left(-\log \dfrac{hits(LCS(a,b))}{total\_webs} \right)}{\left(-\log \dfrac{hits(a)}{total\_webs} - \log \dfrac{hits(b)}{total\_webs} \right)} \tag{7}$$

Jiang & Conrath distance measure can be rewritten as follows.

$$dis_{jcn} \_ IR(a,b) = (IC\_IR(a) + IC\_IR(b)) - 2 \times sim_{res} \_ IR(a,b) =$$

$$= \left(-\log \frac{hits(a)}{total\_webs} - \log \frac{hits(b)}{total\_webs} \right) - 2x\left(-\log \frac{hits(LSC(a,b))}{total\_webs} \right) \tag{8}$$

In any case, the main problem of using such a general repository as the Web to estimate concept's appearance probabilities is language ambiguity. This problem appears when concept probabilities are estimated by means of word's (instead of concept's) web hit counts. On the one hand, different synonyms or lexicalizations of the same concept may result in different IC_IR values, introducing bias. On the other hand, the same word may have different senses and, in consequence, correspondences to several concepts. As a result, ambiguity may lead to inconsistent concept probability estimations. Fortunately, specific domain terms which are the object of this study, due to their concreteness, are rarely ambiguous in contrast to general words. So, our hypothesis is that language ambiguity will not compromise the web statistics when comparing biomedical concepts. This is also evaluated in the next section.

5   Evaluation

The most common way of evaluating similarity measures is by using a set of word pairs whose similarity has been assessed by a group of human experts. Computing the correlation between the computerized and human-based ratings, one is able to obtain a quantitative value of the similarity function's quality, enabling an objective comparison against other measures. In a general setting, the most commonly used benchmark is the Miller and Charles set [17] of 30 ranked domain-independent word pairs.

For the biomedical domain, Pedersen *et al.* [19], in collaboration with Mayo Clinic experts, created a set of word pairs referring to medical disorders whose similarity were evaluated by a group of 3 expert physicians. After a normalization process, a final set of 30 word pairs with the averaged similarity measures provided by physicians in a scale between 1 and 4 were obtained (see Table 1). The correlation between physician judgments was 0.68.

Table 1. Set of 30 medical term pairs with associated averaged expert's similarity scores (extracted from [19]). Note that the term *"lung infiltrates"* is not found in SNOMED-CT.

Term 1	Term 2	Physician score (averaged)
Renal failure	Kidney failure	4.0
Heart	Myocardium	3.3
Stroke	Infarct	3.0
Abortion	Miscarriage	3.0
Delusion	Schizophrenia	3.0
Congestive heart failure	Pulmonary edema	3.0
Metastasis	Adenocarcinoma	2.7
Calcification	Stenosis	2.7
Diarrhea	Stomach cramps	2.3
Mitral stenosis	Atrial fibrillation	2.3
Chronic obstructive pulmonary disease	*Lung infiltrates*	2.3
Rheumatoid arthritis	Lupus	2.0
Brain tumor	Intracranial hemorrhage	2.0
Carpal tunnel syndrome	Osteoarthritis	2.0
Diabetes mellitus	Hypertension	2.0
Acne	Syringe	2.0
Antibiotic	Allergy	1.7
Cortisone	Total knee replacement	1.7
Pulmonary embolus	Myocardial infarction	1.7
Pulmonary fibrosis	Lung cancer	1.7
Cholangiocarcinoma	Colonoscopy	1.3
Lymphoid hyperplasia	Laryngeal cancer	1.3
Multiple sclerosis	Psychosis	1.0
Appendicitis	Osteoporosis	1.0
Rectal polyp	Aorta	1.0
Xerostomia	Alcoholic cirrhosis	1.0
Peptic ulcer disease	Myopia	1.0
Depression	Cellulitis	1.0
Varicose vein	Entire knee meniscus	1.0
Hyperlipidemia	Metastasis	1.0

Pedersen *et al.* [19] used that benchmark to evaluate the IC-based similarity measures described in section 3 exploiting the SNOMED-CT hierarchy (to retrieve the LCS) and the Mayo Clinic Corpus (to statistically assess concept's IC). Term pair *"chronic obstructive pulmonary disease" – "lung infiltrates"* was excluded from the test bed as the later term was not found in SNOMED-CT terminology. The correlation they obtained for the remaining 29 pairs under this testing condition is shown in the second column of Table 2.

In order to compare the performance of the same measures when substituting the domain-specific preprocessed corpus of Mayo Clinical Notes for the Web, we have used the adapted versions of the IC-based measures as presented in section 4. We have also taken SNOMED-CT as the reference ontology and the same benchmark as Pedersen. In order to obtain term appearances from the Web we have used Live

Search[3] as the web search engine as it does not introduce limitations on the number of queries performed per day [23]. Even though, any other general-purpose search engine (such as Google, Altavista or Yahoo) may be employed instead. The correlation results obtained in this case for each adapted measure are presented in the third column of Table 2.

It is worth to note that, due to the syntactical complexity of some of the LCSs extracted from SNOMED-CT (e.g. being *"morphologically altered structure"* the LCS of *"calcification"* and *"stenosis"*) data sparseness may appear because a very few number of occurrences of the exact expression can be found (*"morphologically altered structure"* returned only 48 matches in Live Search). In order to tackle this problem, we simplified syntactically complex LCSs by taking the noun phrase on the right of the expression or removing some adjectives on the left of the noun phrase (i.e. *"morphologically altered structure"* was replaced by *"altered structure"*). In this manner, as we generalize the LCS, the scope of the statistical assessment is increased (more hit counts are considered due to the simpler query expression) without losing much of the semantic context. Other approaches dealing with IC-based measures do not face this problem, as term frequencies of the LCS are manually computed from the background corpus at a conceptual level [11] (i.e. a document may cover the *"morphologically altered structure"* topic even though the exact term expression never appears in the text, a situation which is detected and taken into consideration by a human expert). This contrasts to the strict keyword-matching of standard Web search engines which only seek for the presence or absence of the query expression into the text.

Table 2. Correlations for IC-based measures against physician judgments according to the two different background corpus

Measure	Mayo Clinic Corpus (as presented in [19])	Web via LiveSearch (modified measures of section 4)
Lin	0.60	**0.62**
Jiang and Conrath	0.45	**0.58**
Resnik	0.45	**0.48**

Evaluating the results presented in table 2 and, considering that the correlation between the group of expert physicians involved in the manual evaluation was 0.68 (which represent an upper bound for a computerized approach), as expected, the correlation values obtained by the performed experiments are lower. What it is worth to note is that similarity values computed from the Web (with the formulas presented in section 4) correlate significantly better than those obtained from a domain-specific pre-processed corpus (presented in [19]). In some cases (Jiang & Conrath measure), the improvement is almost a 20% (0.45 vs 0.58) with respect to the upper bound and, in others (Lin measure), results are very close to the human judgments (with a correlation of 0.62 vs 0.68).

---

[3] http://www.live.com/

That is a very interesting conclusion, showing that the Web (accessed by means of publicly available web search engines), despite its generality, noise, lack of structure and unreliability of individual sources is able to provide a robust semantic evidence for concrete domains such as biomedicine. In this case, the size and heterogeneity of the Web aids to provide accurate estimations of domain information distribution at a social scale, which improves the distribution observed in a much more reliable and structured source. In addition, even though polysemy and synonymy may affect the computation of concept probabilities from word web hit counts (due to the limitations of the strict keyword matching algorithms implemented by search engines, as stated in sections 4 and 5), thanks to the reduced ambiguity of domain-specific words, results do not seem to be significantly affected.

6 Conclusions

In this paper, we studied the influence of the background corpus used by ontology-driven IC-based similarity measures when applied to a specific domain such as biomedicine.

In previous approaches, it was argued that, for supervised (ontology-based) similarity measures, a domain-specific corpus was needed to achieve reliable similarity values for domain-specific concepts [19]. Considering the characteristics of the Web and its success in previous attempts of exploiting it to tackle other language processing tasks [24], we adapted IC-based measures to compute concept probabilities from term web search hit counts.

According to the hypothesis stated in this paper, the evaluation has shown that using the Web, despite of being a priori less reliable, noisier and unstructured, similarity values are even more reliable (compared to human judgements) than those obtained from a domain-specific corpus (maintaining the same domain ontology, SNOMED-CT, as background knowledge). Moreover, the limitations of the strict keyword matching implemented by web search engines have not handicapped the results, providing better estimations than appearance frequencies computed from pre-processed domain data. Consequently, in this case, the necessity of having a domain corpus is no more required. This is a very interesting conclusion because, usually, domain corpus lacks of coverage, it has a reduced size or even it is not available due to the confidentiality of data.

From the experiments, we can conclude that the Web (in a raw and unprocessed manner) is a valid corpus from which to compute ontology-based semantic similarities in a concrete domain as biomedicine.

After this work, we plan to evaluate web-based similarity measures in other concrete domains for which ontologies are available (such as chemistry or computer science) and to study also how they perform in a general environment with domain independent -and potentially ambiguous- terms. In that last case, it is expected that the result's quality will be compromised by the inaccurate estimation of web-based concept probabilities due to language ambiguity. So, additional strategies may be needed in order to contextualize queries for concept probability estimation by exploiting available ontological knowledge (such as attaching the LCS to every web query).

Acknowledgements

The work is partially supported by the Universitat Rovira i Virgili (2009AIRE-04) and the DAMASK Spanish project (TIN2009-11005). Montserrat Batet is also supported by a research grant provided by the Universitat Rovira i Virgili.

References

1. Bollegala, D., Matsuo, Y., Ishizuka, M.: WebSim: A Web-based Semantic Similarity Measure. In: The 21st Annual Conference of the Japanese Society for Artificial Intelligence, pp. 757–766 (2007)
2. Brill, E.: Processing Natural Language without Natural Language Processing. In: Proceedings of the 4th International Conference on Computational Linguistics and Intelligent Text Processing, Mexico City, Mexico, pp. 360–369 (2003)
3. Budanitsky, A., Hirst, G.: Evaluating wordnet-based measures of semantic distance. Computational Linguistics 32(1), 13–47 (2006)
4. Burgun, A., Bodenreider, O.: Comparing terms, concepts and semantic classes in WordNet and the Unified Medical Language System. In: Proceedings of the NAACL 2001 Workshop: WordNet and other lexical resources: Applications, extensions and customizations, Pittsburgh, PA, pp. 77–82 (2001)
5. Caviedes, J., Cimino, J.: Towards the development of a conceptual distance metric for the UMLS. Journal of Biomedical Informatics 37, 77–85 (2004)
6. Church, K.W., Gale, W., Hanks, P., Hindle, D.: Using Statistics in Lexical Analysis. In: Proceedings of Lexical Acquisition: Exploiting On-Line Resources to Build a Lexicon, pp. 115–164 (1991)
7. Cilibrasi, R., Vitanyi, P.M.B.: The Google Similarity Distance. IEEE Transaction on Know-ledge and Data Engineering 19(3), 370–383 (2006)
8. Cimiano, P.: Ontology Learning and Population from Text. In: Algorithms, Evaluation and Applications. Springer, Heidelberg (2006)
9. Fellbaum, C.: WordNet: An Electronic Lexical Database. MIT Press, Cambridge (1998)
10. Iosif, E., Potamianos, A.: Unsupervised Semantic Similarity Computation using Web Search Engines. In: Proceedings of the International Conference on Web Intelligence, pp. 381–387 (2007)
11. Jiang, J., Conrath, D.: Semantic similarity based on corpus statistics and lexical taxonomy. In: Proceedings of the International Conference on Research in Computational Linguistics, Japan, pp. 19–33 (1997)
12. Leacock, C., Chodorow, M.: Combining local context and WordNet similarity for word sense identification. In: Fellbaum (ed.) WordNet: An electronic lexical database, pp. 265–283. MIT Press, Cambridge (1998)
13. Lee, J.H., Kim, M.H., Lee, Y.J.: Information retrieval based on conceptual distance in is-a hierarchies. Journal of Documentation 49(2), 188–207 (1993)
14. Lin, D.: An information-theoretic definition of similarity. In: Proceedings of the 15th International Conf. on Machine Learning, pp. 296–304. Morgan Kaufmann, San Francisco (1998)
15. Lord, P., Stevens, R., Brass, A., Goble, C.: Investigating semantic similarity measures across the gene ontology: the relationship between sequence and annotation. Bioinformatics 19(10), 1275–1283 (2003)

16. Miller, G., Leacock, C., Tengi, R., Bunker, R.T.: A Semantic Concordance. In: Proceedings of ARPA Workshop on Human Language Technology, Morristown, USA, pp. 303–308. Association for Computational Linguistics (1993)

17. Miller, G.A., Charles, W.G.: Contextual correlates of semantic similarity. Language and Cognitive Processes 6(1), 1–28 (1991)

18. Neugyn, H.A., Al-Mubaid, H.: New Ontology-based Semantic Similarity Measure for the Biomedical Domain. In: IEEE Conference on Granular Computing, Atlanta, GA, USA, pp. 623–628 (2006)

19. Pedersen, T., Pakhomov, S., Patwardhan, S., Chute, C.: Measures of semantic similarity and relatedness in the biomedical domain. Journal of Biomedical Informatics 40, 288–299 (2007)

20. Rada, R., Mili, H., Bichnell, E., Blettner, M.: Development and application of a metric on semantic nets. IEEE Transactions on Systems, Man and Cybernetics 9(1), 17–30 (1989)

21. Resnik, P.: Semantic similarity in a taxonomy: An information-based measure and its application to problems of ambiguity in natural language. Journal of Artificial Intelligence. Research 11, 95–130 (1999)

22. Resnik, P.: Using information content to evaluate semantic similarity in a taxonomy. In: Proceedings of 14th International Joint Conference on Artificial Intelligence, pp. 448–453 (1995)

23. Sánchez, D., Moreno, A.: Learning non-taxonomic relationships from web documents for domain ontology construction. Data Knowledge Engineering 63(3), 600–623 (2008)

24. Turney, P.D.: Mining the Web for synonyms: PMI-IR versus LSA on TOEFL. In: Flach, P.A., De Raedt, L. (eds.) ECML 2001. LNCS (LNAI), vol. 2167, pp. 491–499. Springer, Heidelberg (2001)

25. Wilbu, W., Yang, Y.: An analysis of statistical term strength and its use in the indexing and retrieval of molecular biology texts. Computers in Biology and Medicine 26, 209–222 (1996)

26. Wu, Z., Palmer, M.: Verb semantics and lexical selection. In: Proceedings of the 32nd annual Meeting of the Association for Computational Linguistics, New Mexico, USA, pp. 133–138 (1994)

An Anytime Algorithm for Computing Inconsistency Measurement*

Yue Ma[1], Guilin Qi[2], Guohui Xiao[3], Pascal Hitzler[4], and Zuoquan Lin[3]

[1] Institute LIPN, Université Paris-Nord (LIPN - UMR 7030), France
[2] School of Computer Science and Engineering, Southeast University, Nanjing, China
[3] Department of Information Science, Peking University, China
[4] Kno.e.sis Center, Wright State University, Dayton, OH, USA
yue.ma@lipn.univ-paris13.fr, gqi@aifb.uni-karlsruhe.de,
{xgh,lz}@is.pku.edu.cn, pascal@pascal-hitzler.de

Abstract. Measuring inconsistency degrees of inconsistent knowledge bases is an important problem as it provides context information for facilitating inconsistency handling. Many methods have been proposed to solve this problem and a main class of them is based on some kind of paraconsistent semantics. In this paper, we consider the computational aspects of inconsistency degrees of propositional knowledge bases under 4-valued semantics. We first analyze its computational complexity. As it turns out that computing the exact inconsistency degree is intractable, we then propose an anytime algorithm that provides tractable approximation of the inconsistency degree from above and below. We show that our algorithm satisfies some desirable properties and give experimental results of our implementation of the algorithm.

1 Introduction

Inconsistency handling is one of the central problems in the field of knowledge representation. Recently, there is an increasing interest in quantifying inconsistency in an inconsistent knowledge base. This is because it is not fine-grained enough to simply say that two inconsistent knowledge bases contain the same amount of inconsistency. Indeed, it has been shown that analyzing inconsistency is helpful to decide how to act on inconsistency [1], i.e. whether to ignore it or to resolve it. Furthermore, measuring inconsistency in a knowledge base can provide some context information which can be used to resolve inconsistency [2,3,4], and proves useful in different scenarios such as Software Engineering [5].

Different approaches to measuring inconsistency are based on different views of atomic inconsistency [3]. Syntactic ones put atomicity to formulae, such as taking maximal consistent subsets of formulae [6] or minimal inconsistent sets [7]. Semantic ones put atomicity to propositional letters, such as considering the conflicting propositional letters based on some kind of paraconsistent model [8,2,3,9,10]. In this paper, we focus on the computational aspect of a 4-valued semantics based inconsistency degree which is among the latter view.

---

* We acknowledge support by OSEO, agence nationale de valorisation de la recherche in the Quaero project.

D. Karagiannis and Z. Jin (Eds.): KSEM 2009, LNAI 5914, pp. 29–40, 2009.

The main contributions of this paper lie in Section 4, 5, and 6 with new proposed interesting theorems (their proofs are omitted due to space limitation). In Section 4, we show that computing exact inconsistency degrees is a computational problem of high complexity (Θ_2^p-complete). In Section 5, we present an anytime algorithm to provide tractable approximations of the inconsistency degree from above and below, by computing *the lower* and *upper bounds* defined in this paper. We show that our algorithm satisfies some desirable properties. Section 6 will give experimental explanations of the algorithm. To the best of our knowledge, this is the first work that (1) analyzes the complexity issues of computing the inconsistency degree and that (2) attempts to alleviate the intractability of computing the exact inconsistency degree for full propositional logic by approximating it from above and from below in an anytime manner. Our results show that the computation of approximating inconsistency degree can be done tractable; and can be performed to full propositional knowledge bases instead of restricting to CNF to design a tractable paraconsistent reasoning [11].

2   Related Work

Most effort has been directed at theoretical accounts of inconsistency measures, i.e. its definitions, properties, and possible applications. But few papers focus on the computational aspect of inconsistency degree. Among the syntactic approaches, [6] shows the possibility to compute inconsistency degrees using the simplex method. Among the semantics methods, [12] and [10] provide algorithms for computing inconsistency degrees that can be implemented. The algorithm in [10] only deals with KBs consisting of first-order formulas in the form $Q_1 x_1, ..., Q_n x_n . \bigwedge_i (P_i(t_1, ..., t_{m_i}) \wedge \neg P_i(t_1, ..., t_{m_i}))$, where $Q_1, ..., Q_n$ are universal or existential quantifiers. In [12], an algorithm is proposed for full FOL logic. Although it can be applied to measure inconsistency in propositional logic, its computational complexity is too high to be used in in general cases. The anytime algorithm proposed in this paper for computing approximating inconsistency degrees can avoid these shortcomings.

Although our algorithm is inspired by the algorithm in [12], It is significantly different from the existing one. Firstly, ours is motivated by the theoretical results of the tractability of *S-4 entailment* (Theorems 3,4,5). In contrast, the algorithm in [12] is based on a reduction to hard SAT instances, which makes it inherently intractable. Secondly, ours is designed towards obtaining an approximation with guaranteed lower and upper bounds that gradually converge to the exact solution. Thirdly, we implement a new strategy to achieve polynomial time approximations. We present the preliminary evaluation results of the implementation of the algorithm in Section 6. Our evaluation results show our algorithm outperforms that given in [12] and the approximating values are reasonable to replace the exact inconsistency degree.

3   Preliminaries

Let $\mathcal{P}$ be a countable set of propositional letters. We concentrate on the classical propositional language formed by the usual Boolean connectives $\wedge$ (conjunction), $\vee$ (disjunction), $\rightarrow$ (implication), and $\neg$ (negation). A propositional knowledge base K consists of a finite set of formulae over that language. We use $Var(K)$ for the set of propositional letters used in K and $|S|$ for the cardinality of S for any set S.

Next we give a brief introduction on Belnap's four-valued (4-valued) semantics. Compared to two truth values used by classical semantics, the set of truth values for four-valued semantics [13,14] contains four elements: *true, false, unknown* and *both*, written by t, f, N, B, respectively. The truth value B stands for contradictory information, hence four-valued logic leads itself to dealing with inconsistencies. The four truth values together with the ordering $\preceq$ defined below form a lattice, denoted by $\mathbf{FOUR} = (\{t, f, B, N\}, \preceq)$: $f \preceq N \preceq t, f \preceq B \preceq t, N \npreceq B, B \npreceq N$. The four-valued semantics of connectives $\vee, \wedge$ are defined according to the upper and lower bounds of two elements based on the ordering $\preceq$, respectively, and the operator $\neg$ is defined as $\neg t = f, \neg f = t, \neg B = B$, and $\neg N = N$. The designated set of $\mathbf{FOUR}$ is $\{t, B\}$. So a four-valued interpretation $\mathfrak{I}$ is a *4-model* of a knowledge base K if and only if for each formula $\phi \in K, \phi^{\mathfrak{I}} \in \{t, B\}$. A knowledge base which has a 4-model is called *4-valued satisfiable*. A knowledge base K 4-valued entails a formula φ, written $K \models_4 \varphi$, if and only if each 4-model of K is a 4-model of φ. We write K for a knowledge base, and $\mathcal{M}_4(K)$ for the set of 4-models of K throughout this paper. Four-valued semantics provides a novel way to define inconsistentcy measurements [1].

Let $\mathfrak{I}$ be a four-valued model of K. The *inconsistency degree of K with respect to $\mathfrak{I}$*, denoted $\mathrm{Inc}_{\mathfrak{I}}(K)$, is a value in $[0, 1]$ defined as $\mathrm{Inc}_{\mathfrak{I}}(K) = \frac{|Conflict(\mathfrak{I},K)|}{|Var(K)|}$, where $Conflict(\mathfrak{I}, K) = \{p \mid p \in Var(K), p^{\mathfrak{I}} = B\}$. It measures to what extent a given knowledge base K contains inconsistencies with respect to its 4-model $\mathfrak{I}$. Preferred models defined below are used to define inconsistency degrees and especially useful to explain our approximating algorithm later.

Definition 1 (Preferred Models). *The set of preferred models, written PreferModel (K), is defined as PreferModel$(K) = \{\mathfrak{I} \mid \forall \mathfrak{I}' \in \mathcal{M}_4(K), Inc_{\mathfrak{I}}(K) \leq Inc_{\mathfrak{I}'}(K)\}$.*

Definition 2 (Inconsistency Degree). *The* inconsistency degree of K, denoted by $ID(K)$, is defined as the value $Inc_{\mathfrak{I}}(K)$, where $\mathfrak{I} \in PreferModels(K)$.

Example 1. *Let $K = \{p, \neg p \vee q, \neg q, r\}$. Consider two 4-valued models $\mathfrak{I}_1$ and $\mathfrak{I}_2$ of K with $p^{\mathfrak{I}_1} = t, q^{\mathfrak{I}_1} = B, r^{\mathfrak{I}_1} = t$; and $p^{\mathfrak{I}_2} = B, q^{\mathfrak{I}_2} = B, r^{\mathfrak{I}_2} = t$. We have $Inc_{\mathfrak{I}_1}(K) = \frac{1}{3}$, while $Inc_{\mathfrak{I}_2}(K) = \frac{2}{3}$. Moreover, $\mathfrak{I}_1$ is a preferred model of K because there is no other 4-model $\mathfrak{I}'$ of K such that $Inc_{\mathfrak{I}'}(K) < Inc_{\mathfrak{I}_1}(K)$. Then $ID(K) = \frac{1}{3}$.*

One way to compute inconsistency degree is to recast the algorithm proposed in [12] to propositional knowledge bases, where *S-4 semantics* defined as follows is used:

Definition 3 (S-4 Model). *For any given set $S \subseteq Var(K)$, an interpretation $\mathfrak{I}$ is called an S-4 model of K if and only if $\mathfrak{I} \in \mathcal{M}_4(K)$ and satisfies the following condition:*

$$\mathfrak{I}(p) \in \begin{cases} \{B\} & \text{if } p \in Var(K) \setminus S, \\ \{N, t, f\} & \text{if } p \in S. \end{cases}$$

For a given $S \subseteq Var(K)$, the knowledge base K is called *S-4 unsatisfiable* iff. it has no S-4 model. Let φ be a formula and $Var(\{\varphi\}) \subseteq Var(K)$. φ is *S-4 entailed* by K, written $K \models_S^4 \varphi$, iff. each S-4 model of K is an S-4 model of φ.

Theorem 1 ([12]). *For any KB K, we have $ID(K) = 1 - A/|Var(K)|$, where $A = \max\{|S| : S \subseteq Var(K), K \text{ is } S\text{-4 satisfiable}\}$.*

Theorem 1 shows that the computation of $ID(K)$ can be reduced to the problem of computing the maximal cardinality of subsets S of $Var(K)$ such that K is S-4 satisfiable.

4   Computational Complexities

Apart from any particular algorithm, let us study the computational complexity of the inconsistency degree to see how hard the problem itself is. First we define following computation problems related inconsistency degrees:

- $ID_{\leq d}$ (resp. $ID_{<d}$, $ID_{\geq d}$, $ID_{>d}$): Given a propositional knowledge base K and a number $d \in [0, 1]$, is $ID(K) \leq d$ (resp. $ID(K) < d$, $ID(K) \geq d$, $ID(K) > d$)?
- EXACT-ID: Given a propositional knowledge base K and a number $d \in [0, 1]$, is $ID(K) = d$?
- ID: Given a propositional knowledge base K, what is the value of $ID(K)$?

We have the complexities of these problems indicated by following theorem.

Theorem 2. *$ID_{\leq d}$ and $ID_{<d}$ are **NP**-complete; $ID_{\geq d}$ and $ID_{>d}$ are **coNP**-complete; EXACT-ID is **DP**-complete; ID is $\mathbf{FP}^{\mathbf{NP}[\log n]}$-complete[1].*

5   Anytime Algorithm

According to results shown in the previous section, computing inconsistency degrees is usually intractable. In this section, we propose an anytime algorithm to approximate the exact inconsistency degree. Our results show that in P-time we can get an interval containing the accurate value of $ID(K)$. We first clarify some definitions which will be used to explain our algorithm.

5.1   Formal Definitions

Definition 4. *(Bounding Values) A real number x (resp. y) is a lower (resp. an upper) bounding value of the inconsistency degree of K, if and only if $x \leq ID(K)$ (resp. $ID(K) \leq y$).*

Intuitively, a pair of lower and upper bounding values characterizes an interval containing the exact inconsistency degree of a knowledge base. For simplicity, lower (resp. an upper) bounding value is called *lower (resp. upper) bound.*

---

[1] A language L is in the class **DP**[15] iff there are two languages $L_1 \in$ **NP** and $L_2 \in$ **coNP** such that $L = L_1 \cap L_2$. Complexity $\mathbf{P}^{\mathbf{NP}[\log n]}$ is defined to be the class of all languages decided by a polynomial-time oracle machine which on input x asks a total of $\mathcal{O}(\log |x|)$ SAT (or any other problem in **NP**) queries. $\mathbf{FP}^{\mathbf{NP}[\log n]}$ is the corresponding class of functions.

Definition 5. *(Bounding Models) A four-valued interpretation $\mathfrak{I}'$ is a lower (resp. an upper) bounding model of K if and only if for any preferred model $\mathfrak{I}$ of K, Condition 1 holds (resp. Condition 2 holds and $\mathfrak{I}' \in \mathcal{M}_4(K)$):*

$$\text{Condition 1: } |Conflict(\mathfrak{I}', K)| \leq |Conflict(\mathfrak{I}, K)|$$
$$\text{Condition 2: } |Conflict(\mathfrak{I}', K)| \geq |Conflict(\mathfrak{I}, K)|$$

Intuitively, the lower and upper bounding models of K are approximations of preferred models from below and above. We call two-valued interpretations $\mathfrak{I}$ trivial lower bounding models since $Conflict(\mathfrak{I}, K) = 0$ and $ID(K) = 0$ always holds. We are only interested in nontrivial bounding models for inconsistent knowledge bases, which can produce a nonzero lower bound of $ID(K)$.

Example 2. *(Example 1 continued) K has a lower bounding model $\mathfrak{I}_3$ and an upper bounding model $\mathfrak{I}_4$ defined as: $p^{\mathfrak{I}_3} = t, q^{\mathfrak{I}_3} = t, r^{\mathfrak{I}_3} = t$; and $p^{\mathfrak{I}_4} = B, q^{\mathfrak{I}_4} = B, r^{\mathfrak{I}_4} = t$.*

Next proposition gives a connection between *lower (resp. upper) bounds* and *lower (resp. upper) bounding models*.

Proposition 1. *If $\mathfrak{I}$ is a lower (an upper) bounding model of K, $Inc_{\mathfrak{I}}(K)$ is a lower (an upper) bounding value of $ID(K)$.*

By borrowing the idea of guidelines for a theory of approximating reasoning [16], we require that an anytime approximating algorithm for computing inconsistency degrees should be able to produce two sequences $r_1, ..., r_m$ and $r^1, ..., r^k$:

$$r_1 \leq ... \leq r_m \leq ID(K) \leq r^k \leq ... \leq r^1, \tag{1}$$

such that these two sequences have the following properties:

- The length of each sequence is polynomial w.r.t $|K|$;
- Computing r_1 and r^1 are both tractable. Generally, computing r_j and r^j becomes exponentially harder as j increases, but it is not harder than computing $ID(K)$.
- Since computing r_i and r^j could become intractable as i and j increase, we need to find functions $f(|K|)$ and $g(|K|)$ such that computing r_i and r^j both stay tractable as long as $i \leq f(|K|)$ and $j \leq g(|K|)$.
- each r_i (r^j) corresponds to a lower (an upper) bounding model, which indicates the sense of the two sequences.

In the rest, we will describe an anytime algorithm which can produce such two sequences.

5.2   Tractable Approximations from Above and Below

We know that S-4 entailment is generally intractable, which makes algorithms based on S-4 semantics to compute inconsistency degrees time-consuming. In this section, we will distinguish a tractable case of S-4 entailment (proportional to the size of input knowledge base), by which we can compute approximating inconsistency degrees.

Lemma 1. *Let $S = \{p_1, ..., p_k\}$ be a subset of $Var(K)$ and φ be a formula such that $Var(\varphi) \subseteq Var(K)$. $K \models_S^4 \varphi$ if and only if*

$$K \wedge \bigwedge_{q \in Var(K) \setminus S} (q \wedge \neg q) \models_4 \varphi \vee (c_1 \vee ... \vee c_k)$$

holds for any combination $\{c_1, ..., c_k\}$, where each c_i is either p_i or $\neg p_i (1 \leq i \leq k)$.

This lemma shows a way to reduce the S-4 entailment to the 4-entailment. Specially note that if φ is in CNF (conjunctive normal formal), the righthand of the reduced 4-entailment maintains CNF form by a little bit of rewriting, as follows: Suppose $\varphi = C_1 \wedge ... \wedge C_n$. Then $\varphi \vee (c_1 \vee ... \vee c_k) = (C_1 \vee c_1 \vee ... \vee c_k) \wedge ... \wedge (C_n \vee c_1 \vee ... \vee c_k)$ which is still in CNF and its size is linear to that of $\varphi \vee (c_1 \vee ... \vee c_k)$.

Lemma 2 ([17]). *For K in any form and φ in CNF, there exists an algorithm for deciding if $K \models_4 \varphi$ in $\mathcal{O}(|K| \cdot |\varphi|)$ time.*

By Lemma 1 and 2, we have the following theorem:

Theorem 3 (Complexity). *There exists an algorithm for deciding if $K \models_S^4 \varphi$ and deciding if K is S-4 satisfiable in $\mathcal{O}(|K||\varphi||S| \cdot 2^{|S|})$ and $\mathcal{O}(|K||S| \cdot 2^{|S|})$ time, respectively.*

Theorem 3 shows that S-4 entailment and S-4 satisfiability can both be decided in polynomial time w.r.t the size of K, exponential w.r.t that of S, though. So they can be justified in P-time if $|S|$ is limited by a logarithmic function of $|K|$.

Next we study how to use Theorem 3 to tractably compute upper and lower bounding values of inconsistency degrees.

Lemma 3. *Given two sets S and S' satisfying $S \subseteq S' \subseteq \mathcal{P}$, if a theory K is S-4 unsatisfiable, then it is S'-4 unsatisfiable.*

By Lemma 3, we get a way to compute upper and lower bounds of $ID(K)$ shown by Theorems 4 and 5, respectively.

Theorem 4. *Given $S \subseteq Var(K)$, if K is S-4 satisfiable, then $ID(K) \leq 1 - |S|/|Var(K)|$.*

Theorems 3 and 4 together show that for a monotonic sequence of sets $S_1, ..., S_k$, where $|S_i| < |S_{i+1}|$ for any $1 \leq i \leq k - 1$, if we can show that K is S_i-4 $(i = 1, ..., k)$ satisfiable one by one, then we can get a sequence of decreasing upper bounding values of the inconsistency degree of K in time $\mathcal{O}(|K||S_i| \cdot 2^{|S_i|})$. If $|S_i| = \mathcal{O}(\log |K|)$, it is easy to see that the computation of an upper bound is done in polynomial time with respect to the size of K. In the worst case (i.e., when $S = Var(K)$), the complexity of the method coincides with the result that ID$_\leq$ is NP-complete (Theorem 2).

Theorem 5. *For a given w $(1 \leq w \leq |Var(K)|)$, if for each w-size subset S of $Var(K)$, K is S-4 unsatisfiable, then $ID(K) \geq 1 - (w - 1)/|Var(K)|$.*

Theorems 3 and 5 together show that for a monotonic sequence of sets $S_1, ..., S_m$ satisfying $|S_i| < |S_{i+1}|$, if we can prove that K is $|S_i|$-4 unsatisfiable[2] for each $i \in [1, m]$,

---

[2] For the sake of simplicity, we say that K is l-4 satisfiable for $l \in \mathbb{N}$, if there is a subset $S \subseteq Var(K)$ such that K is S-4 satisfiable. We say that K is l-4 unsatisfiable if K is not l-4 satisfiable.

then we can get a series of increasing lower bounds of the inconsistency degree of K. For each w, it needs at most $\binom{|Var(K)|}{w}$ times tests of S-4 unsatisfiability. So it takes $\mathcal{O}(\binom{|Var(K)|}{w}|K|w \cdot 2^w)$ time to compute a lower bound $1 - (w-1)/|Var(K)|$. If and only if w is limited by a constant, we have that each lower bound is obtained in polynomial time by Proposition 2.

Proposition 2. *Let $f(n) = \mathcal{O}(\binom{n}{k} \cdot 2^k)$ where $0 \le k \le n$. There exists a $p \in \mathbb{N}$ such that $f(n) = \mathcal{O}(n^p)$ if and only if k is limited by a constant which is independent of n.*

Suppose r_i, r^j in Inequation 1 are defined as follows:

$$r^j = 1 - |S|/|Var(K)|, \text{ where } K \text{ is } |S|\text{-4 satisfiable, } j = |S|;$$

$$r_i = 1 - \frac{|S| - 1}{|Var(K)|}, \text{ where } K \text{ is } |S|\text{-4 unsatisfiable, } i = |S|.$$

By Theorems 3, 4 and 5 and Proposition 2, we get a way to compute the upper and lower bounds of $ID(K)$ which satisfy: if $j \le \log(|K|)$ and $i \le M$ (M is a constant independent of $|K|$), r^j and r_i are computed in polynomial time; Both i and j cannot be greater than $|Var(K)|$. This is a typical approximation process of a **NP**-complete problem $ID_{\ge d}$ (resp. co**NP**-complete problem $ID_{\le d}$) via polynomial intermediate steps, because each intermediate step provides a partial solution which is an upper (resp. lower) bound of $ID(K)$.

Example 3. *Suppose $K = \{p_i \lor q_j, \neg p_i, \neg q_j \mid 1 \le i, j \le N\}$. So $|Var(K)| = 2N$. To know whether $ID(K) < \frac{3}{4}$, by Theorem 4 we only need to find an S of size $\lceil \frac{2N}{4} \rceil$ such that K is S-4 satisfiable. This is true by choosing $S = \{p_i \mid 1 \le i \le \lceil \frac{2N}{4} \rceil\}$. To know whether $ID(K) > \frac{1}{3}$, Theorem 5 tells us to check whether K is S-4 unsatisfiable for all S of size $\lfloor \frac{4N}{3} \rfloor + 1$. This is true also. So $ID(K) \in [\frac{1}{3}, \frac{3}{4}]$.*

An interesting consequence of the above theoretical results is that we can compute the exact inconsistency of some knowledge bases in P-time. Let us first look at an example.

Example 4. *Let $K = \{(p_i \lor p_{i+1}) \land (\neg p_i \lor \neg p_{i+1}), p_{i_1} \land \ldots \land p_{i_{N-5}}, \neg p_{j_1} \land \ldots \land \neg p_{j_{N-10}}, p_{2t}, \neg p_{3j+1} \lor \neg p_{5u+2}, \}(1 \le i \le N-1, 1 \le 2t, 3j+1, 5u+2 \le N)$. $Var(K) = N$. To approximate $ID(K)$, we can check whether K is l-4 satisfiable for l going larger from 1 by one increase on the value each time. Obviously, K's inconsistency degree is close to 1 if $N \gg 10$. By Theorem 3, we can see that all of these operations can be done in P-time before the exact value obtained.*

More formally, we have the following proposition.

Proposition 3. *If $ID(K) \ge 1 - M/|Var(K)|$, where M is an arbitrary constant which is independent of $|K|$, then $ID(K)$ can be computed in polynomial time.*

Given a knowledge base K with $|Var(K)| = n$. By the analysis given after Theorem 4 and Theorem 5, we know that in the worst case, it takes $\mathcal{O}(\binom{n}{w}|K|w2^w)$ time to get an upper (resp. a lower) bounding value. By Fermat's Lemma[3], its maximal value is near $w = \lceil \frac{2n+1}{3} \rceil$ when n is big enough. It means that *to do dichotomy directly on size $\frac{n}{2}$ will be of high complexity.* To get upper and lower bounding values in P-time instead of going to intractable computation straight, we propose Algorithm 1, which consists of two stages: The first one is to localize an interval $[l_1, l_2]$ that contains the inconsistency degree (line 1-8), while returning upper and lower bounding values in P-time; The second one is to pursue more accurate approximations within the interval $[l_1, l_2]$ by binary search (line 9-17).

Algorithm 1 is an anytime algorithm that can be interrupted at any time and returns a pair of upper and lower bounding values of the exact inconsistency degree. It has five parameters: the knowledge base K we are interested in; the precision threshold ε which is used to control the precision of the returned results; the constant $M \ll |Var(K)|$ to guarantee that the computation begins with tractable approximations; a pair of positive reals a, b which determines a linear function $h(l_2) = al_2 + b$ that updates the interval's right extreme point l_2 by $h(l_2)$ during the first stage (line 5). $h(\cdot)$ decides how to choose the sizes l to test l-4 satisfiability of K. For example, if $h(l_2) = l_2 + 2$, line 5 updates l from i to $i + 1$ (suitable for $ID(K)$ near 1); If $h(l_2) = 2l_2$, line 5 updates l from i to $2i$ (suitable for $ID(K)$ near 0.5); While if $h(l_2) = 2(|Var(K)| - M)$, line 5 updates l by $|Var(K)| - M$ (suitable for $ID(K)$ near 0). We remark that $h(l_2)$ can be replaced by other functions.

Next we give detailed explanations about Algorithm 1. To guarantee that it runs in P-time run at the beginning to return approximations, we begin with a far smaller search interval $[l_1, l_2] = [0, M]$ compared to $|Var(K)|$. The *while* block (line 3) iteratively tests whether the difference between upper and lower bounding values is still lager than the precision threshold and whether K is l-satisfiable, where $l = \lceil \frac{l_2}{2} \rceil$. If both yes, the upper bound r_+ is updated, the testing interval becomes $[l, h(l_2)]$, and the iteration continues; Otherwise (line 7), the lower bound r_- is updated and the search interval becomes $[l_1, l]$. This completes the first part of the algorithm to localize an interval. If $r_+ - r_-$ is already below the precision threshold, the algorithm terminates (line 8). Otherwise, we get an interval $[l_1, l_2]$ such that K is l_1-4 satisfiable and l_2-4 unsatisfiable. Then the algorithm turns to the second "while" iteration (line 9) which executes binary search within the search internal $[l_1, l_2]$ found in the first stage. If there is a subset $|S| = l_1 + \lceil \frac{l_2 - l_1}{2} \rceil$ such that K is S-4 satisfiable, then the search internal shortens to the right half part of $[l_1, l_2]$ (line 12), otherwise to the left half part (line 14). *During this stage, K keeps l_2-4 unsatisfiable and l_1-4 satisfiable for $[l_1, l_2]$.* Until $r_+ - r_-$ below the precision threshold, the algorithm finishes and returns upper and lower bounds.

Theorem 6 (Correctness of Algorithm 1). *Let r_+ and r_- be values computed by Algorithm 1. We have $r_- \leq ID(K)$ and $r_+ \geq ID(K)$. Moreover, $r_+ = r_- = ID(K)$ if $\varepsilon = 0$.*

The following example gives a detailed illustration.

Example 5. *(Example 3 contd.) Let $\varepsilon = 0.1, h(l_2) = 2l_2$, and $M = 4 \ll N$. Algorithm 1 processes on K as follows:*

---

[3] V.A. Zorich, *Mathematical Analysis*, Springer, 2004.

Algorithm 1. Approx_Incons_Degree(K, ε, M, a, b)

Input: KB K; precision threshold $\varepsilon \in [0, 1[$; constant $M \ll |Var(K)|$; $a, b \in \mathbb{R}^+$
Output: Lower bound r_- and upper bound r_+ of $ID(K)$
1: $r_- \leftarrow 0; r_+ \leftarrow 1$ {Initial lower and upper bounds}
2: $\bar{\varepsilon} \leftarrow r_+ - r_-; n \leftarrow |Var(K)|; l_1 \leftarrow 0; l_2 \leftarrow M; l \leftarrow \lceil \frac{l_2}{2} \rceil$
3: **while** $\bar{\varepsilon} > \varepsilon$ and K is l-4 satisfiable **do**
4:      $r_+ \leftarrow (1 - l/n); \bar{\varepsilon} \leftarrow r_+ - r_-$ {Update upper bound}
5:      $l_1 \leftarrow l; l_2 \leftarrow h(l_2); l \leftarrow \lceil \frac{l_2}{2} \rceil$ {Update search interval}
6: **end while**
7: $r_- \leftarrow 1 - (l - 1)/n; \bar{\varepsilon} \leftarrow r_+ - r_-; l_2 \leftarrow l$
8: **if** $\bar{\varepsilon} \leq \varepsilon$ **then**   **return** r_+ and r_- **end if**
9: **while** $\bar{\varepsilon} > \varepsilon$ **do**
10:      $l \leftarrow l_1 + \lceil \frac{l_2 - l_1}{2} \rceil$
11:      **if** K is l-4 satisfiable **then**
12:          $r_+ \leftarrow (1 - l/n); \bar{\varepsilon} \leftarrow r_+ - r_-; l_1 \leftarrow l$
13:      **else**
14:          $r_- \leftarrow 1 - (l - 1)/n; \bar{\varepsilon} \leftarrow r_+ - r_-; l_2 \leftarrow l$
15:      **end if**
16: **end while**
17: **return** r_+ and r_-

Denote the initial search interval $[l_1^0, l_2^0] = [0, 4]$. After initializations, $l = 2$ and line 3 is executed. Obviously, K is S-4 satisfiable for some $|S| = l$ (e.g. $S = \{p_1, p_2\}$). So we get a newer upper bound $r_+ = \frac{2N-l}{2N}$. Meanwhile, the difference between upper and lower bounds $\bar{\varepsilon}$ becomes $\frac{2N-l}{2N} > \varepsilon$, and the search interval is updated as $[l_1^1, l_2^1] = [l, 2l_2]$ and $l = 4$.

Stage 1. The while iteration from line 3 is repeatedly executed with double size increase of l each time. After c times such that $2^{c-1} \leq N < 2^c$, $l = 2^c$ and K becomes l-4 unsatisfiable. The localized interval is $[2^{c-1}, 2^c]$. It turns to line 7 to update the lower bound by $1 - \frac{l-1}{2N}$. The newest upper bound is $1 - 2^{c-2}/N$, so $\bar{\varepsilon} = 2^{c-2}/N$. If $\bar{\varepsilon} \leq \varepsilon$, algorithm ends by line 8. Otherwise, it turns to stage 2.

Stage 2. By dichotomy in the interval $[2^{c-1}, 2^c]$, algorithm terminates until $\bar{\varepsilon} \leq \varepsilon$.

Unlike Example 5, for the knowledge base in Example 4, since its inconsistency degree is quite close to 1, it becomes S-4 satisfiable for an S such that $|S|$ is less than a constant M. Therefore, after the first stage of Algorithm 1 applying on this knowledge base, the localized interval $[l_1, l_2]$ is bounded by M. For such an interval, the second stage of the algorithm runs in P-time according to Theorem 3 and Proposition 2. So Algorithm 1 is a P-time algorithm for the knowledge base given in Example 4. However, it fails for other knowledge bases whose inconsistency degrees are far less than 1. Fortunately, the following proposition shows that by setting the precision threshold ε properly, Algorithm 1 can be executed in P-time to return approximating values.

Proposition 4. *Let s be an arbitrary constant independent of $|K|$. If $\varepsilon \geq 1 - \frac{h^s(M)}{2|Var(K)|}$, where $h^s(\cdot)$ is s iterations of $h(\cdot)$, Algorithm 1 terminates in polynomial time with the difference between upper and lower bounds less than ε ($r_+ - r_- \leq \varepsilon$).*

The following proposition shows that r_- and r_+ computed by Algorithm 1 have a sound semantics in terms of *upper* and *lower bounding models* defined in Definition 5.

Proposition 5. *There is a lower (an upper) bounding model $\mathfrak{J}'$ ($\mathfrak{J}''$) of K such that $Inc_{\mathfrak{J}'}(K) = r_-$ ($Inc_{\mathfrak{J}''}(K) = r_+$).*

Summing up, we have achieved an anytime algorithm for approximately computing inconsistency degrees which is:

– *computationally tractable:* Each approximating step can be done in polynomial time if $|S|$ is limited by a logarithmic function for upper bounds (Theorems 3 and 4) and by a constant function for lower bounds (Theorems 3 and 5).
– *dual and semantical well-founded:* The accurate inconsistency degree is approximated both from above and from below (Theorem 6), corresponding to inconsistency degrees of some upper and lower bounding models of K (Proposition 5).
– *convergent:* More computation resource available, more precise values returned (Theorems 4 and 5). It always converges to the accurate value if there is no limitation of computation resource (Theorem 6) and terminates in polynomial time for special knowledge bases (Proposition 3).

6   Evaluation

Our algorithm has been implemented in Java using a computer with Intel E7300 2.66G, 4G, Windows Server 2008. Algorithm 1 gives a general framework to approximate inconsistency degrees from above and below. In our implementation, we set $M = 2, h(l_2) = l_2 + 2$. That is, the first *while* loop (see line 3) keeps testing l-4 satisfiability of K from $l = i$ to $i + 1$. So the interval $[l_1, l_2]$ localized in the first stage of the algorithm satisfies $l_2 = l_1 + 1$ and the second binary search is not necessary. According to our analysis in Section 5, this avoids direct binary search which needs to test all $\frac{n!}{(n/2)!(n/2)!}$ subsets of $Var(K)$, where $n = |Var(K)|$.

There are tow main sources of complexity to compute approximating inconsistency degrees: *the complexities of S-4 satisfiability* and *of search space*. The S-4 satisfiability that we implemented is based on the reduction given in Lemma 1 and the tractable algorithm for 4-satisfiability in [17]. Our experiments told us that search space could heavily affect efficiency. So we carefully designed a truncation strategy to limit the search space based on the monotonicity of S-4 unsatisfiability. That is, if we have found an S such that K is S-4 unsatisfiable, then we can prune all supersets S' of S which makes K S'-4 unsatisfiable. We implemented this strategy in breadth-first search on the binomial tree [18,19] of subsets of $Var(K)$.

Figure 1 shows the evaluation results over knowledge bases[4] in Example 3 with $|K| = N^2 + 2N$ and $|Var(k)| = 2N$ for $N = 5, 7, 8, 9, 10$. The left part of the figure shows how the preset precision threshold ε affects the run time performance of our algorithm: the smaller ε is, the longer it executes. If $\varepsilon \geq 0.7$, the algorithm terminated

---

[4] We use instances of Example 3 because they are the running examples through the paper and meet the worst cases of the algorithm (e.g. the *truncation strategy* discussed later cannot be applied). We want to show the performance of our algorithm in its worst case.

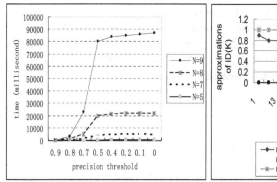
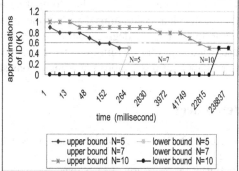

Fig. 1. Evaluation results over KBs in Example 3 with $|K| = N^2 + 2N$ and $|Var(K)| = 2N$ for $N = 5, 7, 8, 9, 10$

easily (at most $18.028s$ for $N = 9$ and much less for $N < 9$). The quality of the approximations at different time points is shown on the right part of the figure. The decreasing (resp. increasing) curves represent upper (resp. lower) bounds for $N = 5, 7, 10$, respectively. Note that the inconsistency degrees of all the three knowledge bases are 0.5.

For large knowledge bases, it is time-consuming to compute the exact inconsistency degrees. For example, for $N = 10$, our algorithm took 239.935s to get the accurate inconsistency degree. In contrast, by costing much less time, approximating values (upper bounds for these examples) can provide a good estimation of the exact value and are much easier to compute. For example, when $N = 10$, the algorithm told us that the inconsistency degree is less than 0.8 at 3.9s; and when $N = 5$, we got the upper bound 0.6 at 0.152s. Note that in these experiments, the lower bounds were updated slowly. In fact, the exact inconsistency degrees were obtained as soon as the first nonzero lower bounding values were returned. This is because we set $M = 2, h(l_2) = l_2 + 2$ in our implementation. If we set M and $h(\cdot)$ differently, the results will be changed, as shown in Example 3 in Section 5.

We need to point out that our truncation strategy cannot be applied to the test data used in the experiments because no subsets can be pruned. Therefore, although our experiments show the benefits of the approximations, our algorithm can increase significantly when the truncation strategy is applicable and if we carefully set M and $h(\cdot)$. Take $\{p_i, \neg p_j \mid 0 \leq i, j < 20, j \text{ is odd}\}$ for example, our optimized algorithm run less than 1s whilst it run over 5min without the truncation strategy.

7 Conclusion

In this paper, we investigated computational aspects of the inconsistency degree. We showed that the complexities of several decision problems about inconsistency degree are high in general. To compute inconsistency degrees more practically, we proposed an general framework of an anytime algorithm which is *computationally tractable, dual and semantical well-founded, and improvable and convergent.* The experimental results of our implementation show that computing approximating inconsistency degrees is much faster than computing the exact inconsistency degrees in general. The approximating inconsistency degrees can be useful in many applications, such as knowledge base evaluation and merging inconsistent knowledge bases. We will further study on the real applications of approximating inconsistency degree in the future work.

References

1. Hunter, A.: How to act on inconsistent news: Ignore, resolve, or reject. Data Knowl. Eng. 57, 221–239 (2006)
2. Hunter, A.: Measuring inconsistency in knowledge via quasi-classical models. In: Proc. of AAAI 2002, pp. 68–73. AAAI Press, Menlo Park (2002)
3. Hunter, A., Konieczny, S.: Approaches to measuring inconsistent information. In: Bertossi, L., Hunter, A., Schaub, T. (eds.) Inconsistency Tolerance. LNCS, vol. 3300, pp. 191–236. Springer, Heidelberg (2005)
4. Hunter, A., Konieczny, S.: Shapley inconsistency values. In: Proc. of KR 2006, pp. 249–259. AAAI Press, Menlo Park (2006)
5. Mu, K., Jin, Z., Lu, R., Liu, W.: Measuring inconsistency in requirements specifications. In: Godo, L. (ed.) ECSQARU 2005. LNCS (LNAI), vol. 3571, pp. 440–451. Springer, Heidelberg (2005)
6. Knight, K.: Measuring inconsistency. Journal of Philosophical Logic 31(1), 77–98 (2002)
7. Hunter, A., Konieczny, S.: Measuring inconsistency through minimal inconsistent sets. In: Proc. of KR 2008, pp. 358–366 (2008)
8. Grant, J.: Classifications for inconsistent theories. Notre Dame J. of Formal Logic 19, 435–444 (1978)
9. Grant, J., Hunter, A.: Measuring inconsistency in knowledgebases. Journal of Intelligent Information Systems 27, 159–184 (2006)
10. Grant, J., Hunter, A.: Analysing inconsistent first-order knowledge bases. Artif. Intell. 172, 1064–1093 (2008)
11. Coste-Marquis, S., Marquis, P.: A unit resolution-based approach to tractable and paraconsistent reasoning. In: Proc. of ECAI, pp. 803–807 (2004)
12. Ma, Y., Qi, G., Hitzler, P., Lin, Z.: An algorithm for computing inconsistency measurement by paraconsistent semantics. In: Mellouli, K. (ed.) ECSQARU 2007. LNCS (LNAI), vol. 4724, pp. 91–102. Springer, Heidelberg (2007)
13. Belnap, N.D.: A useful four-valued logic. In: Modern uses of multiple-valued logics, pp. 7–73. Reidel Publishing Company, Boston (1977)
14. Arieli, O., Avron, A.: The value of the four values. Artif. Intell. 102, 97–141 (1998)
15. Papadimitriou, C.H. (ed.): Computational Complexity. Addison Wesley, Reading (1994)
16. Schaerf, M., Cadoli, M.: Tractable reasoning via approximation. Artificial Intelligence 74, 249–310 (1995)
17. Cadoli, M., Schaerf, M.: On the complexity of entailment in propositional multivalued logics. Ann. Math. Artif. Intell. 18, 29–50 (1996)
18. Malouf, R.: Maximal consistent subsets. Computational Linguistics 33, 153–160 (2007)
19. Mu, K., Jin, Z., Liu, W., Zowghi, D.: An approach to measuring the significance of inconsistency in viewpoints framework. Technical report, Peking University (2008)

Forwarding Credible Information in Multi-agent Systems*

Patrick Krümpelmann[1,**], Luciano H. Tamargo[2,**], Alejandro J. García[2], and Marcelo A. Falappa[2]

[1] Technische Universität Dortmund, Germany
patrick.kruempelmann@udo.edu
[2] CONICET - Universidad Nacional del Sur, Bahía Blanca, Argentina
{lt,ajg,maf}@cs.uns.edu.ar

Abstract. In this work we extend the communication abilities of agents in multi-agent systems by enabling them to reason about the credibility of information to be shared with other agents. We propose a framework in which agents exchange sentences of a logical language enriched by meta-information. We discuss several possible approaches and present an advanced approach overcoming previously shown problems. For this, we make use of a calculation method for the plausibility of information known from approaches to belief dynamics in multi-agent systems. Moreover, we present how this can be implemented in a multi-agent system.

Keywords: Knowledge Representation, Plausibility, Multi-agent System.

1 Introduction

In this work, we focus on enhancing reasoning and communication skills of agents that are part of multi-agent systems (MAS). We consider a group of communicating and collaborating agents that share information and have beliefs about the credibility of their fellow agents. In our proposed framework agents can exchange logical information stored in these beliefs, annotated with meta-information concerning the credibility of this information. Thus, agents can acquire information from multiple sources and incorporate it into their proper beliefs.

In this paper we investigate how an agent can forward information to other agents that could have been acquired from other agents. In particular, we study how to rationally choose meta-information to be sent. At any point in time, each agent has initial beliefs (in form of evidential knowledge) and generic beliefs, as well as knowledge acquired from other agents. The agent has to deal with messages received both from other agents and from the environment itself. Similar to Dragoni [2] and Cantwell [1], in our approach informant agents can have different credibilities. In this article, as in [11], a credibility order among agents is introduced and, based on this order, a comparison criterion among beliefs is

---

* Partially supported by PIP-CONICET and Universidad Nacional del Sur.
** Krümpelmann and Tamargo are the first authors and Phd students.

D. Karagiannis and Z. Jin (Eds.): KSEM 2009, LNAI 5914, pp. 41–53, 2009.

defined. We propose to attach to each piece of information an agent identifier representing the credibility of the transferred information, as in [7] and [11]. The choice of the agent identifier that is to be sent along with the piece of information is crucial, as it influences the decision of the receiver about whether to accept the transmitted information. Thus, it is in the interest of the sending agent, and in fact in the interest of the whole coalition of agents, to choose this meta-information carefully. We present different ways to select this identifier and give a categorization of possible approaches, to later discuss their advantages and disadvantages. This discussion leads to the definition of a reasoning approach using the credibility of information that selects the most appropriate agent identifier to be sent. Later, we also show how we can adapt multi-source belief revision techniques for this selection operator and present the resulting operator by adapting work from [11]. Moreover, we present an implementation of this operator using answer set programming based on [6], using a common technique for the implementation of the reasoning component of agents in MAS. In this way, we are extending the reasoning component towards the computation of meta-information on the one hand, while improving the communication component by giving reasoning capabilities to it.

The reminder of the paper is organized as follows. In Section 2 we introduce the general framework in which we are working. Section 3 discusses diverse approaches to the forwarding of information in this setting. In Section 4 we introduce a forwarding criterion that uses a method of reasoning about the plausibility of information. Section 5 elaborates the implementation of our proposed approach for MAS and Section 6 concludes our work.

2   Framework

For the presentation of our work we adapt and extend the general communication framework from [7] by combining it with methods from [11]. Hence, our approach can easily be integrated in a complex, implemented multi-agent system like [7] and is compatible with many MAS standards. Agents interact with its fellow agents by means of *messages* (M). In this approach we assume that a message is at least a triple *(S,R,I)* where S is the sender, R is the set of receivers and I is the content of the message. As explained in [7], a message can have other components (as a speech act) but these are out of the scope of this paper.

When interacting, agents incorporate the received messages into their knowledge base in form of *information objects* (I). In this paper we focus on the representation of a knowledge base in form of information objects and on the transmission of information objects, abstracting from the actual communication act. Hence, for simplicity we consider only I from M in the following. For the identification of the individual agents we introduce a finite set of agent identifiers that is denoted as $\mathbb{A} = \{A_1, \ldots, A_n\}$.

Definition 1 (Information object [7]). *An* information object *is a tuple $I = (\phi, A_i)$, where ϕ is a sentence of some logical language $\mathcal{L}$ and $A_i \in \mathbb{A}$.*

In this paper we adopt a propositional language $\mathcal{L}$ with a complete set of Boolean connectives, namely $\{\neg, \wedge, \vee, \rightarrow, \leftrightarrow\}$. Also, we assume the existence of a logical consequence operator Cn which determines the logical closure of the set it is applied to. This operator satisfies *inclusion* ($B \subseteq Cn(B)$), *iteration* ($Cn(B) = Cn(Cn(B))$), and *monotonicity* (if $B \subseteq C$ then $Cn(B) \subseteq Cn(C)$) and includes the classical consequence operator. In general, we write $\phi \in Cn(B)$ as $B \vdash \phi$. The information objects defined above are used to represent an agent's belief base.

Definition 2 (Belief base [7,11]). *A belief base of an agent A_i is a set $K_{A_i} = \{I_0, \ldots, I_m\}$ containing information objects received from other agents and proper initial beliefs represented by information objects (ϕ, A_i). The set $\mathcal{K} = 2^{\mathcal{L} \times \mathbb{A}}$ represents the set of all belief bases.*

Next, two auxiliary functions are introduced in order to obtain the set of sentences and the set of agents that belong to a belief base $K \in \mathcal{K}$.

Definition 3 (Sentence function [11]). *The* sentence function*, $Sen : \mathcal{K} \to 2^{\mathcal{L}}$, is a function s.t. for a given belief base $K \in \mathcal{K}$, $Sen(K) = \{\phi : (\phi, A_i) \in K\}$.*

We call a belief base K consistent if $Cn(Sen(K))$ is consistent.

Definition 4 (Agent identifier function [11]). *The* agent identifier function*, $Ag : \mathcal{K} \to 2^{\mathbb{A}}$, is a function s.t. for a base $K \in \mathcal{K}$, $Ag(K) = \{A_i : (\phi, A_i) \in K\}$.*

As stated above, we investigate how an agent can forward information to other agents. In particular, we study how to rationally choose meta-information to be sent. In our approach agents can obtain information by receiving messages with information objects (I). Therefore, agents can use the agent identifier of I to evaluate the truthfulness of this particular information. For this reason we use an *assessment* function to represent the credibilities each agent gives to other agents known to it. These credibilities are kept in a separate structure to allow for a dynamic change of credibilities within time. According to this, we assume a set of credibility labels $\mathcal{C} = \{c_1, \ldots, c_n\}$ (known to all agents) with a strict order $\prec_c$ on them (*i.e.*, $\prec_c$ satisfies *transitivity* and *totality*).

Definition 5 (Assessment [7]). *An assessment c_{A_i} for the agent A_i is a function $c_{A_i} : \mathbb{A} \to \mathcal{C}$ assigning a credibility value from $\mathcal{C}$ to each agent $A_j \in \mathbb{A}$.*

Note, that each agent in the MAS has its own assessment. This means in particular that different agents can have different assessments which can also be dynamic. As informant agents are ranked by their credibilities, a credibility order over the set $\mathbb{A}$ is introduced. This order is total (*i.e.*, satisfies *transitivity*, *totality* and *antisymmetry*) and is based on the assessment of each agent. Hence, each agent has a particular credibility order among agents.

Definition 6 (Credibility order among agents [11]). *A credibility order among agents for an agent A_i, denoted by '$\leq_{Co}^{A_i}$', is a total order over $\mathbb{A}$ where $A_1 \leq_{Co}^{A_i} A_2$ means that according to A_i, A_2 is at least as credible than A_1, and holds if $c_{A_i}(A_1) \prec_c c_{A_i}(A_2)$ or $c_{A_i}(A_1) = c_j$ and $c_{A_i}(A_2) = c_j$ with $c_j \in \mathcal{C}$. The strict relation $A_1 <_{Co}^{A_i} A_2$, denoting that A_2 is strictly more credible than A_1, is defined as $A_1 \leq_{Co}^{A_i} A_2$ and $A_2 \not\leq_{Co}^{A_i} A_1$. Moreover, $A_1 =_{Co}^{A_i} A_2$ means that A_1 is as credible as A_2, and it holds when $A_1 \leq_{Co}^{A_i} A_2$ and $A_2 \leq_{Co}^{A_i} A_1$.*

Example 1. Consider the set of agent identifiers $\mathbb{A} = \{A_1, A_2, A_3, A_4\}$ and the set of credibility labels $\mathcal{C} = \{c_1, c_2, c_3\}$, where $c_1 \prec_c c_2 \prec_c c_3$. Suppose that the belief base of the agent A_1 is $K_{A_1} = \{(\phi, A_2), (\phi, A_4), (\psi, A_3), (\psi \to \phi, A_1)\}$. Observe that K_{A_1} has two tuples with the sentence ϕ. Suppose that according to the assessment of A_1, $c_{A_1}(A_1) = c_1$, $c_{A_1}(A_2) = c_2$, $c_{A_1}(A_3) = c_2$ and $c_{A_1}(A_4) = c_3$. Then, the credibility order, according to A_1, is: $A_1 \leq_{Co}^{A_1} A_2$, $A_1 \leq_{Co}^{A_1} A_3$, $A_1 \leq_{Co}^{A_1} A_4$, $A_2 \leq_{Co}^{A_1} A_3$, $A_3 \leq_{Co}^{A_1} A_2$, $A_2 \leq_{Co}^{A_1} A_4$, and $A_3 \leq_{Co}^{A_1} A_4$. In a straightforward way, this relation implies $A_1 <_{Co}^{A_1} A_2 =_{Co}^{A_1} A_3 <_{Co}^{A_1} A_4$.

In the rest of this paper we directly use the credibility order among agents for simplicity of representation. The belief base structure is based on information objects, which include meta information and in our case in particular the source of the information. The sources again are connected to the current evaluation of the credibility by means of the assessment of the respective agent. This structure enables epistemic operators to include the credibility of information into the reasoning process. The approach adopted here can deal with a variety of epistemic operators potentially varying among different agents as shown in [7]. A belief operation is a function, that revises (in an abstract manner) the logical belief base of an agent appropriately to a newly given evidence, *i.e.*, a newly received message.

Definition 7 (Belief base operation [7]). *A belief base operation is a function $\mathcal{K} \times \mathcal{M} \to \mathcal{K}$ where $\mathcal{M}$ is the set of all possible messages and $\mathcal{K} = 2^{\mathcal{L} \times \mathbb{A}}$.*

As stated above, an agent A_i can receive an information object $I = (\phi, A_p)$ from other agents through communication. In order to incorporate I to its belief base K_{A_i}, the agent can use a belief base operation that considers its current beliefs (K_{A_i}), the logical sentence that has been received (ϕ) and the agent identifier A_p present in I. For example, the non-prioritized revision operator $\circ : \mathcal{K} \times \mathcal{M} \mapsto \mathcal{K}$ defined in [11] can be used. This operator behaves as follows.

If the incoming information I is consistent with K_{A_i} (*i.e.*, $Sen(K_{A_i}) \not\vdash \neg\phi.$)[1], then $K_{A_i} \circ I = K_{A_i} \cup \{I\}$. Hence, a belief base may contain the same belief in multiple tuples but with different agent identifiers (*e.g.*, in Example 1 ϕ is contained in two tuples). Although a sentence ϕ can have several proofs, there is no redundancy because each tuple represents a different informant. The advantage of this is made clear below. Consider for example that $(\phi, A_h) \in K_{A_i}$ and by this operation (ϕ, A_p) is added to K_{A_i}, if A_p if more credible than A_h then the plausibility of ϕ is increased (see Section 4.1 for details).

Since an agent can receive information that is not consistent with its own beliefs, in order to maintain its belief base consistent, it has to decide whether to accept or reject the new information I. For example, the operator "$\circ$" decides whether to accept or reject I considering the *credibility order among agents*. Incoming information can be rejected when the agent has more credible beliefs that contradict the new information. If an agent decides to accept a new belief that is contradictory with its belief base, then it has to select some beliefs from its belief base in order to withdraw them and avoid the contradiction.

---

[1] In this article, we assume that an agent does not receive contradictions.

3   Forwarding Information

In the following, we describe different criteria for forwarding information that determine which agent identifier is considered by the receiver at the moment of reasoning. That is to say, we analyze different alternatives which determine which agent identifier is to be sent in the information object.

As stated above, when an agent sends information to another agent, it sends information objects. Consider Example 1, if A_1 wants to send ϕ to A_2, it should send the tuple $I = (\phi, Agent)$. As we show below, there are several choices for the identifier "$Agent$" of I: it can be the proper sender (A_1); one of the identifiers stored with ϕ in the sender's base (e.g., A_2, A_4) that can be one of them arbitrarily or the more credible of them; or, in order to decide which identifier to send, a deeper analysis of the whole knowledge base can be performed. In this section, three criteria are introduced and analyzed, and in the following section we propose a more elaborated criterion that takes the plausibility of sentences obtained from agents credibility into consideration.

Sender identifier criterion. As in [2], a simple forwarding criterion can be implemented by sending an information object $I = (\phi, A_i)$ where A_i is always the identifier of the sending agent and ϕ is the belief to be forwarded.

Example 2. Let A_1 and A_2 be two agents and $K_{A_1} = \{(\phi, A_3)\}$. If A_1 wants to send ϕ to A_2 then A_1 forwards to A_2 a message with $I = (\phi, A_1)$.

This criterion has as advantage that it is simple, its implementation is easy and as stated by Dragoni in [2] "agents do not communicate the sources of the assumptions, but they present themselves as completely responsible for the knowledge they are passing on". Nevertheless, the original source of the information can be lost. This can lead to a change of the credibility of forwarded information with respect to the sender's belief base. In other words, the credibility of the belief forwarded can be increased or decreased as each agent has its own assessment. For instance, in Example 2 it can be seen that if $A_1 <_{Co}^{A_1} A_3$ then the credibility of ϕ is decreased with respect to the assessment of A_1. In this case, if $(\neg\phi, A_2) \in K_{A_2}$ and $A_1 <_{Co}^{A_2} A_2 <_{Co}^{A_2} A_3$ then A_2 can reject (ϕ, A_1) because $A_1 <_{Co}^{A_2} A_2$ although the original source is A_3 with $A_2 <_{Co}^{A_2} A_3$. Furthermore, if $A_3 <_{Co}^{A_1} A_1$ then the credibility of ϕ is increased. This case can be considered reasonable as the absence of reasons of A_1 against the acceptation of ϕ make this information more credible. Thus, A_1 believes in ϕ and can be considered as another informant for this information. However, this criterion can lead to unnatural settings as we show in the following example.

Example 3. Consider $\{A_1, A_2, A_3, A_4\} \subseteq \mathbb{A}$ where $A_1 <_{Co}^{A_i} A_2 <_{Co}^{A_i} A_3 <_{Co}^{A_i} A_4$ with $1 \leq i \leq 4$. Let $K_{A_1} = \{(\phi, A_1)\}$, $K_{A_2} = \emptyset$, $K_{A_3} = \{(\neg\phi, A_3)\}$ and $K_{A_4} = \emptyset$. Suppose that A_1 sends (ϕ, A_1) to A_4 using the *sender identifier criterion*. Since $K_{A_4} = \emptyset$ then $K_{A_4} = \{(\phi, A_1)\}$. Now A_4 sends (ϕ, A_4) to A_2 and then $K_{A_2} = \{(\phi, A_4)\}$. Then, A_2 has a more credible version of ϕ than its informant $(A_1 <_{Co}^{A_2} A_4)$. This does not make sense since if A_3 sends $(\neg\phi, A_3)$ to A_2 and A_4 then, following the revision process proposed above, A_2 rejects $(\neg\phi, A_3)$ $(A_3 <_{Co}^{A_2} A_4)$ but A_4 accepts $(\neg\phi, A_3)$ withdrawing (ϕ, A_1) from K_{A_4} $(A_1 <_{Co}^{A_4} A_3)$.

Source identifier criterion. Another approach for forwarding information can be implemented by sending an information object $I = (\phi, A_i)$ where A_i is always the identifier of the original informant stored in the belief base of the sender. This criterion has the constraint that the belief ϕ to be sent has to appear explicitly in only one information object in the belief base.

Example 4. Let A_1 and A_2 be two agents and $K_{A_1} = \{(\phi, A_3)\}$. If A_1 wants to send ϕ to A_2 then A_1 forwards to A_2 a message with $I = (\phi, A_3)$.

This criterion, similar to the previous one, has as advantage that it is simple and can be implemented easily, and it also overcomes one disadvantage of the previous criterion. That is, the original informant is maintained such that the credibility of the belief does not change after forwarding with respect to the sender's assessment. For this reason, this criterion does not lead to unnatural settings as showed in Example 3. However, considering Example 4, in the case of $A_3 <_{Co}^{A_1} A_1$ we loose the property of increasing the credibility of non-conflicting information of the former approach. According to this, it would be a good idea to increase the credibility of ϕ and to send it with the agent identifier A_1 as argued in "sender identifier criterion" criterion. That is, in this criterion the sender is not considered as possible information source.

Combined criterion. As stated above, there can be different tuples containing the same sentence. For instance, in Example 1, ϕ can be found in two explicit tuples $((\phi, A_2)$ and $(\phi, A_4))$. In this case, if A_1 wants to forward ϕ, it should decide which agent identifier appears in the content of the message, A_2 or A_4. We can opt for one of the two policies, we can choose the most credible agent or the least credible one. If we select the least credible agent identifier based on the assessment of the sender agent, we loose the most credible informant of the sender after forwarding. However, this does not occur if we select the most credible agent. But in this case, we are facing similar disadvantages as shown for the "source identifier" criterion. That is, if the sender is more credible than any source of the sentence to be forwarded, it would be a good idea to increase the credibility of the sentence by sending it with the agent identifier of the sender as argued before. This can be resolved performing the calculation of the most credible agent identifier with respect to all agent identifiers which are explicitly represented in the tuples and the sender. For instance, in Example 1 A_1 forwards ϕ with the most credible agent identifier from A_1 (sender identifier), A_2 and A_4. If the sender identifier is forwarded with the belief, then this criterion leads to an unnatural setting as has been shown in Example 3.

A particular case occurs when there exist several tuples with the same belief and different agent identifiers which are considered equally credible by the sender. In this case, we assume that the sender sends one of these agent identifiers according to some policy, although the credibility function of the receiver might assign different credibilities to these agents.

4 Plausibility-Based Criterion

In this section we propose a more elaborated criterion that takes the plausibility of sentences obtained from agents credibility into consideration. Note that in the combined criterion explained above, the agent identifier is obtained through a simple calculation from the set of information objects of the sender's belief base which explicitly contain the belief to be forwarded. However, the combined criterion does not consider that a sentence ϕ can be entailed from a set of information objects. For instance in Example 1, if A_1 wishes to send ϕ then A_1 should consider three proofs: $\{(\phi, A_2)\}$, $\{(\phi, A_4)\}$, $\{(\psi, A_3), (\psi \rightarrow \phi, A_1)\}$.

This criterion proposes to calculate the plausibility of a sentence ϕ based on all its proofs before being forwarded. This calculation should return an agent identifier which is used as the agent identifier of ϕ. In the following subsection we show how this calculation can be formalized.

4.1   Plausible Belief Bases Based on Agent Credibility

In this section we introduce a *plausibility function* as defined in [11] which we intend to use for the determination of the agent identifier of information objects to be forwarded. We use the agent identifiers of the agent's beliefs to compute the plausibility of the beliefs, *i.e.*, each of the agent's beliefs have an associated plausibility that depends on the agent identifier and the *credibility order among agents*. The behavior of this form of plausibility is similar to epistemic entrenchment as defined in [3]. That is, if ϕ and ψ are sentences in $\mathcal{L}$, the notation $\phi \preceq_{K_A} \psi$ means "ψ is at least as plausible as ϕ relative to the belief base K of agent A".

A belief base $K \in \mathcal{K}$ supports explicit and entailed sentences. The explicit sentences are those contained in $Sen(K)$, while the entailed sentences are those that are not in $Sen(K)$ but they are entailed by sentences in $Sen(K)$. In order to obtain the entailed sentences from a base K we use the following function:

Definition 8 (Belief function [11]). *The* belief function, $Bel : \mathcal{K} \rightarrow 2^{\mathcal{L}}$, *is a function s.t. for a belief base* $K \in \mathcal{K}$, $Bel(K) = \{\phi : \phi \in \mathcal{L} \text{ and } Sen(K) \vdash \phi\}$.

In general, there may be several proofs for ϕ from K. Therefore, to calculate the plausibility of a sentence (ϕ) we must analyze all of its proofs. For doing that, all the minimal subsets of K entailing ϕ are obtained. For this, we adapt the notion of kernel sets [5].

Definition 9 (Kernel). *Let* $K \in \mathcal{K}$ *and* $\phi \in \mathcal{L}$. *Then* H *is a* ϕ-kernel *of* K *if and only if* 1) $H \subseteq K$; 2) $Sen(H) \vdash \phi$; 3) *if* $H' \subset H$, *then* $Sen(H') \nvdash \phi$.

Thus, a ϕ-kernel is a minimal set of tuples from K that entails ϕ. The set of ϕ-kernels of K is denoted $K^{\perp\!\!\!\perp}\phi$ and is called *kernel set*.

We follow a cautious approach of plausibility calculation, that is, from each ϕ-kernel, we want to obtain the least plausible tuples. This gives us the plausibility of each proof. Then, the plausibility of a derived sentence ϕ is the greatest plausibility among those of each ϕ-kernel. In order to define this, two functions are given next.

Definition 10 (Least credible sources function [11]). *The* least credible sources function, $min : \mathcal{K} \rightarrow 2^{\mathcal{K}}$, *is a function s.t. for a given belief base* $K_A \in \mathcal{K}$, $min(K_A) = \{(\phi, A_i) : (\phi, A_i) \in K_A \text{ and for all } (\varphi, A_j) \in K_A, A_i \leq_{Co}^{A} A_j\}$.

Definition 11 (Most credible sources function [11]). *The* most credible sources function, $max : \mathcal{K} \to 2^{\mathcal{K}}$, *is a function s.t. for a given belief base $K_A \in \mathcal{K}$,* $max(K_A) = \{(\phi, A_i) : (\phi, A_i) \in K_A \text{ and for all } (\varphi, A_j) \in K_A, A_j \leq^A_{Co} A_i\}$.

Next, we introduce a function which returns the plausibility of a sentence ϕ that can be explicitly in K or entailed from K. This plausibility is based on a single agent identifier which is used when ϕ is compared to other beliefs. However, as stated above, with respect to the assessment of an agent A_i there may exist two agent identifiers (A_1 and A_2) such that $A_1 =^{A_i}_{Co} A_2$. For this case of a draw, we introduce a function which returns a single agent identifier given a set agent identifiers based on a given policy (for instance, the policy could be based on a lexicographical ordering among agent identifiers - A_1 is lesser than A_2).

Definition 12 (Selection function). *The* selection function *of an agent A_i,* $\mathcal{S}_{A_i} : 2^{\mathbb{A}} \to \mathbb{A}$, *is a function such that for a given set of agent identifiers with equal credibility with respect to the assessment of A_i, it returns a single agent identifier based on a given policy.*

Definition 13 (Plausibility function). *The* plausibility function, $Pl : \mathcal{L} \times \mathcal{K} \to \mathbb{A}$, *is a function such that for a $\phi \in \mathcal{L}$ and a belief base $K_A \in \mathcal{K}$,* $Pl(\phi, K_A) = \mathcal{S}_A(Ag(max(\bigcup_{X \in K^{\perp\!\!\!\perp}_A \phi} min(X))))$.

Example 5. Consider a set $\mathbb{A} = \{A_1, A_2, A_3\}$ where the credibility order according to A_1 is $A_3 <^{A_1}_{Co} A_2 <^{A_1}_{Co} A_1$. The belief base of agent A_1 is $K_{A_1} = \{(\psi, A_3), (\phi, A_2), (\phi \to \psi, A_2), (\phi \to \psi, A_1), (\omega, A_3), (\omega \to \psi, A_1), (\varphi \to \psi, A_3), (\rho, A_1), (\omega \to \rho, A_2)\}$. Suppose that agent A_1 needs to calculate the plausibility of ψ. In order to do so, A_1 performs the following steps.

- *Step 1.* $K^{\perp\!\!\!\perp}_{A_1}\psi = \{H_a, H_b, H_c, H_d\}$ where $H_a = \{(\psi, A_3)\}$, $H_b = \{(\phi, A_2),$ $(\phi \to \psi, A_2)\}$, $H_c = \{(\phi, A_2), (\phi \to \psi, A_1)\}$, and $H_d = \{(\omega, A_3), (\omega \to \psi, A_1)\}$.
- *Step 2.* $min(H_a) = \{(\psi, A_3)\}$, $min(H_b) = \{(\phi, A_2), (\phi \to \psi, A_2)\}$, $min(H_c) = \{(\phi, A_2)\}$, and $min(H_d) = \{(\omega, A_3)\}$.
- *Step 3.* $max(\{(\psi, A_3), (\phi, A_2), (\phi \to \psi, A_2), (\omega, A_3)\}) = \{(\phi, A_2), (\phi \to \psi, A_2)\}$.
- *Step 4.* $Ag(\{(\phi, A_2), (\phi \to \psi, A_2)\}) = \{A_2\}$.

Therefore, $Pl(\psi, K_{A_1}) = A_2$. Hence, when ψ is compared with other beliefs, A_2 is used as the informant of ψ (the plausibility of ψ is given by A_2).

4.2   Forwarding Criterion Using a Plausibility Function

In this section, we define a new criterion in which the agent identifier to be sent is obtained from the calculation of the plausibility of the sentence to be sent. The plausibility function we proposed in the previous subsection fulfills the criteria as laid out before such that it is suitable in order to obtain the agent identifier to be sent.

Plausibility-based criterion. A new forwarding criterion can be implemented by sending an information object $I = (\phi, A_i)$ where A_i is the agent identifier obtained using the plausibility function defined above, i.e., $A_i = Pl(\phi, K_{A_1})$ where A_1 is the sender of the message.

Example 6. Let us consider Example 5 again. If the agent A_1 wishes to send ψ to agent A_4 then, according to the plausibility based criterion, A_1 sends the tuple $(\psi, Pl(\psi, K_{A_1}))$ to A_4. That is, A_1 sends, based on its belief base K_{A_1} and its credibility order "$\leq_{co}^{A_1}$", (ψ, A_2) to A_4.

An important decision we made is to forward an agent identifier with a sentence rather than a credibility label in order to give additional information to the beliefs. One reason for this decision is that we achieve a more dynamic framework since the evaluation of the credibility of the agent identifiers is separated by use of the assessment function. Note that the assessment function may change in time realizing dynamic assessments and learning processes of agents. Hence, the credibility order among agents can be changed without changing the knowledge base or the operator. That is, if the credibility order among agents changes, then the plausibility of all sentences also changes without having to modify the belief base of the agent. Another reason for this decision is that since each agent has its own *assessment* (as stated in Section 2), is more suitable to send agent identifiers as this way the receiver agent can evaluate the received belief based on the credibility it has according to its own assessment. This means that the sending agent expresses that it considers the information it transmits as credible as it considers the agent identifier in the information object. Now it is up to the receiver to assess how credible it considers each agent from its perspective using its own assessment function. We belief, that this represents an advanced way of communication in multi-agent systems.

We adopted the plausibility function of an approach [11] of multi-source belief revision [9] for the means of improving the communication in MASs. As our proposed setting and framework is one of multi-source belief revision the proposed forwarding criterion blends in perfectly into a system equipped with the corresponding belief operator. This leads to a consistent strategy of belief operations as well as communication and can improve the overall agent performance. Furthermore, the general idea of using a plausibility function in this way can be adopted to other operators. In the following section we demonstrate that our presented extensions of the multi-agent framework can be easily implemented and integrated in existing systems and thus are ready to be used.

5   Implementation in Logic Programming

For the implementation we show how our approach to forwarding can be implemented within an advanced MAS blending in with the belief operations. For this we adapt credibility logic programs [6,7] and sketch how this can be used to implement our proposed approach. We assume here, that our language for knowledge representation is the one of logic programming. In particular, an information object consists of an agent identifier and a logic program $I = (P, A)$ and the agents store sets of these as there knowledge base. We are working with extended logic programs under the answer set semantics [4]. A rule r is written as $H(r) \leftarrow \mathcal{B}(r)$ where the head of the rule $H(r)$ is either empty or consists of a literal L and the body $\mathcal{B}(r)$ is a set of literals $\{L_0, \ldots, L_m, \text{not } L_{m+1}, \ldots, \text{not } L_n\}$. The body consists of a set of literals $\mathcal{B}(r)^+ = \{L_0, \ldots, L_m\}$ and a set of default negated literals denoted by $\mathcal{B}(r)^- = \{\text{not } L_{m+1}, \ldots, \text{not } L_n\}$. Credibility logic programs work on a set of extended logic programs [4] and a total preference relation on this set. This can be seen as a sequence of programs $\mathcal{P} = (P_1, \ldots, P_n)$ whose order reflects the preference order on the programs which enables the assignment of priorities to programs. This sequence of programs can be extracted from our belief base by taking the program part of each tuple, ordering them using the corresponding agent identifiers in combination with the agent assessment function. The knowledge base of an agent A can be seen as $K = \{(P_1, A_1), \ldots, (P_n, A_n)\}$. For the logic program representation of the knowledge base we overload the agent function defined above as $Ag(P) = A$ iff $(P, A) \in K$. Given this, we construct the sequence of logic programs $\mathcal{P} = (P_1, \ldots, P_n)$ with $Ag(P_i) \leq_{Co}^{Ap} Ag(P_{i+1})$, $1 \leq i < n$. Note that in this representation the program identifiers are unique even though they might have the same content.

In credibility logic programs the credibility of a head literal is determined given the inferred credibilities of the body literals. For the propagation of credibilities a cautious approach in which the minimum of the credibilities of the body literals is used to prioritize the head literal is used. When the credibility of a literal L is to be used for propagation purposes the maximal credibility of the existent prioritized literals for L is used. This way, we achieve an implementation of the plausibility criterion presented above using exclusively the language of extended logic programming. The original alphabet $\mathcal{G}$ of $\mathcal{P}$ is extended to the alphabet $\mathcal{G}'$ by adding several new predicates and atoms revealing the inferred credibility, $i.e.$, plausibility, of each literal. For each literal $L(\boldsymbol{x}_L)$ occurring in $\mathcal{P}$ a prioritized version $\hat{L}(\boldsymbol{x}_L, \mu)$ is added. Let $\mathcal{H}(\mathcal{P})$ denote the set of literals occurring in the head of a rule r in $\mathcal{P}$. The achieved implementation is independent of any specific answer set solver like smodels [10] or DLV [8]. For DLV, however, there exists an implementation of credibility logic programs as a front-end. Next, we state the construction of the a program for the generation of the plausibilities for each literal of an answer set which is an adaption of the construction seen in [6].

Definition 14 (Credibility propagation program). *The following rules are generated from a sequence* $\mathcal{P} = (P_1, \ldots, P_n)$ *of programs. For each program* $P_i \in \mathcal{P}$ *with an agent identifier* μ_i *and each rule* $r_j \in P_i : H \leftarrow \mathcal{B}(r_j)$:

$$\hat{H}(\mu) \leftarrow \mathcal{B}(r_j), minr_j(\mu), \text{not } exLowerMinr_j(\mu).$$
$$exLowerMinr_j(\mu) \leftarrow minr_j(\mu), minr_j(\mu'), Prec(\mu', \mu).$$
$$minr_j(\mu_i) \leftarrow .$$
$$\textit{For each } L \in \mathcal{B}(r_j) \text{: } minr_j(\mu) \leftarrow \mathcal{B}(r_j), \hat{L}(\mu), \text{not } exHigherL(\mu).$$

For each $L \in \mathcal{H}(\mathcal{P})$:

$$exHigherL(\mu) \leftarrow \hat{L}(\mu), \hat{L}(\mu'), Prec(\mu, \mu').$$
$$sendL(\mu) \leftarrow \hat{L}(\mu), \text{not } exHigherL(\mu).$$

Furthermore the preference predicate is defined by the following rules: For each program $P_i \in \mathcal{P}: Prec(\mu_{P_i}, \mu_{P_{i+1}}) \leftarrow .$ *And rules for transitivity:*

$$Prec(x, z) \leftarrow Prec(x, y), Prec(y, z).$$

For each rule r_j the credibilities of all body literals are represented by the $minr_j$ predicates which are inferred by the maximal of the respective literals credibilities. Additionally, one $minr_j$ representing the rule credibility is generated. The minimal of the credibility values for the body $\mathcal{B}(r_j)$, which are given by all $minr_j$ predicates is unified with the credibility argument of the head literal of rule r_j. The max and min functions are realized by the introduction of the predicates $exLowerMinr_j$ for each rule $r_j \in \mathcal{P}$ and $exHigherL$ for each literal occurring in the head a rule in $\mathcal{P}$; for more details cf. [6]. The answer set of the constructed credibility propagation program now contains for each literal the encoded credibility as an argument and for each literal a predicate $sendL(\mu)$ expressing the agent identifier to be send for the literal L. This enables the agent to infer the plausibility of each literal and to determine the agent which represents this plausibility from the answer set. Hence, we showed here that there is an implementation of our here proposed approach at hand which can be used in current implementations of multi-agent systems.

6 Conclusion and Related Work

In this paper we introduced a framework which enhances the communication skills of an agent in MAS by combining them with its reasoning abilities to allow the propagation of credible information. We assumed a collaborative MAS where deliberative agents can receive new information from others through communication and in which they have beliefs about the credibilities of their fellow agents. Similar to Dragoni [2] and Cantwell [1], in our approach, informant agents can have different levels of credibility. However, in [1] a relation of *trustworthiness* is introduced over sets of sources and not between single sources, whereas we used an assessment function in order to represent the credibilities each agent gives to other agents known to it. In [2] their tuples contain 5 elements: <Identifier, Sentence, OS, Source, Credibility>, where Origin Set (OS) records the assumption nodes upon which it really ultimately depends (as derived by the theorem prover). In contrast to them, in our model a tuple only stores a sentence and a source, but a tuple does not store the credibility. That is, in our model the plausibility of a sentence is not explicitly stored with it, as [2] does. Thus, when the plausibility of some sentence is needed the *plausibility function* (Definition 13) is applied. As is shown in Example 5, given a sentence ϕ its plausibility depend on its proofs (ϕ-kernels). Therefore, if one of the sentences of these proof changes, then the plausibility of ϕ may change.

We proposed to send an agent identifier with a piece of information which represents the credibility of the transferred information as in [7] and [11]. Based on this setting, we investigated how an agent can forward information to other agents that, in turn, could have been acquired from other agents. In particular, we studied how to rationally choose the meta-information to be forwarded. The choice of the agent identifier that is to be sent along with the piece of information is crucial, as it influences the decision of the receiver about whether to accept the transmitted information. Here, we presented different ways to select this identifier and gave a categorization of possible approaches, to later discussed their advantages and disadvantages. This discussion led to the definition of a criterion that uses a plausibility function which determines the plausibility of a sentence based on all its proofs according to a given base. In [2] the agents do not communicate the sources of the assumptions, but they present themselves as completely responsible for the knowledge they are passing on; receiving agents consider the sending ones as the sources of all the assumptions they are receiving from them.

An important decision we made is to forward an agent identifier with a sentence rather than a credibility label in order to give additional information to the beliefs. One reason for this decision is that since each agent has its own *assessment*, is more suitable to send agent identifiers as this way the receiver agent can evaluate the belief received based on the credibility it has according to its own assessment. Another reason is that we achieve a more dynamic framework since the evaluation of the credibility of the agent identifiers is separated by use of the assessment function. That is, if the credibility order among agents changes, then the plausibility of all sentences also changes without having to modify the belief base of the agent.

In addition to this, we presented an implementation of this operator using answer set programming adapting methods of [6], and thus using a common technique for the implementation of the reasoning component of agents in MASs. To sum up, we extended the reasoning component towards the computation of meta-information and additionally improved the communication component by giving reasoning capabilities to it.

One of the limitations of our communication framework is that a total order among agent is considered for each agent. As future work, we want to relax this assumption and to consider a partial order among agents. We also plan to expand the framework towards the incorporation of temporal information into the information objects.

References

1. Cantwell, J.: Resolving conflicting information. Journal of Logic, Language and Information 7(2), 191–220 (1998)
2. Dragoni, A., Giorgini, P., Puliti, P.: Distributed belief revision versus distributed truth maintenance. In: Proc. of the Sixth IEEE Conference on Tools with AI, pp. 499–505 (1994)
3. Gärdenfors, P., Makinson, D.: Revisions of knowledge systems using epistemic entrenchment. In: Second Conference on Theoretical Aspects of Reasoning about Knowledge Conference, pp. 83–95. Morgan Kaufmann, San Francisco (1988)
4. Gelfond, M., Lifschitz, V.: The stable model semantics for logic programming. In: ICLP/SLP, pp. 1070–1080. MIT Press, Cambridge (1988)
5. Hansson, S.O.: Kernel contraction. Journal of Symbolic Logic 59(3), 845–859 (1994)
6. Krümpelmann, P., Kern-Isberner, G.: Propagating credibility in answer set programs. In: 22nd Workshop on (Constraint) Logic Programming (WLP 2008), pp. 50–59. Martin-Luther-Universität Halle-Wittenberg, Germany (2008)
7. Krümpelmann, P., Thimm, M., Ritterskamp, M., Kern-Isberner, G.: Belief operations for motivated BDI agents. In: Proc. of AAMAS 2008, pp. 421–428 (2008)

8. Leone, N., Pfeifer, G., Faber, W., Eiter, T., Gottlob, G., Perri, S., Scarcello, F.: The DLV system for knowledge representation and reasoning. ACM Trans. Comput. Logic 7(3), 499–562 (2006)
9. Liu, W., Williams, M.-A.: A framework for multi-agent belief revision. Studia Logica 67(2), 291–312 (2001)
10. Niemelä, I., Simons, P.: Smodels - an implementation of the stable model and well-founded semantics for normal lp. In: Fuhrbach, U., Dix, J., Nerode, A. (eds.) LPNMR 1997. LNCS, vol. 1265, pp. 421–430. Springer, Heidelberg (1997)
11. Tamargo, L.H., García, A.J., Falappa, M.A., Simari, G.R.: Consistency maintenance of plausible belief bases based on agents credibility. In: 12th International Workshop on Non-Monotonic Reasoning (NMR), pp. 50–58 (2008)

Convergence Analysis of Affinity Propagation

Jian Yu and Caiyan Jia

Department of Computer Science, Beijing Jiaotong University,
Beijing 100044, P.R. China
{jianyu,cyjia}@bjtu.edu.cn

Abstract. Recently, Frey & Dueck proposed a novel clustering algorithm named affinity propagation (AP), which has been shown to be powerful as it costs much less time and reaches much lower error. However, its convergence property has not been studied in theory. In this paper, we focus on convergence property of the algorithm. The properties of the decision matrix when the affinity propagation algorithm converges are given, and the criterion that affinity propagation without the damping factor oscillates is obtained. Based on such results, we point out that damping factor might be important to alleviate oscillation of the affinity propagation, but it is not necessary to add a tiny amount of noise to a similarity matrix.

1 Introduction

Cluster analysis is an important part of scientific data analysis. When a large data set is given, it is supposed to be divided into several subgroups in order to be analyzed fast and efficiently. Therefore, many clustering algorithms have been proposed in the literature [1].

Recently, Frey & Dueck (2007) proposed a novel clustering algorithm, affinity propagation (AP), using similarity matrix [2]. The algorithm has been proved powerful as it costs much less time, reaches much lower error, does not require to specify the number of clusters which can be tuned by preference value p defaulted as the median value of a similarity matrix, and is suitable for non-symmetrical or sparse similarity matrixes [2]. It invokes many interests among researchers [3-10].

However, as Mezard (2007) [3] pointed out, the convergence property of affinity propagation is one of the main open challenges. Although Frey & Dueck pointed out that the numerical oscillations occur in updating the messages, they thought the affinity propagation oscillates probably since the energy function of AP may have multiple minima with corresponding multiple fixed points of the update rules. They offer two solutions: 1) adding a tiny amount of noise to the similarity to prevent degenerate situations, 2) introducing the damping factor λ. However, they did not clearly answer the following questions. Does affinity propagation without damping factor λ oscillate in a larger number of iterations or in unlimited iterations? If it is the latter, what conditions make affinity propagation without damping factor λ oscillate? Does adding a tiny amount of noise to the similarity avoid oscillations of AP?

D. Karagiannis and Z. Jin (Eds.): KSEM 2009, LNAI 5914, pp. 54–65, 2009.

In this paper, we will focus on the above questions, and study the convergence properties of affinity propagation in some cases and give the condition that the affinity propagation without a damping factor λ oscillates.We found that damping factor might be important to AP, but it is not necessary to add a tiny amount noise to a similarity matrix of AP.

The rest of paper is organized as follows. Section 2 introduces the notations and definitions for AP. Section 3 presents the main results about convergence properties of AP. Section 4 shows some numerical experiments. Section 5 concludes the paper.

2   Preliminaries

Supposed that a data set X consists of n d-dimensional data points $x_1, x_2, \cdots, x_n$, the similarity matrix of AP is defined as

$$s = [s(i,k)]_{n \times n}, \quad where \ \forall i \neq k, s(i,k) = -\|x_i - x_k\|^2, s(k,k) = p.$$

Where p is the preference value mentioned above.

When the similarity matrix s is given as above, let $S = [S(i,k)] = s + p \times I_n$, then the responsibility matrix $R = [r(i,k)]_{n \times n}$ and the availability matrix $A = [a(i,k)]_{n \times n}$ are define by

$$r(i,k) = S(i,k) - max_{j \neq k}\{(a(i,j) + S(i,j)\}, \forall i, k,$$
$$a(i,k) = min\{0, r(k,k) + \sum_{j \notin \{i,k\}} max\{0, r(j,k)\}\}, i \neq k,$$
$$a(k,k) = \sum_{j \neq k} max\{0, r(j,k)\}.$$

They are just the updating equations of AP.

However, Frey & Dueck pointed out that directly using the above updating equations could lead to numerical oscillations in some cases. They introduced a damping factor λ and a tiny amount of noise in the AP algorithm. In other words, $S = s + p \times I_n + \varepsilon_{n \times n}$ is used as the input of AP [2].

The procedure of APF (affinity propagation with the damping factor λ) can be expressed in **Algorithm APF** (see it in the next page).

Remark: In the source code of APF written by Frey & Dueck, the number of exemplars is the number of points such that $r^{(iter)}(i,i) + a^{(iter)}(i,i) > 0$, which is different from the rule of data partition in APF.

For simplicity, we denoted the affinity propagation without damping factor as APW. We call $E = R + A$ a decision matrix of AP. And we refer to S as a similarity matrix, and we need the following definitions.

Definition 1. (The right nearest neighbor) The right nearest neighbor to the data point i in the data set X w.r.t S is denoted as i_*^S, where

$$i_*^S = argmax_{k \in \{1,2,\cdots,n\} \setminus \{i\}}\{S(i,k)\}.$$

Definition 2. (The second right nearest neighbor) The second right nearest neighbor to the data point i in the data set X w.r.t S is denoted as i_{**}^S, where

$$i_{**}^S = argmax_{k \in \{1,2,\cdots,n\} \setminus \{i,i_*^S\}}\{S(i,k)\}.$$

The minimum value of the diagonal elements of similarity matrix S is denoted as α, where $\alpha = min_i\{S(i,i)\} = p + min_i\{\varepsilon(i,i)\}$.

Algorithm APF

Input: $l = 0$, $r^{(0)}(i,k) = 0$, $a^{(0)}(i,k) = 0$ for all i, k. $S = s + p \times I_n + \varepsilon_{n \times n}$ and $\lambda(0 < \lambda < 1)$ is damping factor, *iter* represents the maximum iteration number, and $\varepsilon_{n \times n}$ is a noise matrix.

Step 1: Responsibility updates

$$r^{(l+1)}(i,k) = \lambda r^{(l)}(i,k) + (1-\lambda)(S(i,k) - max_{j \neq k}\{a^{(l)}(i,j) + S(i,j)\}).$$

Step 2: Availability updates

$$a^{(l+1)}(i,k) = \lambda a^{(l)}(i,k) + (1-\lambda)min\{0, r^{(l+1)}(k,k) + \sum_{j \neq k} max\{0, r^{(l+1)}(j,k)\}\}, \ i \neq k,$$
$$a^{(l+1)}(k,k) = \lambda a^{(l)}(k,k) + (1-\lambda)\sum_{j \neq k} max\{0, r^{(l+1)}(j,k)\}\}.$$

Step 3: data partitions
If $l < iter$, the $l = l + 1$ go to Step 1, else if $l = iter$, making partition according to the following formula

$$c_i = argmax_k\{r^{(iter)}(i,k) + a^{(iter)}(i,k)\}.$$

Definition 3. (Self-availability) When a similarity matrix S is given, self-availability of the point k w.r.t S is defined as

$$a_k^S = \sum_{i_*^S = k} (S(i, i_*^S) - S(i, i_{**}^S)).$$

Definition 4. (Maximal availability) When a similarity matrix S is given, maximal availability of the data set w.r.t S is defined as

$$\nabla_a^S = max_k\{\sum_{i_*^S = k} (S(i, i_*^S) - S(i, i_{**}^S))\}.$$

Definition 5. (Neighbor diameter) Neighbor diameter of the point k of the data set w.r.t S is defined as
$$n_k^S = S(k, k_{**}^S).$$

Definition 6. (Maximal neighbor diameter) Maximal neighbor diameter of the data set w.r.t S is defined as

$$N_d^S = min_k\{S(k, k_{**}^S)\}.$$

Definition 7. (Symmetry degree) Symmetry degree of similarity matrix S is defined as
$$\nabla_s^S = max\{0, max_{i \neq k}\{-S(k, k_*^S) + S(i,k)\}\}.$$

Definition 8. (Symmetry deviation) Symmetry deviation of similarity matrix S is defined as

$$\nabla_d^S = max\{max_{i,k}\{|S(i,k) - S(k,i)|\}, max_i\{S(i,i)\} - min_i\{S(i,i)\}\}.$$

Definition 9. (Potential matrix) $\Gamma = S + A$ is called the potential matrix $\Gamma = [\tau(i,k)]_{n \times n}$ of the affinity propagation algorithm.

3   Convergence Properties of APF and APW

Theorem 1. If $0 \leq \lambda < 1$, $lim_{l \to +\infty} r^{(l)}(i,k) = r(i,k)$ and $lim_{l \to +\infty} a^{(l)}(i,k) = a(i,k)$ exist for all i, k, then

$$r(i,k) = S(i,k) - max_{j \neq k}\{(a(i,j) + S(i,j)\}, \forall i, k,$$
$$a(i,k) = min\{0, r(k,k) + \sum_{j \notin \{i,k\}} max\{0, r(j,k)\}\}, \ i \neq k,$$
$$a(k,k) = \sum_{j \neq k} max\{0, r(j,k)\}.$$

Proof. It easy to know by the updating equations of AP.

In other words, if the affinity propagation algorithm converges for $0 \leq \lambda < 1$, then its convergent set satisfies the definitions of the responsibility and the availability of AP. Hence, Theorem 1 offers a way to study the convergence property of the decision matrix of APF.

Corollary 1. *If the affinity propagation algorithm converges for $0 \leq \lambda < 1$, then 1) each line of the decision matrix $E = [e(i,k)]_{n \times n} = R + A$ has at most one positive element, at least $(n-1)$ non positive elements.*
  2) $e(i,k) \leq e(k,k)$.
  3) $max_k\{e(i,k)\} = max_{k', e(k',k') \geq 0}\{e(i,k')\}$.

Proof. 1) By Theorem 1, we know that the decision matrix $E = [e(i,k)]_{n \times n}$ can be expressed as follows.

$$e(i,k) = \tau(i,i_*^\Gamma) - max\{\tau(i,i_{**}^\Gamma), \tau(i,i)\}, \ for \ \ k = i_*^\Gamma,$$
$$e(i,k) = \tau(i,k) - max\{\tau(i,i_*^\Gamma), \tau(i,i)\}, \ for \ \ k \notin \{i_*^\Gamma, i\},$$
$$e(i,i) = \tau(i,i) - \tau(i,i_*^\Gamma).$$

Obviously, if $e(i,i) > 0$, then $\tau(i,i) \geq \tau(i,i_*^\Gamma)$. Therefore, if $k \neq i$, $e(i,k) = \tau(i,k) - \tau(i,i) \leq 0$. If $e(i,i) < 0$, then $\tau(i,i) < \tau(i,i_*^\Gamma)$. If $k = i_*^\Gamma$, $e(i,k) \geq 0$. If $k \neq i_*^\Gamma$ and $k \neq i$, $e(i,k) = \tau(i,k) - \tau(i,i_*^\Gamma) \leq 0$.
  2) By Theorem 1, it is easy to know that if $i \neq k$,

$$a(i,k) = min\{0, r(k,k) + \sum_{j \notin \{i,k\}} max\{0, r(j,k)\}\} \leq r(k,k) + a(k,k) - max\{0, r(i,k)\}.$$

Therefore, we can know the following facts.

$$a(i,k) + max\{0, r(i,k)\} \leq r(k,k) + a(k,k),$$
$$a(i,k) + r(i,k) \leq a(i,k) + max\{0, r(i,k)\},$$
$$a(i,k) + r(i,k) \leq r(k,k) + a(k,k).$$

Thus,

$$e(i,k) \le e(k,k).$$

3) It is easy to be proved by the above analysis.

Corollary 1 offers a clear description of the decision matrix when APF converges, and can help us to judge the oscillation and determine the data partition for the affinity propagation algorithm. In fact, Corollary 1 tells us that if the affinity propagation algorithm converges and the point i takes the point k as its exemplar, then $e(i,k) \ge 0$. In other words, if there exists a row in the decision matrix, in which all elements are negative or at least two elements are positive, AP does not converge at that iteration. And more, when the APF converges, $e(i,i) < 0$ means that the data point i is guaranteed not to be an exemplar.

As for the number of exemplars, Corollary 1 shows that the number of exemplars should be the number of points such that $e(i,i) \ge 0$ when APF converges, which is different from $e(i,i) > 0$ used in the source code written by Frey & Dueck. If APF converges and the elements of the diagonal line of the decision matrix has no element equal to zero, the number of exemplars should be the number of points such that $e(i,i) > 0$. At this condition, the convergence of APF may not be guaranteed. Thus it is recommended to use the original definition of the exemplars in [2], i.e. the exemplars are determined by $i = argmax_k\{e^{(iter)}(i,k)\}$.

As for data partition, Corollary 1 means that, when APF converges, $k = argmax_{k',\,e(k',k')\ge 0}\{e(i,k')\}$ identifies the data point k as the exemplar for the data point i. Such results can reduce the computational complex of APF surely. Similarly, it can only be used when APF converges. Therefore, it is much safer to use $argmax_k\{e^{(iter)}(i,k)\}$ to partition a data set.

Lemma 1. $\forall 0 \le \lambda < 1$, 1) if $\forall i, S(i,i) \ge S(i,i_*^S)$ then

$$
\begin{aligned}
r^{(l)}(i,k) &= (1-\lambda^l)(S(i,k)-S(i,i)) \le 0, for\ i \ne k,\\
r^{(l)}(k,k) &= (1-\lambda^l)(S(k,k)-S(k,k_*^S)) \ge 0,\\
a^{(l)}(i,k) &= 0, \ \forall\ i,k.
\end{aligned}
$$

2)If $\forall i, S(i,i) \le min_{i\ne k}\{S(i,k)\}$,

$$
\begin{aligned}
r^{(l)}(i,k) &\le (1-\lambda^l)(S(i,k)-S(i,i)), for\ i \ne k,\\
r^{(l)}(k,k) &\ge (1-\lambda^l)(S(k,k)-S(k,k_*^S)),\\
a^{(l)}(i,k) &\le (1-\lambda^l - l(1-\lambda)\lambda^l)(S(k,k)-S(k,k_*^S)), for\ i \ne k,\\
a^{(l)}(k,k) &\ge (1-\lambda^l - l(1-\lambda)\lambda^l)\sum_{i=1}^n (S(i,k)-S(k,k)).
\end{aligned}
$$

Proof. It is easy to be proved by standard mathematical induction method.

Apparently, Lemma 1 implies Theorem 2.

Theorem 2. If $\forall i, S(i,i) \ge S(i,i_*^S)$, then APF and APW converge and the number of exemplars will be n.

Corollary 2. If $|\varepsilon(i,k)| \le 0.5\varepsilon$ and $p \ge max_i\{s(i,i_*^s)\}+\varepsilon$, then APF and APW converge when the input is $s + p \times I_n + \varepsilon_{n\times n}$.

Proof. It is easy to prove that $p \geq max_i\{s(i, i_*^s)\} + \varepsilon$ and $|\varepsilon(i, k)| \leq 0.5\varepsilon$ implies that $S(i, i) = p + \varepsilon(i, k) \geq max_i\{s(i, i_*^s)\} + \varepsilon + \varepsilon(i, k) \geq max_i\{s(i, i_*^s)\} + 0.5\varepsilon$. Similarly, the definition of $s(i, i_*^s)$ and ε implies that $s(i, i_*^s) + 0.5\varepsilon \geq s(i, i_*^S) + \varepsilon(i, i_*^S) = S(i, i_*^S)$. Combining the above results, the proof can be finished by Theorem 2.

Corollary 2 and Theorem 2 tell us that the preference value p should be less than $max_i\{s(i, i_*^s)\} + 2max_{i,k}\varepsilon(i, k)$. Otherwise, APF cannot work as a clustering algorithm.

According to the updating equations and Definition 7 and 8, we have Lemma 2 and Lemma 3.

Lemma 2. *If* $\lambda = 0$,

$r^{(1)}(i, k) = S(i, k) - S(i, i_*^S), for\ k \neq i,$
$r^{(1)}(k, k) = S(i, i_*^S) - S(i, i_{**}^S),$
$a^{(1)}(i, k) = min\{0, S(k, k) - S(k, k_*) + \sum_{j_*^S = k, j \neq i}(S(j, j_*^S) - S(j, j_{**}^S))\}, for\ k \neq i,$
$a^{(1)}(k, k) = \sum_{j_*^S = k}(S(j, j_*^S) - S(j, j_{**}^S)).$

Lemma 3. $\nabla_s^S \leq \nabla_d^S$.

Lemma 4. *If* $l \leq 2, \lambda = 0, \alpha \leq N_d^S - 2\nabla_a^S - 6\nabla_d^S$, *then*

$$a^{(2l-1)}(i, k) = S(k, k) - S(k, k_*^S), for\ k \neq i,$$
$$a^{(2l-1)}(k, k) = 0,$$
$$r^{(2l)}(i, k) \geq S(i, k) - \alpha - 2\nabla_d^S, for\ k \neq i,$$
$$r^{(2l)}(k, k) \geq -2\nabla_d^S,$$
$$a^{(2l)}(i, k) = 0, for\ k \neq i,$$
$$a^{(2l)}(k, k) \geq S(k, k_*^S) - \alpha + 2\nabla_a^S,$$
$$r^{(2l+1)}(i, k) \leq S(i, k) - S(i, i_*^S) - 2\nabla_a^S \leq 0, for\ i \neq k,$$
$$r^{(2l+1)}(k, k) = S(k, k) - S(k, k_*^S).$$

Proof. In fact, we only need to calculate $r^{(m)}(i, k), a^{(m)}(i, k), a^{(m)}(k, k)$, m=1, 2, 3, 4, 5, respectively, then the proof can be easily finished by mathematical induction method. Therefore, we use the following six steps to finish this proof.

First Step. Calculating $r^{(1)}(i, k), a^{(1)}(i, k), a^{(1)}(k, k)$.
    By the definitions, we know that the inequalities, (1), (2) and (3), hold.

$$-S(k, k_*^S) \leq -N_d^S \tag{1}$$

$$\sum_{j_*^S = k, j \neq i}(S(j, j_*^S) - S(j, j_{**}^S)) \leq \sum_{j_*^S = k}(S(j, j_*^S) - S(j, j_{**}^S)) \tag{2}$$

$$\sum_{j_*^S = k}(S(j, j_*^S) - S(j, j_{**}^S)) \leq \nabla_a^S \tag{3}$$

Combining (1), (2), (3) and the assumption of this lemma, we can get

$$S(k,k)-S(k,k_*^S)+\sum_{j_*^S=k,j\neq i}(S(j,j_*^S)-S(j,j_{**}^S))\leq\alpha+\nabla_d^S-N_d^S+\nabla_a^S\leq-\nabla_a^S-5\nabla_d^S\leq0 \quad (4)$$

By Lemma 2 and (4), it easy to finish the first step.

Second Step. Calculating $r^{(2)}(i,k), a^{(2)}(i,k), a^{(2)}(k,k)$.

We know that

$$a^{(1)}(i,k')+S(i,k')$$
$$=S(k',k)-S(k',(k')_*^S)+S(i,k')+\sum_{j_*^S=k',j\neq i}(S(j,j_*^S)-S(j,j_{**}^S))$$
$$\leq\alpha+2\nabla_d^S+\nabla_a^S.$$

And

$$r^{(2)}(i,k)\geq S(i,k)-\alpha-2\nabla_d^S-\nabla_a^S, for\ i\neq k \quad (5)$$

By Definition 8, we also know that

$$r^{(2)}(i,k)\geq S(k,i)-\alpha-3\nabla_d^S-\nabla_a^S \quad (6)$$

$$r^{(2)}(k,k)\geq\alpha-\alpha-2\nabla_d^S-\nabla_a^S=-2\nabla_d^S-\nabla_a^S \quad (7)$$

By (6) and $\alpha\leq N_d^S-2\nabla_a^S-6\nabla_d^S$, it is easy to know that

$$r^{(2)}(k_*^S,k)\geq S(k,k_*^S)-\alpha-3\nabla_d^S-\nabla_a^S\geq3\nabla_d^S+\nabla_a^S\geq0 \quad (8)$$

$$r^{(2)}(k_{**}^S,k)\geq S(k,k_{**}^S)-\alpha-3\nabla_d^S-\nabla_a^S\geq3\nabla_d^S+\nabla_a^S\geq0 \quad (9)$$

If $i=k_*^S$, then (7) and (9) can follow (10).

$$r^{(2)}(k,k)+\sum_{i'\notin\{i,k\}}max\{0,r^{(2)}(i',k)\}\geq-\nabla_a^S-2\nabla_d^S+r^{(2)}(k_{**}^S,k)\geq\nabla_d^S\geq0 \quad (10)$$

If $i=k_{**}^S$, then (7) and (8) can follow (11).

$$r^{(2)}(k,k)+\sum_{i'\notin\{i,k\}}max\{0,r^{(2)}(i',k)\}\geq-\nabla_a^S-2\nabla_d^S+r^{(2)}(k_*^S,k)\geq\nabla_d^S\geq0 \quad (11)$$

If $i\neq k_*^S,k_{**}^S$, then we have (12) by combining (7), (8) and (9).

$$r^{(2)}(k,k)+\sum_{i'\notin\{i,k\}}max\{0,r^{(2)}(i',k)\}\geq-2\nabla_d^S-\nabla_a^S+r^{(2)}(k_*^S,k)+r^{(2)}(k_{**}^S,k)$$
$$\geq-2\nabla_d^S-\nabla_a^S+2(3\nabla_d^S+\nabla_a^S)\geq0$$
$$(12)$$

According to the inequalities, (10), (11) and (12), we have if $i\neq k$, $a^{(2)}(i,k)=0$.
Similarly, we can obtain

$$a^{(2)}(k,k)\geq S(k,k_*^S)+S(k,k_{**}^S)-2\alpha-2(\nabla_a^S+3\nabla_d^S)\geq S(k,k_*^S)-\alpha \quad (13)$$

Third Step. Calculating $r^{(3)}(i,k), a^{(3)}(i,k), a^{(3)}(k,k)$.

It's easy to know (13) implies (14).

$$S(i,i) + a^{(2)}(i,i) \geq \alpha + S(i,i_*^S) - \alpha = S(i,i_*^S) \tag{14}$$

If $k = i_*^S$, it is easy to prove that

$$max_{k' \neq k}\{a^{(2)}(i,k') + S(i,k')\} = max\{S(i,i_{**}^S), S(i,i) + a^{(2)}(i,i)\} \geq S(i,i_*^S) \tag{15}$$

If $k \neq i_*^S$, then it is easy to prove that

$$max_{k' \neq k}\{a^{(2)}(i,k') + S(i,k')\} = max\{S(i,i_*^S), S(i,i) + a^{(2)}(i,i)\} \geq S(i,i_*^S) \tag{16}$$

(15) and (16) imply (17) if $i \neq k$

$$r^{(3)}(i,k) = S(i,k) - max_{k' \neq k}\{a^{(2)}(i,k') + S(i,k')\} \leq S(i,k) - S(i,i_*^S) \leq 0 \tag{17}$$

By this way, we also know that

$$r^{(3)}(k,k) = S(k,k) - S(k,k_*^S),$$
$$a^{(3)}(i,k) = S(k,k) - S(k,k_*^S),$$
$$a^{(3)}(k,k) = 0.$$

Fourth Step. Calculating $r^{(4)}(i,k), a^{(4)}(i,k), a^{(4)}(k,k)$.

$$a^{(3)}(i,k') + S(i,k') = S(k',k') - S(k',(k')_*^S) + S(i,k') \leq \alpha + 2\nabla_d^S, \text{ for } i \neq k' \tag{18}$$

By (18), we can know that

$$r^{(4)}(i,k) \geq S(i,k) - \alpha - 2\nabla_d^S, \text{ for } k \neq i \tag{19}$$

$$r^{(4)}(k,k) \geq -2\nabla_d^S \tag{20}$$

By Definition 8, we know that

$$r^{(4)}(k_*^S,k) \geq S(k,k_*^S) - \alpha - 3\nabla_d^S \geq 3\nabla_d^S + 2\nabla_a^S \geq 0 \tag{21}$$

$$r^{(4)}(k_{**}^S,k) \geq S(k,k_{**}^S) - \alpha - 3\nabla_d^S \geq 3\nabla_d^S + 2\nabla_a^S \geq 0 \tag{22}$$

Combining (20), (21) and (22), we can know that

$$a^{(4)}(i,k) = 0, \text{ for } k \neq i$$
$$a^{(4)}(k,k) \geq S(k,k_*^S) + S(k,k_{**}^S) - 2\alpha - 6\nabla_d^S \geq S(k,k_*^S) - \alpha + 2\nabla_a^S \tag{23}$$

Fifth Step. Calculating $r^{(5)}(i,k), a^{(5)}(i,k), a^{(5)}(k,k)$.

By (23), we can follow

$$S(i,i) + a^{(4)}(i,i) \geq S(i,i_*^S) + 2\nabla_a^S \tag{24}$$

If $k = i_*^S$, (24) implies (25).

$$max_{k' \neq k}\{a^{(4)}(i, k') + S(i, k')\} = max\{S(i, i_{**}^S), S(i, i) + a^{(4)}(i, i)\}$$
$$= S(i, i) + a^{(4)}(i, i) \geq S(i, i_*^S) + 2\nabla_a^S \qquad (25)$$

If $k \neq i_*^S$, then it is easy to prove that

$$max_{k' \neq k}\{a^{(4)}(i, k') + S(i, k')\} = max\{S(i, i_*^S), S(i, i) + a^{(4)}(i, i)\}$$
$$= S(i, i) + a^{(4)}(i, i) \geq S(i, i_*^S) + 2\nabla_a^S \qquad (26)$$

Combining (25) and (26), we can know that

$$r^{(5)}(i, k) \leq S(i, k) - S(i, i_*^S) - 2\nabla_a^S \leq 0, \quad for \ i \neq k,$$
$$r^{(5)}(k, k) = S(k, k) - S(k, k_*^S).$$

Therefore, we can know that

$$a^{(5)}(i, k) = S(k, k) - S(k, k_*^S), \quad for \ i \neq k,$$
$$a^{(5)}(k, k) = 0.$$

Sixth Step. The proof can be easily finished by standard mathematical induction method.

Since $\alpha = p + min_i\{\varepsilon(i, i)\}$, Theorem 3 tells us that $p \leq N_d^S - 2\nabla_a^S - 6\nabla_d^S - min_i\{\varepsilon(i, i)\}$, APW oscillates. It is easy to see that $N_d^S - 2\nabla_a^S - 6\nabla_d^S - min_i\{\varepsilon(i, i)\}$ is independent of p. Moreover, any matrix $B = [b(i, k)]_{n \times n}$ can be expressed as the form $s + \varepsilon_{n \times n}$, where $s = 0.5(B + B^T) - diag(B)$, $\varepsilon_{n \times n} = 0.5(B - B^T) + diag(B)$. Obviously, $s = 0.5(B + B^T) - diag(B)$ is symmetry. Therefore, even the preference for each point k takes the different value $b(k, k)$, APW still oscillates for the matrix B if $\alpha = min_i\{b(i, i)\} \leq N_d^B - 2\nabla_a^B - 6\nabla_d^B$.

Notice that $p = \alpha$, then $B = [b(i, k)]_{n \times n}$ can be expressed as the form $s + p \times I_n + \varepsilon_{n \times n}$, where $s = 0.5(B + B^T) - diag(B)$, $\varepsilon_{n \times n} = 0.5(B - B^T) + diag(B) - p \times I_n$. Therefore, it is not necessary to let each point k takes different preference.

Since $\nabla_a^S = max\{max_{i,k}\{|\varepsilon(i, k) - \varepsilon(k, i)|\}, max_i\{\varepsilon(i, i)\} - min_i\{\varepsilon(i, i)\}\}$, by similar analysis, N_d^S and ∇_a^S also have nothing to do with the preference p when $S = s + p \times I_n + \varepsilon_{n \times n}$. They can be denoted as $N_d^{s + \varepsilon_{n \times n}}$, $\nabla_d^{\varepsilon_{n \times n}}$, respectively. So Lemma 4 can follow Theorem 3.

Theorem 3. *If* $p \leq N_d^{s + \varepsilon_{n \times n}} - 2\nabla_a^{s + \varepsilon_{n \times n}} - 6\nabla_d^{\varepsilon_{n \times n}} - min_i\{\varepsilon(i, i)\}$, *then APW oscillates.*

In order to clearly show the impact of the noise matrix on the APW, we can prove the facts as follows.

Lemma 5.

$$N_d^S \geq N_d^s - max_{i,k}\{|\varepsilon(i, k)|\}, and \ \nabla_a^S \leq \nabla^s + 2(n - 1) \times max_{i,k}\{|\varepsilon(i, k)|\}$$

where

$$\nabla^s = \sum_{i=1}^n |s(i, i_*^s) - s(i, i_{**}^s)|$$

and $0 \leq max_{i,k}|\varepsilon(i, k)| \leq 0.5min_{i \neq k, i \neq k', s(i,k) - s(i,k') \neq 0}\{|s(i, k) - s(i, k')|\}$.

Proof. $0 \leq max_{i,k}|\varepsilon(i,k)| \leq 0.5min_{i\neq k, i\neq k', s(i,k)-s(i,k')\neq 0}\{|s(i,k) - s(i,k')|\}$ means that $S(i, i_{**}^S) \geq s(i, i_{**}^s) - max_i\varepsilon(i, i_{**}^s)$, therefore, $N_d^S \geq N_d^s - max_{i,k}\{|\varepsilon(i,k)|\}$.

If $s(i, i_*^s) - s(i, i_{**}^s) > 0$, then $i_*^s = i_*^S$ and

$$S(i, i_*^S) - S(i, i_{**}^S) \leq s(i, i_*^s) - s(i, i_{**}^s) + 2max_{i,k}\{|\varepsilon(i,k)|\}$$

If $s(i, i_*^s) - s(i, i_{**}^s) = 0$, then

$$S(i, i_*^S) - S(i, i_{**}^S) \leq 2max_{i,k}\{|\varepsilon(i,k)|\}$$

So, $\nabla_a^S \leq \nabla_a^s + 2(n - 1) \times max_{i,k}\{|\varepsilon(i,k)|\}$. then it is easy to get the second inequality of the lemma.

By Lemma 4, we have

Corollary 3. *If $max_{i,k}\{|\varepsilon(i,k)|\} \leq 0.5\varepsilon$ and $p < N_d^s - 2\nabla_a^s - 2n\varepsilon - 5\varepsilon$, where $0 \leq \varepsilon \leq 0.5min_{i\neq k, i\neq k', s(i,k)-s(i,k')\neq 0}\{|s(i,k) - s(i,k')|\}$, then APW oscillates when the input is $s + p \times I_n + \varepsilon_{n\times n}$.*

Proof. By Lemma 5 and Theorem 3, we have the result.

By Theorem 3 and Corollary 3, we know the condition that APW does oscillate no matter what properties of the similarity matrix have. They tell us that AP oscillates not because the energy function of AP may have multiple minima. Moreover, Corollary 3 clearly proves that adding a tiny amount of noise to the similarity cannot prevent oscillation of APW. Therefore, it is not recommended to add a tiny amount of noise to a similarity matrix in APW.

According to the above analysis, the convergence property of APW is an open problem for $max_i\{s(i, i_*^s)\} > p > N_d^s - 2\nabla_a^s$ when no adding a tiny amount of noise to the similarity. As a matter of fact, Frey & Dueck (2007) pointed the preference p could be the median or the minimum of the input similarities. Notice the definition of $N_d^s - 2\nabla_a^s$, it is easy to guess that such suggested values of preference do not belong to $max_{i\neq k}\{s(i,k)\} > p > N_d^s - 2\nabla_a^s$ in a greater probability.

As for general data sets, Theorem 2 and 3 tell us that APW experiences change from convergence to oscillation for $max_i\{s(i, i_*^s)\} > p > N_d^s - 2\nabla_a^s$. Apparently, if the similarity matrix satisfies $s(j, j_*^s) - s(j, j_{**}^s) = 0$ for any j, then Theorem 2 and 3 show the convergence property of APW for any value of p, where $p = min_k\{s(k, k_{**}^s)\}$ is the unique point that the convergence property of APW changes from convergence to oscillation. It is an interesting question that there exists a data set satisfying the above assumption? In the next section, we will construct the data sets satisfying $s(j, j_*^s) - s(j, j_{**}^s) = 0$ for any j.

4   Numerical Oscillations of APW and APF

In this section, we choose two small data sets, one is Cube-2k, the other data set is 25 two dimensional data points (Fdat) in [2], to simply show the convergence properties of APW and APF.

Cube-$2k$ consists of $2k$ clusters and each cluster consists of 2^k k-dimensional points located at 2^k vertices of a unit hypercube. Namely,

$$\text{Cube-}2k = \{x \in R^k | x = y + z, \forall y \in A, \forall z \in B\},$$

where

$$A = \{y = [y_1, y_2, \cdots, y_k] \in R^k | y_j = 0 \, or \, 1, \forall 1 \le j \le k\},$$
$$B = \{z = [z_1, z_2, \cdots, z_k] \in R^k | \exists! j \, s.t. \, z_j = \pm 2k, otherwise \, z_j = 0\}.$$

From the definition of Cube-$2k$, Cube-$2k$ is well-separated and suitable for cluster analysis. It is easy to know that $s(j, j_*^s) - s(j, j_{**}^s) = 0$ for any j holds for Cube-$2k$ as $s(j, j_*^s) = s(j, j_{**}^s) = -1$ for any j holds in Cube-$2k$, which also means that $min_k\{s(k, k_{**}^s)\} = max_i\{s(i, i_*^s)\}$. For Cube-$2k$, the convergence property of APW is completely finished by Theorem 2 and 3 when no adding a tiny amount of noise to the similarity. When adding a tiny amount of noise to similarities, Corollary 3 tells us that APW does oscillate when $p < -1 - (2^{2k+1} + 10) * max_{i,k}\{|\varepsilon(i, k)|\}$. Therefore, adding a tiny amount of noise to similarities cannot prevent oscillation of APW.

As for Fdat, the detailed description can be seen in [2]. For Fdat, $max_{i \ne k} s(i, k) = -0.5476$, $N_d^s - 2\nabla_\alpha^s = -10.7346$, numerical experiments show APW converges for $p > -1.76$, oscillates for $p < -1.77$ ($iter$=100000). So, APW changes from convergence to oscillation when p decreases from -0.5476 to -10.7346, which is consistent with our above analysis.

As for cluster analysis, it is unexpected that every point in the data set is clustered as one cluster. Therefore, when APF or APW output n exemplars, it says that APF or APW is invalid as a clustering algorithm. Theorem 3 tell us an interesting fact if $\lambda = 0$ and $p \le N_d^s - 2\nabla_a^s$, the number of exemplars the affinity propagation algorithm outputted is always n when the number of exemplars is determined by $i = max_k\{e^{(iter)}(i, k)\}$ instead of $e^{(iter)}(i, k) > 0$. Corollary 3 also tells us that the number of exemplars the affinity propagation algorithm outputted is always n when $max_i\{s(i, i_*^s)\} \le p$ if no noise matrix is added to a similarity matrix. Therefore, Theorem 2 and 3 illustrate that APW is invalid as a clustering algorithm for most values of the preference. In the extreme case, we know that APW is invalid for the data sets like Cube-$2k$ as a clustering algorithm no matter what value of the preference p is taken according to the above analysis although Cube-$2k$ is well separated as a test data. Moreover, Theorem 3 tells us that the number of exemplars does not monotonically decrease with the decreasing preference p for APW if judging the number of exemplars by $i = max_k\{e^{(iter)}(i, k)\}$.

As for adding a tiny amount of noise to similarities, we has run APF many times on the above two data sets. In this paper, the noise matrix $\varepsilon_{n \times n} = [\varepsilon(i, k)]_{n \times n}$ is supposed to obey the normal distribution $N(0, \varepsilon)$, where δ_{ik} is the Kronecker symbol and

$$\varepsilon = 10^{-12}(max_{i,k}[s(i, k) + p \times \delta_{ik}] - min_{i,k}[s(i, k) + p \times \delta_{ik}]).$$

The numerical experiments tell us when the above noise matrix is added to the similarity matrixes, the oscillation of APF still occurs for $p = -251$ and

$\lambda = 0.9, 0.99, 0.999$ on Cube-4, for $p = -251$ and $\lambda = 0.5$ on Fdat. The results are also consistent with our analysis in Section 3.

However, considering damping factor $0 < \lambda \leq 1$, it is still very challenging to prove the convergence of AP. We have done some numerical experiments on the aspect. Given the space limitation, we did not show the results in the paper.

5 Conclusions

In this paper, we give a clear description of the decision matrix when APF converges, and show the convergence properties of APW. We gave the concrete conditions that APW does oscillate. As for special data sets, like Cube-2k, it is proved that APW does not work as a clustering algorithm although such data sets can be considered friendly to cluster analysis. Moreover, the analysis suggests that it is not necessary to add a tiny amount of noise to a similarity matrix of AP for avoiding oscillation, but it is very important to choose the proper damping factor when using APF.

Acknowledge

This work was supported in part by NSFC (Grant No. 90820013, 60875031 and 60905029), 973 Project (Grant No. 2007CB311002), the Research Fund for the Doctoral Program of Higher Education (Grant No. 2007004038).

References

1. Jain, A.K., Murty, M.N., Flynn, P.J.: Data clustering: a review. ACM Computing Surveys 31(3), 264–323 (1999)
2. Frey, B.J., Dueck, D.: Clustering by passing messages between data points. Science 315, 972–976 (2007); (Supporting material is available on Science Online)
3. Mezard, M.: Where are the exemplars? Science 315, 949–951 (2007)
4. Xiao, J.X., Wang, J.D., Tan, P., Quan, L.: Joint affinity propagation for multiple view segmentation. In: Proc. of ICCV, pp. 1–7 (2007)
5. An, S., Liu, W.Q., Venkatesh, S.: Acquiring critical light points for illumination subspaces of face images by affinity propagation clustering. In: Ip, H.H.-S., Au, O.C., Leung, H., Sun, M.-T., Ma, W.-Y., Hu, S.-M. (eds.) PCM 2007. LNCS, vol. 4810, pp. 647–654. Springer, Heidelberg (2007)
6. Michael, J.B., Hans, F.K.: Comment on clustering by passing messages between data points. Science 319, 726c (2008)
7. Frey, B.J., Dueck, D.: Response to comment on clustering by passing messages between data points. Science 319, 726d (2008)
8. Michele, L.S., Martin, W.: Clustering by soft-constraint affinity propagation: applications to gene-expression data. Bioinformatics 23(20), 2708–2715 (2007)
9. Lai, D.R., Lu, H.T.: Identification of community structure in complex networks using affinity propagation clustering method. Modern Physics Letters B 22(16), 1547–1566 (2008)
10. Jia, Y.Q., Wang, G.D., Zhang, C.H., Hua, H.S.: Finding Image Exemplars Using Fast Sparse Affinity Propagation. In: Proc. of ACM Multimedia, pp. 639–642 (2008)

Propagation of Random Perturbations under Fuzzy Algebraic Operators

Zheng Zheng[*], Shanjie Wu, and Kai-Yuan Cai

Department of Automatic Control, Beijing University of Aeronautics and Astronautics,
Beijing 10083, China

Abstract. Since the introduction of fuzzy sets theory, fuzzy method made successful applications in many areas, such as fuzzy control and intelligent decision systems. In these areas, there are usually random perturbations caused by the constantly changing of real situations, thus the analysis of the stability and robustness is an important issue for the applications. In the side of fuzzy methods, we have a corresponding problem: will a small random perturbation of input cause a big oscillation of output of a fuzzy method? In particular, when the distributions of random perturbations are given, what is the propagation of random perturbations in fuzzy schemes? In this paper, we start to answer the question. We estimate the expectation of the propagated perturbations under different fuzzy algebraic operators with two analysis methods. Some examples are presented to show the effectiveness and features of the analysis methods.

Keywords: Robustness of fuzzy methods; Fuzzy algebraic operators; Fuzzy set; Random perturbation.

1 Introduction

Since the very inception of fuzzy sets [1], many fuzzy schemes have been developed based on different fuzzy operators [2, 3]. Applications of these fuzzy techniques have been successful in various fields especially in fuzzy control [4, 5] and intelligent decision systems [6], where fuzzy sets expressed by quantitative terms. Ideally, there should be no random perturbations on the fuzzy sets in fuzzy schemes. In this way, fuzzy schemes in mathematically quantitative terms will exactly represent or capture the essence of requirement. Unfortunately, in real applications, there are usually random perturbations caused by the constantly changing of real situations. The perturbations will be propagated by fuzzy schemes, and they may cause unexpected results.

Thus, a question arises: will a random perturbation of input cause a big oscillation of output of fuzzy methods? Or: Is a mathematical fuzzy scheme robust or perturbation-resistant against the random perturbations? For perturbations of fuzzy schemes, many researchers have provided their approaches [7-11]. Their approaches to perturbations of input are based on some proximity of fuzzy sets, using proximity measure in [7], or α-similarity measure in [8], or δ-equality in our previous work [9], or interval perturbation in [10]. As extensions, the robustness of many fuzzy operators measured by perturbations is investigated in [11]. Most of the efforts concentrated on

---

[*] Corresponding author.

D. Karagiannis and Z. Jin (Eds.): KSEM 2009, LNAI 5914, pp. 66–77, 2009.

how to estimate the limits of the propagated perturbations in fuzzy schemes if the limits of the perturbations on input fuzzy sets are given. However, for random perturbations, the estimation of the limits is not enough. Other properties are needed to be analyzed, such as the expectation, the variance and so on.

In this paper, we focus on the fuzzy methods based on fuzzy algebraic operators and study the propagation of random perturbations that are independent and uniformly distributed. Given fuzzy premises and relations, two methods are developed to analyze the propagated random perturbations under fuzzy algebraic operators.

The paper is structured as follows: After introducing two lemmas, we obtain the expectation of the propagated perturbations under different algebraic operators. Two methods are used to estimate the random perturbations, which are mathematical analysis method in Section 2 and simulation based analysis method in Section 3. Some examples show the effectiveness and the features of the two analysis methods. Final remarks are made in Section 4.

2 Mathematical Analysis of Random Perturbations under Fuzzy Algebraic Operators

In this section, a mathematical analysis method is shown to analyze the propagated random perturbations under fuzzy algebraic operators. The analysis results can serve as certain criterions for choices of fuzzy algebraic operators in real applications, and they can also be used to estimate the output of fuzzy methods with random perturbations. Unless otherwise stated, suppose:

(1) A_1, A_2, A_1' and A_2' are fuzzy sets defined on U, and $\mu_{A_1}(x)$, $\mu_{A_2}(x)$, $\mu_{A_1'}(x)$ and $\mu_{A_2'}(x)$ are memberships of A_1, A_2, A_1' and A_2' respectively;

(2) $\mu_{A_1'}(x) = \min\left(1, \mu_{A_1}(x) + \delta_1(x)\right)$ and $\mu_{A_2'}(x) = \min\left(1, \mu_{A_2}(x) + \delta_2(x)\right)$, where $\delta_1(x)$, $\delta_2(x)$ are independent random numbers uniformly distributed over the interval $[0, a]$ $(0<a<1)$.

Note, in fuzzy environments, the memberships of fuzzy sets must belong to interval $[0, 1]$, thus functions maximize and minimize are used to limit the ranges of the elements into $[0, 1]$;

(3) $E(\bullet)$ represents the expectation of a random number.

Without loss of generality, suppose $\mu_{A_1}(x) \leq \mu_{A_2}(x)$ when analyzing $E\left(\left|\mu_{A_1' \circ A_2'}(x) - \mu_{A_1 \circ A_2}(x)\right|\right)$.

For convenience of mathematical analysis, two lemmas are proposed first. The proof of the lemmas is not shown because of space limit.

Lemma 2.1. Suppose X is a random number whose probability density function is $f_X(z)$ and b is a real number, then the expectation of $\min(b, X)$ is

$$E_{\min(b, X)}(z) = \int_{-\infty}^{b} z \cdot f_X(z)\, dz + \int_{b}^{+\infty} b \cdot f_X(z)\, dz$$

Lemma 2.2. Suppose X, Y are independent random numbers uniformly distributed over the interval $[0, a]$ ($0 < a < 1$). Besides, let c and d be real numbers and $c \leq d$, where $c, d \in [0, 1]$, then the probability density function of $\max(c + X, d + Y)$ is as follows:

$$
f_{\max(c+X,\,d+Y)}(z) =
\begin{cases}
\dfrac{2z - c - d}{a^2} & \text{if } d \leq z < \max(c + a, d) \\[2mm]
\dfrac{1}{a} & \text{if } \max(c + a, d) \leq z \leq d + a \\[2mm]
0 & \text{else}
\end{cases}
$$

2.1 Basic Fuzzy Algebraic Operators

Proposition 2.1. Suppose $A_1 \cup A_2$ represent the union of A_1 and A_2 and $A_1' \cup A_2'$ the union of A_1' and A_2', i.e.,

$$\mu_{A_1 \cup A_2}(x) = \max\left(\mu_{A_1}(x), \mu_{A_2}(x)\right) \text{ and } \mu_{A_1' \cup A_2'}(x) = \max\left(\mu_{A_1'}(x), \mu_{A_2'}(x)\right).$$

If $\mu_{A_1}(x) \leq \mu_{A_2}(x)$, the expectation of the random perturbation on $\mu_{A_1 \cup A_2}(x)$ is

$$
E\left(\left|\mu_{A_1' \cup A_2'}(x) - \mu_{A_1 \cup A_2}(x)\right|\right) =
\begin{cases}
E_1 & \text{if } \mu_{A_2}(x) \leq 1 \leq \mu_{A_1}(x) + a \\
E_2 & \text{if } \mu_{A_2}(x) \leq \mu_{A_1}(x) + a \leq 1 \leq \mu_{A_2}(x) + a \\
E_3 & \text{if } \mu_{A_2}(x) \leq \mu_{A_1}(x) + a \leq \mu_{A_2}(x) + a \leq 1 \\
E_4 & \text{if } \mu_{A_1}(x) + a \leq \mu_{A_2}(x) \leq 1 \leq \mu_{A_2}(x) + a \\
\dfrac{a}{2} & \text{else}
\end{cases}
$$

where

$$E_1 = \frac{\mu_{A_1}(x)\left(\mu_{A_2}(x) - 1\right)^2}{2a^2} - \frac{\left(\mu_{A_2}(x)\right)^3}{6a^2} + \frac{\mu_{A_2}(x)}{2a^2} + 1 - \mu_{A_2}(x),$$

$$E_2 = \frac{\left(\mu_{A_1}(x) - \mu_{A_2}(x) + a\right)^3}{6a^2} - \frac{\left(\mu_{A_2}(x)\right)^2 - 2\mu_{A_1}(x) + 1}{2a} + 1 - \mu_{A_2}(x),$$

$$E_3 = \frac{\left(\mu_{A_1}(x) - \mu_{A_2}(x) + a\right)^3}{6a^2} + \frac{a}{2},$$

$$E_4 = 1 - \frac{\left(\mu_{A_2}(x) - 1\right)^2}{2a} - \mu_{A_2}(x).$$

Proof. First, we can get

$$E\left(\left|\mu_{A_1'\cup A_2'}(x)-\mu_{A_1\cup A_2}(x)\right|\right)=E\left(\max\left(\mu_{A_1'}(x),\mu_{A_2'}(x)\right)\right)-\max\left(\mu_{A_1}(x),\mu_{A_2}(x)\right)$$

$$=E\left(\max\left(\min\left(1,\mu_{A_1}(x)+\delta_1(x)\right),\min\left(1,\mu_{A_2}(x)+\delta_2(x)\right)\right)\right)-\max\left(\mu_{A_1}(x),\mu_{A_2}(x)\right)$$

$$=E\left(\min\left(1,\max\left(\mu_{A_1}(x)+\delta_1(x),\mu_{A_2}(x)+\delta_2(x)\right)\right)\right)-\max\left(\mu_{A_1}(x),\mu_{A_2}(x)\right).$$

Then, $E\left(\left|\mu_{A_1'\cup A_2'}(x)-\mu_{A_1\cup A_2}(x)\right|\right)$ can be calculated from five cases as follows.

(1) If $\mu_{A_2}(x)\le 1\le\mu_{A_1}(x)+a$, then

$$E\left(\left|\mu_{A_1'\cup A_2'}(x)-\mu_{A_1\cup A_2}(x)\right|\right)$$

$$=\int_{\mu_{A_2}(x)}^{1}z\times\frac{2z-\mu_{A_1}(x)-\mu_{A_2}(x)}{a^2}dz+\int_{1}^{\mu_{A_1}(x)+a}1\times\frac{2z-\mu_{A_1}(x)-\mu_{A_2}(x)}{a^2}dz$$

$$+\int_{\mu_{A_1}(x)+a}^{\mu_{A_2}(x)+a}1\times\frac{1}{a}dz-\mu_{A_2}(x) \hspace{2cm}\text{(by Lemma 2.1 and 2.2)}$$

$$=\frac{\mu_{A_1}(x)\left(\mu_{A_2}(x)-1\right)^2}{2a^2}-\frac{\left(\mu_{A_2}(x)\right)^3}{6a^2}+\frac{\mu_{A_2}(x)}{2a^2}+1-\mu_{A_2}(x);$$

(2) If $\mu_{A_2}(x)\le\mu_{A_1}(x)+a\le 1\le\mu_{A_2}(x)+a$, then

$$E\left(\left|\mu_{A_1'\cup A_2'}(x)-\mu_{A_1\cup A_2}(x)\right|\right)$$

$$=\int_{\mu_{A_2}(x)}^{\mu_{A_1}(x)+a}z\times\frac{2z-\mu_{A_1}(x)-\mu_{A_2}(x)}{a^2}dz+\int_{\mu_{A_1}(x)+a}^{1}z\times\frac{1}{a}dz+\int_{1}^{\mu_{A_2}(x)+a}1\times\frac{1}{a}dz-\mu_{A_2}(x)$$

$$\hspace{8cm}\text{(by Lemma 2.1 and 2.2)}$$

$$=\frac{\left(\mu_{A_1}(x)-\mu_{A_2}(x)+a\right)^3}{6a^2}-\frac{\left(\mu_{A_2}(x)\right)^2-2\mu_{A_1}(x)+1}{2a}+1-\mu_{A_2}(x);$$

(3) If $\mu_{A_2}(x)\le\mu_{A_1}(x)+a\le\mu_{A_2}(x)+a\le 1$, then

$$E\left(\left|\mu_{A_1'\cup A_2'}(x)-\mu_{A_1\cup A_2}(x)\right|\right)$$

$$=\int_{\mu_{A_2}(x)}^{\mu_{A_1}(x)+a}z\times\frac{2z-\mu_{A_1}(x)-\mu_{A_2}(x)}{a^2}dz+\int_{\mu_{A_1}(x)+a}^{\mu_{A_2}(x)+a}z\times\frac{1}{a}dz-\mu_{A_2}(x)$$

$$\hspace{8cm}\text{(by Lemma 2.1 and 2.2)}$$

$$=\frac{\left(\mu_{A_1}(x)-\mu_{A_2}(x)+a\right)^3}{6a^2}+\frac{a}{2};$$

(4)   If $\mu_{A_1}(x)+a \le \mu_{A_2}(x) \le 1 \le \mu_{A_2}(x)+a$, then

$$E\left(\left|\mu_{A_1'\cup A_2'}(x)-\mu_{A_1\cup A_2}(x)\right|\right)$$

$$= \int_{\mu_{A_2}(x)}^{1} z\times\frac{1}{a}dz + \int_{1}^{\mu_{A_2}(x)+a} 1\times\frac{1}{a}dz - \mu_{A_2}(x)$$

$$= \frac{1-\left(\mu_{A_2}(x)\right)^2}{2a} + \frac{\mu_{A_2}(x)+a-1}{a} - \mu_{A_2}(x) \qquad \text{(by Lemma 2.1 and 2.2)}$$

$$= 1 - \frac{\left(\mu_{A_2}(x)-1\right)^2}{2a} - \mu_{A_2}(x);$$

(5)   If $\mu_{A_1}(x)+a \le \mu_{A_2}(x) \le \mu_{A_2}(x)+a \le 1$, then

$$E\left(\left|\mu_{A_1'\cup A_2'}(x)-\mu_{A_1\cup A_2}(x)\right|\right)$$

$$= \int_{\mu_{A_2}(x)}^{\mu_{A_2}(x)+a} z\times\frac{1}{a}dz - \mu_{A_2}(x) \qquad \text{(by Lemma 2.1 and 2.2)} \ \blacksquare$$

$$= \frac{\left(\mu_{A_2}(x)+a\right)^2 - \left(\mu_{A_2}(x)\right)^2}{2a} - \mu_{A_2}(x) = \frac{a}{2}.$$

Note, the proof of the following propositions is not shown for space limit.

Proposition 2.2. Suppose $A_1 \cap A_2$ represent the intersection of A_1 and A_2 and $A_1' \cap A_2'$ the intersection of A_1' and A_2', i.e.,

$$\mu_{A_1\cap A_2}(x) = \min\left(\mu_{A_1}(x), \mu_{A_2}(x)\right) \text{ and } \mu_{A_1'\cap A_2'}(x) = \min\left(\mu_{A_1'}(x), \mu_{A_2'}(x)\right).$$

If $\mu_{A_1}(x) \le \mu_{A_2}(x)$, the expectation of the random perturbation on $\mu_{A_1\cap A_2}(x)$ is

$$E\left(\left|\mu_{A_1'\cap A_2'}(x)-\mu_{A_1\cap A_2}(x)\right|\right) = \begin{cases} E_1' & \text{if } \mu_{A_2}(x) \le 1 \le \mu_{A_1}(x)+a \\ E_2' & \text{if } \mu_{A_2}(x) \le \mu_{A_1}(x)+a \le \min\left(1, \mu_{A_2}(x)+a\right) \\ \dfrac{a}{2} & \text{if } \mu_{A_1}(x)+a \le \mu_{A_2}(x) \end{cases} \quad (1)$$

where

$$E_1' = \frac{\left(\mu_{A_2}(x)-\mu_{A_1}(x)-a\right)^3}{6a^2} - \frac{\mu_{A_1}(x)+a}{2a^2} + \frac{\left(\mu_{A_1}(x)+a\right)^3}{6a^2}$$

$$- \frac{\left(\mu_{A_1}(x)+a-1\right)^2\left(\mu_{A_1}(x)+a\right)}{2a^2} + \frac{a}{2} + \frac{1}{3a^2},$$

$$E_2' = \frac{\left(\mu_{A_2}(x)-\mu_{A_1}(x)-a\right)^3}{6a^2} + \frac{a}{2}.$$

Proof. Omit for space limit.

Proposition 2.3. Suppose $\overline{A_1}$ and $\overline{A_1}'$ represent the complements of A_1 and A_1' respectively, i.e.,

$$\mu_{\overline{A_1}}(x) = 1 - \mu_{A_1}(x) \text{ and } \mu_{\overline{A_1}'}(x) = 1 - \mu_{A_1'}(x).$$

Then the expectation of the perturbation on $\mu_{\overline{A_1}}(x)$ is

$$E\left(\left|\mu_{\overline{A_1}'}(x) - \mu_{\overline{A_1}}(x)\right|\right) = \begin{cases} 1 - \dfrac{\left(\mu_{A_1}(x) - 1\right)^2}{2a} - \mu_{A_1}(x) & \text{if } \mu_{A_1}(x) + a > 1 \\ \dfrac{a}{2} & \text{if } \mu_{A_1}(x) + a \leq 1 \end{cases}$$

Proof. Omit for space limit.

2.2 Extended Fuzzy Algebraic Operators

Proposition 2.4. Suppose $A_1 \cdot A_2$ represent the product of A_1 and A_2 and $A_1' \cdot A_2'$ the product of A_1' and A_2', i.e.,

$$\mu_{A_1 \cdot A_2}(x) = \mu_{A_1}(x)\mu_{A_2}(x) \text{ and } \mu_{A_1' \cdot A_2'}(x) = \mu_{A_1'}(x)\mu_{A_2'}(x).$$

If $\mu_{A_1}(x) \leq \mu_{A_2}(x)$, the expectation of the perturbation on $\mu_{A_1 \cdot A_2}(x)$ is

$$E\left(\left|\mu_{A_1' \cdot A_2'}(x) - \mu_{A_1 \cdot A_2}(x)\right|\right)$$

$$= \begin{cases} \left(1 - \dfrac{\left(\mu_{A_1}(x) - 1\right)^2}{2a}\right)\left(1 - \dfrac{\left(\mu_{A_2}(x) - 1\right)^2}{2a}\right) - \mu_{A_1}(x)\mu_{A_2}(x) \\ \qquad\qquad \text{if } \mu_{A_1}(x) + a > 1 \text{ and } \mu_{A_2}(x) + a > 1 \\[2mm] \left(\mu_{A_1}(x) + \dfrac{a}{2}\right)\left(1 - \dfrac{\left(\mu_{A_2}(x) - 1\right)^2}{2a}\right) - \mu_{A_1}(x)\mu_{A_2}(x) \\ \qquad\qquad \text{if } \mu_{A_1}(x) + a \leq 1 \text{ and } \mu_{A_2}(x) + a > 1 \\[2mm] \left(\mu_{A_1}(x) + \dfrac{a}{2}\right)\left(\mu_{A_2}(x) + \dfrac{a}{2}\right) - \mu_{A_1}(x)\mu_{A_2}(x) \\ \qquad\qquad \text{if } \mu_{A_1}(x) + a \leq 1 \text{ and } \mu_{A_2}(x) + a \leq 1 \end{cases}$$

Proof. Omit for space limit.

Proposition 2.5. Suppose $A_1 \uplus A_2$ represent the bold sum of A_1 and A_2 and the $A_1' \uplus A_2'$ bold sum of A_1' and A_2', i.e.,

$$\mu_{A_1 \uplus A_2}(x) = \min\left(1, \mu_{A_1}(x) + \mu_{A_2}(x)\right), \ \mu_{A_1' \uplus A_2'}(x) = \min\left(1, \mu_{A_1'}(x) + \mu_{A_2'}(x)\right).$$

If $\mu_{A_1}(x) \le \mu_{A_2}(x)$, the expectation of the perturbation on is

(1) If $\mu_{A_2}(x) + a \ge 1$,

$$E\left(\left\|\mu_{A_1' \uplus A_2'}(x) - \mu_{A_1 \uplus A_2}(x)\right\|\right)$$

$$= \begin{cases} \dfrac{\left(\mu_{A_1}(x) + \mu_{A_2}(x) - 1\right)^3}{6a^2} - \left(\mu_{A_1}(x) + \mu_{A_2}(x) - 1\right) & \text{if } \mu_{A_1}(x) + \mu_{A_2}(x) \le 1 \\ 0 & \text{if } \mu_{A_1}(x) + \mu_{A_2}(x) > 1 \end{cases}$$

(2) If $\mu_{A_2}(x) + a < 1$,

$$E\left(\left\|\mu_{A_1' \uplus A_2'}(x) - \mu_{A_1 \uplus A_2}(x)\right\|\right)$$

$$= \begin{cases} 0 & \text{if } \mu_{A_1}(x) + \mu_{A_2}(x) > 1 \\ \dfrac{\left(\mu_{A_1}(x) + \mu_{A_2}(x) - 1\right)^3}{6a^2} - \left(\mu_{A_1}(x) + \mu_{A_2}(x) - 1\right) & \text{if } \mu_{A_1}(x) + \mu_{A_2}(x) \le 1 \le \mu_{A_1}(x) + \mu_{A_2}(x) + a \\ \dfrac{\left(1 - \mu_{A_1}(x) - \mu_{A_2}(x)\right)^3}{6a^2} - \dfrac{\left(1 - \mu_{A_1}(x) - \mu_{A_2}(x) - a\right)^2}{a} + \dfrac{2a}{3} & \text{if } \mu_{A_1}(x) + \mu_{A_2}(x) + a \le 1 \le \mu_{A_1}(x) + \mu_{A_2}(x) + 2a \\ a & \text{if } \mu_{A_1}(x) + \mu_{A_2}(x) + 2a \le 1 \end{cases}$$

Proof. Omit for space limit.

Proposition 2.6. Suppose $A_1 \|_\lambda A_2$ represent the convex linear sum of A_1 and A_2 and $A_1' \|_\lambda A_2'$ the convex linear sum of A_1' and A_2', and $\lambda \le 1$, i.e.,

$$\mu_{A_1 \|_\lambda A_2}(x) = \lambda \min\left(\mu_{A_1}(x), \mu_{A_2}(x)\right) + (1 - \lambda) \max\left(\mu_{A_1}(x), \mu_{A_2}(x)\right),$$

and $\quad \mu_{A_1' \|_\lambda A_2'}(x) = \lambda \min\left(\mu_{A_1'}(x), \mu_{A_2'}(x)\right) + (1 - \lambda) \max\left(\mu_{A_1'}(x), \mu_{A_2'}(x)\right).$

If $\mu_{A_1}(x) \le \mu_{A_2}(x)$, the expectation of the perturbation on $\mu_{A_1 \|_\lambda A_2}(x)$ is

$$E\left(\left\|\mu_{A_1' \|_\lambda A_2'}(x) - \mu_{A_1 \|_\lambda A_2}(x)\right\|\right)$$

$$= \lambda E\left(\left\|\mu_{A_1' \cap A_2'}(x) - \mu_{A_1 \cap A_2}(x)\right\|\right) + (1 - \lambda) E\left(\left\|\mu_{A_1' \cup A_2'}(x) - \mu_{A_1 \cup A_2}(x)\right\|\right),$$

where $E\left(\left|\mu_{A_1' \cap A_2'}(x) - \mu_{A_1 \cap A_2}(x)\right|\right)$ and $E\left(\left|\mu_{A_1' \cup A_2'}(x) - \mu_{A_1 \cup A_2}(x)\right|\right)$ are shown at

Proposition 3.1 and 3.2.

Proof. Omit for space limit.

3 Simulation Based Analysis of Random Perturbations under Fuzzy Algebraic Operators

As in Section 2, a natural way to evaluate the propagation of random perturbations in fuzzy schemes is to define and derive it analytically. However, in some cases, the random perturbations of fuzzy sets are difficult to be analyzed mathematically, especially when the distributions of $\delta_1(x)$ and $\delta_2(x)$ are very complicated or the fuzzy scheme is too complex to be analyzed. Thus, new methods need to be developed to deal with these cases.

In this following, we present a Monte Carlo simulation based analysis method to analyze the propagated random perturbations under fuzzy algebraic operator " $\circ$ ".

Simulation Algorithm 1

Step 1. Obtain fuzzy sets A_1 and A_2, and suppose $\mu_{A_1}(x)$, $\mu_{A_2}(x)$ are the memberships of A_1 and A_2 respectively ;

Step 2. For $\forall x \in U$, generate p sets of random numbers $\left\{\delta_1^j(x), \delta_2^j(x)\right\}$, $j = 1, \cdots, p$, where $\delta_1^j(x), \delta_2^j(x)$ are independent and they uniformly distributed over the interval $[0, a]$ ($0 < a < 1$);

Step 3. Let $\mu_{A_1^{j'}}(x) = \min\left(1, \mu_{A_1}(x) + \delta_1^j(x)\right)$ and $\mu_{A_2^{j'}}(x) = \min\left(1, \mu_{A_2}(x) + \delta_2^j(x)\right)$, where $\mu_{A_1^{j'}}(x)$ and $\mu_{A_2^{j'}}(x)$ be the memberships of $A_1^{j'}$ and $A_2^{j'}$ ($j = 1, \cdots, p$);

Step 4. Obtain the results of $A_1 \circ A_2$ and $A_1^{j'} \circ A_2^{j'}$, where $j = 1, \cdots, p$;

Step 5. For $\forall x \in U$, calculate $\dfrac{1}{p} \sum_{j=1}^{p} \left|\mu_{A_1^{j'} \circ A_2^{j'}}(x) - \mu_{A_1 \circ A_2}(x)\right|$, which is the estimation

of the expectation of random perturbation on $\mu_{A_1 \circ A_2}(x)$;

Step 6. Stop.

Obviously, $E\left(\left|\mu_{A_1' \circ A_2'}(x) - \mu_{A_1 \circ A_2}(x)\right|\right) = \lim_{p \to +\infty} \left(\dfrac{1}{p} \sum_{j=1}^{p} \left|\mu_{A_1^{j'} \circ A_2^{j'}}(x) - \mu_{A_1 \circ A_2}(x)\right|\right)$. From

the algorithm, it is easy to design an algorithm to get the expectation of the propagated perturbations under different fuzzy operators, while it is often difficult for mathematical analysis method to get analysis results. However, because

$\frac{1}{p}\sum_{j=1}^{p}\left|\mu_{A_1^{j'}\circ A_2^{j'}}(x)-\mu_{A_1\circ A_2}(x)\right|$ can be close to $E\left(\left|\mu_{A_1'\circ A_2'}(x)-\mu_{A_1\circ A_2}(x)\right|\right)$ only when p is a huge number, thus the simulation algorithm is usually time consuming. Besides, the analysis results of simulation based method cannot be expressed by mathematical expressions as mathematical method, which is easy to be understood and further used. In summary, mathematical analysis method and simulation based analysis method each has its advantages and disadvantages, so we should choose them according to the requirements.

In the following, an example is given to use the two methods to analyze a case. With Matlab 6.0, we program simulation algorithm 1 and the mathematical expressions analyzed by the mathematical method to calculate the expectations of propagated random perturbations respectively. The simulations are processed on the computer with 2.33GHz CPU and 2G memory space.

Example 1: Let $p=10^6$ (in Simulation Algorithm 1), $\lambda=0.25$ (in Proposition 2.7), and

$$A_1 = \begin{bmatrix} 0.7177 & 0.0859 & 0.9065 & 0.2017 & 0.1525 & 0.6100 & 0.2455 \\ 0.2070 & 0.4858 & 0.4940 & 0.4377 & 0.4213 & 0.3594 & 0.0755 \\ 0.4509 & 0.5675 & 0.2727 & 0.3367 & 0.4410 & 0.7317 & 0.2469 \\ 0.9032 & 0.4927 & 0.5723 & 0.1155 & 0.3425 & 0.9514 & 0.9637 \\ 0.8190 & 0.8764 & 0.5557 & 0.2152 & 0.8946 & 0.7252 & 0.1390 \\ 0.6533 & 0.6099 & 0.0049 & 0.1518 & 0.9203 & 0.4065 & 0.4567 \\ 0.4855 & 0.7520 & 0.2696 & 0.3620 & 0.2125 & 0.9258 & 0.4760 \end{bmatrix},$$

$$A_2 = \begin{bmatrix} 0.7178 & 0.9003 & 0.9067 & 0.7366 & 0.5521 & 0.1658 & 0.0930 \\ 0.1612 & 0.4857 & 0.3355 & 0.2708 & 0.4123 & 0.4683 & 0.1474 \\ 0.4539 & 0.7761 & 0.6814 & 0.3237 & 0.4749 & 0.7208 & 0.9654 \\ 0.0982 & 0.9368 & 0.5745 & 0.6944 & 0.8695 & 0.8654 & 0.9637 \\ 0.8281 & 0.8673 & 0.7513 & 0.2242 & 0.4847 & 0.6960 & 0.4014 \\ 0.8248 & 0.1739 & 0.1692 & 0.2340 & 0.7805 & 0.4169 & 0.4767 \\ 0.9152 & 0.7442 & 0.2846 & 0.7966 & 0.2065 & 0.7497 & 0.6980 \end{bmatrix}.$$

Suppose for $\forall x \in U$, $\delta_1^j(x), \delta_2^j(x)$ are independent random numbers uniformly distributed over the interval [0, 0.025], where $j=1,\cdots,p$. In table 1, $\sum_{x\in U}E$ is the abbreviation of $\sum_{x\in U}E\left(\left|\mu_{A_1'\circ A_2'}(x)-\mu_{A_1\circ A_2}(x)\right|\right)$, where "$\circ$" is an algebraic operator. t_1 and t_2 are the processing times of the two analysis methods respectively.

Table 1. Comparison of analysis methods

No.	Fuzzy Methods	Mathematical Analysis		Simulation based analysis		δ
		$\sum_{x \in U} E$	t_1 (sec.)	$\sum_{x \in U} E$	t_2 (sec.)	Equalities based analysis
1	$A_1 \cup A_2$	0.64482	0.015	0.64484	27.282	0.975
2	$A_1 \cap A_2$	0.58018	0.015	0.58019	27.219	0.975
3	$\overline{A_1}$	0.61250	<0.001	0.61245	21.079	0.975
4	$A_1 \cdot A_2$	0.64737	0.015	0.64737	23.172	0.950
5	$A_1 \uplus A_2$	0.63543	0.016	0.63548	27.187	0.950
6	$A_1 \parallel_\lambda A_2$	0.62866	0.031	0.62872	40.5320	0.975

More than one dozen of simulations with different parameters have been processed, the results of which show that:

(1) When p is huge, simulation based analysis results are close to the mathematical analysis results;
(2) The simulation based method is more time consuming than the mathematical based method;
(3) The analysis results in this paper and those of the δ-equalities based method are different. The reason is that, the analysis results in this paper are the expectations of the propagated random perturbations, while the results of the δ-equalities based method are the limits. The two kinds of analysis results show different robustness aspects of fuzzy methods.

We can easily extend simulation algorithm 1 to analyze random perturbations propagated in more complicated fuzzy methods. For example, we can compare the propagated perturbations of two fuzzy reasoning methods by simply extending Simulation Algorithm 1.

Example 2. Let $A \rightarrow B$ be a fuzzy relation and C be a fuzzy premise, which are

$$A = [0.3987 \quad 0.3584 \quad 0.2853 \quad 0.8686 \quad 0.6264 \quad 0.2412 \quad 0.9781],$$
$$B = [0.6405 \quad 0.2298 \quad 0.6813 \quad 0.6658 \quad 0.1347 \quad 0.0225 \quad 0.2622],$$
$$C = [0.1165 \quad 0.0693 \quad 0.8529 \quad 0.1803 \quad 0.0324 \quad 0.7339 \quad 0.5365].$$

There are two candidate fuzzy reasoning methods. For the first one, suppose the min conjunction and the Dines-Rescher implication operator are used in the fuzzy reasoning, that is

$$\mu_D(y) = \sup_{x \in U} \Big(\min \big(\mu_C(x), \max \big(1 - \mu_A(x), \mu_B(y) \big) \big) \Big),$$

where D is the fuzzy consequence. For the second one, suppose the min conjunction and the Reichenbach implication operator are used in the fuzzy reasoning, that is

$$\mu_D(y) = \sup_{x \in U}\left(\min\left(\mu_C(x), 1 - \mu_A(x) + \mu_A(x)\mu_B(y)\right)\right),$$

where D is the fuzzy consequence. Suppose δ_A, δ_B and δ_C are random functions on A, B and C respectively, where for $\forall x \in U$ $\delta_A(x)$, $\delta_B(x)$ and $\delta_C(x)$ are independent random numbers with normally distribution N(0, 1). Let $p = 10^6$, then we can calculate the propagated perturbations of the two fuzzy reasoning methods by simply extending simulation algorithm 1 and the results are showed in table 2.

Table 2. Comparison of two fuzzy reasoning methods

	Fuzzy reasoning method 1	Fuzzy reasoning method 2
$\sum_{x \in U} E$	2.0735	1.3396

From table 2, it is obvious that fuzzy reasoning method 2 has much less propagated perturbations than method 1, thus we can choose method 2 as the final fuzzy reasoning method.

4 Conclusion

Up to this point, two methods are used to analyze the propagation of the random perturbations. For the mathematical analysis method, the results are mathematical expressions, thus they are easy to be understood and further used. However, the process of mathematical analysis is usually difficult especially when the fuzzy schemes or the random distributions of the perturbations are complicated. For the simulation based analysis method, the analysis algorithm is easy to be designed and processed. However, the analysis results cannot be expressed mathematically, thus it is not easy to be understood and further used. And because the simulation based analysis results are statistical, it is usually slow to be calculated. The analysis results of both methods can serve as certain criterions for choices of fuzzy algebraic operators in real applications, and they can also be used to estimate the output of fuzzy methods with random perturbations.

Acknowledgements

This work was supported by Aviation Science Foundation of China (2008ZG51092).

References

[1] Zadeh, L.A.: The concept of a linguistic variable and its applications to approximate reasoning, I, II, III. Information Science 8, 199–249 (1974); 8, 301-357 (1974); 9, 43-93 (1975)

[2] Dubois, D., Lang, J., Prade, H.: Fuzzy sets in approximate reasoning, parts 1 and 2. Fuzzy Sets and Systems 40, 143–244 (1991)

[3] Wang, G.J.: On the logic foundation of fuzzy reasoning. Information Science 117, 231–251 (1999)

[4] Driankov, D., Hellendoorn, H., Reinfrank, M.: An Introduction To Fuzzy Control. Springer, New York (1993)

[5] Wang, L.X.: A Course In Fuzzy Systems And Control. Prentice-Hall, Englewood Cliffs (1997)

[6] Carlsson, C., Fuller, R.: Fuzzy Reasoning In Decision Making and Optimization. Physica-Verlag, Heidelberg (2002)

[7] Pappis, C.P.: Value approximation of fuzzy systems variables. Fuzzy Sets and Systems 39, 111–115 (1991)

[8] Hong, D.H., Hwang, S.Y.: A note on the value similarity of fuzzy systems variables. Fuzzy Sets and Systems 66, 383–386 (1994)

[9] Cai, K.Y.: Robustness of fuzzy reasoning and δ-equalities of fuzzy sets. IEEE Transactions on Fuzzy Systems 9, 738–750 (2001)

[10] Cheng, G.S., Fu, Y.X.: Error estimation of perturbations under CRI. IEEE Transactions on Fuzzy Systems 14, 709–715 (2006)

[11] Li, Y.M., Li, D.C., Pedrycz, W., Wu, J.J.: An approach to measure the robustness of fuzzy reasoning. International Journal of Intelligent Systems 20, 393–413 (2005)

[12] DeGroot, M.H.: Probability And Statistics. Addison-Wesley Pub. Co., Reading (1986)

Two Approaches to Iterated Belief Contraction

Raghav Ramachandran, Abhaya C. Nayak, and Mehmet A. Orgun

Department of Computing, Macquarie University,
NSW-2109, Australia

Abstract. Iterated Belief Contraction is a relatively less explored area in belief change and intuition for it is often driven by work in the area of iterated belief revision. For many of the iterable belief contraction functions defined in the literature, very little is known about their properties. In this paper we recall two iterable contraction functions, *Natural contraction* and *Priority contraction*, defined by Nayak and colleagues. Here we characterize both these contraction functions via some simple properties of iterated contraction.

1 Introduction

The logic of belief change studies the process of changing the current set of beliefs of an agent upon receiving new information. The received information could be consistent with the existing beliefs or conflict with some of them. When the information is consistent with the existing beliefs, incorporating the information into the set of beliefs is not much of a challenge. This change is called *Belief Expansion*. When the received information contradicts some existing beliefs, believing it along with presently held beliefs leads to inconsistency. Hence including the information into the set of beliefs of the agent becomes a difficult task. Adding this information to the set of beliefs while maintaining the consistency is called *Belief Revision*. There are many accounts detailing the procedure of belief revision [5, 12]. While Expansion and Revision deal with adding the information to the existing set of beliefs, removal of beliefs from the belief set is another form of change. This is called *Belief Contraction*. Just as in revision, there are many accounts of contraction functions found in the literature [1, 6, 10].

Belief expansion, revision and contraction can be considered as belief change functions taking one belief state to another.[1] There is a school of thought that considers belief expansion and belief contraction to be the basic change operations, and belief revision is a composition of these two basic operations given by the Levi Identity.[2] In this paper, we consider belief revision could be performed as a combination of contraction and expansion. Therefore, in this paper, we concern ourselves with belief contraction functions only. We begin by introducing

---

[1] A belief state of the agent gives many aspects of the agent's epistemic state, such as the set of beliefs, degree of trust in these beliefs.

[2] Levi Identity states that Belief Revision can be obtained from a Belief Expansion following a Belief Contraction. $\mathcal{K}_\alpha^* = (\mathcal{K}_{\neg\alpha}^-)_\alpha^+$. Here $*$ denotes revision, $-$ denotes contraction and $+$ denotes expansion.

D. Karagiannis and Z. Jin (Eds.): KSEM 2009, LNAI 5914, pp. 78–89, 2009.

two iterable contraction functions, the Natural contraction function and Priority contraction function, based on the System of Spheres approach [10, 13].[3] Contraction functions that can perform iterated changes are termed Iterable. Our aim is to characterize these two contraction functions.

The organization of this paper is as follows: In section 2, we give the necessary background in the System of Spheres approach and the Epistemic Entrenchment approach to belief change. In section 3, we describe Priority and Natural contraction functions. Section 4 deals with an earlier approach at characterizing Lexicographic contraction function [9]. In section 5, we discuss properties desirable of iterable contraction functions. Section 6 presents representation theorems connecting the properties discussed in section 5 and priority and natural contraction functions.

2   Background

The literature is rich with different approaches to contract a belief α from a belief state $\mathcal{K}$. The AGM belief change theory has been very influential in this topic. In the AGM theory [1,6], the belief state is restricted to a belief set[4] closed under classical consequence operation. Let $\mathcal{K}$ be the belief set of the agent and $\alpha, \beta \in \mathcal{K}$. A function $-$ is called an AGM contraction function if and only if it satisfies the AGM contraction postulates listed below.

1. *Closure*: $\mathcal{K}_\alpha^-$ is a theory whenever $\mathcal{K}$ is a theory.[5]
2. *Inclusion*: $\mathcal{K}_\alpha^- \subseteq \mathcal{K}$.
3. *Vacuity*:   If $\alpha \notin Cn(\mathcal{K})$, then $\mathcal{K}_\alpha^- = \mathcal{K}$.
4. *Success*:   If $\emptyset \nvdash \alpha$, then $\alpha \notin \mathcal{K}_\alpha^-$.
5. *Preservation*: If $Cn(\alpha) = Cn(\beta)$, then $\mathcal{K}_\alpha^- = \mathcal{K}_\beta^-$.
6. *Recovery*:   If $\alpha \in \mathcal{K}$, then $\mathcal{K} \subseteq (\mathcal{K}_\alpha^-)_\alpha^+$.
7. *Intersection*: $\mathcal{K}_\alpha^- \bigcap \mathcal{K}_\beta^- \subseteq \mathcal{K}_{\alpha \wedge \beta}^-$, for any theory $\mathcal{K}$.
8. *Conjunction*: If $\beta \notin \mathcal{K}_{\alpha \wedge \beta}^-$, then $\mathcal{K}_{\alpha \wedge \beta}^- \subseteq \mathcal{K}_\beta^-$, for any theory $\mathcal{K}$.

Motivation behind these postulates can be found in [6]. Along with these postulates a construction procedure for such a contraction function was proposed [1, 6]. The construction of an AGM contraction function requires a selection function, which chooses a subset of the set of all maximal subsets of the belief set that do not entail the belief being contracted. This selection function is defined for a particular belief set $\mathcal{K}$. Because of this the contraction function constructed using this selection function cannot be used to contract a belief β from the contracted belief set $\mathcal{K}_\alpha^-$. To contract a belief β from the belief set $\mathcal{K}_\alpha^-$, a different AGM contraction function needs to be constructed using a selection

---

[3] In [13] natural contraction is termed as AGM contraction or Conservative contraction and Priority contraction is termed as Moderate Contraction.

[4] A belief set is a set of the beliefs of the agent, where the beliefs are represented as sentences in the language $\mathcal{L}$.

[5] A theory is a set of sentences closed under the classical consequence operation.

function defined for that belief set. Thus, the belief change functions given by the AGM approach are not iterable. This problem has been addressed in [2] where the authors assume a selection function over the language $\mathcal{L}$ instead of over the belief set $\mathcal{K}$. This selection function is then restricted to any particular belief set. In this way, the AGM approach has been extended to iterable belief change functions. This approach is very similar to the *superselectors* used by Hansson in early '90s [8].

Another approach to the belief change theory is the epistemic entrenchment approach. In epistemic entrenchment based on [6], the belief state of the agent is given by a binary relation $\leq$ over the set of sentences in the language, $\mathcal{L}$. For any $\alpha, \beta \in \mathcal{L}$, $\alpha \leq \beta$ is read as β *is at least as entrenched as* α. Given an entrenchment relation $\leq$ on $\mathcal{L}$ which reflects the epistemic state of the agent, the set of beliefs of the agent can be derived from $\leq$, as shown below:

$$\mathcal{K}_{\leq} = \begin{cases} \{\alpha : \bot < \alpha\} \text{ if } \bot < \beta, \text{for any } \beta \\ \{\alpha : \bot \leq \alpha\} \text{ otherwise.} \end{cases} \tag{E1}$$

To perform changes to the belief set of the agent, the entrenchment relation needs to be changed. In [11] belief revision is performed by changing the entrenchment relation. The initial entrenchment relation $\leq$ is changed to another entrenchment relation $\leq'$, which results in changing the belief set $\mathcal{K}_{\leq}$ to $\mathcal{K}_{\leq'}$. In [11] it is shown that changing a given entrenchment relation to another entrenchment relation ensures that the belief change operations defined this way are iterable.

Another well known approach to belief change is based on System of Spheres [7]. Here the belief state of the agent is given by a total pre-order relation $\sqsubseteq$ on the set of all possible worlds Ω over the language $\mathcal{L}$.[6] Given a $\sqsubseteq$ the belief set of the agent is,

$$\mathcal{K}_{\sqsubseteq} = \{\beta \in \mathcal{L} \mid \sqsubseteq\text{-min}(\Omega) \subseteq [\beta]\}, \tag{G1}$$

where $\sqsubseteq\text{-min}(\Omega) = \{\omega \in \Omega \mid \omega \sqsubseteq \omega', \text{ for every } \omega' \in \Omega\}$ and $[\beta] = \{\omega \in \Omega \mid \omega \models \beta\}$. In this approach, changes to the agent's beliefs are performed by changing the total pre-order relation $\sqsubseteq$ on Ω to another total pre-order relation $\sqsubseteq'$. Since $\sqsubseteq$ is changed to another total pre-order relation, the change function is iterable. In [11], a clear relation between the system of spheres approach and the epistemic entrenchment approach is provided.[7]

3   Contraction Functions Based on System of Spheres

Let $\sqsubseteq$ be a total pre-order relation on Ω and $\mathcal{K}$ denote the belief set of the agent corresponding to $\sqsubseteq$. In order to contract a belief α from $\mathcal{K}$ the total pre-order $\sqsubseteq$ on Ω needs to be changed. From equation (G1), it is understood that

---

[6] A total pre-order relation is a reflexive, connected and transitive binary relation.

[7] A total pre-order relation $\sqsubseteq$ on Ω is a counter-part of an epistemic entrenchment relation $\leq$ on $\mathcal{L}$ if and only if $\forall \alpha, \beta \in \mathcal{L}$, $\alpha \leq \beta$ if and only if either $[\beta] = \Omega$ or there exists $\omega_1 \in \sqsubseteq\text{-min}[\neg\alpha]$ such that $\omega_1 \sqsubseteq \omega_2$ for every world $\omega_2 \in \sqsubseteq\text{-min}[\neg\beta]$.

the changed total pre-order $\sqsubseteq_\alpha^\ominus$ on Ω should contain at least one model of $\neg\alpha$, i.e., $\sqsubseteq_\alpha^\ominus$-$\min(\Omega) \cap [\neg\alpha] \neq \emptyset$, so that $\alpha \notin \mathcal{K}_\alpha^\ominus$.[8] In [7], Grove proved that for a contraction function, based on the System of Spheres approach, to satisfy the postulates given in the AGM theory, $\sqsubseteq_\alpha^\ominus$ has to be such that

$$\sqsubseteq_\alpha^\ominus\text{-}\min(\Omega) = \sqsubseteq\text{-}\min(\Omega) \bigcup \sqsubseteq\text{-}\min[\neg\alpha] \tag{G2}$$

This condition on $\sqsubseteq_\alpha^\ominus$ does not guarantee a unique way of changing $\sqsubseteq$ to $\sqsubseteq_\alpha^\ominus$. Different ways of changing $\sqsubseteq$ have been proposed, each giving different contraction functions. Here we recall two such contraction functions proposed in [10].

3.1   Moderate or Priority Contraction Function

The basic idea behind priority contraction is simple. Suppose we want to remove a belief α. The content of α can be viewed to be the same as the set of sentences $\{\beta \to \alpha \mid \beta \in \mathcal{L}\}$. Furthermore, since $\alpha \vdash \beta \to \alpha$ for every such $\beta \to \alpha$, each of them can be viewed as contributing to the content of α. When we remove α from a belief set, under usual circumstances, α becomes a non-belief; its degree of entrenchment gets reduced to minimum. One would expect that something along that line would happen to all the sentences $\beta \to \alpha$. That would happen to a number of them, since out of every pair $\gamma \to \alpha$ and $\neg\gamma \to \alpha$, at least one will need to be removed and its entrenchment will become minimal. The question is, what happens to those sentences $\beta \to \alpha$ which are not removed from the belief set. One might argue that even if some sentence $\beta \to \alpha$ is not removed via removal of α, its degree of entrenchment should be reduced – its status should be demoted. How this idea can be formally captured is a contentious issue.[9] One way to go about it is the following: the degree of entrenchment of $\beta \to \alpha$ for some particular β is reflected by the minimally plausible worlds in $[\beta] \cap [\neg\alpha]$. So, by demoting all the worlds in $[\neg\alpha]$ we can expect to reduce the degree of entrenchment of every $\beta \to \alpha$. This is captured by the following formalization.

Given a total pre-order relation $\sqsubseteq$ on Ω, and a sentence α to be removed from the associated belief set, priority contraction function $\ominus$ changes the relation $\sqsubseteq$ to a new relation $\sqsubseteq_\alpha^\ominus$. If α is *not* believed in the state represented by the prior $\sqsubseteq$, then $\sqsubseteq$ is not changed. In the principal, non-trivial case, where $\alpha \in \mathcal{K}$ and $\nvdash \alpha$, the new total pre-order $\sqsubseteq_\alpha^\ominus$ satisfies the following conditions: For any $\omega, \omega' \in \Omega$,

1. When $\omega \models \alpha$ and $\omega' \models \alpha$ then $\omega \sqsubseteq_\alpha^\ominus \omega'$ if and only if $\omega \sqsubseteq \omega'$.
2. When $\omega \models \neg\alpha$ and $\omega' \models \neg\alpha$ then $\omega \sqsubseteq_\alpha^\ominus \omega'$ if and only if $\omega \sqsubseteq \omega'$.
3. When $\omega \models \alpha$, $\omega \notin \sqsubseteq$-$\min(\Omega)$ and $\omega' \models \neg\alpha$ then $\omega' \sqsubset_\alpha^\ominus \omega$.[10]
4. When $\omega \in \sqsubseteq$-$\min(\Omega)$ or $\omega \in \sqsubseteq$-$\min[\neg\alpha]$, then $\omega \sqsubseteq_\alpha^\ominus \omega'$, for any $\omega' \in \Omega$.

---

[8] Here, $\ominus$ denotes a contraction function acting on the relation $\sqsubseteq$.

[9] Since the entrenchment relation is changing in the process, it is not immediately obvious how to capture it in a relational setup. It is akin to the comparison of inter-personal utility functions. We will take up this issue on another occasion.

[10] From here onwards we use $\sqsubset$ to denote the strict part of $\sqsubseteq$, that is, $\omega \sqsubset \omega'$ iff both $\omega \sqsubseteq \omega'$, and $\omega' \not\sqsubseteq \omega$.

It is clear from conditions (1) and (2) above that relative plausibility of worlds in $[\alpha]$ (and respectively in $[\neg\alpha]$) are not affected. Condition (3) captures the uniform demotion of worlds in $[\neg\alpha]$. Condition (4) reflects (G2).[11]

3.2   Natural or Conservative Contraction Function

In [3](pg 522) Boutilier defines a simple belief revision function called *Natural Revision*. He states that,

> If $\omega \in$ min(M, A) by the basic requirement ω must be minimal in R′, and these must be the only minimal worlds in R′,

where R and R′ are binary relations on Ω before and after revision, respectively, ϕ is the valuation function, A is the proposition being revised and M is $\langle \Omega, R, \phi \rangle$. This definition states that when revising by A, the minimally plausible worlds in [A] become the only minimally plausible worlds in Ω in the revised relation R′. There is no other change made to the relation R. Natural contraction follows Natural revision in changing the relation on Ω only with respect to the minimally plausible worlds in $[\neg\alpha]$ when contracting by α. Given a total pre-order relation $\sqsubseteq$ on Ω, and a sentence α, a natural contraction function $\ominus$ changes the relation $\sqsubseteq$ when contracting by α to $\sqsubseteq_\alpha^\ominus$.[12] The changed total pre-order relation $\sqsubseteq_\alpha^\ominus$ satisfies the following conditions: For any $\omega, \omega' \in \Omega$,

1. When $\omega \in \sqsubseteq\text{-min}(\Omega)$ or $\omega \in \sqsubseteq\text{-min}[\neg\alpha]$, then $\omega \sqsubseteq_\alpha^\ominus \omega'$, for any $\omega' \in \Omega$.
2. When $\omega, \omega' \notin (\sqsubseteq\text{-min}(\Omega) \bigcup \sqsubseteq\text{-min}[\neg\alpha])$, $\omega \sqsubseteq_\alpha^\ominus \omega'$ if and only if $\omega \sqsubseteq \omega'$

It is clear that a natural contraction function changes $\sqsubseteq$ just enough to satisfy the AGM contraction postulates. The relative orderings of worlds in $[\alpha]$ (and respectively in $[\neg\alpha]$) are not changed.

These two contraction functions do not show any difference in a single contraction because in the first contraction they mirror an AGM contraction function. But their differences surface when performing the second contraction. To illustrate the differences, we make use of the well known example given by Darwiche and Pearl in [4], which was extended to the following in [10].

Example 1. Let the agent believe that *Mr. Craig is rich(r) and Mr. Craig is smart(s)*. $\Omega = \{ \{s, r\}, \{s, \neg r\}, \{\neg s, r\}, \{\neg s, \neg r\}\}$, and The epistemic state of the agent can be given by any total pre-order $\sqsubseteq$ on Ω. Let $\sqsubseteq$ be $\{s, r\} \subsetneq \{s, \neg r\} \subsetneq \{\neg s, r\} \subsetneq \{\neg s, \neg r\}$ and $\sqsubseteq\text{-min}(\Omega) = \{\{s, r\}\}$. Suppose the agent doubts the smartness of Mr. Craig and decides not to believe s. According to both Priority and Natural contraction functions, the agent will continue to believe r, but no longer believe s. After the contraction, suppose the agent decides to withdraw

---

[11] The definition of Moderate contraction function is analogous to Lexicographic revision, [11], but the name Lexicographic contraction is used for a more complicated contraction function in [10].

[12] The following definition of natural contraction holds only for non-trivial cases, just as in the definition of Priority contraction.

its belief in r, then according to Priority Contraction function, the agent will end up believing *If s, then r*. And according to Natural Contraction function, the agent will end up believing *Either s or r*.

The literature on the system of spheres approach and the epistemic entrenchment approach give a construction of iterable belief contraction functions. It is also necessary to study the properties of these iterable belief contraction functions because that would bring out the differences between these contraction functions.

4   Postulates for Iterated Contraction

In [9] the authors define a Principled iterative contraction operation, which is an AGM contraction function, that satisfies Principled Factored Insertion, Given $\beta \in \mathcal{K}_\alpha^-$,

1. If $\alpha \rightarrow \beta \in (\mathcal{K}_\alpha^-)_{\bar{\beta}}^-$, then $(\mathcal{K}_\alpha^-)_{\bar{\beta}}^- = \mathcal{K}_\alpha^- \cap \mathcal{K}_{\alpha \vee \beta}^-$
2. If $\alpha \vee \beta \in (\mathcal{K}_\alpha^-)_{\bar{\beta}}^-$, then $(\mathcal{K}_\alpha^-)_{\bar{\beta}}^- = \mathcal{K}_\alpha^- \cap \mathcal{K}_{\alpha \rightarrow \beta}^-$
3. If $\alpha \rightarrow \beta, \alpha \vee \beta \notin (\mathcal{K}_\alpha^-)_{\bar{\beta}}^-$, then $(\mathcal{K}_\alpha^-)_{\bar{\beta}}^- = \mathcal{K}_\alpha^- \cap \mathcal{K}_{\alpha \rightarrow \beta}^- \cap \mathcal{K}_{\alpha \vee \beta}^-$.

The authors believed that Principled Factored Insertion would be enough to characterize what they call the Lexicographic contraction operation[13] but only showed that a Lexicographic contraction operation satisfies Principled Factored Insertion. Here we show that both Priority and Natural contraction functions satisfy Principled Factored Insertion.

Theorem 1. *Given a total pre-order relation $\sqsubseteq$ on Ω, when contracting by α any contraction function that satisfies (G2) and preserves the ordering of worlds within $[\alpha]$ and $[\neg\alpha]$ satisfies Principled Factored Insertion(PFI).*

Proof: Let $\alpha, \beta \in \mathcal{K}$ be the two beliefs to be contracted. Let $\ominus$ be a contraction function which satisfies (G2) and preserves the ordering of worlds in $[\alpha]$ and $[\neg\alpha]$. $\sqsubseteq_\alpha^\ominus$ denotes the changed pre-order relation after contraction by α. We have that $[(\mathcal{K}_\alpha^\ominus)_{\bar{\beta}}^\ominus] = \sqsubseteq\text{-min}(\Omega) \cup \sqsubseteq\text{-min}[\neg\alpha] \cup \sqsubseteq_\alpha^\ominus\text{-min}[\neg\beta]$. $\beta \in \mathcal{K}_\alpha^\ominus$ gives that $\sqsubseteq\text{-min}[\neg\alpha] = \sqsubseteq\text{-min}[\neg\alpha \wedge \beta]$.

Suppose $\alpha \vee \beta \in (\mathcal{K}_\alpha^\ominus)_{\bar{\beta}}^\ominus$, then we have $\sqsubseteq_\alpha^\ominus\text{-min}[\neg\beta] = \sqsubseteq_\alpha^\ominus\text{-min}[\neg\beta \wedge \alpha]$. Since $\ominus$ preserves ordering of worlds in $[\alpha]$, $\sqsubseteq_\alpha^\ominus\text{-min}[\neg\beta \wedge \alpha] = \sqsubseteq\text{-min}[\neg\beta \wedge \alpha]$. Therefore, $[(\mathcal{K}_\alpha^\ominus)_{\bar{\beta}}^\ominus] = (\sqsubseteq\text{-min}(\Omega) \cup \sqsubseteq\text{-min}[\neg\alpha]) \cup (\sqsubseteq\text{-min}(\Omega) \cup \sqsubseteq_\alpha^\ominus\text{-min}[\neg\beta \wedge \alpha])$. Hence, $(\mathcal{K}_\alpha^\ominus)_{\bar{\beta}}^\ominus = \mathcal{K}_\alpha^\ominus \cap \mathcal{K}_{\alpha \rightarrow \beta}^\ominus$.

Suppose $\alpha \rightarrow \beta \in (\mathcal{K}_\alpha^\ominus)_{\bar{\beta}}^\ominus$, then we have $\sqsubseteq_\alpha^\ominus\text{-min}[\neg\beta] = \sqsubseteq_\alpha^\ominus\text{-min}[\neg\beta \wedge \neg\alpha]$. It is assumed that $-$ is a contraction function that preserves the ordering of worlds in $[\neg\alpha]$. Therefore we have, $\sqsubseteq_\alpha^\ominus\text{-min}[\neg\beta \wedge \neg\alpha] = \sqsubseteq\text{-min}[\neg\beta \wedge \neg\alpha]$. This gives $[(\mathcal{K}_\alpha^\ominus)_{\bar{\beta}}^\ominus] = (\sqsubseteq\text{-min}(\Omega) \cup \sqsubseteq\text{-min}[\neg\alpha]) \cup (\sqsubseteq\text{-min}(\Omega) \cup \sqsubseteq_\alpha^\ominus\text{-min}[\neg\beta \wedge \neg\alpha])$. Hence, $(\mathcal{K}_\alpha^\ominus)_{\bar{\beta}}^\ominus = \mathcal{K}_\alpha^\ominus \cap \mathcal{K}_{\alpha \vee \beta}^\ominus$.

---

[13] The Lexicographic contraction operation [10] is defined on a total pre-order relation on the set of possible worlds Ω. It differs from Natural and Priority contraction in how the worlds in $[\alpha]$ and $[\neg\alpha]$ are related to each other after contraction by α.

When neither $\alpha \vee \beta \in (\mathcal{K}_\alpha^\ominus)_\beta^\ominus$ nor $\alpha \to \beta \in (\mathcal{K}_\alpha^\ominus)_\beta^\ominus$, then $\sqsubseteq_\alpha^\ominus$-min$[\neg\beta]{=}\sqsubseteq_\alpha^\ominus$-min$[\neg\alpha \wedge \neg\beta]\cup \sqsubseteq_\alpha^\ominus$-min$[\alpha \wedge \neg\beta]$. Since the contraction function $\ominus$ preserves the ordering of worlds in $[\alpha]$ and $[\neg\alpha]$, we have $\sqsubseteq_\alpha^\ominus$-min$[\neg\alpha \wedge \neg\beta]{=}\sqsubseteq$-min$[\neg\alpha \wedge \neg\beta]$ and $\sqsubseteq_\alpha^\ominus$-min$[\alpha \wedge \neg\beta]{=}\sqsubseteq$-min$[\alpha \wedge \neg\beta]$. Hence $(\mathcal{K}_\alpha^\ominus)_\beta^\ominus = \mathcal{K}_\alpha^\ominus \cap \mathcal{K}_{\alpha\vee\beta}^\ominus \cap \mathcal{K}_{\alpha\to\beta}^\ominus$. ∎

When contracting $\mathcal{K}$ by α, both Priority and Natural contraction functions satisfy (G2) and preserve the ordering of worlds within $[\alpha]$ and $[\neg\alpha]$. Therefore, Priority and Natural contraction functions satisfy PFI. It is clear that we need properties that capture the changes to ordering between the worlds of $[\alpha]$ and $[\neg\alpha]$. In the following section, we extend PFI and discuss properties that could be expected from an iterable contraction function. We also present results reflecting the changes to the ordering of worlds in Ω caused by any contraction function satisfying these properties.

5   Properties of Iteratable Contraction Functions

Let $\mathcal{K}$ be a theory denoting the belief set of the agent. Let $\alpha, \beta \in \mathcal{K}$. The following discussion is about the problem of iterated contraction of α followed by β from $\mathcal{K}$. The agent always believes in tautologies. Hence, it is not possible to successfully contract any valid sentence from the belief set. Thus in the following discussion we assume that both α, β are not logical tautologies. There are two cases to be studied in iterated contraction, either $\beta \in \mathcal{K}_\alpha^-$ or $\beta \notin \mathcal{K}_\alpha^-$. When $\beta \notin \mathcal{K}_\alpha^-$, contracting β from $\mathcal{K}_\alpha^-$ becomes a trivial problem. In the following discussion we assume that $\beta \in \mathcal{K}_\alpha^-$.

In [11] the PFI is derived by using the identity

$$(\alpha \vee \beta) \wedge (\alpha \to \beta) \equiv \beta \tag{I}$$

Suppose the agent believes in $\alpha \vee \beta$ and $\alpha \to \beta$, then the agent believes in β. Contracting β from $\mathcal{K}_\alpha^-$ is related to contracting $\alpha \vee \beta$ from $\mathcal{K}_\alpha^-$ or $\alpha \to \beta$ from $\mathcal{K}_\alpha^-$ or both. Therefore, $\alpha \vee \beta$ and $\alpha \to \beta$ form important factors when contracting β from $\mathcal{K}_\alpha^-$.

Suppose the agent has justification to believe in $\alpha \vee \beta$ independent of α or β, then the agent may not withdraw its belief in $\alpha \vee \beta$ when withdrawing β from $\mathcal{K}_\alpha^-$. Thus, $\alpha \vee \beta \in (\mathcal{K}_\alpha^-)_\beta^-$ and PFI gives, $(\mathcal{K}_\alpha^-)_\beta^- = \mathcal{K}_\alpha^- \cap \mathcal{K}_{\alpha\to\beta}^-$. Here, we denote independent justification for $\alpha \vee \beta$ as $\alpha < \alpha \vee \beta$ and $\beta < \alpha \vee \beta$, where $<$ denotes strict epistemic entrenchment relation. From the definition of contraction functions based on entrenchment relations, $\alpha < \alpha \vee \beta$ and $\beta < \alpha \vee \beta$ we get $\alpha \in \mathcal{K}_\beta^-$ and $\beta \in \mathcal{K}_\alpha^-$. Hence,

$$\alpha \in \mathcal{K}_\beta^- \Rightarrow (\mathcal{K}_\alpha^-)_\beta^- = \mathcal{K}_\alpha^- \cap \mathcal{K}_{\alpha\to\beta}^-. \tag{1}$$

Lemma 1. *If $-$ is a contraction function that satisfies (1) and let $\omega \in [\alpha]$, $\omega' \in [\neg\alpha]$, $\omega, \omega' \notin \sqsubseteq$-$min(\Omega)\cup \sqsubseteq$-$min[\neg\alpha]$, $\omega \sqsubsetneq \omega'$ then $\omega \sqsubsetneq_\alpha^- \omega'$.*

Proof: Proofs for the lemmas have not been provided in this paper due to space constraints. Interested readers can contact the authors for the same. ∎

Suppose α and β are such that $\vdash \alpha \vee \beta$, this could be considered as a strong justification for the agent to believe in $\alpha \vee \beta$. Therefore, $\vdash \alpha \vee \beta$ is a special case of $\alpha \in \mathcal{K}_\beta^-$. Thus we have,

$$\vdash \alpha \vee \beta \Rightarrow (\mathcal{K}_\alpha^-)_\beta^- = \mathcal{K}_\alpha^- \bigcap \mathcal{K}_{\alpha \to \beta}^-. \tag{2}$$

Lemma 2. *Let $\omega, \omega' \in [\alpha]$. Suppose $-$ satisfies (2) then $\omega \sqsubseteq_\alpha^- \omega'$ if and only if $\omega \sqsubseteq \omega'$.*

When the agent has explicit evidence for believing in $\alpha \to \beta$ then the agent might still believe in $\alpha \to \beta$ after the iterated contraction of β following α from $\mathcal{K}$. Hence, upon contraction β, we have $\alpha \to \beta \in \mathcal{K}_\beta^-$. Thus from PFI,

$$\alpha \to \beta \in \mathcal{K}_\beta^- \Rightarrow (\mathcal{K}_\alpha^-)_\beta^- = \mathcal{K}_\alpha^- \bigcap \mathcal{K}_{\alpha \vee \beta}^-. \tag{3}$$

Lemma 3. *Given $-$ satisfies (3), for any ω, ω' such that $\omega \sqsubsetneq \omega'$ and $\omega \in [\neg \alpha]$, $\omega' \in [\alpha]$ with $\omega, \omega' \notin \sqsubseteq\text{-}min(\Omega) \cup \sqsubseteq\text{-}min[\neg\alpha]$ then $\omega \sqsubsetneq_\alpha^- \omega'$.*

Given $\vdash \alpha \to \beta$, the agent always retains its belief in $\alpha \to \beta$ after any contraction. Therefore, $\vdash \alpha \to \beta$ is a special case of the agent having explicit evidence for believing in $\alpha \to \beta$. This gives us

$$\vdash \alpha \to \beta \Rightarrow (\mathcal{K}_\alpha^-)_\beta^- = \mathcal{K}_\alpha^- \bigcap \mathcal{K}_{\alpha \vee \beta}^-. \tag{4}$$

Lemma 4. *Suppose $-$ satisfies (4), given $\omega, \omega' \in [\neg \alpha]$ and $\omega \sqsubseteq \omega'$ then $\omega \sqsubseteq_\alpha^- \omega'$.*

Suppose $\nvdash \alpha \vee \beta$. In [9] there is a lot of discussion in how the agent could believe in $\alpha \vee \beta$ after withdrawing its belief in both α and β. The agent might withdraw its belief in $\alpha \vee \beta$ more readily than in $\alpha \to \beta$ when contracting β from $\mathcal{K}_\alpha^-$. Hence,

$$\nvdash \alpha \vee \beta \Rightarrow (\mathcal{K}_\alpha^-)_\beta^- = \mathcal{K}_\alpha^- \bigcap \mathcal{K}_{\alpha \vee \beta}^-. \tag{5}$$

Lemma 5. *Consider a contraction function $-$ satisfying (5). For $\omega, \omega' \in \Omega$ if $\omega \in [\alpha] \setminus \sqsubseteq\text{-}min[\alpha]$ and $\omega' \in [\neg \alpha]$ then $\omega' \sqsubsetneq_\alpha^- \omega$.*

When the agent has no explicit justification for either $\alpha \vee \beta$ or $\alpha \to \beta$, the agent might withdraw its belief in both instead of making a choice between them. Hence,

$$(\mathcal{K}_\alpha^-)_\beta^- = \mathcal{K}_\alpha^- \bigcap \mathcal{K}_{\alpha \vee \beta}^- \bigcap \mathcal{K}_{\alpha \to \beta}^-. \tag{6}$$

Lemma 6. *Let $-$ satisfy (6). Given $\omega \in [\alpha] \setminus \sqsubseteq\text{-}min[\alpha], \omega' \in [\neg \alpha] \setminus \sqsubseteq\text{-}min[\neg \alpha]$ if $\omega \sqsubseteq \omega'$ and $\omega' \sqsubseteq \omega$ then $\omega \sqsubseteq_\alpha^- \omega'$ and $\omega' \sqsubseteq_\alpha^- \omega$.*

The equations (1) to (6) are some possiblities that come about in iterated contraction. But no iterable contraction function can satisfy all the above listed properties. The properties (5) and (1) are contradictory. (5) says that when

$\nvdash \alpha \vee \beta$, the agent may not believe in $\alpha \vee \beta$ when no longer believing in α and β. The agent is indifferent to any independent justification that $\alpha \vee \beta$ might have whereas (1) says that even though $\nvdash \alpha \vee \beta$, the agent might still believe in $\alpha \vee \beta$ when it has sufficient reason to do so. Equation (5) does not sit well along with equation (6) too. (6) says that when neither $\alpha \vee \beta$ nor $\alpha \rightarrow \beta$ have extra evidence, the agent withdraws its belief in both of them, but according to (5), it is deemed enough to withdraw the belief in just $\alpha \vee \beta$. Thus, no contraction function satisfying (5) can satisfy (1) and (6) and vice-versa.

Let a contraction function satisfy (3), the agent has extra evidence supporting its belief in $\alpha \rightarrow \beta$ then $\alpha \rightarrow \beta \in (\mathcal{K}_\alpha^-)_\beta^-$. To avoid triviality, we have $\nvdash \alpha \vee \beta$. Since $\alpha \rightarrow \beta \in (\mathcal{K}_\alpha^-)_\beta^-$, to contract by β, $\alpha \vee \beta$ has to be contracted. Hence both (3) and (5) are consistent with each other. As (4) is a special case of (3), we have (5) and (4) are consistent with each other. From the above discussion we have that, if any contraction function satisfies (3) and (4) then it satisfies (5).[14] And if a contraction function satisfies (5) it does not satisfy (1) and (6) and vice-versa. Thus, the properties presented above can be listed as two sets: $\{(2)$ and $(5)\}$ and $\{(1), (3)$ and $(6)\}$.

The following lemmas are converses of lemmas 1 to 6.

Lemma 7. *Let* $-$ *be a contraction function such that for all* $\omega, \omega' \in [\alpha]$, $\omega \sqsubseteq_\alpha^- \omega'$ *iff* $\omega \sqsubseteq \omega'$. *Then* $-$ *satisfies (2).*

Lemma 8. *Let* $-$ *be a contraction function such that for all* $\omega, \omega' \in [\neg\alpha]$, *we have* $\omega \sqsubseteq_\alpha^- \omega'$ *iff* $\omega \sqsubseteq \omega'$. *Then the contraction function satisfies (4).*

Lemma 9. *Let* $-$ *preserve the ordering of worlds in* $[\neg\alpha]$ *after contraction by* α. *Suppose* $-$ *satisfies the following* $\forall \omega \in [\alpha] \backslash \sqsubseteq$-$min[\alpha]$ *and* $\omega' \in [\neg\alpha]$, $\omega' \sqsubset_\alpha^- \omega$, *then* $-$ *satisfies (5).*

Lemma 10. *If* $-$ *is a contraction function that satisfies the following* $\forall \omega, \omega' \notin (\sqsubseteq$-$min(\Omega) \cup \sqsubseteq$-$min[\neg\alpha])$, $\omega \sqsubseteq_\alpha^- \omega'$ *iff* $\omega \sqsubseteq \omega'$, *then* $-$ *satisfies(1), (3) and (6).*

6   Representation Results

In the previous section we listed some desirable properties in an iterable contraction function. In this section, we give some important results connecting these properties and Priority and Natural contraction functions. Let $\mathcal{K}$ be a belief set and $\sqsubseteq$ be a total pre-order relation on Ω such that $[\mathcal{K}] = \sqsubseteq$-$min(\Omega)$. Let $\alpha, \beta \in \mathcal{K}$, such that $\nvdash \alpha$ and $\nvdash \beta$, also $\beta \in \mathcal{K}_\alpha^-$. In this section we assume that $-$ is a contraction function that satisfies AGM postulates, hence $-$ satisfies (G2).

Theorem 2. *Any AGM contraction function satisfying the properties (2) and (5) is a Priority contraction function.*

---

[14] It has to be noted here that the discussion holds only for non-trivial cases of when $\nvdash \alpha, \nvdash \beta, \beta \in \mathcal{K}_\alpha^-$ and $\gamma \notin \mathcal{K}_\gamma^-$ for any $\gamma \in \mathcal{K}$.

Proof Sketch: Let $-$ be a contraction function that satisfies the properties (2) and (5). For any $\alpha, \beta \in \mathcal{K}$, if $\vdash \alpha \vee \beta$, then $(\mathcal{K}_\alpha^-)_\beta^- = \mathcal{K}_\alpha^- \cap \mathcal{K}_{\alpha \to \beta}^-$ and $\nvdash \alpha \vee \beta$, then $(\mathcal{K}_\alpha^-)_\beta^- = \mathcal{K}_\alpha^- \cap \mathcal{K}_{\alpha \vee \beta}^-$.

Let $\sqsubseteq$ be a total pre-order relation on the set of all possible worlds Ω such that $\sqsubseteq$-$\min(\Omega) = [\mathcal{K}]$. From *lemma 2 and 5* we have

$\forall \omega, \omega' \in [\alpha] \setminus \sqsubseteq$-$\min(\Omega)$, $\omega \sqsubseteq_\alpha^- \omega'$ if and only if $\omega \sqsubseteq \omega'$ and $\forall \omega \in [\alpha] \setminus \sqsubseteq$-$\min[\alpha]$ and $\omega' \in [\neg \alpha]$, then $\omega' \sqsubsetneq_\alpha^- \omega$.

Let α, β be such that $\vdash \alpha \to \beta$. This is a special case of $\nvdash \alpha \vee \beta$ because if we have $\vdash \alpha \to \beta$ and $\vdash \alpha \vee \beta$ then this gives $\vdash \beta$ which makes contraction by β a trivial operation. Hence, when $\vdash \alpha \to \beta$, for contraction by β to be non-trivial, we need $\nvdash \alpha \vee \beta$. Therefore, property (5) also includes in itself the following:

$\vdash \alpha \to \beta$ then $(\mathcal{K}_\alpha^-)_\beta^- = \mathcal{K}_\alpha^- \cap \mathcal{K}_{\alpha \vee \beta}^-$. Thus from *lemma 4* we have $\forall \omega, \omega' \in [\neg \alpha]$ $\omega \sqsubseteq_\alpha^- \omega'$ if and only if $\omega \sqsubseteq \omega'$.

Thus $-$ is a priority contraction function. ∎

The above theorem shows that when an AGM contraction function, $-$, which satisfies (2) and (5) is used to contract any α from $\mathcal{K}_\sqsubseteq$, the relation $\sqsubseteq$ is changed to $\sqsubseteq_\alpha^-$. And $\sqsubseteq_\alpha^-$ is the same relation as $\sqsubseteq_\alpha^\ominus$, where $\ominus$ is a priority contraction function, as defined in section 3.1 . In other words, the above result gives the sufficiency condition for a priority contraction function.

Theorem 3. *Every Priority Contraction function satisfies the properties (2) and (5).*

Proof Sketch: Let $-$ be a priority contraction function and $\sqsubseteq$ be a total pre-order relation on Ω.

Then we have that $\forall \omega, \omega' \in [\alpha]$, $\omega \sqsubseteq_\alpha^- \omega'$ if and only if $\omega \sqsubseteq \omega'$. Thus from *lemma 7* for any $\beta \in \mathcal{K}_\alpha^-$ such that $\vdash \alpha \vee \beta$, $-$ satisfies property (2).

Given $\omega \in [\alpha] \setminus \sqsubseteq$-$\min(\Omega)$ and $\omega' \in [\neg \alpha]$, we have that $\omega' \sqsubsetneq_\alpha^- \omega$. From *lemma 9* we have that for any $\beta \in \mathcal{K}_\alpha^-$ such that $\nvdash \alpha \vee \beta$, $-$ satisfies property (5). ∎

The above result gives the necessary conditions for a priority contraction function. To summarise the above two theorems, a contraction function based on the system of spheres approach is a priority contraction function if it satisfies the AGM contraction postulates along with

1. $\vdash \alpha \vee \beta \Rightarrow (\mathcal{K}_\alpha^-)_\beta^- = \mathcal{K}_\alpha^- \cap \mathcal{K}_{\alpha \to \beta}^-$
2. $\nvdash \alpha \vee \beta \Rightarrow (\mathcal{K}_\alpha^-)_\beta^- = \mathcal{K}_\alpha^- \cap \mathcal{K}_{\alpha \vee \beta}^-$.

And an AGM contraction function that satisfies the above properties is a priority contraction function acting on a total pre-order relation $\sqsubseteq$ on Ω.

Theorem 4. *An AGM contraction function satisfying the properties (1), (3) and (6) is a Natural contraction function.*

Proof Sketch: Let $-$ be an AGM contraction function that satisfies (1), (3) and (6). From *lemma 1, 3 and 6* we have that $\forall \omega, \omega' \notin \sqsubseteq$-$\min(\Omega) \cup \sqsubseteq$-$\min[\neg\alpha]$, $\omega \sqsubseteq_\alpha^- \omega'$ if and only if $\omega \sqsubseteq \omega'$ and (G2) gives $\forall \omega \in \sqsubseteq$-$\min(\Omega)$ or $\omega \in \sqsubseteq$-$\min[\neg\alpha]$. Thus $\omega \sqsubseteq_\alpha^- \omega'$, for any $\omega' \in \Omega$. Therefore $-$ is a natural contraction function. ∎

The above theorem shows that when an AGM contraction function, $-$, which satisfies (1), (3) and (6), changes the total pre-order relation $\sqsubseteq$ to $\sqsubseteq_\alpha^-$. And $\sqsubseteq_\alpha^-$ is the same as relation $\sqsubseteq_\alpha^\ominus$, where $\ominus$ is a natural contraction function as defined in section 3.2. The above result gives the sufficiency condition for a natural contraction function.

Theorem 5. *Every Natural Contraction function satisfies the properties (1), (3) and (6).*

Proof Sketch: Let $-$ be a natural contraction function. Then from (G2) $-$ is an AGM contraction function. And from *lemma 10* we have $-$ satisfies (1), (3) and (6).   ∎

The above result gives the necessary conditions for a natural contraction function. Summarizing the above two theorems, a contraction based on the system of spheres approach is a natural contraction function if it satisfies the AGM contraction postulates along with

1. $\alpha \in \mathcal{K}_\beta^- \Rightarrow (\mathcal{K}_\alpha^-)_\beta^- = \mathcal{K}_\alpha^- \bigcap \mathcal{K}_{\alpha \to \beta}^-$
2. $\alpha \to \beta \in \mathcal{K}_\beta^- \Rightarrow (\mathcal{K}_\alpha^-)_\beta^- = \mathcal{K}_\alpha^- \bigcap \mathcal{K}_{\alpha \vee \beta}^-$
3. $\alpha, \alpha \to \beta \notin \mathcal{K}_\beta^- \Rightarrow (\mathcal{K}_\alpha^-)_\beta^- = \mathcal{K}_\alpha^- \bigcap \mathcal{K}_{\alpha \vee \beta}^- \bigcap \mathcal{K}_{\alpha \to \beta}^-$

And an AGM contraction satisfying the above properties is a Natural contraction function acting on a total pre-order relation $\sqsubseteq$ on Ω.

7   Concluding Remarks

In this paper, we examined two intuitive approaches to iterated belief contraction proposed by Nayak and his colleagues in [10], namely *Natural Contraction* and *Priority Contraction*. The proposals were couched in semantic terms, and their properties have not been studied before. We showed that, perhaps somewhat surprisingly, both of these operations satisfy the condition called *Principled Factored Insertion* which was used in [9] to motivate *Lexicographic Contraction*, and was initially thought to characterize it. Furthermore, we proposed a list of simple and arguably acceptable properties that one could expect an iterable contraction function to satisfy. We have demonstrated that different subsets of these properties fully characterize Priority and Natural contraction functions. We leave the characterization of Lexicographic contraction for future work.

Acknowledgements. The work presented in this paper has been supported in part by a Macquarie University Research Development Grant (MQRDG) and Macquarie University Research Excellence Scholarship (MQRES).

References

1. Alchourrón, C.E., Gärdenfors, P., Makinson, D.: On the logic of theory change: Partial meet contraction and revision functions. Journal of Symbolic Logic 50, 510–530 (1985)
2. Areces, C., Becher, V.: Iterable agm functions. In: Williams, M.-A., Rott, H. (eds.) Frontiers in Belief Revision, pp. 261–277. Kluwer, Dordrecht (2001)
3. Boutilier, C.: Revision sequences and nested conditionals. In: International Joint Conferences on Artificial Intelligence, pp. 519–525 (1993)
4. Darwiche, A., Pearl, J.: On the logic of iterated belief revision. Artifical Intelligence 89, 1–29 (1997)
5. Friedman, N., Halpern, J.Y.: Belief revision: A critique. In: Knowledge Representation and Reasoning, pp. 421–431 (1996)
6. Gärdenfors, P.: Knowledge in Flux: Modeling the Dynamics of Epistemic States. Bradford Books/ MIT Press, Cambridge (1988)
7. Grove, A.: Two modellings for theory change. Journal of Philosophical Logic 17, 157–170 (1988)
8. Hansson, S.O.: Belief Base Dynamics. Uppsala University, PhD thesis (1991)
9. Nayak, A.C., Goebel, R., Orgun, M.A.: Iterated belief contraction from first principles. In: International Joint Conferences on Artificial Intelligence, pp. 2568–2573 (2007)
10. Nayak, A.C., Goebel, R., Orgun, M.A., Pham, T.: Iterated belief change and the levi identity. In: Delgrande, J.P., Lang, J., Rott, H., Tallon, J.-M. (eds.) Belief Change in Rational Agents. Dagstuhl Seminar Proceedings. Internationales Begegnungs- und Forschungszentrum für Informatik (IBFI), Schloss Dagstuhl, Germany, vol. 05321 (2005)
11. Nayak, A.C., Pagnucco, M., Peppas, P.: Dynamic belief revision operators. Artif. Intell. 146(2), 193–228 (2003)
12. Rott, H.: Change, Choice and Inference: A study of belief revision and nonmonotonic reasoning. Oxford Science Publications, Clarendon Press (2001)
13. Rott, H.: Shifting priorities: Simple representations for twenty-seven iterated theory change operators. In: Lagerlund, H., Lindstrom, S., Sliwinski, R. (eds.) Modality Matters: Twenty-five Essays in Honour of Krister Segerberg, pp. 359–384. Uppsala University Press (2006)

The Dual Spatial Connectives of Separation Logic

Yuming Shen[1,2], Yuefei Sui[1], and Ju Wang[3]

[1] Key Laboratory of Intelligent Information Processing, Institute of Computing Technology, Chinese Academy of Sciences 100190, China
shenyuming@ict.ac.cn
[2] Graduate University of Chinese Academy of Sciences, 100049, China
[3] School of Computer Science and Information Engineering,
Guangxi Normal University 541004, China

Abstract. Separation logic has two spatial connectives $*$ and $-*$. It is known that $*$ and $-*$ are not dual each other, like 'and' and 'or', 'for all' and 'there exists', 'necessarily' and 'possibly', etc. To define the dual connectives of $*$ and $-*$ there are two choices: one is to take $*$ and $-*$ as special logical connectives; another is to take $*$ and $-*$ as binary modalities. Correspondingly, the dual modalities of $*$ and $-*$ are represented as the dual connectives of $*$ and $-*$, and as the dual modalities of $*$ and $-*$, where the latter can be represented by unary modalities in the case that the formulas are defined in a special form.

Keywords: Separation logic, modal logic, spatial connectives, dual modalities.

1 Introduction

Separation logic is a spatial logic for reasoning about mutable heap structure (see, e.g., [1],[2],[3],[12]), which is an extension of Hoare logic to describe the applications of programs on the heap structures and the reasoning about memory update(see, e.g., [7],[11]).

Except the logical implication and the negation, separation logic has two spatial connectives: the separating conjunction $\varphi_1 * \varphi_2$ which means the existence of a split of the current heap into two disjoint sub-heaps that satisfy φ_1 and φ_2 respectively; and its adjunct implication $\varphi_1 - *\varphi_2$ which means that whenever a fresh heap that satisfies φ_1 is composed with the current heap, the result satisfies φ_2, where there is an implicit universal quantification over fresh heaps in $*$; and an implicit universal quantification over fresh heaps in $- *$.

To consider the dual spatial connectives of $*$ and $-*$, being taken as connectives, the dual spatial connectives $\twoheadleftarrow$ and $\leftharpoondown$ of $*$ and $-*$ are defined as follows:

$$\varphi_1 \twoheadleftarrow \varphi_2 \equiv \neg(\neg\varphi_1 * \neg\varphi_2);$$
$$\varphi_1 \leftharpoondown \varphi_2 \equiv \neg(\neg\varphi_1 - *\neg\varphi_2).$$

Being taken as modalities, there is no modal logic containing binary modalities, as $*$ and $-*$, and the definitions of the dual binary modalities.

D. Karagiannis and Z. Jin (Eds.): KSEM 2009, LNAI 5914, pp. 90–99, 2009.

Logically, every logical connective has the corresponding connective in semantics. For example, $\varphi_1 \lor \varphi_2$ is true in a model if and only if either φ_1 or φ_2 is true in the model. And the semantics of logical quantifiers and modalities are explained by the quantifiers in semantics. For example, $\forall x \varphi(x)$ is true in a model if and only if for each element a in the universe of the model, $\varphi(x/a)$ is true in the model; $\Box \varphi$ is true in a model if and only if for each possible world w in the model, φ is true at possible world w.

By the semantical definitions of $*$ and $-*$, $*$ and $-*$ may be taken as modalities. To define the dual modalities, we first notice the dual modalities of the known unary modalities.

	$\Box(\varphi)$
alethic	φ is necessarily true
doxastic	The agent knows that φ is true
	The agent believes that φ is true
temporal	φ is true at any time
	$\Diamond(\varphi)$
alethic	φ is possibly true
doxastic	The agent doesn't know that φ is false
	The agent doesn't doubt that φ is true
temporal	φ is true at some time

Given a unary modality $\Box$, the dual modality $\Diamond$ of $\Box$ is defined as

$$\Diamond \varphi \equiv \neg \Box \neg \varphi.$$

How to define the dual modalities of binary modalities is the problem we shall discuss in the paper, and a way is given to define the binary modalities in terms of unary modalities is used to represent the binary spatial connectives in terms of unary modalities.

The paper is organized as follows: the next section gives the basic definition of the basic separation logic; the third section defines the dual modalities of $*$ and $-*$, according to the connective reading of $*$ and $-*$; the fourth section defines the dual modalities of $*$ and $-*$, according to the modality reading of $*$ and $-*$; the fifth section defines the dual modalities of $*$ and $-*$ by the unary modalities; and the last section concludes the paper.

Our notation is standard, a reference is [2]. To distinguish the symbols used in syntax and semantics, we shall use $\rightsquigarrow$ and $\multimap$ to denote $*$, $-*$ in [2].

2   The Basic Separation Logic

The logical language for propositional separation logic contains the following symbols:

- a constant symbol **0**;
- variable symbols: $\mathbf{x_0}, \mathbf{x_1}, \ldots$;
- a binary predicate symbol: $=$;

- a triple predicate symbol: $\mapsto$;
- logical connectives: $\top$, emp, $\neg$, $\rightarrow$;
- spatial connectives: $\rightsquigarrow$, $-\!\circ$.[1]

Remark. Notice that the propositional separation logic is a fragment of the first-order separation logic, in which the quantifiers do not occur, and the propositions are the atomic formulas, that is, $E = E|\top|\text{emp}|E \mapsto E_1, E_2$. Moreover, there is no negation symbol $\neg$, just as in databases. □

The syntax of propositional separation logic: let E and φ be the expressions and formulas, respectively.

$$E = \mathbf{0}|\mathbf{x};$$
$$\varphi = E = E|E \mapsto E_1, E_2|\top|\text{emp}|\neg\varphi_1|\varphi_1 \rightarrow \varphi_2|\varphi_1 \rightsquigarrow \varphi_2|\varphi_1 -\!\circ \varphi_2.$$

Intuitively, $\varphi_1 \rightsquigarrow \varphi_2$ means that the current heap can be split into two disjoint sub-heaps satisfying φ_1 and φ_2, respectively; and $\varphi_1 -\!\circ \varphi_2$ means that for any heap disjoint from the current heap and satisfying φ_1, the heap composing of the heap with the current heap satisfies φ_2.

A partial function $f : X \rightarrow_{\texttt{fin}} Y$ is a finite map f from X to Y, such that $\text{dom}(f) \subseteq X, \text{range}(f) \subseteq Y$, and $|f|$ is finite. Given two partial functions $f, g : X \rightarrow_{\texttt{fin}} Y, f\#g$ is defined iff $\text{dom}(f) \cap \text{dom}(g) = \emptyset$, and for any $x \in X$,

$$f * g(x) = \begin{cases} f(x) \text{ if } x \in \text{dom}(f) \\ g(x) \text{ if } x \in \text{dom}(g) \\ \uparrow \quad \text{ if } x \notin \text{dom}(f) \cup \text{dom}(g) \end{cases}$$

Let Loc be the set of locations, Var be the set of (program) variables, and Val be the set of all the values. Values, stacks, heaps and states are defined as follows:

$$v \in Val = Loc \cup \{0\};$$
$$s \in Stack = Var \rightarrow_{\texttt{fin}} Val;$$
$$h \in Heap = Loc \rightarrow_{\texttt{fin}} Val \times Val;$$
$$(s, h) \in State = Stack \times Heap,$$

where value 0 is the null location.

A model M for the propositional separation logic is a tuple $(Val, Stack, Heap, State, \#, *)$, where

o $Val = Loc \cup \{0\}$ is the set of all the values;
o $Stack$ is the set of all the finite assignments such that for any $s \in Stack$ and variable $\mathbf{x}, s(\mathbf{x}) \in Val$;
o $Heap$ is the set of all the finite functions from Loc to Val;
o $State$ is the set of all the pairs (s, h), where $s \in Stack, h \in Heap$;
o $\#$ is a binary relation on $Heap$ such that for any $h, h' \in Heap, h\#h'$ iff $\text{dom}(h) \cap \text{dom}(h') = \emptyset$; and

---

[1] To distinguish the semantics from the syntax, we use $\#, *$ as the functions in the semantics, and use $\rightsquigarrow$, $-\!\circ$ as the spatial connectives in the syntax to represent $*$, $-\!*$, respectively.

$\circ \ * : Heap \times Heap \to Heap$ is a binary partial function such that for any $h, h' \in Heap, (h, h') \in \mathrm{dom}(*)$ iff $h \# h'$, and for any $x \in Loc$,

$$h * h'(x) = \begin{cases} h(x) & \text{if } x \in \mathrm{dom}(h) \\ h'(x) & \text{if } x \in \mathrm{dom}(h') \\ \uparrow & \text{otherwise.} \end{cases}$$

For any $(s, h) \in State$, the interpretation $E^{(s,h)}$ of E is defined as follows:

$$E^{(s,h)} = \begin{cases} s(\mathbf{x}) & \text{if } E = \mathbf{x} \\ 0 & \text{if } E = 0 \end{cases}$$

and the satisfaction $(s, h) \models \varphi$ of φ in (s, h) is defined as follows:[2]

$$\begin{cases} E_1^{(s,h)} = E_2^{(s,h)} & \text{if } \varphi = E_1 = E_2 \\ \mathrm{dom}(h) = \{E^{(s,h)}\} \& h(E^{(s,h)}) = (E_1^{(s,h)}, E_2^{(s,h)}) & \text{if } \varphi = E \mapsto E_1, E_2 \\ (s, h) \models \varphi_1 \Rightarrow (s, h) \models \varphi_2 & \text{if } \varphi = \varphi_1 \to \varphi_2 \\ \mathrm{dom}(h) = \emptyset & \text{if } \varphi = \mathrm{emp} \\ (s, h) \not\models \varphi_1 & \text{if } \varphi = \neg\varphi_1 \\ \mathbf{E}h_1, h_2(h_1 \# h_2 \& h_1 * h_2 = h \& (s, h_1) \models \varphi_1 \& (s, h_2) \models \varphi_2) & \text{if } \varphi = \varphi_1 \rightsquigarrow \varphi_2 \\ \mathbf{A}h_1(h \# h_1 \& (s, h_1) \models \varphi_1 \Rightarrow (s, h * h_1) \models \varphi_2) & \text{if } \varphi = \varphi_1 \multimap \varphi_2 \end{cases}$$

φ is satisfied in M, denoted by $M \models \varphi$, if for each $(s, h) \in State, (s, h) \models \varphi$.

3   The Spatial Connectives Taken as Logical Connectives

Define the triple relations $R^{\rightsquigarrow}$ and $R^{\multimap}$ on $Heap$ as follows: for any $h, h_1, h_2 \in Heap$,

$$R^{\rightsquigarrow}(h, h_1, h_2) \text{ iff } h_1 \# h_2 \& h = h_1 * h_2;$$
$$R^{\multimap}(h, h_1, h_2) \text{ iff } h \# h_1 \& h_2 = h * h_1.$$

To define the dual operators of $\rightsquigarrow$ and $\multimap$, from the semantical point of view, we first review the dual quantifiers and dual modalities.

$$M \models \forall x \varphi(x) \text{ iff } \mathbf{A}a(M \models \varphi(x/a))$$
$$M \models \exists x \varphi(x) \text{ iff } \mathbf{E}a(M \models \varphi(x/a))$$
$$M, w \models \Box\varphi \text{ iff } \mathbf{A}w'((w, w') \in R \Rightarrow M, w' \models \varphi)$$
$$M, w \models \Diamond\varphi \text{ iff } \mathbf{E}w'((w, w') \in R \& M, w' \models \varphi).$$

From the point of syntactical view, we compare the dual operators of $\rightsquigarrow$ and $\multimap$ with the logical connectives $\vee$ and $\wedge$.

$$\varphi_1 \vee \varphi_2 \equiv \neg(\neg\varphi_1 \wedge \neg\varphi_2);$$
$$\varphi_1 \wedge \varphi_2 \equiv \neg(\neg\varphi_1 \vee \neg\varphi_2).$$

---

[2] In syntax we use $\neg, \wedge, \to, \forall, \exists$ to denote the logical connectives and quantifiers; in semantics we use $\sim, \&, \Rightarrow, \mathbf{A}, \mathbf{E}$ to denote the corresponding connectives and quantifiers.

Let $\leftarrow\!\!\!\cdot$ and $\llcorner\!\!\!\!-$ be the dual operators for $\rightsquigarrow$ and $-\!\circ$, respectively. Then, we define

$$\varphi_1 \leftarrow\!\!\!\cdot \varphi_2 \equiv \neg(\neg\varphi_1 \rightsquigarrow \neg\varphi_2);$$
$$\varphi_1 \llcorner\!\!\!- \varphi_2 \equiv \neg(\neg\varphi_1 -\!\circ \neg\varphi_2),$$

and correspondingly, we have the following definition of satisfaction.

$(s, h) \models \varphi_1 \leftarrow\!\!\!\cdot \varphi_2$
iff $(s, h) \models \neg(\neg\varphi_1 \rightsquigarrow \neg\varphi_2)$
iff $(s, h) \not\models \neg\varphi_1 \rightsquigarrow \neg\varphi_2$
iff $\sim \mathbf{E}h_1, h_2(R^\rightsquigarrow(h, h_1, h_2)\&(s, h_1) \models \neg\varphi_1\&(s, h_2) \models \neg\varphi_2)$
iff $\mathbf{A}h_1, h_2(\sim R^\rightsquigarrow(h, h_1, h_2)$ or $(s, h_1) \not\models \neg\varphi_1$ or $(s, h_2) \not\models \neg\varphi_2)$
iff $\mathbf{A}h_1, h_2(R^\rightsquigarrow(h, h_1, h_2) \Rightarrow (s, h_1) \models \varphi_1$ or $(s, h_2) \models \varphi_2)$;
$(s, h) \models \varphi_1 \llcorner\!\!\!- \varphi_2$
iff $(s, h) \models \neg(\neg\varphi_1 -\!\circ \neg\varphi_2)$
iff $(s, h) \not\models \neg\varphi_1 -\!\circ \neg\varphi_2)$
iff $\sim \mathbf{A}h_1, h_2(R^{-\circ}(h, h_1, h_2)\&(s, h_1) \models \neg\varphi_1 \Rightarrow (s, h_2) \models \neg\varphi_2)$
iff $\mathbf{E}h_1, h_2(R^{-\circ}(h, h_1, h_2)\&(s, h_1) \models \neg\varphi_1\&(s, h_2) \not\models \neg\varphi_2)$
iff $\mathbf{E}h_1, h_2(R^{-\circ}(h, h_1, h_2)\&(s, h_1) \models \neg\varphi_1\&(s, h_2) \models \varphi_2)$.

Hence, we define

$(s, h) \models \varphi_1 \rightsquigarrow \varphi_2$ iff $\mathbf{E}h_1, h_2(R^\rightsquigarrow(h, h_1, h_2)\&(s, h_1) \models \varphi_1\&(s, h_2) \models \varphi_2)$;
$(s, h) \models \varphi_1 \leftarrow\!\!\!\cdot \varphi_2$ iff $\mathbf{A}h_1, h_2(R^\rightsquigarrow(h, h_1, h_2) \Rightarrow (s, h_1) \models \varphi_1$ or $(s, h_2) \models \varphi_2)$;
$(s, h) \models \varphi_1 -\!\circ \varphi_2$ iff $\mathbf{A}h_1, h_2(R^{-\circ}(h, h_1, h_2)\&(s, h_1) \models \varphi_1 \Rightarrow (s, h_2) \models \varphi_2)$
$(s, h) \models \varphi_1 \llcorner\!\!\!- \varphi_1$iff $\mathbf{E}h_1, h_2(R^{-\circ}(h, h_1, h_2)\&(s, h_1) \models \neg\varphi_1\&(s, h_2) \models \varphi_2)$.

A good dual definition should satisfy the following condition:

$$\forall x\varphi(x) \equiv \neg\exists x\neg\varphi(x);$$
$$\exists x\varphi(x) \equiv \neg\forall x\neg\varphi(x);$$
$$\Box\varphi \equiv \neg\Diamond\neg\varphi;$$
$$\Diamond\varphi \equiv \neg\Box\neg\varphi.$$

Hence, for $\rightsquigarrow$ and $-\!\circ$, we should have the following equivalences:

$$\varphi_1 \rightsquigarrow \varphi_2 \equiv \neg(\neg\varphi_1 \leftarrow\!\!\!\cdot \neg\varphi_2);$$
$$\varphi_1 -\!\circ \varphi_2 \equiv \neg(\neg\varphi_1 \llcorner\!\!\!- \neg\varphi_2).$$

Proposition 1. *For any formulas φ_1 and φ_2,*

$$\varphi_1 \rightsquigarrow \varphi_2 \equiv \neg(\neg\varphi_1 \leftarrow\!\!\!\cdot \neg\varphi_2);$$
$$\varphi_1 -\!\circ \varphi_2 \equiv \neg(\neg\varphi_1 \llcorner\!\!\!- \neg\varphi_2),$$

that is, for any model M for the propositional separation logic and state $(s, h), (s, h) \models \varphi_1 \rightsquigarrow \varphi_2$ iff $(s, h) \models \neg(\neg\varphi_1 \leftarrow\!\!\!\cdot \neg\varphi_2)$; and $(s, h) \models \varphi_1 -\!\circ \varphi_2$ iff $(s, h) \models \neg(\neg\varphi_1 \llcorner\!\!\!- \neg\varphi_2)$.

Proof. Given a model M for the propositional separation logic, for any $(s, h) \in$ *State*, we have the following equivalences:

$$(s, h) \models \neg(\neg\varphi_1 \leftharpoonup \neg\varphi_2) \text{ iff } \sim ((s, h) \models \neg\varphi_1 \leftharpoonup \neg\varphi_2)$$
$$\text{iff } \sim \mathbf{A}h_1, h_2(R^{\rightsquigarrow}(h, h_1, h_2) \Rightarrow (s, h_1) \models \neg\varphi_1 \text{ or } (s, h_2) \models \neg\varphi_2)$$
$$\text{iff } \mathbf{E}h_1, h_2(R^{\rightsquigarrow}(h, h_1, h_2)\& \sim ((s, h_1) \not\models \varphi_1 \text{ or } (s, h_2) \not\models \varphi_2))$$
$$\text{iff } \mathbf{E}h_1, h_2(R^{\rightsquigarrow}(h, h_1, h_2)\&(\sim (s, h_1) \not\models \varphi_1 \& \sim (s, h_2) \not\models \varphi_2))$$
$$\text{iff } \mathbf{E}h_1, h_2(R^{\rightsquigarrow}(h, h_1, h_2)\&(s, h_1) \models \varphi_1\&(s, h_2) \models \varphi_2))$$
$$\text{iff } (s, h) \models \varphi_1 \rightsquigarrow \varphi_2;$$
$$(s, h) \models \neg(\neg\varphi_1 \leftharpoondown \neg\varphi_2) \text{ iff } \sim ((s, h) \models \neg\varphi_1 \leftharpoondown \neg\varphi_2)$$
$$\text{iff } \sim \mathbf{E}h_1, h_2(R^{\multimap}(h, h_1, h_2)\&(s, h_1) \models \neg\neg\varphi_1\&(s, h_2) \models \neg\varphi_2)$$
$$\text{iff } \mathbf{A}h_1, h_2(R^{\rightsquigarrow}(h, h_1, h_2)\&(s, h_1) \models \varphi_1 \Rightarrow\sim (s, h_2) \not\models \varphi_2))$$
$$\text{iff } \mathbf{A}h_1, h_2(R^{\rightsquigarrow}(h, h_1, h_2)\&(s, h_1) \models \varphi_1 \Rightarrow (s, h_2) \models \varphi_2))$$
$$\text{iff } (s, h) \models \varphi_1 \multimap \varphi_2. \qquad \square$$

4   The Spatial Connectives Taken as Binary Modalities

Another choice to define the dual spatial connectives in separation logic is to take $\rightsquigarrow$ and $\multimap$ dual each other. The fact is that we cannot define as so, because $R^{\rightsquigarrow}$ and $R^{\multimap}$ are two different triple relations.

For the simplicity of notations, we use $\leftharpoonup$ and $\leftharpoondown$ as the dual spatial connectives of $\rightsquigarrow$ and $\multimap$, respectively. We define

$$\varphi_1 \leftharpoonup \varphi_2 = \neg(\varphi_1 \rightsquigarrow \neg\varphi_2);$$
$$\varphi_1 \leftharpoondown \varphi_2 = \neg(\varphi_1 \multimap \neg\varphi_2).$$

The reason of this definition can be seen in the following reasonings:

$$(s, h) \models \neg(\varphi_1 \rightsquigarrow \neg\varphi_2) \text{ iff } \sim ((s, h) \models \varphi_1 \rightsquigarrow \neg\varphi_2)$$
$$\text{iff } \sim (\mathbf{E}h_1, h_2(R^{\rightsquigarrow}(h, h_1, h_2)\&(s, h_1) \models \varphi_1\&(s, h_2) \models \neg\varphi_2)$$
$$\text{iff } \mathbf{A}h_1, h_2(R^{\rightsquigarrow}(h, h_1, h_2)\&(s, h_1) \models \varphi_1 \Rightarrow\sim ((s, h_2) \models \neg\varphi_2))$$
$$\text{iff } \mathbf{A}h_1, h_2(R^{\rightsquigarrow}(h, h_1, h_2)\&(s, h_1) \models \varphi_1 \Rightarrow (s, h_2) \models \varphi_2);$$
$$(s, h) \models \neg(\varphi_1 \multimap \neg\varphi_2) \text{ iff } \sim ((s, h) \models \varphi_1 \multimap \neg\varphi_2)$$
$$\text{iff } \sim (\mathbf{A}h_1, h_2(R^{\multimap}(h, h_1, h_2)\&(s, h_1) \models \varphi_1 \Rightarrow (s, h_2) \models \neg\varphi_2)$$
$$\text{iff } \mathbf{E}h_1, h_2(R^{\multimap}(h, h_1, h_2)\&(s, h_1) \models \varphi_1\& \sim ((s, h_2) \models \neg\varphi_2))$$
$$\text{iff } \mathbf{E}h_1, h_2(R^{\multimap}(h, h_1, h_2)\&(s, h_1) \models \varphi_1\&(s, h_2) \models \varphi_2).$$

Notice the duality of the forms of the following equivalences:

$$(s, h) \models \neg(\varphi_1 \rightsquigarrow \neg\varphi_2) \text{ iff } \mathbf{A}h_1, h_2(R^{\rightsquigarrow}(h, h_1, h_2)\&(s, h_1) \models \varphi_1 \Rightarrow (s, h_2) \models \varphi_2);$$
$$(s, h) \models \varphi_1 \multimap \varphi_2 \text{ iff } \mathbf{A}h_1, h_2(R^{\multimap}(h, h_1, h_2)\&(s, h_1) \models \varphi_1 \Rightarrow (s, h_2) \models \varphi_2);$$

and

$$(s, h) \models \neg(\varphi_1 \multimap \neg\varphi_2) \text{ iff } \mathbf{E}h_1, h_2(R^{\multimap}(h, h_1, h_2)\&(s, h_1) \models \varphi_1\&(s, h_2) \models;$$
$$(s, h) \models \varphi_1 \rightsquigarrow \varphi_2 \text{ iff } \mathbf{E}h_1, h_2(R^{\rightsquigarrow}(h, h_1, h_2)\&(s, h_1) \models \varphi_1\&(s, h_2) \models \varphi_2).$$

Hence, from the semantical duality, we can define the dual spatial connectives as follows:

$$\varphi_1 \rightsquigarrow \varphi_2 = \neg(\varphi_1 \leftsquigarrow \neg\varphi_2);$$
$$\varphi_1 \leftsquigarrow \varphi_2 = \neg(\varphi_1 \rightsquigarrow \neg\varphi_2);$$
$$\varphi_1 \multimap \varphi_2 = \neg(\varphi_1 \rightsquigarrow \neg\varphi_2);$$
$$\varphi_1 \multimapinv \varphi_2 = \neg(\varphi_1 \multimap \neg\varphi_2).$$

To verify that $\varphi_1 \rightsquigarrow \varphi_2 = \neg(\varphi_1 \leftsquigarrow \neg\varphi_2)$ and $\varphi_1 \multimap \varphi_2 = \neg(\varphi_1 \rightsquigarrow \neg\varphi_2)$, we have the following equivalences:

$(s,h) \models \neg(\varphi_1 \leftsquigarrow \neg\varphi_2)$ iff $\sim ((s,h) \models \varphi_1 \leftsquigarrow \neg\varphi_2)$
iff $\sim (\mathbf{A}h_1, h_2(R^{\rightsquigarrow}(h, h_1, h_2)\&(s, h_1) \models \varphi_1 \Rightarrow (s, h_2) \models \neg\varphi_2)$
iff $\mathbf{E}h_1, h_2(R^{\rightsquigarrow}(h, h_1, h_2)\&(s, h_1) \models \varphi_1 \& \sim ((s, h_2) \models \neg\varphi_2))$
iff $\mathbf{E}h_1, h_2(R^{\rightsquigarrow}(h, h_1, h_2)\&(s, h_1) \models \varphi_1 \&(s, h_2) \models \varphi_2)$
iff $(s,h) \models \varphi_1 \rightsquigarrow \varphi_2;$
$(s,h) \models \neg(\varphi_1 \multimapinv \neg\varphi_2)$ iff $\sim ((s,h) \models \varphi_1 \multimapinv \neg\varphi_2)$
iff $\sim (\mathbf{E}h_1, h_2(R^{\multimap}(h, h_1, h_2)\&(s, h_1) \models \varphi_1 \&(s, h_2) \models \neg\varphi_2)$
iff $\mathbf{A}h_1, h_2(R^{\multimap}(h, h_1, h_2)\&(s, h_1) \models \varphi_1 \Rightarrow \sim ((s, h_2) \models \neg\varphi_2))$
iff $\mathbf{A}h_1, h_2(R^{\multimap}(h, h_1, h_2)\&(s, h_1) \models \varphi_1 \Rightarrow (s, h_2) \models \varphi_2)$
iff $(s,h) \models \varphi_1 \multimap \varphi_2.$

Hence, from the logical point of view, we can define the dual spatial connectives of $\rightsquigarrow$ and $\multimap$ as follows:

$$\varphi_1 \leftsquigarrow \varphi_2 = \neg(\varphi_1 \rightsquigarrow \neg\varphi_2);$$
$$\varphi_1 \multimapinv \varphi_2 = \neg(\varphi_1 \multimap \neg\varphi_2).$$

and we have the following equivalences:

$$\varphi_1 \multimap \varphi_2 \equiv \neg(\varphi_1 \multimapinv \neg\varphi_2);$$
$$\varphi_1 \rightsquigarrow \varphi_2 \equiv \neg(\varphi_1 \leftsquigarrow \neg\varphi_2).$$

5   The Unary Modalities to Represent $\rightsquigarrow$ and $\multimap$

To use unary modalities to represent $\rightsquigarrow$ and $\multimap$, we need the following modalities:

$$L, N^L;$$
$$\Box, N^\Box,$$

and their dualities:

$$M, N^M;$$
$$\Diamond, N^\Diamond,$$

where the dual pairs are given in the following table:[3]

$$(L, M)\ (N^L, N^M);$$
$$(\Box, \Diamond)\ (N^\Box, N^\Diamond).$$

---

[3] In the history of the modal logic, L and M were/are used to denote $\Box$ and $\Diamond$, respectively.

The formulas φ are defined as follows:

$$\varphi = E = E|E \mapsto E_1, E_2|\top|\text{emp}|\neg\varphi_1|\varphi_1 \to \varphi_2$$
$$|L(\varphi_1 \to N^L\varphi_2)|M(\varphi_1 \wedge N^M\varphi_2)|\square(\varphi_1 \to N^\square\varphi_2)|\lozenge(\varphi_1 \wedge N^\lozenge\varphi_2).$$

Given a model M and $(s,h) \in State$, the satisfaction of φ is defined as follows:

$$(s,h) \models L(\varphi_1 \to N^L\varphi_2) \text{ iff } \mathbf{A}h_1(R^L(h,h_1)\&(s,h_1) \models \varphi_1$$
$$\Rightarrow \mathbf{A}h_2(R^{N^L}(h,h_1,h_2) \Rightarrow (s,h_2) \models \varphi_2));$$
$$(s,h) \models M(\varphi_1 \wedge N^M\varphi_2) \text{ iff } \mathbf{E}h_1(R^L(h,h_1)\&(s,h_1) \models \varphi_1$$
$$\&\mathbf{E}h_2(R^{N^L}(h,h_1,h_2)\&(s,h_2) \models \varphi_2));$$
$$(s,h) \models \square(\varphi_1 \to N^\square\varphi_2) \text{ iff } \mathbf{A}h_1(R^\square(h,h_1)\&(s,h_1) \models \varphi_1$$
$$\Rightarrow \mathbf{A}h_2(R^{N^\square}(h,h_1,h_2) \Rightarrow (s,h_2) \models \varphi_2));$$
$$(s,h) \models \lozenge(\varphi_1\&N^\lozenge\varphi_2) \text{ iff } \mathbf{E}h_1(R^\square(h,h_1)\&(s,h_1) \models \varphi_1$$
$$\&\mathbf{E}h_2(R^{N^\square}(h,h_1,h_2)\&(s,h_2) \models \varphi_2)),$$

where for any $h, h_1, h_2 \in H$,

$$R^L(h,h_1) \text{ iff } \mathbf{E}h_2(h_1\#h_2\&h = h_1 * h_2);$$
$$R^{N^L}(h,h_1,h_2) \text{ iff } R^{\rightsquigarrow}(h,h_1,h_2);$$
$$R^\square(h,h_1) \text{ iff } h\#h_1;$$
$$R^\square(h,h_1,h_2) \text{ iff } R^{-\circ}(h,h_1,h_2).$$

Define the following correspondence:

$$\varphi_1 \rightsquigarrow \varphi_2 \equiv M(\varphi_1 \wedge N^M\varphi_2);$$
$$\varphi_1 \leftharpoondown \varphi_2 \equiv L(\varphi_1 \to N^L\varphi_2);$$
$$\varphi_1 \multimap \varphi_2 \equiv \square(\varphi_1 \to N^\square\varphi_2);$$
$$\varphi_1 \leftharpoonup \varphi_2 \equiv \lozenge(\varphi_1\&N^\lozenge\varphi_2).$$

Then, we have the following equivalences:

$(s,h) \models \varphi_1 \rightsquigarrow \varphi_2$
iff $\mathbf{E}h_1, h_2(h_1\#h_2\&h = h_1 * h_2\&(s,h_1) \models \varphi_1\&(s,h_2) \models \varphi_2)$
iff $\mathbf{E}h_1(R^L(h,h_1)\&(s,h_1) \models \varphi_1\&\mathbf{E}h_2(R^{N^L}(h,h_1,h_2)\&(s,h_2) \models \varphi_2))$
iff $(s,h) \models M(\varphi_1 \wedge N^M\varphi_2)$;

and

$(s,h) \models \varphi_1 \multimap \varphi_2$
iff $\mathbf{A}h_1(h_1\#h\&(s,h_1) \models \varphi_1 \Rightarrow (s,h * h_1) \models \varphi_2)$
iff $\mathbf{A}h_1(R^\lozenge(h,h_1)\&(s,h_1) \models \varphi_1 \Rightarrow \mathbf{A}h_2(R^{N^\lozenge}(h,h_1,h_2) \Rightarrow (s,h_2) \models \varphi_2))$
iff $(s,h) \models \lozenge(\varphi_1 \wedge N^\lozenge\varphi_2)$.

By the functionality of $*$, for any h, h_1, if there is an $h_2 \in H$ such that $h_1\#h_2$ and $h = h_1 * h_2$ then h_2 is unique. Hence, $\mathbf{E}h_2$ is equivalent to $\mathbf{A}h_2$.

Then, we have the following

Proposition 2. *Given a model M for the propositional separation logic and (s, h) $\in$ State, for any formula φ in the propositional separation logic, let $\sigma(\varphi)$ be the corresponding formula of φ in the modalized the propositional separation logic,*

$$(s, h) \models \varphi \text{ iff } (s, h) \models \sigma(\varphi). \qquad \square$$

6   Conclusions and Future Work

To give a logic, the logical properties of the symbols in the logic should be considered firstly. In separation logic, taken as the binary modalities, the spatial connectives $\rightsquigarrow$ and $\multimap$ have two kinds of the dual modalities: from the connective reading or semantic reading of $\rightsquigarrow$ and $\multimap$,

$$\varphi_1 \leftarrow \varphi_2 \equiv \neg(\neg\varphi_1 \rightsquigarrow \neg\varphi_2);$$
$$\varphi_1 \leftharpoonup \varphi_2 \equiv \neg(\neg\varphi_1 \multimap \neg\varphi_2),$$

and from the duality reading of $\rightsquigarrow$ and $\multimap$,

$$\varphi_1 \leftarrow \varphi_2 \equiv \neg(\varphi_1 \rightsquigarrow \neg\varphi_2);$$
$$\varphi_1 \leftharpoonup \varphi_2 \equiv \neg(\varphi_1 \multimap \neg\varphi_2).$$

The spatial connectives $\rightsquigarrow$ and $\multimap$ can be represented by unary modalities $M, N^M, \square, N^\square$ as follows:

$$\varphi_1 \rightsquigarrow \varphi_2 \equiv M(\varphi_1 \wedge N^M \varphi_2);$$
$$\varphi_1 \multimap \varphi_2 \equiv \square(\varphi_1 \rightarrow N^\square \varphi_2).$$

The further work is that according to the dual spatial connectives of $\rightsquigarrow$ and $\multimap$, there is an inference system that should be built, so that the automatic reasoning in separation logic becomes possible.

Acknowledgements

The work was supported by the National Natural Science Foundation of China under Grant Nos.60496326, 60573063, 60573064, 60773059 and the National High-Tech Research and Development Plan of China under Grant No.2007AA01Z325.

References

1. Brookes, S.D.: A semantics of concurrent separation logic. Theoretical Computer Science 375, 227–270 (2007)
2. Calcagno, C., Gardner, P., Hague, M.: From separation logic to first-order logic. In: Sassone, V. (ed.) FOSSACS 2005. LNCS, vol. 3441, pp. 395–409. Springer, Heidelberg (2005)

3. Calcagno, C., O'Hearn, P., Yang, H.: Local action and abstract separation logic. In: 22nd LICS, pp. 366–378 (2007)
4. Cardelli, L., Caires, L.: A spatial logic for concurrency. In: Kobayashi, N., Pierce, B.C. (eds.) TACS 2001. LNCS, vol. 2215, pp. 1–37. Springer, Heidelberg (2001)
5. Cardelli, L., Gordon, D.: Anytime, anywhere. modal logics for mobile ambients. In: 27th POPL (2000)
6. Chaochen, Z., Hoare, C.A.R., Ravn, A.P.: A calculus of durations. Inform. Proc. Letters 40, 269–276 (1991)
7. Hoare, C.A.R.: Communicating Sequential Processes. Prentice Hall, Englewood Cliffs (1985)
8. Hoare, T., O'Hearn, P.: Separation logic semantics for communicating processes. Electronic Notes in Theoretical Computer Science 212, 3–25 (2008)
9. Lozes, E.: Separation logic preserves the expressive power of classical logic (2004), http://www.diku.dk/topps/space2004/space.nal/etienne.pdf
10. O'Hearn, P.W.: Resources, concurrency and local reasoning. Theoretical Computer Science 375, 271–307 (2007)
11. Parkinson, M., Bierman, G.: Separation logic and abstraction. In: POPL 2005, pp. 12–14 (2005)
12. Reynolds, J.C.: Separation logic: a logic for shared mutable data structures. In: LICS, pp. 55–74. IEEE, Los Alamitos (2002)

Knowledge Engineering in Future Internet

Vedran Hrgovcic, Wilfrid Utz, and Robert Woitsch

BOC Asset Management GmbH, Baeckerstrasse 5,
1010 Vienna, Austria
{Vedran.Hrgovcic,Wilfrid.Utz,Robert.Woitsch}@boc-eu.com

Abstract. The business and IT alignment acknowledged by enterprises as an important factor in achieving competitive advantage in the market is undergoing slight changes as it faces the challenges and requirements imposed by the Future Internet initiative. The goal to support the adaptation to the challenges and requirements is achieved by applying knowledge engineering (KE) to all three alignment layers: business, the "IT-Socket" and IT layer. Paper introduces approaches to support model transformation and translation – enabling multilingual environments on the business layer, the semantic workflow approach – enabling next generation IT infrastructure layer, and the IT-Socket approach for alignment.

Keywords: Knowledge Engineering, Ontologies, BPM, Meta Models, Semantic Workflows, Future Internet, Business and IT Alignment.

1 Introduction

Enterprises, no matter how well established they may be, are subject to major changes based on the alterations of the important factors such as technology used, status in the market, and law and regulations in order to stay competitive. ICT sector is a prominent example of how such changes may alter the enterprise but as well the industry on the whole. One of the most important changes that will have a major consequence on the IT industry and enterprises within this sector in the next few years are the goals and requirements imposed by the Future Internet (FI) initiative.

The current major issue with the FI is that it is not being recognized as a "must be considered" but as a "just another hype" by majority of the IT managers. The reasons for such passivity include the fact that everything is working just fine. That is, the internet has been successful in the last years and has grown steadily without risking the stability of the worldwide and company internal networks. Much more effective backbones, wireless access, mobile access, etc have been introduced without having a negative impact on most enterprises. The issue itself lies within the tremendous success. The growth rate between 2000 and 2008 was 342,2% summing to more than 1 billion of users with approximately 1600 pages per month viewed by each user on average [1]. By 2012 the amount of data transferred will pass the 20.000 Petabytes[1] per month with video files being the major transportation medium for information

---

[1] 1 Petabyte equals to 1000 terabytes.

D. Karagiannis and Z. Jin (Eds.): KSEM 2009, LNAI 5914, pp. 100–109, 2009.

averaging 50% of the overall traffic [2], the amount of devices connected to the internet will grow to 1 trillion [3], etc. In order to be ready for the aforementioned changes architectural problem that came in the spotlight affecting following areas such as Networks, Things and Objects, Content and Media, Software and Services, Security and Privacy, Governance, and Applications has to be solved. Due to fact that IT has changed its role from a supporting force to an industry of its own and without it some of the core processes in the enterprise wouldn't work anymore or would jeopardize the current position of the enterprise in the market it is more then ever important to support the business and IT alignment. The importance of such alignment lies first in the fact that it reduces the idle running and unnecessary overpowering of the IT infrastructure – according to Gartner "8 out 10 dollars invested in IT are dead money as they are not contributing directly to business growth [4] – by allowing better estimations regarding investment in and measuring the value of the IT. Second, an enterprise having aligned its business and IT (internal as well as IT provided by external 3rd parties) can react to changes in environment (e.g. law and regulation) much faster and keep advantages over the competitors.

The obstacles in the business and IT alignment are manifold. First of all, the languages "spoken" by different actors in the alignment process are different. IT manager is usually happy with UML [5] but such language may not provide enough information for someone from the business level, and vice versa. Therefore a framework that allows modelling and translation between different languages of the stakeholders is required for the alignment.

Further, based on the fact that even enterprises coming from the same cluster will impose different demands and requirements on the IT, successful alignment requires a "plug-it" ability allowing connecting and using the IT resources without having to worry about the underlying infrastructure. An approach currently developed in the plugIT EC research project [6] is tackling this issue in order to provide an alignment between the business and IT by allowing them to connect using the "IT-Socket".

Finally, following the initiatives on the translation and transformation of the stakeholder languages – allowing even the steering of the workflows on the IT level directly from the business level as outlined in [15] and the alignment over the "IT-Socket" an additional requirement for smart and adaptive IT infrastructure featuring e.g. Semantic Workflows rose. The goal is to create an infrastructure that would provide a reliable backend for the FI and benefit from introduction of the semantic technologies to the IT layer. This issue has been tackled in the LD-CAST [7] and FIT [8] EC research projects focused on design, deployment and execution of the smart and adaptive workflows.

This paper focuses on the aspect of the development and evolution of the business and IT alignment during time and the goal to enable seamless adaptation to changes in the environment by applying the knowledge engineering to support the evolution of the alignment on all three levels: Business, IT-Socket and IT Level.

The structure of this paper is as follows: after the introductory section describing the environment and overall challenges of the work presented, second chapter provides an overview on the Knowledge Engineering (KE) aspects in the Business Process Management (BPM). Third chapter presents the "IT-Socket" approach for the business and IT alignment. Fourth chapter introduces the idea of the Semantic IT Infrastructure based on the KE concepts. Finally the paper concludes with the short summary of the work done and outlook on the open research questions.

2 Knowledge Engineering in the Business Layer

BPM, which has become a commodity nowadays, is being used in the majority of larger enterprises in order to define, externalize and manage business processes. A specific enterprise BPM is seen as an instantiation of the BP-Framework (as described in the [9]) which defines (1) Business Models – defining the basics of the BPM approach, e.g. available elements and their relations, (2) Domain – defining the specifics imposed by the selected scenario, e.g. process documentation, (3) Regulations – defining the legal, business or technological regulations and (4) Model Processing – defining automated operations applicable for the available models, for a particular enterprise.

In order to apply the introduced BP framework (see [9] for more details) in a specific enterprise scenario the involvement of a BPM method is required. One of the currently used BPM methods applied in the aforementioned EC research projects is the Business Process Management Systems (BPMS) which consist out of five processes [10]:

- Strategic Decision Process – identifying the selected process and defining required resources
- Re-Engineering Process – change and adaptation of models to increase efficiency
- Resource Allocation Process – alignment of the business with the available IT infrastructure
- Workflow Management Process – concrete execution of a workflow
- Performance Evaluation Process – collecting and analysing the process-execution data

The alignment of the business with the IT layer is supported by application of the KE in order to foster (1) execution of the knowledge intensive activities (out of activities composing the business process) and (2) extraction of the available knowledge (in order to formalise it).

In order to carry out the aforementioned activities first the knowledge available has to be analysed applying the four dimensions of the knowledge space (as outlined in [9]):

- Form – defines the syntax and semantics of the knowledge
- Content – defines the domain of the application
- Use – defines the usage of the models
- Interpretation – machine (KE) or human (Knowledge Management)

Next step is the integration of the KE and BPM which is carried out on meta model level (second layer as shown in Fig. 1), according to the requirements imposed by the specific scenario. Hence, the integration can be carried out by applying meta model integration patterns (as introduced in [11]):

- Reference Pattern – linking one element from the KE to the one element in the BPM meta model, e.g. in order to describe specific element in more detail by providing additional semantic information in form of ontologies,
- Extension Pattern – integration of additional concepts from KE to BPM model, e.g. extending the model library with business rules
- Transformation Pattern – transformation of the parts of BPM meta model into the KE model, e.g. for automated generation of domain ontologies

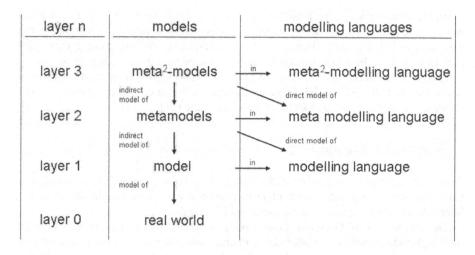

layer n	models	modelling languages
layer 3	meta²-models	meta²-modelling language
layer 2	metamodels	meta modelling language
layer 1	model	modelling language
layer 0	real world	

Fig. 1. An overview on the Meta Model Layers to depict the level chosen for KE/BPM integration, based on Strahinger [12] adapted by Karagiannis [13]

The business and IT alignment based on the Reference Pattern integration approach was applied in the LD-CAST EC research project [7] by linking the elements of the business process (activities) with corresponding concepts available in the LD-CAST ontology in order to enable creation and management of the smart workflows situated within the IT layer. The elements composing the business process and workflow were annotated with the concepts from the ontology, which in earlier step were applied to annotate web-services used in execution of the workflow. This approach allowed on the one hand dynamic population of the workflows and their execution based on inputs by the end users (selecting the desired web-service compositions) and on the other hand allowed direct interaction of the stakeholders from the business layer with workflows based in the IT layer. More detailed information on this approach can be found in [14], [15] and [16].

The Extension Pattern integration approach was used in the FIT EC research project [8] where KE concepts such as SWRL² – in order to express business rules, and parts of OWL³ – in order to extend the SWRL definition, were applied as an extension to the BP-Framework ADOe*Gov*® (a BP-Framework for e-Government)[25]. This approach provided the adaptability to the business and IT alignment, as it used the KE to externalize parts of the processes that were subject to frequent changes (e.g. changes in law and local regulations are common in the e-Government area) and therefore enable faster reaction and eliminate the requirement for the re-deployment of the processes – only business rules had to be changed. More details on this integration approach can be found in [15], [17] and [18].

The Transformation Pattern integration approach was applied in the BREIN EC research project [19] in order to extract the ontology concepts from the available business processes, hence creating a KE model which was used to support the usage of

² SWRL, http://www.w3.org/Submission/SWRL, Access [13.6.2009].
³ OWL, http://www.w3.org/2004/OWL, Access [16.03.2009].

semantic technologies in the project (e.g. dynamic discovery of knowledge resources based on the selected task and the experience level of the requestor, or supporting the work of grid-based agents). The ontology generation is carried out using a bootstrapping approach. To be more specific KE model is used to provide a support when creating new BP models (e.g. taxonomy or granularity check) which are then used to add new concepts to the KE model. More on this approach can be found in [19], [20] and on the dynamic discovery and knowledge provision in [21].

3   Knowledge Engineering in the Business and IT Alignment

As introduced in the first section of this paper, the middle layer between the business layer – presented in chapter 2 and IT layer presented in chapter 4 is the so-called IT-Socket layer which supports the business and IT alignment.

The idea behind the IT-Socket is to provide an access point to businesses searching for IT providing desired services (both internal and external IT service providers), where they will be able to connect to without having to think about the underlying IT infrastructure. That is, they will be able to plug-in and receive IT services as they would plug-in to a power socket to receive electrical power.

In order to allow this, having in mind the different languages "spoken" by the stakeholders and different requirements that business could impose when requesting and buying IT services – e.g. car manufacturer seeking a CPU hours for simulation and company performing off-site backups – one have to (1) solve the translation / transformation issue (different modelling languages) and consequently to (2) provide a holistic modelling framework capable of supporting many different modelling languages.

In order to solve the first issue the conceptual challenge, as outlined in [22], the first step is the creation of the modelling language ontologies in order to gather the formal representation of the modelling principles of each modelling language considered for the alignment.

The second step is the integration and translation between the created modelling language ontologies followed by the creation of the reference ontologies for modelling languages in order to foster meta-model integration.

Finally based on the aforementioned results the creation of domain ontologies out of graphical modelling languages and translation between them is the last step to provide a KE supported alignment of the business and IT layers.

The second issue is the provision of the modelling framework supporting many different modelling languages and hence allowing creation of the models using different tools is the technical challenge of the IT-Socket. Currently over 140 different modelling tools are available on the market [23], which can be classified according to following layers: (1) the graphical user interface – GUI, (2) the business logic and (3) model repository. This implies that any approach considering integration of different modelling tools within one modelling framework will have to take into account these tree layers. The architecture of the Next Generation Modelling Framework (NGMF) as described in [22] supports these tree layers by providing following architecture: (1) web modelling portal – the top layer providing GUI functionality, (2) service oriented modelling framework – providing business logic, e.g. service registration, discovery orchestration, etc, (3) open model repository – handling the storage of created models and (4) additionally to the traditional three layer architecture of modelling tools (as outlined in [24]), there is the fourth layer housing the Semantic Modelling Kernel which is responsible for the integration and transformation between the modelling tools, languages and models.

In order to fully support the business and IT alignment in aspect of the FI IT-Socket relies on application of the KE on both business and IT level. In the following chapter the next generation smart and adaptive IT infrastructure required to ensure FI compliance is presented.

4 Knowledge Engineering in the IT Infrastructure

Being an essential part of the IT world, the infrastructure layer has undergone some major revisions and changes in the last 50 years. Some of the dominant hardware architecture approaches have vanished to make place for more advanced solutions. Hence the evolution starting from mainframe, minicomputers, client/server and web/virtualization has currently landed in the cloud computing area.

In the software part the stunning success of the SOA based applications changed the way how business deploy and manage their IT. The ability to provide access to the legacy systems using web service layers along with novel fully web service based applications provided easy and in most cases best value for the money approach for the enterprises.

The SOA allowed enterprises to outsource parts of their workflows to external parties and thus to increase availability and capacity of their own services, raising their position in the market.

The major issue with the most implementations is that the workflows are mainly deployed as static items, allowing almost no changes in the services used for execution of specific tasks once they are released by workflow manager.

Although this may appear to be a stable and proven approach as opposed to for instance allowing interference during the execution, it goes against the principles of FI – access everything and everywhere, and in not so distant future it will lower the competitive advantage of the enterprises.

The goal for the new IT infrastructure capable of handling the demands and requirements imposed by the FI would be deployment of semantically enriched, easily adaptable and smart workflows.

The concept of the semantically enriched workflows describes the workflows that have been enhanced with semantic concepts, created by applying the KE on both the workflow level itself and the business process level in order to ensure the alignment.

Easily adaptable and smart workflows are able to react to changes in the environment. This includes both the IT layer – e.g. self-healing capabilities in case of failure and the external layer – e.g. adapting to changes in the market, law or regulations.

In order to create this smart IT infrastructure a concept of the abstract workflows has to be introduced. The abstract workflows are defined as abstract as they have no services bound to the activities but just the descriptions of the requirements in order to satisfy the activity parameters. Abstract workflows are not executable in the sense of the traditional workflow but they can be deployed to the running systems and started.

After the start, specific mechanisms (based on the results of the KE applied previously – see Integration patterns in chapter 2) such as semantic search and discovery (1) seek out services based on the semantic descriptions available within the workflow tasks and available services, (2) take notice of the specified parameters by the end users – who start the workflow, such as cost, time, etc, (3) applies the available business rules in order to satisfy internal and external regulations, and based on the execution type – e.g. manual, automatic, proceed with workflow execution.

The modelling of the abstract workflows – and as final result the deployable semantic workflows – is structured in three layers:

- Simple Abstract Workflows (AW-1) – This layer consist of descriptions and models of the workflows having different abstraction levels and structure, e.g. text documents, visio models, implicit knowledge, etc. They have in common that they do not follow any workflow standard; they are in most cases not executable and deployable and require major effort to transform into the next layer's workflows – the Structured Abstract Workflows. AW-1 workflows are usually created by persons that are not that familiar with the workflow technology but have vast knowledge of the business processes, market environment, IT infrastructures, etc which are used in order to design the workflows. This layer may be avoided in a scenario where team consisting of experienced workflow designer and business manager and enough resources (mainly time) is available. If these constraints are fulfilled it is possible to start directly from AW-2 level.

- Structured Abstract Workflows (AW-2) – the second layer as opposed to the AW-1 models involves the usage of specific notations for workflow languages (e.g. BPEL[4]) in order to design deployable workflows. The design of AW-2 workflows is either started from the scratch using only the knowledge provided by the first level as an input, or by extending already available AW-1 workflows. In both cases KE provides support to create AW-2 compliant workflows. AW-2 workflows can be used in two ways, first they can be used as traditional workflows (requires definition and assignment of the required web-services) and executed in the runtime systems or they can be used as input in the next layer in order to create Semantic Workflows. AW-2 workflows are designed by experienced workflow designers based on the input from the business managers and are highly structured and designed following the standards and internal enterprise regulations.

- Semantic Workflows (SEMWF) – the third and final layer introduces the semantic workflows. These workflows are built on top of the results from the first two layer, and as opposed to AW-1 and AW-2 they are fully deployable and executable – as described in this chapter and not in sense of traditional workflows. Modelling of the SEMWF workflows involves another stakeholder in the process, namely the ontology expert is required in order to apply the results of the previously applied KE. The ontology used

---

[4] BPEL, http://www.oasis-open.org/committees/wsbpel, Access [13.06.2009].

to semantically enrich the workflow – the AW-2 workflow, can be obtained from various sources: (1) as a result of the KE on two previous levels, (2) using enterprise ontology, if it includes the domain in question or (3) ontologies from 3rd party provider. Currently (1) is used in most cases as it provides required concepts for the workflow/service description and it is generated during the process, (2) is mostly not existent as generation of enterprise ontology implies cooperation of many internal and external business, IT and ontology experts – considered too costly by most enterprises today and (3) will gain more importance through the FI initiative as external providers whose core competence is development and management of ontologies will be able to provide reference ontologies for the fraction of cost compared to developing enterprise ontology in-house – semantics as a service approach.

As described previously the execution of the SEMWF is carried out using additional components which take over the task of finding appropriate services according to the requirement-descriptions provided within the deployed workflow instance. When speaking of execution of the SEMWF, we have to consider following questions:

How is the SEMWF execution started?

- Manual Execution – based on the input parameters or combination thereof provided by the end user at the runtime, e.g. required time of completion, maximum costs, etc., end user is presented with possible selections of executable workflows fulfilling the requirements. After the selection of one combination workflow execution is started.
- Automatic Execution – automatic execution is carried out automatically as soon as one executable workflow is available. This is usually done by agents. In order not to waste resources additional parameters can be predefined, and can be marked either mandatory, optional or combination thereof.

How is the SEMWF execution carried out?

- Static Execution – is used in such scenarios where workflow execution is marked as mission critical and costs or time do not play major role and SLA agreements can not be negotiated on the fly. This means that once the available web services have been discovered and selected the whole workflow is populated with exactly one web service per activity (given that each activity is carried out by one web service or additional one in case of fail-over) and then executed. The advantage of this approach is that the stability is increased but it lowers the adaptability and the flexibility of the SEMWF and thus does not use all improvements introduced by the SEMWF.
- Dynamic Execution – is more advanced approach as opposed to the Static Execution as it allows dynamic population of the services at the runtime. To be more specific this approach performs initial search and discovery to find best available candidates out of available web services which are then pre-marked and will be used in case nothing better is found when their

workflow activity is being executed. Therefore during the execution of the activity 1, system tries to find better services for following activities then the one which have been found when the workflow was deployed / started. This approach implies usage of SLA negotiation on the fly in order to guarantee the execution of the workflow. It has numerous advantages over the Static Execution, but currently it can not guarantee the same level of stability.

Based on the three-layer design approach and application of the KE technologies used on the business layer, the SEMWF provides ability to create semantic IT infrastructure that can be aligned with business layer applying the IT-Socket approach and thus act as one of the core components of the Future Internet initiative for enterprises.

5 Summary and Outlook

This paper introduced the research goal of enabling the business and IT alignment using the IT-Socket approach, providing answers to the challenges of the Future Internet and making the IT-Socket approach Future Internet compliant by applying knowledge engineering on all three alignment levels. Some of the technologies and approaches presented have been already tested in EC research and commercial projects yielding important insights on how to position the alignment approach within the Future Internet.

Some of the technologies such as the IT Socket are currently developed within the plugIT EC Research project and will be tested in three different scenarios: Certification, Virtual Organization and Governance. More information can be found at [25]. And finally some of the technologies such as dynamic execution of the SEMWF are still open research questions that will become significant as the Future Internet becomes more important.

References

1. Internet World Stats, Homepage: http://www.internetworldstats.com (access: 13.06.2009)
2. Da Silva, J.: Towards the Future Internet, ISI - European Satcom Day, http://ftp.cordis.europa.eu/pub/fp7/ict/docs/ ch1-jds-isi24april2009short_en.pdf (access: 13.06.2009)
3. Bayou, R., Guirao, F.: The Future Internet in FP7, European Commission
4. Gartner: Gartner Says Eight of Ten Dollars Enterprises Spend on IT is "Dead Money", http://www.gartner.com/it/page.jsp?id=497088 (access: 13.06.2009)
5. Unified Modelling Language Homepage, http://www.uml.org (access 13.06.2009)
6. plugIT Homepage, http://www.plug-it-project.eu (access: 13.06.2008)
7. LD-CAST Project Homepage, http://www.ldcastproject.com (access: 13.06.2008)
8. FIT Project Homepage, http://www.fit-project.org (access: 13.06.2008)
9. Karagiannis, D., Woitsch, R.: Knowledge Engineering in Business Process Management. In: Handbook on Business Process Management. LNCS. Springer, Heidelberg (2009) (in press)

10. Karagiannis, D., Junginger, S., Strobl, R.: Introduction to Business Process Management Systems Concepts. In: Scholz-Reiter, B., Stickel, E. (Hrsg.) Business Process Modelling, pp. 81–106. Springer, Heidelberg (1996)
11. Kühn, H.: PhD Thesis: Methodenintegration im Business Engineering, University of Vienna (2004)
12. Strahringer, S.: Metamodellierung als Instrument des Methodenvergleichs: eine Evaluierung am Beispiel objektorientierter Analysemethoden, Shaker, Aachen (1996)
13. Karagiannis, D., Kühn, H.: Metamodelling Platforms. In: Bauknecht, K., Tjoa, A.M., Quirchmayr, G. (eds.) EC-Web 2002. LNCS, vol. 2455, p. 182. Springer, Heidelberg (2002)
14. Catapano, A., D'Atri, A., Hrgovcic, V., Ionita, A.D., Tarabanis, K.: LD-CAST: Local Development Cooperation Actions Enabled by Semantic Technology. In: Eastern Europe eIGov Days, Prague, Czech Republic (2008)
15. Hrgovcic, V., Woitsch, R., Utz, W., Leutgeb, A.: Adaptive and Smart e-Government Workflows - Experience Report from the Projects FIT and LDCAST. In: eChallenges e-2008 Conference (eChallenges 2008), Stockholm, Sweden. IOS Press, Amsterdam (2008)
16. Karagiannis, D., Woitsch, R., Utz, W., Hrgovcic, V., Eichner, H.: Business Episodes and Workflow Integration: A Use Case in LD-CAST. In: 3rd International Conference on Internet and Web Applications and Services (ICIW 2008), Athens, Greece, pp. 84–89. IEEE, Los Alamitos (2008)
17. Karagiannis, D., Utz, W., Woitsch, R., Leutgeb, A.: Business Processes and Rules - An eGovernment Case Study. In: AAAI 2008 Spring Symposium - AI Meets Business Rules and Process Management, Stanford, USA. AAAI, Menlo Park (2008)
18. Leutgeb, A., Utz, W., Woitsch, R., Fill, H.-G.: Adaptive Processes in E-Government - A Field Report about semantic-based approaches from the EU-Project "FIT". In: Proceedings of the International Conference on Enterprise Information Systems (ICEIS 2007), Funchal, Madeira - Portugal, pp. 264–269. INSTICC (2007)
19. Karagiannis, D., Utz, W., Woitsch, R., Eichner, H.: BPM4SOA Business Process Models for Semantic Service-Oriented Infrastructures. In: eChallenges e-2008 Conference (eChallenges 2008), Stockholm, Sweden. IOS Press, Amsterdam (2008)
20. Woitsch, R., Eichner, H., Hrgovcic, V.: Modelling Interoperability: The Modelling Framework of BREIN. In: STASIS/BREIN Workshop, part of 4th International Conference on Interoperability for Enterprise Software and Applications (I-ESA 2008), Berlin, Germany. Workshop Proceedings (2008)
21. Hrgovcic, V., Woitsch, R.: Enhancing Semantic E-Government Workflows through Service Oriented Knowledge Provision. In: 4th International Conference on Internet and Web Applications and Services (ICIW 2009), Venice, Italy. IEEE, Los Alamitos (2009)
22. Woitsch, R., Karagiannis, D., Plexousakis, D., Hinkelmann, K.: Plug your Business into IT: Business and IT-Alignment using a Model-Based IT-Socket. In: eChallenges e-2009 Conference (eChallenges 2009), Istanbul, Turkey (in press, 2009)
23. plugIT Deliverable, D 2.2: State of the Art in Modelling, restricted access
24. Junginger, S., Kühn, H., Strobl, R., Karagiannis, D.: Ein Geschäftsprozessmanagement-Werkzeug der nächsten Generation-ADONIS: Konzeption und Anwendungen, pp. 392–401. Wirtschaftsinformatik, Vieweg (2000)
25. Palkovits, S., Wimmer, M.: Processes in e-Government – A Holistic Framework for Modelling Electronic Public Services. In: Traunmüller, R. (ed.) EGOV 2003. LNCS, vol. 2739, pp. 213–219. Springer, Heidelberg (2003)

Developing Diagnostic DSSs Based on a Novel Data Collection Methodology

Kaya Kuru[1,2], Sertan Girgin[1], Kemal Arda[2,3], Uğur Bozlar[1],
and Veysel Akgün[1]

[1] Gülhane Military Medical Academy
kkuru@gata.edu.tr
[2] Middle East Technical University
[3] Ankara Oncology Hospital

Abstract. Although necessary information on prognostic implications is missing and reliable data are available in very few areas of medicine, there is an increasing demand for diagnostic decision support systems (DDSS), mainly due to the multitude of variables involved and highly complex relations between them. Unfortunately, existing approaches seem inadequate for providing accurate and high quality data – a prerequisite for establishing a successful DDSS. In this paper, we demonstrate how SISDS methodology that aims to remedy the deficiencies of current systems in use can be utilized to ease the data collection process and provide opportunities to construct DDSSs without tedious pre-processing and data preparation steps. We also provide empirical results on a real-world testbed application in the field of radiology.

1 Introduction

Diagnostic decision support systems (DDSS) are computer programs that are designed to assist health professionals in making proper diagnosis at the point of care; a DDSS would take the patients clinical data and propose a set of appropriate diagnoses. However, due to the multitude of variables and complex relationship between them developing a DDSS is a non-trivial task[1]. One possible approach is to define rule sets based on experts' opinion. Although this approach may be feasible in certain cases, it has certain drawbacks and prone to errors. It is practically impossible for health care professionals to keep track of all relevant medical knowledge, and they have limited ability to deal effectively with large amounts of information[2]. Personal bias may distort an objective judgment; furthermore, even agreement among the experts does not always guarantee correctness: empirically, researchers have found that experts in some domains have been correct only 40-60% of the time [3]. Constructing the rule set can also be labor intensive.

---

[1] A physician may be confronted with more than 200 variables in critical cases [1].
[2] Humans may not be able to develop a systematic response to any problem involving more than seven variables [2].

D. Karagiannis and Z. Jin (Eds.): KSEM 2009, LNAI 5914, pp. 110–121, 2009.

An alternative approach is to employ *machine learning* (ML), in particular *classification*, techniques to predict possible diagnoses automatically based on existing medical data, which has the potential to build more accurate and reliable models. Harnessing this potential, however, depends on two interrelated factors: the structure and the quality of the data, and the chosen classification method. Unfortunately, reliable data is available in very few areas of medicine [4]; medical reports, that constitute the main source of data, are almost always in unstructured format, mainly due to the cognitive overload imposed by existing structured data entry approaches, and incomplete since details are assumed to be common knowledge and left out [5]; this, consequently, necessitates tedious data preparation/preprocessing steps.

In our previous work, we proposed a novel methodology called "Structured, Interactive, Standardized, Decision Supporting" (SISDS) that aims to ease the data entry process and avoid these additional steps [6]. SISDS combines most of the favorable features of the exiting methods while removing their deficiencies; it enables *transparently* collecting structured and *well-formed* data while the health professionals edit the corresponding medical report in a *natural free-text like* style. In this paper, we extend that work and show how SISDS can be combined with off-the-shelf classification methods to develop a real-world DDSS with high accuracy, as a proof-of-concept of its effectiveness and to verify the viability of the proposed approach. We also discuss the particular characteristics of collected medical data and how it can be leveraged to improve the diagnosis predictions. The rest of the paper is organized as follows: in the next section, we briefly describe the SISDS method and our testbed in the field of radiology, together with the properties of the data set. In Section 3, we present and discuss empirical results obtained by using various classifiers on the data set. Finally, Section 4 concludes with possible future work.

2   The SISDS Method and the Data Set

Professionals frequently declare the need for improvements in medical report quality at their institutions [7,8]. These declarations indicate a necessity for new methods that are both effective and with less cognitive pressure – in between free-text reporting and sophisticated menu-driven structured approaches, which would provide a through communication among professionals, and also facilitate high level operations, such as population based inferences and diagnosis/decision support. In this section, we will first discuss some essential characteristics of such a method, and then formally describe a particular solution that provides them.

Cognitive load and hierarchical structure. Reducing the cognitive overload is of utmost importance in medical reporting. There exists a direct relationship between the amount and complexity of information that need to be entered/processed by users and the cognitive load; hence, *reducing the amount and complexity of information would also reduce the cognitive load*. Let us consider a typical esophagus radiology report which would, among other things, contain observations about the shape of the mucosal relief, the section, length

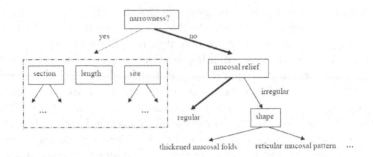

Fig. 1. The hierarchical structure. Boxes correspond to data entries and line labels indicate possible answers. The dashed box groups a set of data entries that are activated when there is narrowness. The normal values are shown with thick edges.

and the site of the narrowness of the esophagus etc. When entering data for a particular case, only a subset of this information may actually be relevant. For example, one of the questions to be answered in this report would be the following: *"Is there any narrowness without a clear expansion in the esophagus during the transition of the contrast media?"*. Usually, the answer to this question is *no*, and in this case the mucosal relief should be entered, which can be either *regular* or *irregular*; if the mucosal relief is irregular then the shape of the irregularity should also be specified; otherwise, this information is not required. In case there is a narrowness that is the answer is yes, mucosal relief is not important and a completely different set of information should be entered including and depending on the properties of the narrowness, such as its section, length and site. Note that, this inherently leads to a *nested and hierarchical* structure as depicted in Figure 1, in which data entered at a certain point determines the information flow, and consequently, the related data that should be entered: note that this may lead to a combinational expansion.

In medical reporting, we can identify two main goals: (i) to make medical reports easily accessible, complete and comprehensible by all users, and (ii) to be able to extract medical data out of them for further analysis. In order to accomplish these goals, abstraction seems essential; we can single out three main layers with the following properties:

Data. The data fields, or *variables*, that constitute a report must be consistent and *well defined* having specific types (such as nominal, numeric, measurement unit etc.). By assigning *default values* to variables and defining *constraints* over them (such as a permissible value range) cognitive overload can be reduced and erroneous input can be prevented.

Logic. A data entry may *encapsulate* multiple data fields. Furthermore, there may exist dependencies and relations between data entries; assigning a specific value to a particular variable may necessitate filling other data entries as well, or invalidate them. This inherently leads to a *nested and hierarchical* structure, in which data entered at a certain point determines the information flow. By initially starting from the default values for the normal cases and *interactively walking on the necessary steps* while completing the report, the number of data entries, and hence cognitive overload, can be reduced considerably. This dynamic hierarchical structure emerges as a common feature of almost all kinds of medical reports, and can be realized by assigning triggers, defined in terms of boolean expressions, to the data entries.

Presentation. Data and logic level can be regarded as the backend that defines the structure of the report; presentation level, on the other hand, is the frontend that defines how the report is rendered for data collection and viewing. The separation of presentation from data and logic would enable to generate different views of the same data based on user requirements.

Now, starting from the data level we will describe the SISDS method and discuss how it possesses the listed properties. The complete formal definition can be found in [6]. The building block in SISDS is a data field, or *variable*, defined by a tuple $\langle var, type, val, opts \rangle$ where var is the name and $type$ is the type of the variable, val is its initial value, and $opts$ is a set of pairs of the form $\{\langle name_1, val_1 \rangle, \ldots \}$ where $name_i$ denotes the name of the ith option and val_i is its value; typical options include the minimum, maximum and normal values of a variable. $type$ is either one of predefined types or if it is a nominal variable it is a set of possible values, ex. $\{yes, no\}$. A *data entry* is a unit of data request and encapsulates one or more variables; it is defined by a tuple $\langle label, vars, defs, triggers \rangle$ where $label$ is a unique identifier denoting the data entry, $vars$ is a set of variable definitions, $defs$ is a set of data request/view definitions (DRVDs) and $triggers$ is a set of triggers that activate related data entries. Each trigger in $triggers$ is a pair of the form $\langle cond, action \rangle$ where $cond$ is a boolean expression with embedded variable references and $action$ specifies an action to be executed when the condition holds. The variable references in the boolean expression are of the form $\langle label, var \rangle$ where var is the name of the variable, $label$ is the identifier of the data entry that the variable belongs to or $\emptyset$ if it belongs to the current data entry. While evaluating the boolean expression, the variable references are replaced with their current values[3]. An *action* can be a set of labels that denote the data entries to be activated, a message to be displayed, or a diagnosis prediction; it is important to note that cyclic activations are not allowed, that is a descendant of a data entry cannot re-activate it. Each DRVD is a tuple of the form $\langle type, lang, def \rangle$ where $type$ denotes the type of the DRVD, $lang$ denotes the language of the definition, and def is the body of the definition. The body of the definition is an arbitrary string with embedded variable references of the form $\langle label, var, vals, opts \rangle$ where $label$ and var are defined as above, $vals$ is a set of mappings for nominal variables to map possible values of the variable to string counterparts, and $opts$ is a set of options as in the definition of variables. Typical options include format specifiers to determine the rendering of the variable. DRVDs are used by the presentation layer to render

---

[3] The values of the variables with measurement data types are automatically converted into a common unit before evaluation since their units may be altered by the user.

data entry forms or reports based on their *type*; this gives rise to a unified view in which data collection and viewing are handled similarly.

Finally, a *report* is tuple $\langle E, M, triggers \rangle$ where E is a set of consistent data entries, that is all data entries referred in the trigger conditions of these data entries exists in the report, M is an ordered list of data entry identifiers denoting the *main data entries*, and *triggers* is a set of report-wide triggers; for each identifier in M there must be a corresponding data entry in E. The main data entries constitute the initial skeleton of the report. The report-wide triggers enable to provide diagnosis and other suggestions to the user based on the entered data. From a conceptual point of view, our structured design with interactivity looks like a tree with branches growing from a stem such that the branches collapse and expand as needed in terms of the request from the user as depicted in Figure 4, main data entries being the initially expanded branches. A dynamic hierarchy of *sections* is built as related data entries logically follow-up depending on the defined conditions. This effectively enables the user to focus on problematic parts and record them in more detail while eliminating other parts to save time, thus avoiding inefficiency and cognitive overload. The general elements of formal specification of SISDS are summarized in Table 3 as well as the relationship of data elements are presented in Figure 2.

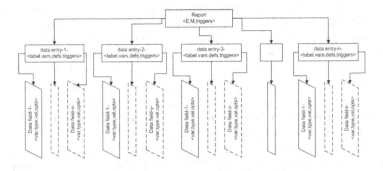

Fig. 2. The relationship of data definitions

The interrelation among the layers of data, logic and presentation, and the user is depicted in Figure 5. This figure summarizes the back-end and the front-end of application in computer model rather than the formal specification. We would like to point out that several existing design patterns, most notably *model-view-controller(MVC)* architectural pattern fits well to this layering: MVC isolates business logic from user interface, resulting in an application where it is easier to modify either the visual appearance of the application or the underlying business rules without affecting the other. The essential purpose of MVC is to bridge the gap between the human user's mental model and digital computer model. More detailed information about MVC design patterns can be found in Hunt's book [9].

In order to verify the viability of the proposed approach, we implemented a Web-based prototype. The main novelties of this particular implementation

definition	description of variables	examples
data field		
<var,type,val,opts>	var :the name of the data field type:the type of the variable val :the initial value of the variable opts:a list of options(name1,val1;...)	patient sex nominal(M,F) male male,M;female,F
data entry		
<label,vars,defs,triggers>	label :a unique identifier vars :a set of variable definitions, defs :a set of data DRVDs[type; lang; def(label; var; vals; opts)], triggers:a pair of the form [cond(label; var)],[action],	relief normal,not normal details are in the paragraphs [relief ==" not normal"], [trigger DRVD]
report		
<E,M,triggers>	E :a set of consistent data entries M :an ordered list of data entry identifiers triggers:a set of report-wide triggers	all data entries main data entries(initial skeleton) triggers

Fig. 3. The formulas of definitions in tabular

are a simple syntax to realize the formal definition given above, visual clues that prevent users from missing other pertinent details while concentrated on a specific subject, and a natural free-text like data entry mechanism such that the entered data directly corresponds to the content of the final report. Although the last item can be accomplished by following a split view approach, it is not cognitively appealing as the user has to go back and forth between different views. The solution that we offer is to use inline editing, that is to present the report in a single view but allow the users to directly manipulate the data on the screen simply by clicking on data fields which are displayed as hyperlinks (Figure 6): The predefined nominal values are displayed for data fields when

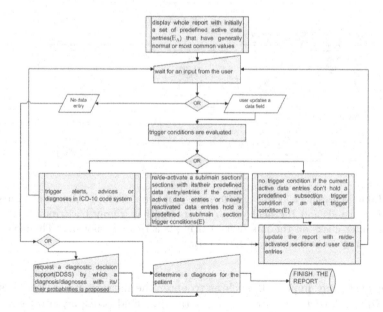

Fig. 4. The interaction of the user with the front-end presentation layer to generate medical reports

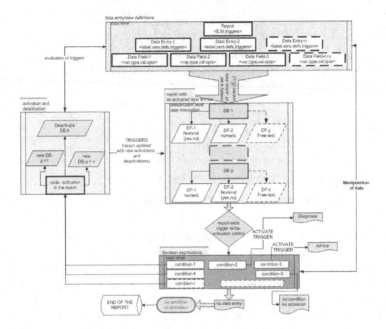

Fig. 5. The interrelation among the layers of data, logic and presentation, and the user. All triggers are evaluated to update the report when the user enter a value for any data entry.

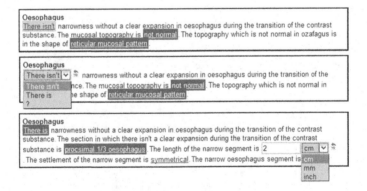

Fig. 6. Inline data entry in free-text format. (top) Initial state. Abnormal values are highlighted in red, and the field yet to be entered has a gray background. (middle) When the user clicks on the link inline editing is activated. (bottom) The new value "There is" triggers another set of data entries.

the user clicks any hyperlink, such as to enter the narrowness as "there is" or "there isn't"'; A text entry or numeric data entry field is displayed if there isn't any predefined nominal value as "the length of the narrow segment" entered as 2 cm. As the user changes the values of variables, the contents of the report

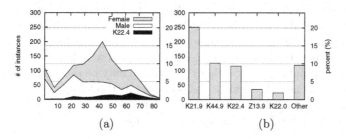

Fig. 7. (a) The age distribution of the data set for different sexes and the target diagnosis. (b) The distribution of ICD-10 diagnoses in the data set.

is also rearranged automatically according to predefined trigger conditions; the trigger conditions are implemented by javascript codes and runs effectively on the client site: report-wide trigger conditions are evaluated by javascript and current values of variables are changed with the definitions of conditions, and furthermore, the mapping of some operators are implemented by javascript.

As a real-world testbed for the SISDS methodology, we chose the field of radiology and a sample esophagus report format is constructed by radiology experts from several hospitals on our prototype, mainly based on the book of Weissleder [10]. The report consists of 13 main and 59 auxiliary data entries in a hierarchy having a maximum depth of 4. Each main data entry has a single nominal variable, and the report contains a total of 72 variables (53 nominal and 19 numerical) making it a fine example of a moderate sized medical report. After filling in the report, at least one diagnosis (up to 4) must be entered according to the ICD-10 coding scheme[4]. In a period of six months, health care professionals from the radiology departments of four different hospitals retrospectively entered real patient esophagus reports using the web-based prototype. The resulting data set contains 1240 instances spanning a period of six years from 2003 to 2008[5]. The instances have missing values for some attributes due to the trigger based dynamic activation of (sub-)sections, or simply because the value was not known by the user. Out of 1240 instances, 717 instances (57.8%) belong to healty patients and remaining 540 instances (42.4%) were tagged by one or more diagnoses. The number of distinct ICD-10 codes was 39. Among them only three were significant: K21.9 (250 instances, 20%), K44.9 (126 instances, 10%) and K22.4 (116 instances, 9.4%); the remaining ones have an average of 4.9 instances (Figure 7b). After applying a conjunctive rule learner using K21.9 and K44.9 diagnoses as target classes, we found out that both of them can be predicted with a high accuracy (98% and 98.8% respectively) depending on the answers of two particular main data entries. Therefore, we opted for the

---

[4] The International Statistical Classification of Diseases and Related Health Problems 10th Revision (ICD-10) is a coding of over 155000 diseases and signs, symptoms, abnormal findings, complaints, social circumstances and external causes of injury or diseases, as classified by the World Health Organization (WHO).

[5] 47.87% male, 52.13% female with a minimum age of 1 and a maximum age of 87.

non-trivial case of K22.4 as our target diagnosis, in which the prediction rate of the conjunctive rule learner is low (71.6%) for the patients having the corresponding health problem[6]. The diagnosis of K22.4 appears in all age intervals and sexes, most notably common for older people and female sexes (Figure 7a). We would like to point out that, although we will mainly be presenting the results on this data set in the next section, the proposed methodology and the followed procedure is more general and our aim here is to accomplish a proof-of-concept that similar studies can be conducted on other domains as well.

3   Experimental Results

Our goal, that is to predict the diagnosis of K22.4 for new esophagus report instances with a high accuracy, is a classical binary classification problem; hence, it is possible to employ various well-known classification techniques (such as Bayesian networks, decision trees, neural network, support vector machines or other functional classifiers) that are compatible with the properties of the data set. In this paper, we will focus on four specific representatives of different approaches: a Bayesian network that uses hill-climbing and a simple estimator that estimates probabilities directly from the data, a multinomial logistic regression model with a ridge estimator, a support vector classifier with sequential minimal optimization algorithm, and an alternating decision tree. The implementations of all these classifiers are available (BayesNet, Logistic, SMO and ADTree) in the Weka application suite developed at the University of Waikato[7]. Our web-based prototype has the capability to export the collected data in a format that can be directly imported by Weka, so that these (and other) classifiers can be tested with ease and a decision support system can be developed rapidly. In our experiments, we used the default parameters of the classifiers and applied 10-fold cross-validation to prevent overfitting[8]. To reduce variability, multiple rounds (in our case, again 10) of cross-validation are performed using different partitions, and the validation results are averaged over the rounds. In Logistic and SMO classifiers, the nominal attributes are transformed into binary numeric attributes and normalized.

The results are depicted in Table 1. We first applied the algorithms to the data set normally without any additional processing steps: the overall prediction rates are high for all classifiers (> 93.5%) and the SMO algorithm has the best accuracy rate (96.7%). However, when we analyse the results in detail, it can be observed that true positive rates (TPR), i.e. correct prediction rate for patients

---

[6] Results are obtained by 10-fold cross-validation.

[7] Waikato Environment for Knowledge Analysis (WEKA) contains a collection of visualization tools and algorithms for data analysis and predictive modeling, and is freely available from http://www.cs.waikato.ac.nz/ml/weka/

[8] In k-fold cross-validation, the data set is partitioned into k equally sized subsets; the analysis are performed on $k - 1$ subsets (training set), and then validated on the remaining one (testing set). 10-fold cross-validation is known to perform well for moderate sized data sets.

Table 1. Average results of the learning algorithms, with some meta learning methods, applied to the data

	ADTree			BayesNet			Logistic			SMO		
	TNR	TPR		TNR	TPR		TNR	TPR		TNR	TPR	
	0.982	0.762	0.962	0.973	0.563	0.935	0.976	0.777	0.958	0.986	0.784	0.967
Cost Sen.	0.933	0.841	0.924	0.953	0.854	0.944	0.958	0.821	0.945	0.949	0.866	0.941
Info. Gain (8)	0.982	0.774	0.963	0.986	0.727	0.962	0.985	0.751	0.963	0.987	0.740	0.963
(16)	0.985	0.771	0.965	0.982	0.756	0.961	0.985	0.773	0.965	0.985	0.779	0.965
(32)	0.983	0.769	0.963	0.974	0.716	0.949	0.980	0.783	0.961	0.986	0.770	0.965
PCA (8)	0.992	0.676	0.963	0.930	0.856	0.923	0.987	0.559	0.946	0.994	0.724	0.927
(16)	0.991	0.731	0.967	0.916	0.888	0.913	0.987	0.744	0.964	0.988	0.724	0.964
(32)	0.985	0.710	0.959	0.877	0.929	0.882	0.980	0.761	0.960	0.984	0.749	0.962

with K22.4 diagnosis, are low ($< 78.4\%$); in our case, TPR (K22.4) is as important as the true negative rate (TNR), i.e. correct prediction rate for healthy patients (in the sense that not suffering from K22.4 diagnosis). TPR is calculated by the evaluation of 116 instances as opposed to 1124 instances for TNR (with a ratio of 1/10.3), which means that the data set is *unbalanced* and prone to bias in the class-wise classification results. This situation also emerges as a common feature of most diagnostic related medical data sets [11]. One possible way to deal with this problem is to use *cost-sensitive classification*. In cost-sensitive classification, classes have different costs associated with them and the training instances are reweighted according to the total cost assigned to each class using a cost matrix. The classes with less number of instances can be assigned higher costs to reduce the number of false predictions, and consequently increase the accuracy, for that class (in our case, TPR). Note that, this also means that the prediction rates for other class(es) will inevitably fall as they will relatively have lower costs (and thus the number of false predictions in those classes will increase). In our experiments, we tested several cost matrices and the best results have been obtained by a cost matrix that assigns a weight of 10.0 to instances with K22.4 diagnosis and 1.0 otherwise. As it can be seen from Table 1, this leads to a significant increase in TPRs for all classifiers (almost 30% increase for the BayesNet and $\approx 8\%$ for the others) despite a small loss of 2%-4.9% in TNRs. Although BayesNet and Logistic classifiers have higher TNRs, SMO is better in TPR and has a similar but slightly lower TNR, and can be a better choice[9].

The data set under study consists of over 70 attributes; experiments show that useless attributes cause the performance of learning schemes to deteriorate [12]. A possible way to prevent this situation is to apply attribute selection techniques to the data set as a pre-processing step, and reduce the number of attributes. In our experiments, we tested two such techniques and determined a set of 8, 16 and 32 (transformed) attributes: *information gain attribute evaluation* (IG) that

---

[9] Other meta learning methods, such as bagging or boosting, couldn't produce comparable results. The details can be found and the established DDSS on SISDS Methodology can be tested at http:/www.gata.edu.tr/sisds

evaluates the worth of an attribute by measuring the information gain with respect to the target class, and *principal component analysis* (PCA) in which attribute reduction is accomplished by choosing eigenvectors that account for a specified percentage of the variance in the data set. For IG the results stay almost the same for ADtree, Logistic and SMO classifiers, and are better for the BayesNet classifier both in terms of TPRs and TNRs. For PCA the results also seem similar (slightly higher TNRs and lower TPRs), except BayesNet in which TPRs increase dramatically to 85.6%, 88.8%, and 92.9% with much sacrifices for TNRs, 93.0%, 91.6%, and 87%7 for 8 to 16 and 32 attributes respectively. Overall, including more attributes increases the TPRs for all algorithms which signifies that all attributes, rather than small subset, add a value to the classification results (probably, due to small number of instances with K22.4 diagnosis in certain age groups).

4   Conclusion and Discussion

As pointed out by Berner [13], health professionals' diagnostic performance can be strongly influenced by the quality of information the system produces and the type of cases on which the system is used. The accuracy and predictive power of the classifiers derived from data depends on the quality of the data. Information systems should enable the capturing of more complete, accurate, specific and timely medical information. Current reporting methods are insufficient to serve robust data collection for establishing DDSS. New methods of generating medical reports are required to avoid errors, decrease variations, enable research, support decisions and provide high quality health services. Within this context, the proposed SISDS methodology, that aims to remove the deficiencies of existing methods and introduces and promises new advantages, emerges as a viable candidate. The implemented web-based prototype and the accurate prediction results obtained on the testbed application that are presented in this paper demonstrate its effectiveness on certain aspects; most notably, facilitating the data collection phase. It also provides a proof-of-concept that similar studies can be conducted and DDSSs can be developed rapidly for other (medical) domains by incorporating SISDS with off-the-shelf solutions; although, discovering novel and more advanced classifiers is still needed, with the appropriate pre-processing steps (such as adding cost-sensitivity) remarkable results can be obtained.

Data collection in medical reporting using systems like SISDS will definitely help health professionals to practice better medicine. Further studies will concentrate on optimizing the parameters of classifiers to achieve better accuracy, and also wide-scale deployment of the system.

References

1. Morris, A.: Applications, R.G.C. In: Hall, j., schmidt, g., wood, l. (eds.) Principles of critical care, pp. 500–514. McGraw-Hill, New York (1992)
2. Miller, G.: The magical number seven, plus of minus two: Some limits to our capacity for processing information. Psychol. Rev. 63(1), 81–97 (1956)

3. Sorenson, J., Grove, H., Selto, F.: Detecting management fraud: An empirical approach. In: Symposium on Auditing Research, pp. 72–116. University of Illinois (1982)
4. Delaney, B.C., Fitzmaurice, D.A., Riaz, A., Hobbs, F.D.R.: Can computerised dss deliver improved quality in primary care? BMC 319(7220), 1 (1999)
5. Taira, R.K., Soderland, S.G., Jakobovits, R.M.: Automatic structuring of radiology free-text reports. Radiographics 21(1), 237–245 (2001)
6. Kuru, K., Girgin, S., Arda, K.: A novel multilingual report generation system for medical applications. In: Proceedings of the 12th International Conference on Artificial Intellegenge in Medicine (extended version is available as technical report METU-MIN-TR-2009-001-KK, Infomatics Inst., METU (2009), http://www.ii.metu.edu.tr)
7. Naik, S., Hanbidge, A., Wilson, S.: Radiology reports: examining radiologist and clinician preferences regarding style and content. American Journal of Roentgenology 176(3), 591–592 (2001)
8. Sistrom, C.L., Langlotz, C.: A framework for improving radiology reporting. J. Am. Coll. Radiol. 2(1), 61–67 (2005)
9. Hunt, J.: Java and object orientation, 2nd edn. Springer, Heidelberg (2002)
10. Weissleder, R., Jones, S.E., Wittenberg, J., Harisinghani, M.G., Harisinghani, M.G.: Primer of Diagnostic Imaging, 3rd edn. Mosby (2003)
11. Waegemann, C.P., et al.: Healthcare documentation: A report on information capture and report generation, p. 6 (2002)
12. Witten, I.H., Frank, E.: Data Mining, 2nd edn. Elsevier, Amsterdam (1997)
13. Berner, E.S., Maisiak, R.S., Cobbs, C.G., Taunton, O.D.: Effects of a decision support system on physicians' diagnostic performance. J. Am. Med. Inform. Assoc. 6(5), 420–427 (1999)

Data Integration for Business Analytics: A Conceptual Approach

Wilfried Grossmann

Institute for Scientific Computing, University Vienna
A 1010 Vienna
Wilfried.grossmann@univie.ac.at

Abstract. We present first and basic ideas for a modeling environment for business analytics. Main emphasis is on modeling components for data preparation, in particular data integration. The model is based on combination of knowledge and techniques from statistical metadata management and from workflow processes. All modeling concepts are presented in a problem oriented formulation. The approach is embedded into an open model framework which aims for a modeling platform for all kinds of models useful in business applications.

Keywords: Business Analytics, Data Preparation, Data Integration, Open Models.

1 Introduction

The empirical foundation of business analytics are data of high quality usually represented in a data warehouse. These data are obtained in an ETL-process by integration and aggregation of data from different sources and used further on either as input for data mining activities or as database for producing OLAP cubes. There are nowadays a number of open source as well as commercial tools available for supporting this process originating either from models for data integration (for example Pentaho-Kettle [14]) or emphasizing more the role of ETL activities within a process model for data mining (for example the CRISP format [2], which is the theoretical foundation of the data mining software PASW, formerly Clementine [13]). Main advantage of all these products is visualization of the activities, which helps users to understand data processing, supports reuse of analyses sequences, and allows to some extent automatic translation of visual description into executable code.

Despite the progress in the development of tools for building data warehouses and data mining, a number of problems are still unsolved, mainly due to the fact that the systems use metadata mainly for representation of syntactic knowledge about the data. Semantic knowledge about the structure of the problem, which plays a key role in defining the analysis model and in the assessment of the results, is represented only in rudimentary form. Consequently the analysis process is defined often in a more or less ad hoc manner according to the informal knowledge and the experience of the business analyst, who knows about possible pitfalls and side effects of procedures. Let us consider the following three simple examples, which show different types of problems occurring in data integration.

D. Karagiannis and Z. Jin (Eds.): KSEM 2009, LNAI 5914, pp. 122–133, 2009.

Example 1. A school administration maintains three databases: (i) A student database informing about grades of students in different subjects in different classes; (ii) A teacher database containing information about teaching the different subjects in the classes; (iii) A book database informing about the books ordered for the different subjects in the classes. Main question of interest is at what level we can integrate these pieces of information into one coherent data cube.

Example 2. A community maintains a housing register for all persons living in the community and a second register informing about social expenditures and working status of the persons. For planning purposes it is of interest to combine the information in the two registers. However, due to administrative reasons both databases do not have accurate information. For example in case of the housing register it is rather difficult to force people to inform the administration if they leave the community; in case of the social register information is many times incomplete and it may happen that there are multiple entries for one persons.

Example 3. A company wants to define one database of all its customers registered at different branches. Combining these different pieces of information has to take into account that some customers are registered in more than one branch and that quality of the information varies from branch to branch.

In order to support the business analyst in solving such problems and inform users about quality of the data extensions of the existing approaches in the following two directions is of interest.

- Incorporation of more detailed knowledge about the data used in business analytics in such way that this knowledge contains not only the data schema but also information about the semantics of the data necessary for correct interpretation and assessment of results.
- Developing of a more structured description of the transformation processes occurring in data integration, which contains not only application of the basic algorithmic building blocks but also information about dependencies and side effects of the algorithms, i.e. the fact that transformations imply often other transformations and need a check about the feasibility of the activities.

Looking at possible solutions for these two points we can find strategies in two different areas. For representation of semantic knowledge about data ideas developed in the area of statistics under the heading statistical metadata are of interest. Using these ideas is not surprising because business analytics itself can be understood as a specific type of statistical analysis. With respect to the second point it seems reasonable to analyze the transformation processes from a workflow perspective. In particular it is worth to think about different business analytic activities in terms of actions and states obtained by the actions.

Elaborating all these issues in full detail is beyond the scope of this paper. What we want to present are some basic conceptual ideas how to combine statistical metadata models with concepts of workflow management. Bringing together such different modeling tools needs a flexible framework open for integration of different concepts and techniques. Recently the Open Models initiative started at the University Vienna which aims for a framework for all kinds of modeling activities useful for business applications [12]. Hence it seems quite natural to formulate the above mentioned ideas within such a framework.

The above described approach towards use of data sources in business analytics is based on the idea of defining one persistent knowledge base for all data mining activities. Under some circumstances such an approach is not appropriate for specific investigations (in particular in problems like example 3), or it is not feasible due to the huge amount of data. Methods for knowledge discovery from multiple databases as described for example in [20] offer an alternative in such cases.

The organization of the paper is as follows. In section two we present some basic ideas about the open model approach, in particular we stress the need for a so called meta^2-view inside the modeling hierarchy, which allows formulation of a language for describing the logic of information processing for a specific modeling domain. Section three outlines basic ideas of the application logic for business analytics and in section 4 we present some basic ideas for a process model for data integration. We conclude with an outlook for further activities in this direction.

2 The Open Model Approach

The Open Model Initiative proposes a platform for all kind of modeling activities useful in business applications and emphasizes usage of different languages and techniques for solving real world problems. In order to allow high flexibility and representation of models in a language close to the actual modeling activity the OMG modeling hierarchy is enlarged by an additional level, the so called Meta^2-level. The basic idea of the Meta^2-level is definition of a language for the logic of information processing for a certain class of applications, which allows afterwards the definition of a metamodel. In our case the application class is defined by business analytics, other examples of application classes are workflows or optimization. Description of such an application logic has to be done at three different levels:

- Domain Logic: This logic is responsible for setting up the semantics of the domain model; most important is the definition of the structures of interest in the domain together with the relationships between the structures.
- Information Logic: The information logic is responsible for the syntactical part of the metamodel and defines the basic information structures used in the model together with the relationships between the information structures.
- Processing Logic: The processing logic defines the modeling procedures and the mechanisms and algorithms used. It comprises all kinds of activities for the information structures necessary for obtaining new knowledge and information about the domain.

Note that such a description is usually not bound to any type of traditional modeling tool and may be different for the different types of logics. For example in case of domain logic one can use techniques for representation of a terminology or more advanced an ontology, but also traditional data modeling techniques like ER-models may be useful. For the information logic a representation in narrative UML may be many times useful and also understood by users, but also XML-like structures may be appropriate. The processing logic is usually based on algorithms taken either from Computer Science or from Mathematics and encompasses also strategies how such algorithms have to be combined. Consequently we can use different tools for representation of the processing logic ranging from mathematical formulas up to pseudo code for algorithms.

After definition of the logic of information processing in a terminology close to the envisaged modeling application we can think about a translation of the logic into a formal specification. This means that we perform the specification in two steps. In the first step we describe the information logic and the resulting metamodel in a user or application oriented language and afterwards we give a formal specification. Usually such a translation into a formal specification is done in a semi automated way, in ideal case even an automated translation may be possible. Fig. 1 taken from [10] summarizes this modeling process.

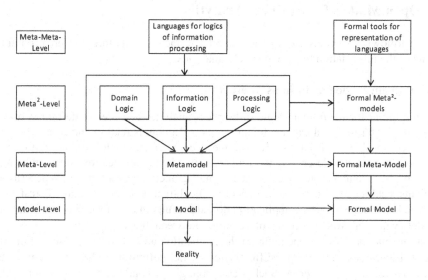

Fig. 1. Formal representation of the modeling process

Building a model in such stepwise manner helps many times to break down the complexity of the modeling process and allows the combination of knowledge from different experts. Users interested in an application have usually high expertise in the domain knowledge. Depending on the domain they formulate their knowledge in a way which is most appropriate for their domain and their understanding. For example in electrical engineering experts are quite familiar with graphical representation of circuits and know quite well how to interpret such diagrams, whereas in economics experts prefer more formulas and graphical representations of functions for analyzing problems. The building of the model itself is often done by experts for a specific modeling class, for example mathematicians, statisticians or experts in optimization. They use their own more general language and know how to map the domain logic into their own modeling language, which allows afterwards the application of sophisticated algorithms. For example mathematical finance uses a special vocabulary which is by no means limited to the domain of finance. In some sense we can understand this translation of the application domain into the information logic as a kind of typing of the objects of the domain into the logic of a specific kind of information processing. This gives many times a much richer information model which is easier to interpret. For example in business analytics the concept of a summary measure is typically an element of the information logic and may be calculated in different ways, e.g. an estimate according to a complex analytical model or simply the result of counting. Such different methods of calculation have to be taken into account in the interpretation of the results. In the first case we have to know details about the accuracy of the model and the calculation and can think about alternatives for calculation, whereas in the second case we only can judge whether the underlying population is really the intended one.

3 Open Models for Business Analytics

In the following sections we outline some basic facts for the three different components of the information logic in business analytics.

3.1 Domain Logic for Business Analytics

Starting point for the domain logic in business analytics is a precise definition of the objects of interest and attributes for the objects. In commercial or industrial applications typical objects of interest are customers, enterprises, employees, sales, products and so on; attributes describing these objects may be name, location, size, number of employees, revenues, and prices. In health management applications terms describing patients, health care structure, illnesses or medical treatments are of main interest.

The knowledge about the application domain results in a first step in a terminology model, which allows definition of relations between the terms. Different types of relations may be useful, depending on the application area. The basic case is thinking about a hierarchy between the different terms, i.e. definition of broader terms, narrower terms and related terms. Such ideas about generalization and restriction may be sometimes the foundation for sophisticated classification systems and also ontologies, for example products may be classified according to specific groups. Related terms may lead to problems in data integration which are often very difficult to handle. Take as example the integration of information about prices from two databases. If one contains prices of goods or a services according to an official price list and a second data base shows only prices effectively paid by customers it is not obvious how such information can be combined.

Besides such terminology oriented models we can define many times more detailed models using the traditional relations of data modeling. For example in commercial applications we can think about a model for an enterprise which has a number of stores located at different places, different types of employees working for the companies, customers classified according to different regions and so on. In such cases the domain model may be formulated as an ER-model.

3.2 Information Logic for Business Analytics

Every model considers only specific facets of a real world problem. The information logic describes the formal structure of the model which allows precise formal analysis of some features. In order to apply this model for the data we have to map the domain logic onto the information logic in an appropriate way. Correct application of a model requires knowledge of this mapping process, hence a transparent representation of essential features of the domain logic in the information logic is of utmost importance. In that sense the information logic gives a syntactic representation of the semantic of the domain. But keep in mind that there is a modeling semantic inside the information logic different from the domain semantic.

Due to the fact that business analytics is based mainly on empirical information and centers on methods developed in statistics, it seems quite natural to start with an information logic designed for statistical data processing. For more than 20 years such a model for representation of knowledge about data processing has been developed under then heading statistical metadata. A general feature of this information logic is that for the analysis itself the data structure is very simple, basically a flat file or summary table. Hence, we have to map the domain logic into such a structure, no matter how elaborated and complex the data model is defined in the domain logic. Hence, the complexity of the information logic in business analytics is not the data model in the traditional sense but an appropriate representation of a number of objects which have to be considered in connection with such a simple data model. Based on ideas of statistical metadata models we sketch in the following the main entities of an information logic for business analytics. Details may be found in [6, 8]. Keeping in mind the explanations in connection with figure 1 we will not use a formal specification language but a more content oriented description.

Empirical information always starts with *observable units* about which information is sought. Typical examples in business intelligence are enterprises, customers, or products. The observable units are elements of a collection of such observation units usually called a *population*. In most cases such a population cannot be observed completely, for example one may have information only about the behavior of registered customers during a specific time period but we are interested to model and understand the behavior of all potential customers in the future. Note that this consideration is different from the concept of a closed world many times assumed in database applications. The difference between entire population and observed part of the population plays an important role if we want to predict the behavior of all units in the population from the observed behavior. Such a generalization is only possible if we have a precise idea about the relation between population and observed units from the population.

Empirical information about the observation units is collected as attributes for the observable units and called in our context *variables*. (We prefer the term variable, because it is more open for an interpretation that the values vary, either in that sense that different units in the population have different values or in that sense that the values may vary for the units itself, for example sales in time.) Variables formalize the properties of interest of the observation units. Most important for business analytics is a clear description of the *value domain* of the variables, which have to be specified and analyzed carefully in case of data integration. The structure of a value domain is defined by measurement units and the relationships between possible measurement units. Typical examples of such relationships between measurement units are conversion formulas or relationships defined by a hierarchical classification scheme.

All data collected for observable units build the basic information entity for all kinds of activities in business analytics and are called *dataset*. Due to the fact that we have in general not complete information about the population we must be aware that the data resemble only the intended knowledge about the population but must not be confused with this knowledge. The relationship between these two bodies of knowledge depends essentially on the way how the data are collected. In business intelligence we are usually not so much interested in the process of data collection itself, we need only basic facts necessary for correct interpretation of the results of our investigations, for example we have to know what proportion of the population has been included in the dataset. Such information has to be represented in a separate information entity called *Additional Attributes*.

Summing up we can state that for business intelligence the basic information entities are observable units, population, variables, value domains, datasets and additional attributes for the dataset. Let us call them altogether *information categories* for this application. The above described relations between the categories are shown schematically in Fig. 2.

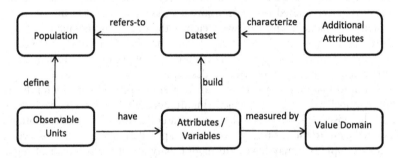

Fig. 2. Relation between information categories in business analytics

The description of the categories has to be given from different points of view. In general we can distinguish between the following four views.

- A *Conceptual Category View* building the bridge from real world problems to statistical representations, i.e. mapping of the domain logic into the information logic.
- A *Structural Category View* describing the categories within the structure defined in Fig. 2.
- A *Representation Category View* necessary for accessing and manipulation of the Category instances.
- An *Administration Category View* responsible for management and bookkeeping of the structures.

With respect to the formal representation of such models different languages have been used. A well known example is the model of the Data Documentation Initiative (DDI) which can be found in [4].

3.3 Processing Logic for Business Analytics

The processing logic for business analytics is basically the processing model of data mining and comprises the following five processing activities (cf. the CRISP processing model [2]): Business Understanding, Data Understanding, Data Preparation, Modeling, and Evaluation. Keeping in mind the model for the information logic outlined in 3.2 and taking a more formally oriented point of view, we can state that each processing activity is composed of a number of **basic operations**. A basic operation may be described as a transformation, which takes a number of instances of categories as input and maps it into a number of new category instances:

$$\left(C_1, C_2, \ldots C_p\right) \rightarrow \left(T_1(C_1, C_2, \ldots C_p), \ldots, T_k(C_1, C_2, \ldots C_p)\right)$$

Although there is usually one type of the categories described in 3.2 which is the main category of interest for the transformation, but in order to handle side effects of a transformation on other categories we have to consider categories of different types in a transformation (cf. [7, 8]). In the following we will outline some important transformations together with the side effects for basic transformations in case of Data Preparation, which encompasses the following types of operations: Data Selection, Data Cleaning, Data Construction and Integration.

Data Selection is an activity which is mainly an operation for objects of the category dataset producing a new dataset. It occurs in two different guises: Either as horizontal selection or as vertical selection. Horizontal selection is basically dropping of a number of variables, i.e. a projection. Vertical selection corresponds to the selection of a specific number of observation units. Obviously both types of selections have side effects on other categories. In case of horizontal selection we have mainly a restriction of the variables, whereas in case of vertical selection we have to adapt the description of the underlying population.

Data cleaning comprises two different basic transformations: basic transformations for detection of erroneous data, so called edits, and basic transformations for correction of these edits. The main categories involved are either variables or data sets. A classification of types of edits with respect to the main categories involved was given for example in [15]. Starting from the pioneering work of Fellegi and Holt [5], there are nowadays many possible strategies for detection of errors, based either on logical conditions defined according to the domain logic or on methods for empirical data analysis, for example finding outliers. A systematic treatment can be found in [3]. The crucial problem is that automated routines are hardly possible. What can be achieved are methodological guidelines. Basic transformations for correction of errors involve mainly instances of the category variables and are well known under the heading imputation. There are a number of possible specifications for algorithms and in [11] one can find a concise systematic survey. Looking for side effects on other categories we have mainly to consider value domains.

Construction of data is a topic which is based on many different methods. The simplest one is construction of new variables by logical or mathematical formulas.

Practically every data manipulation tool and every statistical or data mining programming environment offers such tools under the headings recoding and transformation. More sophisticated is the issue of construction of values for a specific variable. These techniques are sometimes also called imputation procedures, but better is the term statistical matching [17]. With respect to side effects mainly appropriate modifications of the value domains are necessary.

Similar to data selection we can distinguish in case of data integration between horizontal and vertical integration. Horizontal integration amounts to merging at least partially overlapping datasets (in terms of the underlying subpopulations) via common matching variables in order increase the set of available variables. Methods range from standard database joins on a unique key variable over string matching on character variables such as names and addresses to sophisticated probabilistic record linkage [19] or statistical matching procedures [17]. Vertical integration adds cases to a data source by the simple juxtaposition (set union) of (at least partially) equally structured data sources in terms of available variables and value domains that represent disjoint populations or population segments, at least in the ideal case. Besides these operations for instances of the category dataset we have to consider as in case of selection primarily side effects on populations.

Defining a process model from these basic operations according to concepts from workflow analysis has to take into account the specific features mentioned above, which show some similarities to scientific workflow: (i) Compared to usual workflow computation and algorithms play a more important role; (ii) More data are involved; (iii) The number of activities is usually higher. Besides these similarities with scientific workflow there is one special feature in business analytics: control of the entire workflow is determined by rather complex evaluation functions based on various quality criteria for the process. In the next section we will sketch the basic ideas for such a process model for data integration.

4   The Data Integration Process

Having defined a number of basic operations we can generate a process model for a business analytics activity taking into account the following two principles:

- All processes have to be defined not only for one category but have to take into account the side effects on other categories.
- The result of each basic operation has to be evaluated with respect to different evaluation functions which determine further processing activities.

The issue of side effects has been discussed briefly in connection with the description of the basic operations in 3.3 and can be realized in a workflow model by defining swim lanes for the different categories. The evaluation functions summarize the knowledge of the business analysis expert about the result of a basic operation in an operational form and derive rules for further processing. The necessary knowledge for defining such evaluation functions is usually treated under the heading data quality. Such data quality considerations have to take into account the information logic outlined in section 3.2. This means that we have to evaluate not only the dataset itself but also all other categories. Most important are the relation between the dataset and the underlying population and accuracy of the data. Besides traditional quality criteria as discussed in [1] a number of specific quality criteria have been defined in connection with statistical data quality aspects ([18]). For example the relation between population and dataset is analyzed under the heading coverage and for measuring accuracy a number of measures are defined.

In the following we will show how basic operations have to be combined in case of data integration. The data integration process is defined as the process combining different datasets under consideration of side effects onto the category objects associated with the datasets and of evaluation functions. It consists of the following four processing activities:

1. Determination of the observation unit for integration;
2. Alignment procedures for pre-processing the datasets under consideration;
3. Data Integration operations;
4. Alignment procedures for post-processing the integrated dataset.

The core processing activities are of course the integration operations in the third processing step consisting either of a vertical or a horizontal integration. All other processing activities depend on these activities. The first step is necessary in cases where domain models for the data are combined which have no obvious common observation unit as it is the case in Example 1. If we want to integrate information from the different databases a natural candidate for the observation unit for integration are the classes. Obviously for these classes we can obtain only summary information about the students. This example was considered in some detail in [9] using a composite OLAP-Object data model described in [16].

In order to get an idea of the necessary alignment procedures in pre- and post-processing let us consider the prototypic examples 2 and 3 for horizontal and vertical integration as shown in Fig. 3. The left part of Fig. 3 illustrates horizontal integration of datasets D1 and D2 with a matching variable defined according to step 1 of the integration process. The pre-processing steps encompass the necessary database operations in order to obtain the flat file structures D1 and D2. Afterwards we have to perform a number of cleaning operations for identifying precisely the overlap areas for the matching variables. Main basic operations for this processing step are matching operations. Often a mix of different matching methods may be necessary for a number of candidate sets. As a result we obtain the division of dataset D1 into D1a and D1b dataset D2 into D2a and D2b. The evaluation function after this processing step is defined by quality criteria for matching discussed in [17, 19] and evaluate the fit in the dark grey area. Such evaluation has to take into account also the definition of the underlying populations. Based on this evaluation function we have to decide whether the matching is feasible or not. If the matching is feasible a horizontal integration operation is done and we obtain a new dataset D1⊕D2. As side effects we produce the other category objects for the new dataset. Post-processing activities consist mainly in construction operations for the missing values shown in the figure as hatched areas. Different strategies are possible: we can either apply statistical matching for imputing the values or we can decide to drop some parts.

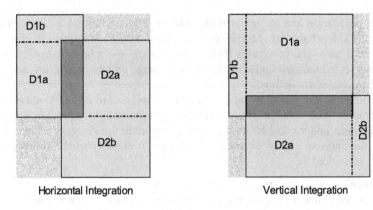

Horizontal Integration   Vertical Integration

Fig. 3. Detailed data view for horizontal and vertical integration

The case of vertical integration shown in the right part of Fig. 3 is similar to horizontal integration. Main pre-processing is identification of possible overlaps in the datasets D1a and D2a shown as dark grey area. This identification is based mainly on the analysis of the underlying populations. Afterwards cleaning operations have to be done because it may be possible that the two datasets contain contradictory information for some common observation units. Again evaluation functions decide about the feasibility of the integration and after integration again data cleaning and construction operations are necessary.

5   Conclusions and Further Work

The conceptual model for data integration outlined in this paper is only a first step in this direction. We have put main emphasis on a description of the structures and processing activities in a language close to business analysis. In a next step we plan to develop a formal model within the open model framework, in particular formulating the integration process in a workflow language. Such a language will support the construction of the integration process in a modular manner. Starting from the core operations for data integration the pre-processing and post-processing activities can be added in a stepwise manner taking into account the evaluation functions. In that sense the envisaged environment will support the process of data integration in obtaining data with high quality.

References

1. Batini, C., Scannapieco, M.: Data Quality - Concepts, Methodologies and Techniques. Springer, Heidelberg (2006)
2. Chapman, P., Clinton, J., Kerber, R., Khabaza, T., Reinartz, T., Shearer, G., Wirth, R.: CRISP-DM1.0 Cross Industry Standard Process for Data Mining (2000), http://www.crisp-dm.org/CRISPWP-0800.pdf
3. Dasu, T., Johnson, T.: Exploratory Data Mining and Data Cleaning, Wiley Series in Probability and Statistics. Hoboken, New Jersey (2003)

4. Data Documentation Initiative (DDI) (2009),
   `http://www.icpsr.umich.edu/DDI/`
5. Fellegi, I.P., Holt, D.: A systematic approach to automatic edit and imputation. Journal of the American Statistical Association 71, 17–35 (1976)
6. Froeschl, K.A., Grossmann, W., Del Vecchio, V.: The concept of metadata: A report on the nature of metadata and how these concepts can be used in practice. Deliverable 5 of the METANET project (2003), `http://www.epros.ed.ac.uk/metanet/`
7. Grossmann, W.: Metadata Usage in Statistical Computing. In: Braverman, A., Hesterberg, T., Minotte, M., Symanizik, J. (eds.) Proceedings of the 35th Symposium on the Interface, pp. 648–663. Interface Foundation of North America (2003)
8. Grossmann, W., Moschner, M.: Towards an Ontology for Data in Business Decisions. In: Karagiannis, D., Reimer, U. (eds.) PAKM 2004. LNCS (LNAI), vol. 3336, pp. 397–407. Springer, Heidelberg (2004)
9. Grossmann, W., Moschner, M.: Knowledge Integration from Mulidimensional Data Sources. In: Moreno Díaz, R., Pichler, F., Quesada Arencibia, A. (eds.) EUROCAST 2007. LNCS, vol. 4739, pp. 345–351. Springer, Heidelberg (2007)
10. Karagiannis, D., Grossmann, W., Höfferer, P.: Open Models A feasibility Study,
    `http://www.openmodels.at/web/omi/home`
11. Nordholt, E.S.: Imputation: Methods, Simulation, Experiments and Practical Examples. International Statistical Review 66, 157–180 (1998)
12. Open Models, `http://www.openmodels.at/web/omi/home`
13. PASW Modeler (formerly Clementine®),
    `http://www.spss.com/software/modeling/modeler/`
14. Pentaho/Kettle, `http://kettle.pentaho.org/`
15. Petrakos, G., Kalogeropoulos, K., Farmakis, G., Stavropoulos, P.: A Classification Scheme of Validation Rules Applied to Statistical Data Bases. In: Nanoupoulos, P., Wilkinson, D. (eds.) Proc. ETK_NTTS_2001. EUROSTAT, pp. 685–693 (2001)
16. Pourabbas, E., Shoshani, A.: The Composite OLAP–Object Data Model: Removing an Unnecessary Barrier. In: Froeschl, K.A., Grossmann, W. (eds.) Proc. 18th Int. Conf. on Scientific and Statistical Database Management – SSDBM 2006, pp. 291–300. IEEE, Los Alamitos (2006)
17. Raessler, S.: Statistical Matching. Springer, Heidelberg (2002)
18. Vardaki, M., Papageorgiou, H.: Statistical Data and Metadata Quality Assessment. In: Handbook of Research on Public Information Technology. Information Science Reference. IGI Global, New York (2007)
19. Winkler, W.E.: Overview of Record Linkage and Current Research Directions. Report RRS2006/02, Washington, D.C.: U.S. Bureau of the Census (2006)
20. Zhang, S., Zhang, C., Wi, X.: Knowledge Discovery from Multiple Databases. Springer, Heidelberg (2004)

New Labeling Strategy for Semi-supervised Document Categorization

Yan Zhu, Liping Jing*, and Jian Yu

School of Computer and Information Technology
Beijing Jiaotong University
Beijing, China
yasmine_zhu@hotmail.com, {lpjing,jianyu}@bjtu.edu.cn

Abstract. Usually, semi-supervised learning requires a number of prior knowledge to supervise the learning process, such as, seeds in Seeded-Kmeans, pair-wise constraints in COP-Kmeans, and labeled data for training an initial useful classifier in S3VM. Such prior knowledge is generally provided by the domain expert, so it is very expensive. In this paper, we propose a new automatical document labeling strategy to derive much more prior knowledge based on the very limited labeled data and the whole data set. Experimental results on 20-Newsgroup text data have shown that the new strategy is helpful for semi-supervised document categorization and improves the learning performance.

Keywords: semi-supervised learning, labeling, prior knowledge, keyword.

1 Introduction

With the abundance of text documents through internet, corporate management systems and digital libraries, the automatic partitioning of documents into sensible groups ranks top on the priority list for all intelligence systems. Various challenges in document categorization(including supervised classification and unsupervised clustering), such as, curse of dimensionality, complex semantics and large volume, must be addressed. At the same time, different users have quite different needs regarding to document categorization because they may view the documents from completely different perspectives [1]. All these made document categorization be a hot and challenging research topic.

A lot of text categorization algorithms have been proposed in the literature, including Bayes classifier [2], SVM method [3], Scatter/Gather [4], Bisecting Kmeans [5], SuffixTree clustering [6], Subspace clustering [7] and etc.. However, such traditional learning has its own drawbacks: supervised classification can not be used in many applications as it requires too much prior knowledge, while unsupervised clustering always performs poorly as they can not capture any information from users. Therefore, more and more researchers paid attention to

---

* Author Yan Zhu and Liping Jing have the same contribution on this paper.

D. Karagiannis and Z. Jin (Eds.): KSEM 2009, LNAI 5914, pp. 134–145, 2009.

semi-supervised learning [8,9], which can effectively use a few prior knowledge to supervise the categorization process. This learning method has been applied to many applications, such as text classification [10], image retrieval [11] and computer-aided diagnosis [12].

Zhu [9] divided the existing semi-supervised learning algorithms into four categories. The first category is based on generative models. Nigam et al. [13] designed a generative model by combining EM framework and Naive Bayes theory. Later, Fujino et al. [14] extended this EM model by introducing a 'bias correction' term and a discriminative training process with the aid of maximum entropy principle. Also, some researchers proposed Kmeans type semi-supervised learning approaches, such as, Seeded-Kmeans [15] and pair-wise constraints Kmeans (COP-Kmeans [16] and HMRF-Kmeans [17]). Graph-based method is the second category which used the graph structure to spread labels from labeled data to the whole data set [18,19,20]. The third category is low-density separation method, for example, S3VM (including TSVM [21], Laplacian SVM [22] and meanS3VM [23]). S3VM adopted SVM on a fixed number of labeled data to provide more useful prior information, and then used unlabeled data to regularize the decision boundary which was in turn to exploit the unlabeled data. The last category is co-training method proposed by Blum and Mitchell [24]. Later, Zhou et al. [25] adopted canonical correlation analysis to make co-training algorithm available on only one labeled point. Similarly, tri-training algorithm [26] with three learners was proposed for semi-supervised learning.

The above semi-supervised learning algorithms are generally assumed that the prior knowledge is provided by domain experts. Here, the prior knowledge is represented in the form of labeled data or pair-wise constraints (must-link and cannot-link constraints). The method to obtain prior knowledge is very expensive. To date, various semi-supervised algorithms were proposed, but there is limited work on automatically deriving prior knowledge from the data [27]. Low-density separation and co-training methods can be taken as automatically deriving approach to some extent, however, these kinds of algorithms are cost-consuming and thus unpractical [9,23]. Recently, researchers tried to use domain ontology (WordNet, Gene Ontology and etc.) [28,29] and other knowledge resources (e.g., Wikipedia) [30] to aid the document clustering, but, their prior information constraints only focus on features, without extending to instance constrains or labels. The methods using word knowledge to derive the labels of some unlabeled data were also devised [31,32], however, these methods can only be carried out under the condition that the annotator can access to the list of categories. The method of extracting topic words from the result of LDA (Latent Dirichlet Allocation) clustering can overcome the above problem [33], the drawback is that the annotator will find it hard to label candidate category words when LDA performs poorly.

In this paper, we present a new labeling strategy to automatically derive more prior knowledge from data set for semi-supervised document categorization. The proposed strategy consists of three parts. In the first part, the keywords will be extracted from the labeled documents. Then the document category keywords will be collected according to the extracted keywords and the known document labels in the second part. Finally, additional unlabeled documents will be labeled based on the statistical information of the category keywords in the unlabeled documents. Using this strategy, more labeled documents will be obtained and input to the existing semi-supervised learning algorithms as the prior knowledge. Experimental results on real-world text data have shown that the proposed strategy is helpful for semi-supervised document categorization and improves the learning performance.

The rest of the paper is organized as follows. In Section 2, we described the proposed labeling strategy in detail. Section 3 showed three semi-supervised categorization algorithms (EM-based Naive Bayes, Linear Neighborhood Propagation and Seeded-Kmeans) and how to combine the new labeling strategy into such algorithms. A series of experimental results on real text data were given and discussed in Section 4. In Section 5, we made a conclusion and showed our future work in brief.

2   Proposed Labeling Strategy

Let $D = \{d_1, d_2, ..., d_L, d_{L+1}, ..., d_N\}$ be a text corpus with N documents. D consists of two parts, $D_L = \{d_1, d_2, ..., d_L\}$ and $D_U = \{d_{L+1}, d_{L+2}, ..., d_N\}$ ($L << N$). Among them, D_L is the set of documents which are labeled manually, and D_U is the set of unlabeled documents. Different from the traditional obtaining prior information methods, our proposed labeling strategy simultaneously asks the experts to mark the keywords describing the document contents when labeling the corresponding documents. Let l_i and w_i be the label and the keyword set of the ith document respectively. For example, a document about space in science is labeled and its keywords are extracted as shown in Fig.1.

Based on such initial prior information, we proposed a labeling strategy as shown in Fig.2. Firstly, the category keywords $\{W_1, W_2, ..., W_K\}$ are collected according to $W = \{w_1, w_2, ..., w_L\}$ and the corresponding document labels, where W_j indicates the keyword set of the jth category, K is the number of topics. Meanwhile, all words appear in more than one category will be ignored.

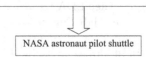

A doucment paragraph about sci:space

...We will assume you mean a NASA astronaut, since it's probably impossible for a non-Russian to get into the cosmonaut corps (paying passengers are not professional cosmonauts), and the other nations have so few astronauts (and fly even fewer) that you're better off hoping to win a lottery. Becoming a shuttle pilot requires lots of fast-jet experience, which means a military flying career; forget that unless you want to do it anyway. So you want to become a shuttle "mission specialist"...

NASA astronaut pilot shuttle

The words extracted from the above paragraph

Fig. 1. From document to extracted words

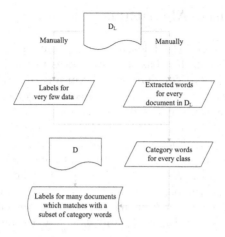

Fig. 2. Flow of the new labeling strategy

Before deriving more prior information, we preprocessed the document collection by selecting the words, removing the stop words, stemming words, and calculating the words' $tf \cdot idf$ value in the corresponding document. Then, the text corpus can be represented as a vector space model, $X = [x_1^T, x_2^T, ..., x_N^T]^T$, where the ith document is $x_i = [x_{i1}, x_{i2}, ..., x_{iM}]$, here M is the total number of words in the collection, x_{it} is $tf \cdot idf$ value of the tth word in the ith document which can be calculated by

$$x_{it} = tf_{it} \times \log \frac{N}{(df_t + 0.01)}. \tag{1}$$

In (1), tf_{it} represents the frequency of the tth word appears in the ith document, df_t is the number of documents in which the tth word occurs. Finally, the data matrix will be normalized so that $\sum_{t=1}^{M} x_{it}^2 = 1, i = 1, 2, ..., N$.

For each document x_i, we check whether it contains the category keywords by

$$z_{ij} = \frac{\Sigma_{p \in idx(W_j)} x_{ip}}{|W_j|} \tag{2}$$

where, $|W_j|$ is the number of keywords the jth category contains, $idx(W_j)$ is the jth category keywords indices in the document collection. Thus, z_{ij} indicates how the ith document is related to the jth category in terms of content. In other words, if a document covers more words which are labeled in the jth category, this document is more similar to jth category in terms of content [34]. As a result, matrix $Z = [z_1^T, z_2^T, ..., Z_N^T]^T$ can be built for N documents and K categories, and $z_i = [z_{i1}, z_{i2}, ..., z_{iK}]$. After normalizing Z (subject to $\sum_{j=1}^{K} z_{ij} = 1, i = 1, 2, ..., N$), its cell value can be used to label the document. In this paper, we assign the ith document into the jth category if $z_{ij} > 0.5 + 1/(K+1)$. With this strategy, more unlabeled documents can be labeled as the prior information for the later semi-supervised learning algorithms.

3   Semi-supervised Algorithms

There are many semi-supervised learning algorithms as mentioned in Introduction. Three algorithms, EM-based Naive Bayes method [13], LNP (Linear Neighborhood Propagation) [20] and Seeded-Kmeans [15] were used to test our proposed labeling strategy, because these three algorithms are efficient. Their computational complexity is linear to the number of data points N. For example, EM-based Naive Bayes and Seeded-Kmeans, generative model methods, have $O(KNT)$ computing time, where K and T represent the number of clusters and the iterations respectively. For LNP, a graph-based method, the major step is to calculate matrix $E_{N \times N}$ by solving N dual optimization problems, thus, the computational complexity of LNP is $O(NR^4)$, where R represents the number of neighbors. R was set 5 in this paper. Next, we will briefly introduce these three semi-supervised algorithms and show how to combine our proposed labeling strategy into them.

3.1   EM-Based Naive Bayes Method

EM-based Naive Bayes method firstly learns an initial classifier from the limited labeled data, and then refines the classifier iteratively until its likelihood function $l(\theta|X, Y)$ reaches to a local maximum value. Here, θ reprensents the parameters of the classifier and Y represents the labels of some data. The classifier is trained with Naive Bayes theory:

$$p(y_i = c_j | x_i; \theta) = \frac{p(x_i | c_j; \theta) \times p(c_j | \theta)}{p(x_i | \theta)} \tag{3}$$

where the jth topic is labeled by c_j. As the denominator in (3) for each topic is same, $p(y_i = c_j | x_i; \theta)$ can be denoted as $p(x_i | c_j; \theta) \times p(c_j | \theta)$. Naive Bayse always assumes that all terms are independent and identical distributed, thus, $p(x_i | c_j; \theta)$ can be expressed as:

$$p(x_i | c_j; \theta) = p(wd_1 | c_j; \theta)^{x_{i1}} \times p(wd_2 | c_j; \theta)^{x_{i2}} \times \ldots \times p(wd_M | c_j; \theta)^{x_{iM}} \tag{4}$$

where, wd_t is the tth term in X. Then the parameter θ of the classifier can be trained by

$$\theta_{c_j} = p(c_j | \theta) = \frac{1 + \Sigma_{i=1}^N \delta_{ij}}{K + N} \tag{5}$$

and

$$\theta_{wd_t | c_j} = p(wd_t | c_j; \theta) = \frac{1 + \Sigma_{i=1}^N \delta_{ij} x_{it}}{M + \Sigma_{s=1}^M \Sigma_{i=1}^N \delta_{ij} x_{is}} \tag{6}$$

where, if the ith data belongs to the jth class, $\delta_{ij} = 1$, otherwise $\delta_{ij} = 0$.

Then, the trained Naive Bayes model is iteratively refined on the unlabeled data until $l(\theta|X, Y)$ reaches to the local maximum value.

3.2   LNP

LNP is a graph-based semi-supervised learning algorithm, which used the graph structure to spread labels from labeled examples to the whole data set. The graph is expressed as $G = \langle V, E \rangle$, where $V = X$ is the vertex set and E is the edge set. Each edge e_{ij} represents the relationship between data x_i and x_j. Unlike the traditional computing methods for e_{ij}, LNP derives the graph edges with neighborhood information by minimizing ξ which is defined as:

$$\xi = \Sigma_{i=1}^{N} \| x_i - \sum_{j, x_j \in N(x_i)} e_{ij} x_j \|^2 \tag{7}$$

$$s.t. \quad \sum_{j, x_j \in N(x_i)} e_{ij} = 1$$

$$e_{ij} > 0$$

Among them, $N(x_i)$ is the neighborhood of x_i. Let $Y^{N \times K}$ be the initial label matrix, and its component $y_{ij} = 1$ if x_i is labeled by the jth class, otherwise, $y_{ij} = 0$.

LNP algorithm is an iteration process to update the prediction model vector F. It begins with $F^{(0)} = Y$, $Y^{N \times K}$ is the initial label index matrix. Then, the matrix F will be updated by

$$F^{(t+1)} = \alpha E F^{(t)} + (1 - \alpha) F^{(0)} \tag{8}$$

where, $0 < \alpha < 1$ is the fraction of label information that data receives from its neighbors. The final label of x_i can be obtained with $\arg\max_{j \leq K} F_{ij}$.

3.3   Seeded-Kmeans

Actually, Seeded-Kmeans is a kind of EM-based semi-supervised learning method, but it is different from EM-based Naive Bayes method on the way of using the prior information. Seeded-Kmeans used the prior information (labeled data) to construct its initial cluster centers. Let y_i be the label of x_i, the initial cluster centers can be formed with

$$u_k = \frac{1}{N_k} \sum_{i=1, y_i=k}^{N} x_i, \quad k = 1, 2, ..., K \tag{9}$$

where, N_k is the number of labeled data belonging to the kth cluster. Once obtaining the initial cluster centers, the standard Kmeans clustering can be applied on the whole data set by the following EM iteration steps:

M-step: assign data x_i to cluster k^* with $k^* = \arg\max_k \| x_i - u_k \|^2$.
E-step: update the cluster centers with (9).

3.4   CW-Type Semi-supervised Learning Methods

In Section 2, we presented a new labeling strategy, i.e., a new method to find more prior information for semi-supervised learning. So far, the existing semi-supervised algorithms, EM-based Naive Bayes, LNP and Seeded-Kmeans have been described. How to combine the new labeling strategy into the existing semi-supervised algorithms will be introduced in this subsection. In order to easy remember, we call the combining process as CW-Type methods, such as, CW-EM-NaiveBayes, CW-LNP and CW-Seeded-Kmeans, which can be summarized as follows.

Table 1. CW-Type semi-supervised method

Input: Document matrix $X = [x_1^T, x_2^T, ..., x_N^T]^T$, number of clusters K, a set of labels $\{l_1, l_2, ..., l_L\}$ for the L documents, and their corresponding keywords $W = \{w_1, w_2, ..., w_L\}$, where w_i is the keyword set for document x_i with label l_i.

Output: K partitions $\{C_1, C_2, \cdots, C_K\}$ of X

Method:

Step1. Calculating $\{W_1, W_2, ..., W_K\}$ according to $W = \{w_1, w_2, ..., w_L\}$ and the corresponding labeled documents.

Step2. Comparing each document x_i with the category keywords W_j by
$$z_{ij} = \frac{\Sigma_{p \in idx(W_j)} x_{ip}}{|W_j|}.$$

Step3. Normalizing Z so that $\sum_{j=1}^{K} z_{ij} = 1, i = 1, 2, ..., N$.

Step4. Assigning the label j to x_i if $z_{ij} > 0.5 + 1/(K+1)$

Step5. Taking the documents with labels as the prior information to run the existing semi-supervised algorithms:
- Training the model parameters for EM-based Naive Bayes with (5,6).
- Initializing the prediction model vector for LNP in (8).
- Finding the initial cluster centers for Seeded-Kmeans with (9).

4   Experimental Results and Discuss

In this paper, we conducted a series of experiments on six subsets of 20-Newsgroup dataset[1] as illustrated in Table 2. As EM-based Naive Bayes method and LNP can be used for prediction, each subset was split into two parts, 80% as training data and 20% as testing data. In the training set, 1% of documents will be previously labeled. Also, some keywords were extracted from these labeled documents. The average number of words for each data set is illustrated in Table 3. Both labeled documents and keywords would be the initial prior information for our labeling strategy. From Table 3, we can see the average number of keywords is very small, in other words, our proposed strategy doesn't need too much external manual work.

Three semi-supervised learning algorithms, EM-based Naive Bayes, LNP and Seeded-Kmeans, were used to test our proposed labeling strategy. The labeled documents (1% of the training data) were the prior information for the original

---

[1] http://people.csail.mit.edu/jrennie/20Newsgroups/

Table 2. Summary of Data Sets

Dataset	Topic	N	Dataset	Topic	N
TA2	alt.atheism	1000	TB2	talk.politics.mideast	1000
	comp.graphics	1000		talk.politics.misc	1000
TA4	comp.graphics	1000	TB4	comp.graphics	1000
	rec.sport.baseball	1000		comp.os.ms-windows.misc	1000
	sci.space	1000		rec.autos	1000
	talk.politics.mideast	1000		sci.electronics	1000
TA4_U	comp.graphics	1000	TB4_U	comp.graphics	1000
	rec.sport.baseball	800		comp.os.ms-windows.misc	800
	sci.space	500		rec.autos	500
	talk.politics.mideast	200		sci.electronics	200

Table 3. Average number of keywords per document in six data sets

TA2	TA4	TA4_U	TB2	TB4	TB4_U
4.56	3.40	2.90	4.50	3.69	3.85

EM-based Naive Bayes, LNP and Seeded-Kmeans algorithms. Therefore, we can compare them with our proposed methods CW-EM-NaiveBayes, CW-LNP and CW-Seeded-Kmeans. Accuracy and normalized mutual information (NMI) were used here as evaluation measures, the two evaluation measures are defined by

$$Accuracy = \frac{\sum_{k=1}^{K} D_k}{N} \tag{10}$$

and

$$NMI(T,C) = \frac{MI(T,C)}{\max\{H(T), H(C)\}}$$

$$MI(T,C) = \sum_{t_i \in T, c_j \in C} p(t_i, c_j) \times \log \frac{p(t_i, c_j)}{p(t_i) \times p(c_j)} \tag{11}$$

$$H(C) = -\sum_{h=1}^{N} p(c_h) \times \log p(c_h)$$

where D_k is the number of documents correctly grouped in cluster k, T represents the set of the documents' true topics, C represents the categorizing results.

Before comparing accuracy and NMI, we listed the error rate of our labeling strategy based on the existing labels of 20-Newsgroup data (because 20-Newsgroup is a benchmark data with document categories). As shown in Section 2 and Section 3, more documents will be labeled and taken as the prior information. Table 4 illustrates the error labeling rates of our proposed strategy for the six datasets. As we expected, our strategy can derive much more labeled data with low error rate.

Table 4. Automatically labeled data

	TA2	TA4	TA4_U	TB2	TB4	TB4_U
Number	1464	3090	1842	1162	2511	1624
Error rate	0.0397	0.0660	0.0868	0.2062	0.2404	0.2523

Next, semi-supervised learning algorithms were applied on the whole training set with the aid of prior information. Figure 3 shows the comparison results of EM-based Naive Bayes and CW-EM-NaiveBayes. Similarly, the experimental results for LNP and CW-LNP are displayed in Figure 4, Seeded-Kmeans and CW-Seeded-Kmeans are listed in Figure 5. All these experimental results on training data have shown that the new labeling strategy improves the learning performance.

Additionally, the inductive results on the testing set for EM-based Naive Bayes type methods and LNP type methods are displayed in Figure 6 and 7.

From the above experimental results, we can say that our proposed labeling method makes the semi-supervised algorithms obtain higher learning accuracy and NMI, i.e., manually extracting a few words when labeling a document and previously automatically assigning part of unlabeled data into categories can provide more supervised information. Thus, it is a promising research direction and necessary, as Jain mentioned in [27], to design an efficient labeling strategy for semi-supervised learning.

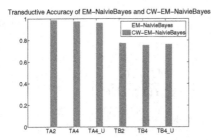 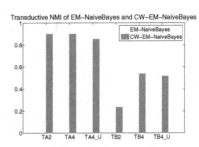

Fig. 3. Transductive results of EM-NaiveBayes type methods

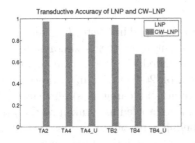 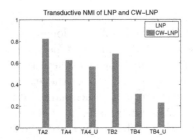

Fig. 4. Transductive results of LNP type methods

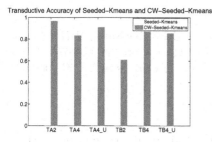
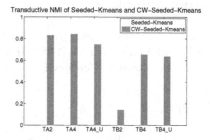

Fig. 5. Transductive results of Seeded-Kmeans type methods

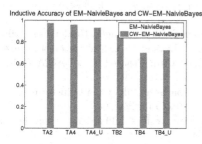
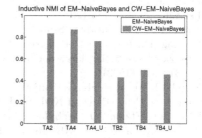

Fig. 6. Inductive results of EM-NaiveBayes type methods

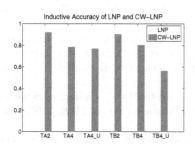
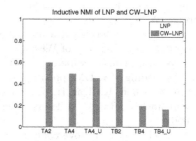

Fig. 7. Inductive results of LNP type methods

5   Conclusions and Future Work

A novel labeling strategy to derive more prior information with the limited labeled data is presented in this paper. Making full use of the extracted words,
the proposed strategy can provide semi-supervised learning algorithms more
prior information. Numerical experiments have shown that the new strategy
is helpful for semi-supervised categorization and can finally improve the learning performance. In the future, we will try to use more background knowledge
(i.e., Ontology) to automatically extract prior information for semi-supervised
learning.

Acknowledgments

The research work in this paper was supported by the National Natural Science Foundation of China (90820013,60875031), 973 project (2007CB311002) and Program for New Century Excellent Talents in University (NCET-06-0078).

References

1. Macskassy, S., Banerjee, A., Davison, B., Hirsh, H.: Human performance on custering web pages: a preliminary study. In: Proceedings of ACM SIGKDD, pp. 264–268 (1998)
2. Friedman, N., Geiger, D., Goldszmidt, M.: Bayesian Network classifier. Machine Learning 29, 131–163 (1997)
3. Vapnik, V.: The Nature of Statistical Learning Theory. Springer, Heidelberg (1995)
4. Cutting, D., Karger, D., Pedersen, J., Tukey, J.: Scatter/Gather: a cluster-based approach to browsing large document collection. In: Proceedings of ACM SIGIR, pp. 318–329 (1992)
5. Steinbach, M., Karypis, G., Kumar, V.: A comparison of document clustering techniques. In: Proceedings of KDD workshop on text mining (2000)
6. Zamir, O., Etzioni, O.: Web document clustering: a feasability demonstration. In: Proceedings of ACM SIGIR, pp. 46–54 (1998)
7. Jing, L., Ng, M., Huang, J.: An entropy weighting k-means algorithm for subspace clsutering of high-dimensional sparse data. IEEE transactions on knowledge and data engineering 19(8), 1026–1041 (2007)
8. Chapelle, O., Zien, A., Scholkopf, B.: Semi-supervised learning. MIT Press, Cambridge (2006)
9. Zhu, X.: Semi-supervised learning literature survey. Computer Sciences Technical Report 1530, University of Wisconsin-Madison (last modified on July19, 2008)
10. Hotho, A., Staab, S., Stumme, G.: Ontologies Improve Text Document Clustering. In: Proceeding of ICDM, pp. 19–22 (2003)
11. Zhou, Z., Chen, K., Dai, H.: Enhancing relevance feedback in image retrieval using unlabeled data. ACM Transactions on information systems 24(2), 219C–244C (2006)
12. Li, M., Zhou, Z.: Improve computer-aided diagnosis with machine learning techniques using undiagnosed samples. IEEE Transactions on systems, man and cybernetics-part A: systems and humans 37, 1088C–1098C (2007)
13. Nigam, K., McCallum, A., Thrun, S., Mitchell, T.: Text classification from labeled and unlabeled documents using EM. Maching Learning 39, 103–134 (2000)
14. Fujino, A., Ueda, N., Saito, K.: A hybrid generative/discriminative approach to semi-supervised classifier desigen. Proceedings of the 20th AAAI (2005)
15. Basu, S., Banerjee, A., Mooney, R.: Semi-supervised clustering by seeding. In: Proceedings of the 19th ICML, Sydney, Australia, pp. 27–34 (2002)
16. Wagstaff, K., Cardie, C., Rogers, S., Schroedl, S.: Constrained K-means Clustering with Background Knowledge. In: Proceedings of the 18th ICML, pp. 577–584 (2001)
17. Basu, S., Bilenko, M., Mooney, R.: A probabilistic framework for semi-supervised clustering. eattle, WA, pp. 59–68 (2004)
18. Zhou, D., Bousquet, O., Lal, N., Weston, J., Scholkopf, B., Olkopf, B.: Learning with Local and Global Consistency. Advances in Neural Information Processing Systems 16 (2004)

19. Zhu, X., Ghahramani, Z., Lafferty, J.: Semi-Supervised Learning Using Gaussian Fields and Harmonic Functions. In: Proceedings of the 20th ICML (2003)
20. Wang, F., Zhang, C.: Label propagation through linear neighborhoods. IEEE Transactions on Knowledge and Data Engineering 20(1) (2008)
21. Joachims, T.: Transductive Inference for Text Classification using Support Vector Machines. In: Proceedings of the 16th ICML (1999)
22. Belkin, M., Niyogi, P., Sindhwani, V.: Manifold regularization: a geometric framework for learning from labeled and unlabeled examples. Journal of machine learning research 7, 2399–2434 (2006)
23. Li, Y., Kwok, J., Zhou, Z.: Semi-supervised learning using label mean. In: Proceedings of the 26th ICML, Canada (2009)
24. Blum, A., Mitchell, T.: Combining labeled and unlabeled data with co-training. In: Proceedings of the 11th COLT, Wisconsin, MI, pp. 92–100 (1998)
25. Zhou, Z., Zhan, D., Yang, Q.: Semi-supervised learning with very few labeled training examples. In: Proceedings of the 22th AAAI (2007)
26. Zhou, Z., Li, M.: Tri-training: Exploiting unlabeled data using three classifiers. IEEE Transactions on Knowledge and Data Engineering 17(11), 1529–1541 (2005)
27. Jain, K.: Data Clustering: 50 Years Beyond K-Means. Springer, Berlin (2008)
28. Hotho, A., Staab, S., Stumme, G.: Wordnet improves text document clustering. In: Proceedings of ACM SIGIR workshop on semantic Web (2003)
29. Jing, L., Zhou, L., Ng, M., Huang, J.: Ontology-based distance measure for text clustering. In: Proceedings of SIAM DM workshop on text mining (2006)
30. Banerjee, S., Ramanathan, K., Gupta, A.: Clustering short texts using wikipedia. In: Proceedings of the 30th ACM SIGIR, pp. 787–788 (2007)
31. Schapire, R., Rochery, M., Rahim, M., Gupta, N.: Incorporating prior knowledge into boosting. In: Proceedings of 19th ICML, pp. 538–545 (2002)
32. Wu, X., Srihari, R.: Incorporating prior knowledge with weighted margin support vector machines. In: Proceedings of 10th ACM SIGKDD, pp. 326–333 (2004)
33. Druck, G., Mann, G., McCallum, A.: Learning from labeled features using generalized expectation criteria. In: Proceedings of 31th annual international ACM SIGIR, pp. 595–602 (2008)
34. Berry, M., Castellanos, M.: Survey of Text Mining II: Clustering, Classification, and Retrieval. Springer, Heidelberg (2008)

A Competitive Learning Approach to Instance Selection for Support Vector Machines

Mario Zechner* and Michael Granitzer

Know-Center and Graz University of Technology
Inffeldgasse 21a
8010 Graz, Austria
{mzechner,mgrani}@know-center.at

Abstract. Support Vector Machines (SVM) have been applied success-fully in a wide variety of fields in the last decade. The SVM problem is formulated as a convex objective function subject to box constraints that needs to be maximized, a quadratic programming (QP) problem. In order to solve the QP problem on larger data sets specialized algorithms and heuristics are required. In this paper we present a new data-squashing method for selecting training instances in support vector learning. Inspired by the growing neural gas algorithm and learning vector quantization we introduce a new, parameter robust neural gas variant to retrieve an initial approximation of the training set containing only those samples that will likely become support vectors in the final classifier. This first approximation is refined in the border areas, defined by neighboring neurons of different classes, yielding the final training set. We evaluate our approach on synthetic as well as real-life datasets, comparing run-time complexity and accuracy to a random sampling approach and the exact solution of the support vector machine. Results show that runtime-complexity can be significantly reduced while achieving the same accuracy as the exact solution and that furthermore our approach does not not rely on data set specific parameterization of the sampling rate like random sampling for doing so. Source code, binary executables as well as the reformatted standard data sets are available for download at http://www.know-center.tugraz.at/forschung/ knowledge_relationship_discovery/downloads_demos/ sngsvm_source_executables

1 Introduction

Support vector machines (SVM) [1] have been applied successfully in a wide variety of fields in the last decade. In their simplest form SVMs are essentially linear classifiers that try to find a maximum margin hyperplane which separates training samples of two different classes. The SVM problem is formulated as a convex objective function subject to box constraints that needs to be maximized, a quadratic programming (QP) problem. The final solution is based on just a subset of the training samples called support vectors.

---

* Student at Graz University of Technology.

D. Karagiannis and Z. Jin (Eds.): KSEM 2009, LNAI 5914, pp. 146–157, 2009.

Solving this QP Problem is usually quadratic in time and storage. This is especially prohobitive in large scale domains such as data mining where millions of documents have to be processed. Various attempts at improving the runtime and space requirements have been made. One line of work focuses on reducing the training set to samples which will likely become support vectors of the final solution. Shin and Cho [2] select training samples based on their neighborhood. They exploit the fact that training samples with heterogeneous neighborhoods in terms of a neighbor's class label are likely to be near the hyperplane induced by the SVM and therefore being potential support vectors. As searching the k-nearest neighbors for all training samples has a time complexity of $O(n^2)$ they introduce an approximation schema that brings the time complexity down to $O(b * n)$ where b is the number of samples with heterogeneous neighborhoods. Their approach seems to have problems with well separated training sets. In [3] Yu et al. use clustering as a preprocessing step. The training set is clustered and the resulting centroids provide the training set for the initial SVM. The results are refined by replacing centroids near the decision surface with their associated training samples and retraining the SVM on those training samples. This procedure is repeated until all training samples have been used. Due to the cluster expansion criterion used their approach only works with linear kernels. Another similar but more general algorithm based on clustering was presented by Boley and Cao [4].

In this paper we follow the notion of speeding up SVM training by reducing the number of training samples. Our focus lies on a fast method that has better runtime complexity than $O(n^2)$ and is robust to parameter selection. Our sparsifying neural gas algorithm (SNG), a combination of learning vector quantization [5] and the growing neural gas algorithm, [6] automatically grows a topological network to reconstruct the input data set. The number of final neurons (i.e. nodes in the network) is determined on the fly. To select the relevant training examples for the SVM we first partition the training set with one SNG for each class. This takes a single pass over the complete training set. Next we merge the two sets of neurons and each neuron receives the label of the class it was trained on. In a second pass over the complete training set we establish connections between the closest two neurons for each training sample. This induces a subset of the Delaunay triangulation which we exploit to determine the border regions between classes. By expanding neurons of different class labels with the training samples in their "receptive field", border regions become densely populated while non border regions are thinned out.

The resulting number of training samples becomes the final SVM training set. Results show that this approach yields faster training time while maintaining most if not all of the support vectors of the full SVM solution. In particular, the contributions of our work are as follows:

1. A new, parameter robust neural gas variation called sparsifying neural gas (SNG) that reliably approximates the topology of the dataset in only 2 passes.
2. An instance selection schema based on the SNG for speeding up support vector machine learning while retaining the SVM's accuracy
3. An empirical study on the performance of our approach comparing it to random sampling as a baseline as well as the full solution of the SVM.

We introduce the SNG in section 2 followed by a review of the SVM formalism in section 3. Section 4 discusses the SVM instance selection in detail while we report results of our experiments in section 5. We conclude and give an outlook on future work in section 6.

2 Sparsifying Neural Gas

In its simplest form data reduction can be accomplished by random sampling and results in [7] indicate that this is the best overall model independent sample selection method. Random sampling is capable to reliably remove noise from a data set. However, due to its stochastic nature random sampling may yield unreliable results and might ignore small clusters of important samples. Furthermore, the sampling rate has to be provided as a parameter highly dependent on the specific data set.

A general approach to data reduction is classical vector quantization (VQ) [8]. In vector quantization a set of codebook vectors is determined maximizing a cost function of the quantization error. Usually a the number of code book vectors and thus the level of compression has to be specified a-priori. The growing neural gas (GNG) [9] by Fritzke overcomes this and other problems by adaptively increasing the number of codebook vectors. However, the decision on the final number of codebook vectors is still necessary as the GNG has a fixed insertion rate and other parameter specific properties. Also, a proof of convergence for the GNG is non-trivial given it's adaptive nature.

With the SNG we try to combine some of the favorable features of various VQ methods. For reducing the time complexity and applying the SNG to large data set our goal is to minimize the number of passes over the data. Competitive learning as used in many online vector quantization methods is our starting point. It has the flexibility to quickly adapt to relevant regions of the data set. For the SNG we utilize the Winner-Takes-All strategy starting with a given set $\mathcal{W}$ of k neurons each represented by a weight vector $\mathbf{w_i} \in \mathcal{W} \subset \mathbb{R}^d, 1 \leq i \leq k$ and a data set of n samples $\mathcal{X} = \{\mathbf{x_j}, 1 \leq j \leq n\}$. For each sample $\mathbf{x_s}$ drawn form $\mathcal{X}$ we select the nearest (winning) neuron $\mathbf{w_{win}}$ by

$$\mathbf{w_{win}} = \min_j ||\mathbf{w_j} - \mathbf{x_s}|| \tag{1}$$

and update it's weight vector as $\mathbf{w'_{win}} = \mathbf{w_{win}} + \mathbf{w_\Delta}$ with

$$\mathbf{w_\Delta} = \eta(\mathbf{x_s} - \mathbf{w_{win}}) \tag{2}$$

where η is the so called learning rate driving the adaptiveness of neurons to new input samples.

As we want a growing network which adapts to the data set topology an insertion criteria is needed. In its original version, Fritzke's GNG keeps a local error statistic for each neuron by tracking its online squared error. This error gets updated for a winning neuron by

$$error_{new}(\mathbf{w_{win}}) = error_{old}(\mathbf{w_{win}}) + ||\mathbf{w_{win}} - \mathbf{x_s}||^2 \tag{3}$$

Additionally, each time a fixed number of samples has been presented to the network a new neuron is inserted halfway between the neuron with the maximum accumulated error and its topological neighbor with the highest error. As outlined above the fixed insertion rate is reducing the flexibility of the network especially when only a single pass over the data is performed. Hence, the GNG parameterization becomes very data set specific. The newly created neuron does not explore new space but rather gives its "parent" neurons the opportunity to adapt to new regions. We therefore introduce a simple heuristic that removes the fixed insertion rate and lets the new neuron instantaneously explore new regions.

In the SNG we extend the fixed insertion rate by introducing another local statistic for each neuron, namely the number of samples a neuron was the closest to. We call this the hit statistic. Each time a neuron is determined as the winner for an input sample we update its hit statistic simply by

$$hits_{new}(\mathbf{w_{win}}) = hits_{old}(\mathbf{w_{win}}) + 1 \tag{4}$$

Combined with the accumulated squared error in equation 3 we can calculate an approximate mean squared error:

$$mse(\mathbf{w_i}) = error(\mathbf{w_i})/hits(\mathbf{w_i}) \tag{5}$$

Of course we have to account for the case $hits(\mathbf{w_i}) = 0$. As a winning neuron is changing its position and therefore the distance to previously seen samples the mse is a very rough approximation typically underestimating the true mse.

We interpret the online mse as the approximate squared radius of the perceptive field of a neuron, i.e. the average distance from the neuron to the samples it won. A new neuron is inserted if

$$mse(\mathbf{w_{win}}) < ||\mathbf{w_{win}} - \mathbf{x_s}||^2 \tag{6}$$

or in words: if the sample lies outside the perceptive field of the winning neuron. The perceptive field of a new neuron as well as its weight vector have to be initialized. We chose a very simple mechanism here and initialize the position $\mathbf{w_{new}}$ of the new neuron to the current samples position. The new neuron inherits the perceptive field of its "parent" neuron $\mathbf{w_{win}}$:

$$\mathbf{w_{new}} = \mathbf{x_s}$$
$$error(\mathbf{w_{new}}) = error(\mathbf{w_{win}}) \tag{7}$$
$$hits(\mathbf{w_{new}}) = hits(\mathbf{w_{win}})$$

The network topology is initialized by randomly selecting two samples as neurons and setting their $error$ and $hits$ to zero. We consider the perceptive field of a winning neuron only if it has seen at least ν samples to avoid growing the network in an to early stage. Note that the perceptive field will only grow while the neuron has not seen ν samples yet and shrink otherwise. For avoiding overly dense regions of neurons we introduce a repulsive behavior between neighboring neurons based on their perceptive field. In the course of determining the winning neuron $\mathbf{w_{win}}$ for a sample we also calculate the second nearest neuron $\mathbf{w_{sec}}$. If the receptive fields of the two neurons overlap we want the second winning neuron to move away from the winning neuron by a small fraction called the repulsion rate ρ. The repulsion criterion is given by:

$$mse(\mathbf{w_{win}}) + mse(\mathbf{w_{sec}}) > ||\mathbf{w_{win}} - \mathbf{w_{sec}}||^2 \tag{8}$$

In case the repulsion criterion is meet we update the second neuron vector as $\mathbf{w'_{sec}} = \mathbf{w_{sec}} + \mathbf{r_\Delta}$ with

$$\mathbf{r_\Delta} = -\rho(\mathbf{w_{win}} - \mathbf{w_{sec}}) \tag{9}$$

This step reduces the overlap of the perceptive fields of two neurons and improves the distribution of the neurons over the data set. It should be noted that occasionally neurons can exist having no sample associated with them. We simply remove those neurons in the final step of the SNG. When all samples of the data set have been presented to the network we perform a second pass over all samples to determine the induced Delaunay triangulation of the network. We again determine the nearest and second nearest neuron for each sample and construct an edge between those neurons. The emerging graph then represents the overall topology of the network. Additionally we note for each sample to which neuron it belongs to. The final result of the SNG is therefore the set of final neurons $\mathcal{W}$, a set of edges $e_{ij} \in \mathcal{E}$ denoting the induced Delaunay triangulation as well as a set $\mathcal{C}_i$ for each neuron $\mathbf{w_i}$ containing the samples $\mathbf{x_s}$ the neuron is closest to. Algorithm 1 depicts the complete procedure of the SNG.

Due to the nature of the SNG there are a couple of caveats. From the insertion rule in inequality 6 it becomes clear that the SNG will aggressively insert new nodes. The SNG does not strictly optimize a cost function as in the case of LVQ or NG and is similar to GNG in that respect.

One downside currently is the absence of a strict theoretical convergence proof of our approach. Additionally, the SNG is order dependent to a small degree. Some areas might be unneccesarily more populated than others due to the approximative nature of the insertion criterion. Also, due to the insertion mechanism the network will adapt to noise. This is a feature shared by many VQ methods and is rather hard to overcome, especially in the case of online VQ.

Finally, three parameters have to be specified, namely the learning rate η, the repulsion rate ρ and the number of samples ν to be seen minimally by a neuron to consider its perceptive field for insertion. In our experiments we have tested various combinations of these parameters ranging from 0.005 to 0.1 for η resp. ρ and from 2 to 20 for ν. We observed that the results of the algorithm do not depend strongly on the choice of parameters. We chose $\eta = 0.05$, $\rho = 0.005$ and $\nu = 5$ for all of our experiments.

Algorithm 1.. Simplifying Neural Gas

Input: $\mathcal{X}, \eta, \rho, \nu$
Output: $\mathcal{W}, \mathcal{C}, \mathcal{E}$ with
  $\mathcal{W} = \{\mathbf{w_1}, ..., \mathbf{w_k}\}$
  $\mathcal{C} = \{\mathcal{C}_1, ..., \mathcal{C}_k\}$ with $\mathcal{C}_i = \{\mathbf{x_j} \in \mathcal{X} : ||\mathbf{w_i} - \mathbf{x_j}|| < ||\mathbf{w_s} - \mathbf{x_j}||, i \neq s\}$
  $\mathcal{E} = \{e_{ij} : \exists \mathbf{x_s} \in \mathcal{X} s.t.$
    $||\mathbf{w_i} - \mathbf{x_s}|| < ||\mathbf{w_u} - \mathbf{x_s}||, \forall u \neq i, \mathbf{w_i}, \mathbf{w_u}, \in \mathcal{W}$ *and*
    $||\mathbf{w_j} - \mathbf{x_s}|| < ||\mathbf{w_v} - \mathbf{x_s}||, \forall v \neq j, \mathbf{w_j}, \mathbf{w_v}, \in \mathcal{W} \setminus \{\mathbf{w_i}\}\}$

Begin SNG
  $\mathcal{C} = \emptyset, \mathcal{E} = \emptyset, \mathcal{W} = \{$ chose 2 random samples from $\mathcal{X} \}$
  $\forall_{1 \leq i \leq 2} error(\mathbf{w_i}) = 0, hits(\mathbf{w_i}) = 0$
  for all $\mathbf{x_s} \in \mathcal{X}$ **do**
    $\mathbf{w_{win}} = argmin_{w_i}||\mathbf{w_i} - \mathbf{x_s}||, \mathbf{w_i} \in \mathcal{W}$
    $\mathbf{w_{sec}} = argmin_{w_j}||\mathbf{w_j} - \mathbf{x_s}||, \mathbf{w_j} \in \mathcal{W} \setminus \{\mathbf{w_{win}}\}$
    if $hits(\mathbf{w_{win}}) > \nu \wedge mse(\mathbf{w_{win}}) < ||\mathbf{w_{win}} - \mathbf{x_s}||^2$ **then**
      $\mathbf{w_{new}} = \mathbf{x_s}$
      $error(\mathbf{w_{new}}) = error(\mathbf{w_{win}})$
      $hits(\mathbf{w_{new}}) = hits(\mathbf{w_{win}})$
    else
      $\mathbf{w_{win}} = \mathbf{w_{win}} + \eta(\mathbf{x_s} - \mathbf{w_{win}})$
      $error(\mathbf{w_{win}}) = error(\mathbf{w_{win}}) + ||\mathbf{w_{win}} - \mathbf{x_s}||^2$
      $hits(\mathbf{w_{win}}) = hits(\mathbf{w_{win}}) + 1$
      if $mse(\mathbf{w_{win}}) + mse(\mathbf{w_{sec}}) > ||\mathbf{w_{win}} - \mathbf{w_{sec}}||^2$ **then**
        $\mathbf{w_{sec}} = \mathbf{w_{sec}} - \rho(\mathbf{w_{win}} - \mathbf{w_{sec}})$
      end if
    end if
  end for
  for all $\mathbf{x_s} \in \mathcal{X}$ **do**
    $\mathbf{w_i} = argmin_{w_i}||\mathbf{w_i} - \mathbf{x_s}||, \mathbf{w_i} \in \mathcal{W}$
    $\mathbf{w_j} = argmin_{w_j}||\mathbf{w_j} - \mathbf{x_s}||, \mathbf{w_j} \in \mathcal{W} \setminus \{\mathbf{w_i}\}$
    $\mathcal{E} = \mathcal{E} \cup e_{ij}, \mathcal{C}_i = \mathcal{C}_i \cup x_s$
  end for
End SNG

---

3 Support Vector Machine Learning

In this section we will briefly review SVM learning, first from a geometrical view followed by its formulation as an optimization problem. Our work is motivated by the geometrical interpretation of SVMs.

In supervised learning scenarios for binary classification we are typically given a set of vectors $\mathcal{X} = \{\mathbf{x_1}, ..., \mathbf{x_n}\}$ with $\mathbf{x_i} \in \mathbb{R}^d$ accompanied by a set of labels $\mathcal{Y} = \{y_1, ..., y_n\}$ with $y_i \in \{-1, 1\}$ where $1 \leq i \leq n$ and d is the number of dimensions. This two sets combined are called the training set from which we want to learn a discriminative function $f(\mathbf{x})$ that returns the correct label for a sample $\mathbf{x}$.

SVMs arrive at the function $f(\mathbf{x})$ by constructing a hyperplane that separates the two classes in the training set with the largest margin. Depending on which side of the hyperplane $\mathbf{x}$ lies $f(\mathbf{x})$ will return the predicted label of the sample $\mathbf{x}$. The solution of the hyperplane is based on vectors in the border regions between the two classes. This vectors are called support vectors. Additionally they can cope with non-separable cases by relaxing the constraints of the optimization problem for separable cases and achieve non-linear classification behavior by incorporating the so called kernel-trick.

A large number of training samples does not contribute to the final solution of the SVM. The discriminating hyperplane is located in the border regions of the two classes. Therefore it is natural to only provide those training samples to the SVM that are within this border region and omit the rest of the samples or replace large sets of them by a representative. The solution of the SVM should not differ provided we present it all support vectors.

4   SNG Instance Selection

The goal of our work is it to effectively reduce the size of the training set for SVM learning. Reflecting section 3, SVM's chose only samples along the discriminative hyperplane to become support vectors. Also, the hyperplane effectively separates the convex hulls of the two classes of the training set. We therefore conclude that only areas of neighboring classes close to each other have to be considered. After discussing how to determine those areas in input space $\mathbb{R}^d$ by using neighborhood properties, we show that the derived properties also hold in feature space when using non-linear kernels.

In our approach we model each class of the training set with a separate instance of SNG, called class specific SNG in the following. We name the set of neurons grown on the positive examples $\mathcal{W}^+$ and the set of neurons generated on the negative training examples $\mathcal{W}^-$. We merge the retrieved sets W^+ and W^- into a final set $\mathcal{W}$ for which we calculate the induced Delaunay triangulation $\mathcal{E}$ in a second pass over the complete training set. As we are only interested in the triangulation induced by $\mathcal{W}$ we omit this calculation step when we first grow the class specific SNGs. While calculating the induced Delaunay triangulation we also note for each sample which neuron of the combined set $\mathcal{W}$ it was closest to resulting in a set $\mathcal{C}_i$ of training samples $\mathbf{x_i}$ for each neuron $\mathbf{w_i} \in \mathcal{W}$. We then find the set of all edges $e_{ij} \in \mathcal{E}$ connecting neurons $\mathbf{w_i}$ and $\mathbf{w_j}$ that do not originate from the same class specific SNG. Topologically those neurons and their respective sets of samples $\mathcal{C}_i$ and $\mathcal{C}_j$ make up the part of the border region between the two classes. We merge $\mathcal{C}_i$ and $\mathcal{C}_j$ to construct the final training set $\mathcal{T}$. Additionally, neurons not in the border region are also added to the training set, but their associated examples are ignored. Finally we train the SVM on the reduced training set $\mathcal{T}$. Figure 1 depicts this process on a two-dimensional toy example. Note that our approach is currently restricted to the binary case. However, an extension to the multi-class case should be straightforward.

Due to space constraints we refrain from giving a proof that this schema holds for various kernels and refer the reader to [10]. Therein the authors show that the neighborhood properties in input space also hold when radial basis function kernels or polynomial kernels are used.

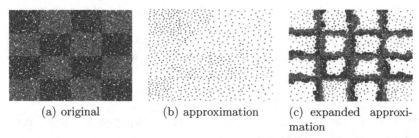

(a) original   (b) approximation   (c) expanded approximation

Fig. 1. (a) The checkerboard dataset. (b) A sparse representation of the checkerboard dataset produced by growing one SNG for each class. (c) The expanded representation of the checkerboard dataset. Neurons of opposite classes that are the nearest neighbor to a sample get replaced by all samples they are nearest to respectively.

5   Experiments

We tested our approach on synthetic as well as standard test data sets using 10-fold cross validation for every experiment. Each data set was split into 10 stratified subsets and all classifiers were trained on the same fold combinations. We compare our approach with the complete SVM solution on the full dataset as well as to stratified random sampling with a sampling rate of 10%,30%, 50% and 70%. In addition we add a sampling rate R* for each run. R* is determined based on the number of examples selected by our SNG based approach and thus should provide a direct comparison between SNG and Random Sampling.

As evaluation criterions we use accuracy (*Accuracy*), the number of support vectors (*SVs*) generated and the time it took to train (*Total Time*) the SVM. We present the mean of these values for each data set as well as their standard deviation. Additionally we report the time taken for growing the SNGs (*SNG Time* and the percentage of the original data set taken for training (*Samples*). We exclude I/O times from all of our timings.

For all data sets we used the RBF kernel. Instead of specifying sigma we replace the term $\frac{1}{2\sigma^2}$ with γ. Table 1 gives an overview of the data sets and the parameters used for training. For the SNGs we used $\eta = 0.05$, $\rho = 0.005$ and $\nu = 5$ for all data sets.

The system was implemented and tested in Java using the Sun Server VM at version 1.6.0.7. As an SVM solver we utilized the Java implementation of LIBSVM [11] at version 2.88. We did not enable shrinking. Our test system consisted of an eight-core Intel Xeon E5420 CPU at 2.5Ghz with 32GB of RAM using only a single core for the test runs. Data sets and source code are available for download[1].

We tested on the following data sets:

– **Adult** a census income data set with 48842 samples and 14 dimensions. We only used the training set for cross validation consisting of 32560 samples.

---

[1] http://en.know-center.at/forschung/knowledge_relationship_discovery/
downloads_demos

- **Banana** a synthetic two-dimensional dataset with a lot of noise and overlap of the class boundaries.
- **Checkboard** a synthetic two-dimensional data set as depicted in figure 1(a)
- **Pendigits** a handwritten digits data set. We tested digit 2 against the rest in our experiments.
- **USPS** a postal zip code data set consisting of images with 16x16 pixels. We tested class "0" against the rest
- **Waveform noise** a synthetic data set with 20 dimensions of real data and 20 dimension of uniform noise.

Table 1. The data sets experiments were performed on and the parameters used

Name	Instances	Dimensions	γ	C
Adult	32560	122	0.005	100
Banana	5299	2	0.5	316
Checkerboard	100000	2	50	100
Pendigits	7493	16	0.0005	100
USPS	9298	256	2	10
Waveform	4999	40	0.05	1

Table 2 presents our experimental results on the above described data sets.

On the Adult data set our approach reduced the data set to about 65% of the original size and outperformed random sampling as well as the full solution in terms of precision. In respect to the full SVM our approach was around 400 seconds faster on average, a 22% speedup. The number of support vectors roughly equals that of the full SVM solution indicating that the reduced data set contained nearly all of them. Our preprocessing approach via the SNGs takes up around 5% of the total runtime which is acceptable.

In case of the banana data set we find a similar behavior. The data set is reduced to around 55% of the original size while preserving all support vectors. We slightly outperform the full SVM in terms of precision while being around 30% faster. Again the preprocessing only takes roughly 5% of the total runtime.

For the checkerboard data set we are again on par with the full SVM solution. The data set is reduced to around 30% of its original size and still preserves all of the support vectors. The runtime is reduced to roughly a third of that of the full SVM solution. Interestingly 70% random sampling does not comes close to this result and has a bigger total runtime than our approach. The preprocessing stage does take up around 23% of the total runtime which is due to the SVM converging faster as the data set is more easily seperable than the adult or banana data set.

On the Pendigits data set we can observe a very interesting behavior. Our approach comes in second in terms of precision, however the the difference is nearly neglect able. The data set is reduced to 18% of its original size yielding a much sparser solution than the full SVM. The number of support vectors is

Table 2. Results on various data sets. R10 to R70 are the results for random sampling from 10% to 70%, R* is the result for random sampling with the same number of samples as the SNG sampling. SSVM is the result for the SNG based sampling approach and SVM is the full solution.

	Accuracy	*Total Time*	*SVs*	*Samples (%)*	*SNG Time*
Adult					
R10	0.8414 ± 0.0059	7.3610	1061 ± 35	10	-
R30	0.8432 ± 0.0064	61.3676	3161 ± 38	30	-
R50	0.8461 ± 0.0058	243.4072	5222 ± 50	50	-
R70	0.8474 ± 0.0058	545.9942	7234 ± 54	70	-
R*	0.8463 ± 0.0060	593.8760	7430 ± 126	65,1296	-
SSVM	$\mathbf{0.8482 \pm 0.0059}$	857.0005	10115 ± 39	65,1296	47,2251
SVM	0.8479 ± 0.0060	1245.3783	10222 ± 28	100	-
Banana					
R10	0.8883 ± 0.0078	0.0506	107 ± 11	10	-
R30	0.9008 ± 0.0081	0.1826	312 ± 14	30	-
R50	0.9038 ± 0.0092	0.4291	527 ± 19	50	-
R70	0.9038 ± 0.0080	0.8235	730 ± 21	70	-
R*	0.9034 ± 0.0107	0.6363	639 ± 32	55,1293	-
SSVM	$\mathbf{0.9060 \pm 0.0074}$	1.0889	1045 ± 8	55,1293	0,0622
SVM	0.9059 ± 0.0073	1.5763	1046 ± 8	100	-
Checkerboard					
R10	0.9912 ± 0.0007	0.9182	441 ± 11	10	-
R30	0.9955 ± 0.0007	4.7641	933 ± 21	30	-
R50	0.9970 ± 0.0006	13.1338	1348 ± 10	50	-
R70	0.9976 ± 0.0006	25.4859	1687 ± 16	70	-
R*	0.9960 ± 0.0010	6.0691	1000 ± 28	29,6086	-
SSVM	$\mathbf{0.9980 \pm 0.0003}$	17.5875	2156 ± 5	29,6086	4,0748
SVM	0.9980 ± 0.0004	50.6111	2156 ± 6	100	-
Pendigits					
R10	0.9939 ± 0.0052	0.0875	372 ± 11	10	-
R30	0.9975 ± 0.0028	0.3271	704 ± 16	30	-
R50	0.9983 ± 0.0013	0.6674	878 ± 16	50	-
R70	0.9981 ± 0.0015	1.0493	1012 ± 32	70	-
R*	0.9969 ± 0.0019	0.1790	564 ± 37	18,2717	-
SSVM	0.9983 ± 0.0012	0.6843	606 ± 38	18,2717	0,4887
SVM	$\mathbf{0.9984 \pm 0.0013}$	1.8018	1151 ± 11	100	-
USPS					
R10	0.9922 ± 0.0028	0.2732	90 ± 6	10	-
R30	0.9949 ± 0.0025	1.3710	150 ± 12	30	-
R50	0.9963 ± 0.0022	2.8140	183 ± 8	50	-
R70	0.9959 ± 0.0024	4.6877	217 ± 8	70	-
R*	0.9940 ± 0.0029	0.8177	121 ± 5	19,6429	-
SSVM	0.9957 ± 0.0017	4.8140	207 ± 9	19,6429	3,3970
SVM	$\mathbf{0.9964 \pm 0.0017}$	9.1343	261 ± 7	100	-
Waveform					
R10	0.8499 ± 0.0154	0.0771	421 ± 3	10	-
R30	0.8868 ± 0.0139	0.5651	1057 ± 9	30	-
R50	0.8928 ± 0.0106	1.4878	1561 ± 16	50	-
R70	0.8970 ± 0.0079	2.9207	2023 ± 11	70	-
R*	0.8986 ± 0.0107	4.2489	2345 ± 86	76,4933	-
SSVM	$\mathbf{0.9014 \pm 0.0085}$	5.3144	2466 ± 28	76,4933	0,5186
SVM	0.8992 ± 0.0089	6.6437	2642 ± 8	100	-

nearly halfed while the runtime is around a third of that of the full SVM. Interestingly randomly sampling 50% of the data set yielded a comparable precision but a more complex model in terms of the number of support vectors. The preprocessing stage takes a lot of the total runtime, in this case 71%. This is again due to the SVM being able to converge rapidly as the classes are easily seperable.

Our approach performed the worst on the USPS data set. Even tough it only took half of the time it took to solve the full SVM problem it was outperformed by random sampling at 50% and 70% both in terms of runtime and precision. The SNG takes up around 70% of the total runtime as calculating the euclidian distance in such highdimensional spaces is expensive. We are however still outperforming random sampling at 10% and 30% in terms of precision as well as random sampling with the same number of samples as our approach.

Finally on the Waveform data set our approach performs again better than the full SVM solution both in terms of precision and runtime. This data set is very noisy so the SNG approach only reduces the data set by around 24%. The number of support vectors is slightly less than that of the worse performing full SVM. Random sampling does not perform very well on this data set overall. As with the adult data set our approach can cope very well with the noise and even produces a solution outperforming the full SVM. The runtime is reduced by around a sixth of that of the full SVM which is not enormous but acceptable given that the solution performs better.

6   Conclusion and Future Work

We could demonstrate that our approach reduces the runtime for solving SVM problems while maintaining or improving the classification performance of the resulting classifier. Especially noisy data sets are handled very well by our approach. The SNG has a faster training time than the full SVM while achieving similar accuracy levels. Random Sampling is usually faster in training, but the sampling rate has to be known a-priori in order to reliably achieve a similar accuracy than the full SVM. Given the results on the USPS data set we consider adapting the SNG to remove the costly distance measurements which increase the runtime for high dimensional problems. One approach would be similar in spirit to spherical k-means clustering where the dot product is used instead of the euclidean distance. This would also open up the domain of text classification to our approach. To further reduce the number of samples provided to the SVM we could try to exploit the induced Delaunay triangulation better. Instead of expanding the neuron by all its associated samples we could further divide this set of samples. However, this adds additional runtime cost which would have to be balanced. The SNG itself could also be used for other machine learning related tasks. We will investigate its use as a fast partitioning method to speed up k-NN classification. This approach would fall in the domain of approximative k-NN search. Hierarchies of SNGs could further speed up the the process. Yet another use for the SNG could be in clustering applications and approximate singular value decomposition which also suffers from a high runtime complexity.

Acknowledgement

The Know-Center is funded within the Austrian COMET Program - Competence Centers for Excellent Technologies - under the auspices of the Austrian Ministry of Transport, Innovation and Technology, the Austrian Ministry of Economics and Labor and by the State of Styria. COMET is managed by the Austrian Research Promotion Agency FFG.

References

1. Vapnik, V.N.: Statistical Learning Theory. Wiley, Chichester (1998)
2. Shin, H., Cho, S.: Fast pattern selection for support vector classifiers. In: Whang, K.-Y., Jeon, J., Shim, K., Srivastava, J. (eds.) PAKDD 2003. LNCS (LNAI), vol. 2637, pp. 376–387. Springer, Heidelberg (2003)
3. Yu, H., Yang, J., Han, J.: Classifying large data sets using svms with hierarchical clusters (2003)
4. Boley, D., Cao, D.: Training support vector machine using adaptive clustering. In: Proceedings of the SIAM International Conference on Data Mining, April 2004, pp. 126–137 (2004)
5. Kohonen, T.: Learning vector quantization. Neural Networks 1(suppl. 1), 3–16 (1988)
6. Fritzke, B.: A growing neural gas network learns topologies. In: Advances in Neural Information Processing Systems, vol. 7, pp. 625–632. MIT Press, Cambridge (1995)
7. Syed, N.A., Liu, H., Sung, K.K.: A study of support vectors on model independent example selection. In: KDD, pp. 272–276 (1999)
8. Gersho, A., Gray, R.M.: Vector quantization and signal compression. Kluwer Academic Publishers, Dordrecht (1992)
9. Fritzke, B.: A growing neural gas network learns topologies. In: Neural Information Processing Systems, pp. 625–632 (1994)
10. Shin, H., Cho, S.: Invariance of neighborhood relation under input space to feature space mapping. Pattern Recognition Letters 26(6), 707–718 (2005)
11. Chang, C.C., Lin, C.J.: LIBSVM: a library for support vector machines (2001), http://www.csie.ntu.edu.tw/~cjlin/libsvm

Knowledge Discovery from Academic Search Engine

Ye Wang[1,2], Miao Jiang[2], Xiaoling Wang[3], and Aoying Zhou[3]

[1] Network Information Center, Second Military Medical University, Shanghai 200433
[2] Department of Computer Science and Engineering, Fudan University, Shanghai 200433
[3] Shanghai Key Laboratory of Trustworthy Computing, Software Engineering Institute,
East China Normal University, Shanghai 200062
{wangee,jiangmiao}@fudan.edu.cn, {xlwang,ayzhou}@sei.ecnu.edu.cn

Abstract. The purpose of a search engine is to retrieve information relevant to a user's query from a big textual collection. However, most vertical search engines, such as Google Scholar and Citeseer, only return the flat ranked list without an efficient result exhibition and knowledge arrangement for given users. This paper considers the problem of knowledge discovery in the literature of search computing. We design some search and ranking strategies to mining potential knowledge from returned search results. A vertical search engine prototype, called Dolphin, is implemented where users are not only getting the results from search engine, but also the knowledge relevant to given query. Experiments show the effectiveness of our approach.

Keywords: Vertical Search engine, Academic Search, Clustering, Ranking, Knowledge discovery.

1 Introduction

Search computing science [20] is a new topic where knowledge discovery and arrangement is a key. Search computing focuses on how to combine and utilize all the search services to get knowledge discovered. We've already have a lot of search services on the web such as different kinds of search engines including general search engines and vertical search engines. In this paper we are discussing how to combine different resources to discover knowledge from a kind of vertical search engine (academic search engine).

Vertical search engine focuses on specialized domain. It integrates a certain kind of information crawled from Web, parse the information intelligently to retrieve the meaningful data, and then present formal and easy-to-understand results to the user. General search engine such as Google or Yahoo could give user a great number of searching results according to the query. However, too many results without good order and organization may puzzle the user. Vertical search engine could bring user more accurate results focused on certain domain. Moreover, it provides more professional templates which could bring user more valuable knowledge not only a flat list of searching results.

D. Karagiannis and Z. Jin (Eds.): KSEM 2009, LNAI 5914, pp. 158–167, 2009.
© Springer-Verlag Berlin Heidelberg 2009

As a kind of vertical search engine, academic search engine is frequently used by researchers. There are many successful academic search engines on WEB already such as GoogleScholar [10], CiteSeer [11], DBLP [12], Rexa [13] and so on. However, most academic search engines return the results to users by a list and users must scan each item/webpage one by one in order to collect and re-organize information based on users' requirements. This procedure is time-consuming even for very experienced researchers. And how to get more useful knowledge from many of these keywords relevant results is still an unsolved problem.

In this paper, we study the knowledge discovery problem in academic search engine and developed a vertical search engine system – Dolphin, which focuses on Academic Web Search. Given a query from user, Dolphin can return the definition of the subject, the categories of the query, and most important, all of this is built online. The Dolphin system supports advanced query template which is used for advanced querying, such as "what are the search trends about the research area; who is the most important researcher in the area".

The rest of this paper is organized as follows. System architecture and components are described in Section 2. Our strategies about obtaining knowledge from academic data are given in Section 3. Related experiments are also given in Section 3, and it's followed by the conclusion in Section 4.

2 Architecture and Data Structure

Dolphin system [15] [19] could provide professional information intelligently according to user's query. System architecture in figure 1 shows that the system is mainly composed by the following four modules.

1. Module of data collecting is responsible for collecting useful information from web, and preprocessing them based on knowledge unit as storage form.
2. Module of data storage stores all the data that collected from web, including information of the papers and the web pages.
3. Module of data center is the bridge of the other modules. It transforms the data format and exchanges data among modules.
4. Query and user interactive module get the input from user, handle the result and return them to the user, and customize the advanced user template.

Most of the data that Dolphin system used is collected from web using a newly implemented topic focused web crawler and all stored data are represented by XML. Currently, in our document collection, there are more than 700000 papers, published during the period from 1970 to 2006. All the data are stored not in web pages but in well-parsed knowledge unit.

We are not using the normal EFS (Extent First Search) to implement our WEB crawler. In our system, we use the topic focused crawling instead of EFS. Different from EFS, in focused crawling we should determine whether a webpage is or not relevant to a specific topic [1] [16], and even we should predict webpage's relevance to a specific topic from its' URL and anchor text [2].

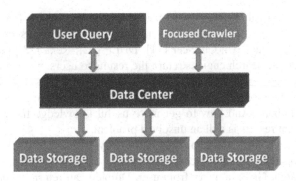

Fig. 1. System Architecture

Generally, we use VSM and TF-IDF to present topic of a document. Weight of each word in the document is evaluated as below.

$$\text{Weight(word)} = \text{tf}_{ik} * \text{IDF}_i = \text{tf}_{ik} * \log(N/n_k + 1)$$

tf_{ik} stands for the times of word Tk's appearance in document Di. IDFi stands for the inverted frequency of documents which has the word Tk. N stands for the numbers of all documents and n_k stands for the numbers of documents which has the word Tk. Considering the factor of length of word, the formula of weight(word) can be normalized as below.

$$\text{Weight(wor d)} = \text{tf}_{ik} * \log(N/n_k + 1) / \sqrt{\sum_{k=1}^{n} (tf_{ik})^2 * \log^2 (N / n_k + 1)}$$

Most of the documents in our system are Web pages (html documents). Html documents are not like normal text documents, they have a lot of format information such as all kinds of tags. All these tags show us some important information of this Web page. For example, tag of "<title>...... </title>" depicts the title of Web page and tag of "<Meta name="keywords" ...>" depicts the keywords of this Web page. Hence we take all these tags into consideration to present the topic of Web page better. We give these tags such as <B>, <Meta name= "description>, <Meta name= "keywords">, <Title>, <A href=>, <h1>... different scores of weight (shown as in Table 1). If a word appears in the important tags, this word should have a "heavier" weight.

For a word Tk, tf_{ik} is evaluated as below.

$$\text{tf}_{ik} = freq_{comm} \times M_{comm} + freq_b \times M_b + freq_{link} \times M_{link} + freq_{title} \times M_{title}$$
$$+ (freq_{m\_dscp} + freq_{m\_keyw}) \times M_{meta} + \sum_{i=1...5} freq_{hi} \times M_{hi}$$

In our system, we are also using a novel score function to evaluate the URLs' correlation about the specific topic. Only the URL whose score is greater than a given threshold is fetched. Three factors contribute to the score. The first one is the content of given web pages, including title, keyword, text and description. The second one is the anchor text of the URL and the third one is the link structure of the connected URLs and pages.

Table 1. Different weight scores of html tags

Tag	Weight	Score
<title>......</title>	M_{title}	1
<Meta name="description" ...>	M_{meta}	0.9
<Meta name="keywords" ...>	M_{meta}	0.9
<h1>......<h1>	M_{h1}	0.8
<h2>......<h2>	M_{h2}	0.7
<h3>......<h3>	M_{h3}	0.6
<h4>......<h4>	M_{h4}	0.5
<h5>......<h5>	M_{h5}	0.4
<A href=...>......</A>	M_{link}	0.3
<B>......</B>	M_b	0.2
Other kinds of tags	M_{comm}	0.1

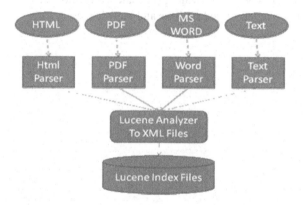

Fig. 2. Data Parsing and Storage

For those satisfied WEB pages, we access them, analyze the content inside, organize them with XML format, and store them into data storage.

Moreover, the crawler should do some preprocessing jobs on the fetched data for further usage which may include erasing the noises, parsing the multi-type data (including html, txt, pdf, msword, etc.) and normalizing the output (as shown in Figure 2). After preprocessing, we'll get much more purified data to build the system's infrastructure. Our system highly depends on the topic related data; hence the selection of the data sources is of great importance. We managed to implement a multi-threaded focused web crawler to fetch the web pages on internet mainly from GoogleScholar, CiteSeer, CSB and DBLP. The paper titles in the DBLP data are used as inputs to query the search engines above, and then we get a large quantity of snippet pages. Our crawler utilizes these snippet pages as seed urls from which the focused crawler begins to get all the information related to the papers. The crawler maintains a link table in database to keep the links' information.

After fetching the urls on internet, the parsed content can be organized into several kinds of XML file, then be stored and indexed. We design 7 types of XML files: Paper Document, Book Document, Thesis Document, WWW Document, Organization Document, People Document and Project Document. Lucene [8] are used to index our xml files (as shown in Figure 2) and to provide a query processor for the users. Lucene is an open source text search engine library which is high-performance and scalable. Crawler continually collects new web pages into system. Thus incremental index are needed. We use Lucene to index some small incremental files into segment index files to speed up the process of indexing.

3   Obtaining Knowledge from Search Results

After crawled and stored a large quantity of academic documents, we provide a simple Google-like query interface in our system. Users could use keywords to query the system to get results he/she may be interested in. Different from traditional academic search engine, we are not only providing the matched results in a ranked list, but also trying to bring users more refined analysis and knowledge.

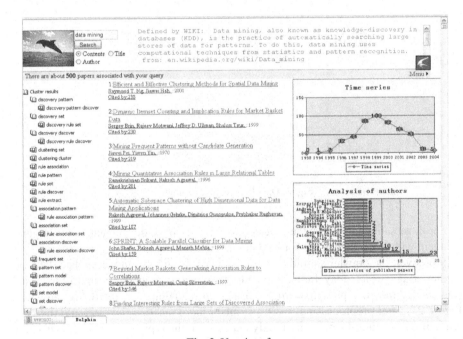

Fig. 3. User interface

3.1   Exhibition of Searching Results

After submitting query keywords, the user will get a comprehensive result shown in Figure 3 which is composed of 5 major parts—the **document list** (middle column in the figure) ranked by the times of being cited, the **clustering result** (left column in the figure), **definition** of the query, **time series** of the published papers on this topic and the **top 20 authors** who are the most prominent ones in this field. And by clicking each label of clustering results, users can get some analysis for each sub-topic respectively, the topics are clustered hierarchically.

3.2 Hierarchical Clustering of Relevant Documents

Hierarchical clustering [7] of relevant documents for a given query is one of the most significant features of the system. Although the open directory project (ODP) is very popular for classification, it is hard to tell the potential clusters below the bottom cluster of ODP. An intelligent automatic hierarchical cluster strategy is expected to resolve the problem here. Document clustering is an automatic grouping of text documents into clusters so that documents within a cluster have high similarity in comparison to one another, but are dissimilar to documents in other clusters. Although many clustering algorithms have been proposed in the literature, most of them do not satisfy the special requirements for clustering documents, suffering from high dimensionality, scalability and easy to browse with meaningful cluster description.

In our system a global frequent itemset hierarchical clustering (FIHC) strategy is used to build clusters, matching the requirement closely. A global frequent itemset is a set of items that appear together in more than a minimum fraction of the whole document set. A global frequent item refers to an item that belongs to some global frequent itemset. A global frequent itemset containing k items is called a global frequent k-itemset. A global frequent item is cluster-frequent in a cluster C if the item is contained in some minimum fraction of documents in C. FIHC uses only the global frequent items in document vectors; thus, the dimensionality is significantly reduced.

The FIHC algorithm can be summarized in three phases: First, construct initial clusters. Second, build a cluster (topic) tree. Finally, prune the cluster tree in case there are too many clusters.

1. Constructing clusters: For each global frequent itemset, an initial cluster is constructed to include all the documents containing this itemset. Initial clusters are overlapping because one document may contain multiple global frequent itemsets. FIHC utilities this global frequent itemset (GFI) as the cluster label to identify the cluster. For each document, the "best" initial cluster is identified and the document is assigned only to the best matching initial cluster. The goodness of a cluster C for a document doc_j is measured by some score function using cluster frequent items of initial clusters. After this step, each document belongs to exactly one cluster. The set of clusters can be viewed as a set of topics in the document set.
2. Building cluster tree: In the cluster tree, each cluster (except the root node) has exactly one parent. The topic of a parent cluster is more general than the topic of a child cluster and they are "similar" to a certain degree. Each cluster uses a global frequent k-itemset as its cluster label. A cluster with a k-itemset cluster label appears at level k in the tree. The cluster tree is built bottom up by choosing the "best" parent at level k-1 for each cluster at level k. The parent's cluster label must be a subset of the child's cluster label. By treating all documents in the child

cluster as a single document, the criterion for selecting the best parent is similar to the one for choosing the best cluster for a document.

3. Pruning cluster tree: The cluster tree can be broad, which becomes not suitable for browsing. The goal of tree pruning is to efficiently remove the overly specific clusters based on the notion of inter-cluster similarity. The idea is that if two sibling clusters are very similar (*Similarity>1*), they should be merged into one cluster.

Using FIHC algorithm, our system could give user not only the cluster list shown in Fig.3, but also a very detailed clusters/hotspots analysis.

3.3 Hits Based Ranking and Searching Results Extension

The goal of our system is to obtain meaningful results from system. For example, when submitting a query of "data mining", users not only want to find the papers which contain the keyword "data mining", but also want to get some more information related to the research area of "data mining". It is a fact that the keyword "data mining" does not appear in some important papers, such as the paper of Rakesh Agrawal, "Mining Association Rules between Sets of Items in Large Databases". Hence the outputted ranked document list from keyword searching method is not good enough. Based on the searching results, we propose and implement HITS-based ranking method for Paper document results.

Considering the citations between two papers as the links, all the linked documents can be described as a directed graph $G = (V;E)$ where node v represents a document and edge e represents a citation in the graph. With this data model, we implement a result ranking algorithm based on HITS [4]. We use ai stands for the authority value of document node i, hi for the hub value and k for the number of iterations.

$$a_i^{(k)} = \sum_{j:e_{ji}\in E} h_j^{(k-1)}$$

$$h_i^{(k)} = \sum_{j:e_{ji}\in E} a_j^{(k-1)}, k = 1,2,3 \ldots$$

Based on the calculated authority value by formula above, we obtain a new ranking order of the result list. For instance, when querying "XML", Dolphin first finds all the paper documents containing the given keywords, and then the top 50 documents which have high citations are picked out to be the root set S in our extended HITS algorithm (as shown in Fig 4).

Then we add papers which are cited by (or citing) these 50 papers to a set T. Finally we got a directed graph which has 3312 nodes (articles). Because the new ranking method is applied on the extended set T not on the root set S, the new ranking result may bring us some new documents which do not contain the keyword "XML". This kind of phenomenon is topic drift. As we discussed before, users may want to get the important papers related to "XML" even if the papers do not contain the queried keyword "XML". So the topic drift may be somewhat valuable in this application. When inputting the

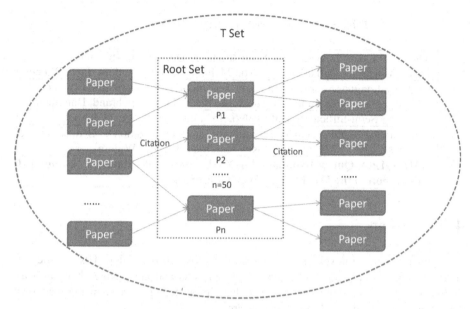

Fig. 4. Graph of Root Set S and Extended Set T

keywords of "XML" as a query, two ranked list are automatically generated. The first one is ordered by the numbers of citations, and the second list is ranked using the extended HITS algorithm. To space restrictions, here we only list the top 5 articles in these two lists from system output. The list of articles ordered by citation is as follows.

As shown in above two lists, the top 4 articles in the **second list** do not contain the keyword "XML" in title or abstract while they are all very important papers in the XML research area. However, the 4th article in the **first list** which contains the keyword "XML" in the abstract is not a "eXtended Markup Language" related article. It's about a kind of functional programming language. This is an example to show that the Dolphin is a more suitable tool for academic searching and analysis.

Table 2. Paper list ordered by the numbers of citations

1	**XML-QL: A Query Language for XML (2000, 238 citations). A standard submission to the World Wide Web Consortium.**
2	A Query Language for XML (1999, 152 citations). Alin Deutsch, Mary Fernandez, Daniela Florescu, Alon Levy, Dan Suciu
3	An Architecture for a Secure Service Discovery Service (1999, 135 citations). Steven E. Czerwinski, Ben Y. Zhao, Todd D. Hodes, Anthony D. Joseph, Y H. Katz
4	Higher-Order Modules and the Phase Distinction (1990, 108 citations) Robert Harper, John C. Mitchell, Eugenio Moggi
5	Quilt: An XML Query Language for Heterogeneous Data Sources (2000, 92 citations). Don Chamberlin, Jonathan Robie, Daniela Florescu

Table 3. Paper list ranked by extended HITS algorithm

1	The Lorel Query Language for Semistructured Data (1996, 467 citations). Serge Abiteboul, Dallan Quass, Jason Mchugh, Jennifer Widom, Janet Wiener
2	A Query Language and Optimization Techniques for Unstructured Data (1998, 321 citations). Peter Buneman, Susan Davidson, Gerd Hillebrand, Dan Suciu
3	Querying Semi-Structured Data (1996, 362 citations). Serge Abiteboul
4	DataGuides: Enabling Query Formulation and Optimization in Semistructured Databases (1997, 224 citations). Roy Goldman, Jennifer Widom
5	**XML-QL: A Query Language for XML (2000, 238 citations). A standard submission to the World Wide Web Consortium.**

4 Conclusion

This paper presents a novel vertical search engine, which is developed as an efficient searching tool to help user get more refined analysis and knowledge about an academic topic or area. Experiments show that the system proposed in this paper is effectiveness and suitable for academic search.

Acknowledgement

This work is supported by NSFC grants (No. 60673137 and No. 60773075), National Hi-Tech 863 program under grant 2008AA01Z1470967 and Shanghai Education Project (Project Number 10ZZ33).

References

[1] Bharat, K., Henzinger, M.R.: Improved algorithms for topic distillation in a hyperlinked environment. In: 21st Annual International ACM SIGIR Conference on Research and Development in Information Retrieval (1998)

[2] Borodin, A., Roberts, G.O., Rosenthal, J.S., Tsaparas, P.: Finding authorities and hubs from link structures on the World Wide Web. In: Tenth International Conference on the World Wide Web, December 2001, pp. 415–429 (2001)

[3] Brin, S., Page, L.: The anatomy of a large-scale hyper textual (web) search engine. In: Seventh International Conference on the World Wide Web (1998)

[4] Lempel, R., Moran, S.: The stochastic approach for link-structure analysis(SALSA)and the TKC effect. In: 9th International World Wide Web Conference (2000)

[5] Kleinberg, J.M.: Authoritative Sources in a Hyperlinked Environment. Journal of the ACM 46(5) (1999)

[6] Farahat, A., LoFaro, T., Miller, J.C., Raeand, G., Ward, L.A.: Existence and Uniqueness of Ranking Vectors for Linear Link Analysis Algorithms

[7] Fung, B.C.M., Wang, K., Ester, M.: Hierarchical document clustering

[8] Lucene, http://lucene.apache.org

 [9] WordNet, http://wordnet.princeton.edu
[10] GoogleScholar, http://scholar.google.com
[11] CiteSeer, http://citeseer.ist.psu.edu/
[12] DBLP, http://www.informatik.uni-trier.de/
[13] Rexa, http://rexa.info/
[14] Fung, B., Wang, K., Ester, M.: Hierarchical document clustering using frequent items. In: SDM 2003, San Francisco, CA (May 2003)
[15] Wang, Y., Geng, Z., Huang, S., Wang, X., Zhou, A.: Academic Web Search Engine — Generating a Survey Automatically. In: WWW 2007, Banff, Alberta, Canada (2007)
[16] De Bra, P., Houben, G., Kornatzky, Y., Post, R.: Information Retrieval in Distributed Hypertexts. In: The 4th RIAO Conference, New York. pp. 481–491 (1994)
[17] Kleinberg, J.M.: Authoritative Sources in a Hyperlinked Environment. In: Tarjan, R.E., Baecker, T. (eds.) Proceedings of the 9th ACM-SIAM Symposium on Discrete Algorithms, pp. 668–677. ACM Press, New Orleans (1997)
[18] Xue, G., Dai, W., Yang, Q., Yu, Y.: Topic-bridged PLSA for Cross-Domain Text Classification. In: Proceedings of the 31st Annual International ACM SIGIR Conference (ACM SIGIR 2008), Singapore, July 20-24, pp. 627–634 (2008)
[19] Dolphin Academic Search Engine, http://dolphin.smmu.edu.cn
[20] Ceri, S.: Search Computing. In: IEEE International Conference on Data Engineering (ICDE 2009), Shanghai, March 29-April 2, pp. 1–3 (2009)

Interactive Visualization in Knowledge Discovery for Academic Evaluation and Accreditation

Anastasios Tsolakidis, Ioannis Chalaris, and George Miaoulis

Department of Informatics, Technological Education Institute, Athens, Greece
atsolakid@cs.teiath.gr, {ixalaris,gmiaoul}@teiath.gr

Abstract. Academic evaluation is a process that gains continuously attention. As a process it requires complex associations discovery and data processing that most of the times are performed by an ad-hoc, manual process by experts. The creation of a theoretical model and support by means of software is a challenging task. In this paper we present VIS-ACC (visualization – accreditation) a methodology that focuses on the processing and extraction of semantic representations from accreditation data. We also present a tool that facilitates the visualization thus automating the evaluation process.

Keywords: Visualization, Accreditation, Knowledge discovery, Data representation, Academic evaluation.

1 Introduction

There is a strong need to ensure the integrity of the results of an accreditation process for academic institutions which undergo such a process in accordance with the principles of the Bologna Declaration. That role was handed to ENQA (**European Association for Quality Assurance in Higher Education**). ENQA standards are currently used as a landmark for regional educational evaluation comities in many European countries such as ADIP – GREECE / AERES-FRANCE / ACCREDITATION COUNCIL – GERMAN etc

National organizations have the privilege to enforce evaluation methods to public or private institutes in order to achieve useful conclusions regarding their educational quality and propose methods of improvement.

Our research first of all aims to develop an application which will process all the useful data regarding each institute's evaluation process and incorporate this processing to the concluding results. In order to come to a neutral conclusion regarding the educational quality of a institute, all data must be accurate, precise and they should meet all the evaluation criteria. Secondly, through that application we will try to visualize those data in a way that will lead to simple and easily understanding conclusions. The data base for our application is referring to Technological Institute of Athens for academic year 2008, taking into consideration the following

- Educational Program
- Learning
- Research

D. Karagiannis and Z. Jin (Eds.): KSEM 2009, LNAI 5914, pp. 168–179, 2009.
© Springer-Verlag Berlin Heidelberg 2009

- Relations with Social/Cultural/Productive parts
- Academic development strategy
- Administrative services and infrastructures

Our application emphasizes in Research -Bachelor's study program - Evaluation process (View State). The main aim is to be able to make useful conclusions in respect to research synergies between the faculty members, research activities etc. We also provide a visualization approach regarding information correlating each lesson's objectives and methods in respect to student's performances. Finally try to help the user to form a quick and accurate opinion regarding the different evaluation tasks, in order to be able to create the reports easily and with less human, subjective intervention.

2 Previous Work

Our research focuses on academic institution's evaluation activities. So far, the accreditation process is performed by experts and without tools support. We focused in providing a concise, easy to comprehend, way to represent the evaluation data regarding the evaluation process. It was very clear that there are two components that are vital, regarding visualization procedure. Therefore we were focused on

- Which ways of representations exist and which of them are suitable for our application's visualization (Representation methods)
- Which limitations or constraints must be met in order to visualize our data precisely without loosing information and help the evaluator to come to right conclusions. (Decision Making using data visualization, Representation & Visualization)

2.1 Existing Representation Methods

The Radial View Tree–map [1] was designed in order to represent a "focus-plus-context style" graph. This method allows the user to select a node and then that node set as the main, in a manner that reduces the confusions.

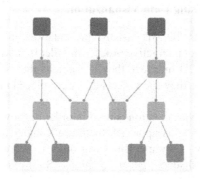

Fig. 1. Tree layout

Fig. 2. Radial View Tree–map

The Tree–map [2][3] shows the size of the elements that take part in a structure that the user wants to represent. For example, a graphical structure can be used to represent the files of a system [3]; that kind of representation gives the ability to a user to observe that file with the largest size in order to delete it.

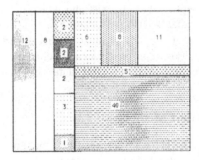

Fig. 3. Tree–map

2.2 Decision Making Using Data Visualization

Voinea et al [4] based on 2D, 3D visual representations use a rule mining process which provides to a user an interactive process in order to discover associations in the represented data. At the 2D interface they use graphs that represent the rules represented and at 3D interface they use objects that were placed in virtual worlds in order to represent that rules.

In another method [5] visualization is used in order to show the dynamic behaviour of a memory allocator. At this approach there are x and y axis that show time and memory correspondingly. They use Blocks in order to represent the process id and they draw them using shaded plateau cushions. So from this representation the observer could easily extract the necessary information. They also show the importance to combine data mining with visualization activities in order to take decisions or to observe a problem.

Another approach is to use a graph in order to extract information for cross cited papers to verify which ones contributed the most to a specific research area [6].

2.3 Representation and Visualization

The HistCite system [7] was designed to help selectively identify the significant (most cited) papers retrieved in topical searches of the Web of Science (SCI, SSCI and/or AHCI). Once a marked list of papers has been created, the resulting Export file is processed by HistCite to create tables ordered by author, year, or citation frequency as well as historiographs which include a small percentile of the most-cited papers and their citation links. A comparative evaluation of different visualization approaches (Treemap, Sequoiaview, BeamTrees, StarTree, botanical trees) can be found in [8]. In [9] matrix-based representations and nodelink diagrams are compared.

Our approach aims in creating an efficient tool for decision making. Another challenge was to use large hierarchically document collections, such as the ones described in [10]. Therefore, we introduce a method to represent the same data from different points of view. In our methodology we will combine those representations in order the viewer to be able to extract conclusions without loosing any information. Also the user does not get confused even when processing a great amount of data and he/she will have the ability to get in more depth only at those information that he/she is interested in.

3 System Description and Architecture

The purpose of our application is to visualize the results of an evaluation from different points of view and different perspectives in order these results to be more clear for the evaluator or the user in general. Based on the need to describe all the factors which are forming the quality of education we have taken into consideration the quality of the following criteria: Educational Program, Learning for bachelor's and masters degree programs, Research, Relations with Social/Cultural/Productive parts, Academic development strategy, Administrative services and infrastructures.
The need for a quality assessment of the current status of an institute is estimated on a 0-5 scale.

In order to develop our application we used: Netbeans 6.1~6.5, J2EE, Mysql - Mysql Query Browser, GlassFish V2, The Google appi, The Tulip Framework, The JavaScript information Visualization Toolkit (JIT).

In order to represent all the amount of data that took place in the evaluation process, we designed our database in a way that gives the ability to each user to extract all the necessary information in an automated way. In most of the situations the user could observe them using flexible representation tools.

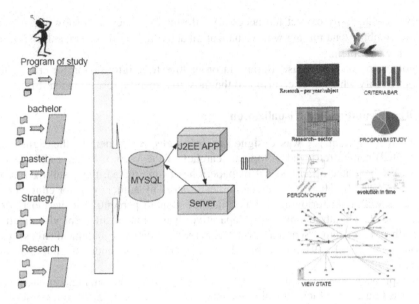

Fig. 4. The architecture of the application and the purpose of our application are shown at the following figure

3.1 View-State

As we mentioned, the main purpose of our application is to visualize the sectors that the educational department will be evaluated in such a way that would help the evaluator to come to useful conclusions and suggestions for future improvement. We

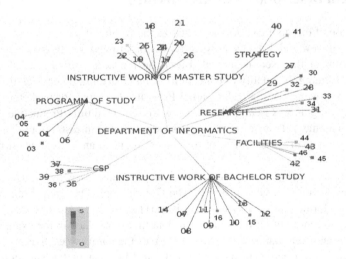

Fig. 5. View- State representation that provide a general view of the categories and the criteria that belongs to each category

concluded that the most appropriate method should be the graph with a balloon view which gives a general view at the observer without to need to get in more depth. From that kind of representation we view the state (good (5), bad (0)) of each sector and moreover for the entire criterion. In order to create that graph we create a module that generates .tlp using retrieval data from the database that can be displayed by the Tulip framework one of the most suitable frameworks for those kind of representations that we need.

In Fig 5 we have a general view of the categories and the criteria related to each category. As we can observe the colour of each node is the same or differs from the others. The colour depends on rate that each criterion has evaluated. The scale is 0- 5 and colour's intensity depends at that scale as shown at the left corner. Also the numbers at the nodes represent the criteria and the parent node the category (sector).

Additional to the representation of the state for each factor of the department we focus on the research sector and Bachelor's study program (quantitative elements) of the evaluated department.

3.2 Research

In order to be able to get knowledge about the research activities of the educational staff of the department we suggest that we should examine

1. The time that they illustrate them and the limit up/down that they have during the period that we examine(**Research –evolution in time**)
2. The research activities that they have create and the collaborations that may be exist in order to publish one of them. (**Research – per year/subject**)
3. The research areas that belongs their activities.(**Research– sector**)
4. The personal information for every professor (**Research – personal**)

As time we have taken into account the period 2003-2008.

As research activities we have take into account: 1) Patents- 2) Journals- 3) Conference papers. As a resource we used the data from an ongoing evaluation at the informatics department.

3.2.1 Research – Evolution in Time
In order to project the evolution of research activities in time (Fig 6) we represent as x 'x the time-related data and at the vertical axis we present research related activities, such as patents, publications etc.

We provide a facility that creates reports from the input data. This tool has the following characteristic: in order to present the evolution of each member of faculty activities in time, we present each faculty member as circle. By choosing the appropriate name from the list of faculty members, we can observe the corresponding circle to move towards the x axis. The movement depends on the number of the activities that each faculty member has achieved, in respect to the time needed.

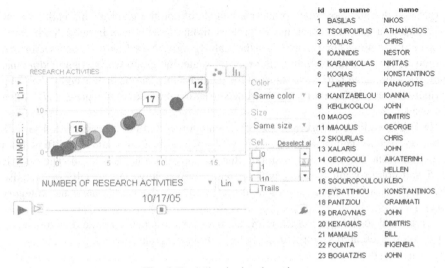

id	surname	name
1	BASILAS	NIKOS
2	TSOUROUPLIS	ATHANASIOS
3	KOILIAS	CHRIS
4	IOANNIDIS	NESTOR
5	KARANIKOLAS	NIKITAS
6	KOGIAS	KONSTANTINOS
7	LAMPIRIS	PANAGIOTIS
8	KANTZABELOU	IOANNA
9	KEKLIKOGLOU	JOHN
10	MAGOS	DIMITRIS
11	MIAOULIS	GEORGE
12	SKOURLAS	CHRIS
13	XALARIS	JOHN
14	GEORGOULI	AIKATERINH
15	GALIOTOU	HELLEN
16	SGOUROPOULOU	KLEIO
17	EYSATTHIOU	KONSTANTINOS
18	PANTZIOU	GRAMMATI
19	DRAGVNAS	JOHN
20	KEXAGIAS	DIMITRIS
21	MAMALIS	BILL
22	FOUNTA	IFIGENEIA
23	BOGIATZHS	JOHN

Fig. 6. Evolution in time in action

3.2.2 Research – Per Year/Subject

The user can choose based on time and subject criteria the representation of data on a graph. The graph's structure is as follows:

- In the centre the year is displayed
- At the perimeter of the second ring the professors of research programs are displayed
- At the perimeter of the next ring the research activities of each professor are displayed

In that way the parameters year-professors-research are represented. So any chosen professor becomes the centre of the next graph and the rest of the initial data displayed are placed at the perimeter of the rest rings respectively.

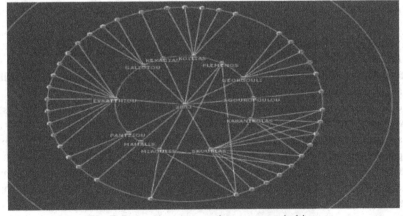

Fig. 7. Research representation – per year/subject

3.2.3 Research – Sector

At this task we evaluated the research activities according to subject area. A research area with more papers than some other will cover a larger surface, therefore it is easy to identify which areas are more active using the graphical representation.

In Fig. 8 we can see the name of the authors with different colours relevant with the number of the papers that they have due the numbers of the others

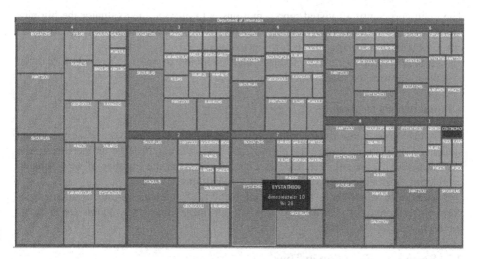

Fig. 8. Representation of research sectors

3.3 Bachelor Degree

Fig. 9 represents a table with student related information for every academic year; for example it shows the number of new enlisted students at the department of

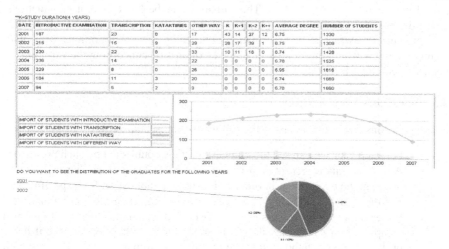

Fig. 9. Representation of the number of students at each year with the information relevant for each one

informatics in a graphical manner. Finally we have the ability to get informed for the duration for graduation of the graduated students for a specific academic year, using a pie chart.

4 Results and Discussion

The main purpose of all visual representations is to provide the evaluator the big picture formulated by evaluating data in order to make up some useful and realistic conclusions and provide suggestions for improvement.

The main sectors where we focused and examined in order to provide a realistic visualization were the following

4.1 Bachelor's Degree

The visual representations are aiming on showing the total number of new enlisted students. For example during the period 2001-2004 we can observe an increase in the total number of new enlisted students and a decrease in the number of new enlisted originating from other high educational level institutes.

Furthermore we observe that the total amount of students of the informatics department is increasing year by year which might means that less students are graduating each year.

4.2 Research – Evolution in Time

That kind of representation helps the user to view the research achieved by several professors through a period of time. Moreover we can take a general view for the research tasks achieved by the faculty members; specifically we concluded that the majority of them were not particularly active, or that some of them do not change their status. In addition we can see that a certain members show a limit up and all the other period their activities are too low. This situation maybe has different explanations. Using the next representation (graph), provided from the application "YEAR-KIND" it is possible to extract more information concerning the collaborations with other faculty members which might provide a good explanation concerning the variations in his/her research performance.

4.3 Research – Year/ Subject

The conclusions that we came using the following representation are:

- For each year we can see the number of articles published by each professor
- Also from this representation we can see the collaborations that were established at the department in order to illustrate a publication for a specific year. For the year of 2003 (Fig 7) we can observe that there are 2 collaborations. For example the professors P(M) – P(P) collaborated once in order to publish a paper. Then we display the publications for the other years to see if those professors had collaborated in the past in order to publish any other research activity. If something like that exists then we could understand that those

professors may be belonging in a common research team. Also using the representation "research area" we can see if those professors (P(M) − P(P)) belong in the same research area. If we observe that they also belong to same area then we understand that they certainly have created a team and they collaborated. Also if we observe that a professor collaborates with different professors each time then we have to take into account if those professors belong to same research area or not ("Research– sector"). If that representation gives to us the knowledge that they collaborate with professors from different areas then the conclusion that we get about the professor is that he/she has not got any specific area that he works. This situation has as a result that this professor is impossible to take part in a team because he has not already defined the area of his interesting.

- Another conclusion that we can extract is the professors that have the most publications at all the years that we examine.
- The application gives to us the ability to examine the results for each year for each research domain. So we have the ability to inform for the kind of the research and moreover for the quality of his/her activities.

4.4 Research – Sector

This representation gives to us the opportunity to observe the different research areas in which the faculty is active. We may also request more results for a specific research area, according to our interests and observe also which areas are the most active. Finally we can analyze its area (square) and to see the participation of each author. The colour of the square that corresponds to each author depends to the rate of attendance to the examined research area.

The main purpose of our implementation is to provide to user all the necessary tools (representative) in order to be able to extract the appropriate conclusions from the evaluation's data. We believe that the combination of the provided different tools will guide the user to get all the necessary information that a user needs.

4.5 View-State

The implementation of the evaluation process in a department is an action that needs a lot of human effort to process all the vast amount of data. The result is the demand for a great amount of manual work and a large percentage of introducing faults in the evaluation process. Our efforts concentrate on the provision of an efficient graphical representation tool in respect to the different aspects of the accreditation process. With that representation the user take a general view of the categories and the criteria that we examine at the evaluated department. As we said before the user should set a rate at each criteria in order to evaluate it. The rates were set on a 1-5 scale. In our representation we have chosen to set those values as metrics for each node. Also as we can see each node has a specific colour. The intensity of the colours depends on the value that we have set for that node. In the case study that we performed the evaluation outcome was rather satisfactory and this is obvious from the very beginning by observing the coloured areas (each colour represents a situation in our 5-degrees scale). In addition the user has the ability to focus on specific categories or criteria in order to inform the reasons that made the experts to set that value. So for example we observe that node 28 was blue. This means that the criteria that represent that node were rated with 5. So the user could go back to the main application to see the reason that made the experts to set that value to 5. In the other hand we observe some nodes like 33 have yellow colour. The kind of colour depends on the value that the user has set for the particular criteria. So the node 33 or the criteria 33 seems to be of low rate. In this situation we can go again into the application on previous pages to view this criteria in order to understand the reason for that rate and maybe to suggest a solution.

5 Conclusions – Further Work

We have presented an application that provides a graphical representation tool to facilitate the accreditation process for academic institutions. The large number of data and the associations between the data make this process difficult and error prone if performed manually. Also, the graphical results help the evaluator to assess the different domains and also to make useful conclusions in respect to the associations hidden behind the rating of a task. Also the user could pay more attention only at these categories that he was interesting depending on the value (colour) that the experts have set.

We believe that this framework provides to user a lot of information relevant with the state of a department in general. We consider a next step the ability to examine all the sectors in more depth and moreover to use a standard model (ISO) and the criteria related with that model in order to evaluate a department or an organisation.

Acknowledgment

We would like to thank Dr. Petros Belsis for useful comments on a previous version of our paper.

References

1. Lamping, J., Rao, R., Pirolli, P.: A focus+context technique based on hyperbolic geometry for visualizing large hierarchies, pp. 401–408 (1995), ISBN:0-201-84705-1
2. Shneiderman, B.: Tree visualization with Tree-maps: A 2-d space-filling approach. ACM Transactions on Graphics (TOG) 11(1), 92–99 (1992)
3. Yee, K.P., Fisher, D., Dhamija, R.: Marti Hearst Animated Exploration of Dynamic Graphs with Radial Layout. In: IEEE Symposium on Information Visualization 2001. University of California, Berkeley (2001)
4. Voinea, L., Telea, A.: Visual Analytics: Visual data mining and analysis of software repositories. Computers and Graphics 31(3), 410–428 (2007)
5. Chena, C., Chenb, Y., Horowitza, M., Houb, H., Liub, Z., Pellegrino, D.: Towards an explanatory and computational theory of scientific discovery. Journal of Informetrics 3(3), 191–209 (2009)

6. Bruls, M., Huizing, K., van Wijk, J.J.: Squarified treemaps. In: de Leeuw, W., van Liere, R. (eds.) Proceedings of Joint Eurographics and IEEE TCVG Symposium on Visualization, pp. 33–42. Springer, Vienna (2000)
7. Burkhard, R., Eppler, M.: Knowledge Visualisation. In: Keller, T., Tergan, S.O. (eds.) Knowledge and Information Visualisation. Springer, Germany (2005)
8. Ghoniem, M., Fekete, J.-D., Castagliola, P.: A comparison of the readability of graphs using node-link and matrix-based representations, pp. 17–24. IEEE Computer Society, Los Alamitos (2004)
9. Granitzer, M., Kienreich, W., Sabol, V., Andrews, K., Klieber, W.: Evaluating a system for interactive exploration of large, hierarchically structured document repositories, pp. 127–134. IEEE Computer Society, Los Alamitos (2004)
10. Chen, C., Morris, S.: Visualizing evolving networks: Minimum spanning trees versus pathfinder networks. In: Proceedings of the IEEE Symposium on Information Visualization (INFOVIS 2004), Washington, DC, USA, pp. 1–8 (2004)

Aggregation Models for People Finding in Enterprise Corpora

Wei Zhang[1,*], Lei Chang[2], Jianqing Ma[1], and Yiping Zhong[1]

[1] Network and Information Engineering Center, Fudan University
{072021095,jqma,ypzhong}@fudan.edu.cn
[2] EMC Research China
chang_lei@emc.com

Abstract. Finding authoritative people of a given field automatically within a large organization is quite helpful in various aspects, such as problem consulting and team building. In this paper, a novel aggregation model is proposed to solve the problem of finding authoritative people. Various kinds of related information in an enterprise repository is assembled to model the knowledge and skills of a candidate, for instance, such information may be the profile which gives a personal description of the candidate, documents related with the candidate, people with similar intellectual structure and so on. Then the candidate is modeled as a multinomial probability distribution over these collected evidence of expertise and candidates are ranked according to the probability of the topic generated by their models. Experimental results on TREC benchmark enterprise corpora demonstrate that our model outperforms current state-of-the-art approaches by a large margin.

1 Introduction

Managing skills and knowledge of employees in a large organization so that the question "who knows what" can be identified is of great value in various aspects [13][10]. For instance, when an employee comes across a problem in a specific field, it is meaningful to use a recommendation system to find experts in this field for consultation. Or, when it comes to team-building for a project which may involve knowledge and skills across several fields, selecting appropriate people and assigning them to the appropriate position are crucial to the success of this project. Thus, people finding system is of great value for large organizations which may have a lot of branches across a country or around the globe. The typical scenario of authoritative people (i.e., expert) search task is like this: given a topic Q, expert finding system will return a list of candidates who are considered to be the most relevant in this field to the users.

Traditional solutions to expert finding are to manually create and maintain a database housing associated information that describes expertise of each individual. Obviously, these methods are quite effort-consuming due to the dynamic

---

* Currently, Wei Zhang is a student at Fudan University.

D. Karagiannis and Z. Jin (Eds.): KSEM 2009, LNAI 5914, pp. 180–191, 2009.

change of personnel and their ever-increasing skills and knowledge. Therefore, automated methods which can analyze enterprise repositories to collect expertise of people and then find out potential experts have dominated in solutions to expert finding. In general, there are two kinds of principle approaches for automatic expert finding: document models and candidate models [6]. They are also called query-dependent and query-independent models respectively in [16].

Both of these two kinds of models concentrate only on a single kind of evidence resource collected from repository, either a profile or a collection of associated documents, which may not well describe the knowledge and skills of an individual. In this paper, a novel and effective method is used to model the expertise of candidates to solve the problem. More specifically, our model represents the knowledge and skills of a candidate by assembling various kinds of expertise evidences collected from enterprise corpora. Compared with document models and candidate models, our method of modeling can represent the knowledge of a candidate in a more accurate and specific manner. In this paper, we focus on the following three kinds of evidence: personal profile, associated documents and similar people. During evidence recognizing and collecting, we adopt several techniques to prevent noise from being incorporated and express the relationship between candidates and collected evidence in a better way, such as sliding window extraction, IDF filtering, weighted association and so on. In addition, our proposed model is flexible to incorporate more expertise evidences besides the information that have been utilized. Finally, the effectiveness of our model is demonstrated by using CECR collection provided by Text Retrieval Conference (TREC) compared with other existing models.

The remainder of this paper is organized as follows. We give a brief overview of related work on expert finding system in Section 2. Then, in Section 3, we describe our aggregation models in detail. Experimental setup is presented in Section 4 and experimental results are reported in Section 5. Finally, Section 6 concludes this work and gives future work.

2 Related Work

Initial approaches to people finding are database based which include Microsoft's SPUD project [9], Hewlett-Packard's CONNEX KM system [1], and the SAGE expert finder [2]. Due to considerable efforts required to set up and maintain the related information, these methods are replaced by automatic approaches. The introduction of Enterprise track by TREC since 2005 has led to various new solutions [8][19][4]. Most of solutions can be classified as either candidate model or document model.

For document models, association between candidates and their associated documents is firstly built. Given a topic, documents are scored using a document retrieval model. Then, the probability of a candidate being an expert is estimated according to the pre-built association and document scores. For example, MITRE's ExpertFinder [15] first finds out documents containing the given topic, then, candidates are ranked by the association which is built based on the distance between occurrences of candidate and query terms in documents. Since ad hoc retrieval plays a key part in document model, in [21], the authors tries to find out to what extent does the performance of ad hoc document retrieval affect the final results of expert finding. Experiments shows that the improvements of ad hoc retrieval does not always result in improvements of expert retrieval. Besides, document-candidate association plays an important part in document model too. In [11], a weighted language model based on the number of citations of a document is employed in a real world academic field.

Candidate models collect useful and related information in the repository and then model the knowledge of candidates on the basis of collected evidence. Then, the probability for a candidate being an expert is directly estimated as the probability of the given topic generated from this expertise model. A famous published system, P@NOPTIC, collects associated documents in the repository and then forms a description document by concatenating all these documents together [7]. In [12], the authors build a description document in a more accurate way by adopting visual page rank and expert home page detection techniques. Visual page rank excludes useless and irrelevant information by analyzing the structure of web pages while expert home page detection, in the opposite, tries to find out the most relevant evidence by matching names of candidates in the title field of a page.

Many other methods are proposed to improve the performance of expert finding system. Perkpva and Croft [16] uses pseudo-relevance feedback to expand the original query and improve the performance of both text retrieval and expert finding. Resources other than repository within an organization are also used as evidences of expertise for candidates, especially evidences collected from WWW. It is reported in [17] that it is beneficial to take Web evidences into account and the combination of local resources and web evidences has resulted in a performance improvement of expert finding.

3   Aggregation Model

In aggregation models, each individual is represented as an *information fragment* set which consists of evidences collected from an enterprise repository. Intuitively, an information segment can be a profile (or resume) of a person, the information about its related people or other information. Various information fragments about an individual are firstly built and then aggregated to model the knowledge of the individual. More formally, the information fragment set of an individual can be formulated as follows:

$$I = \{I_1, I_2, \cdots, I_j\} \tag{1}$$

Where I_k $(1 \leq k \leq j)$ denotes an information fragment of the individual collected from repository. After each candidate is represented with information fragments, the probability of the candidate related to a given topic can be estimated by using the likelihood the fragments relates to the topic. In this work, we focus on the following three information fragments: personal profile, related documents of an individual and similar people. A profile is a single document which gives general description of a candidate's knowledge structure, and relate documents contain those documents which considered to be associated with a candidate. Similar people fragment contains the information about other candidates with similar intellectual structure. The fragment set is expandable. Some other segments which would bring gains to description of expertise for candidates will be incorporated in future work.

After the information fragments of a candidate is constructed, the candidate can be modeled as a multinomial probability distribution over the information fragments. Given a topic Q, the probability of candidate ca being an expert of Q can be estimated by combining the relatedness of the topic to the individual fragment. There are various methods to do this aggregation. Here, we use a weighted aggregation method by allocating different weight to different fragments which can be formulated as follows (query terms t_i in Q are assumed to be generated independently):

$$P(Q|ca) = P(Q|\{I_k\}_{1 \leq k \leq j}) = \prod_{1 \leq k \leq j} P(Q|I_k)^{\lambda_k} = \prod_{1 \leq k \leq j} \{\prod_i^{|Q|} P(t_i|I_k)\}^{\lambda_k} \quad (2)$$

In this work, we use I_1, I_2, I_3 to denote profile, related documents, and similar people information fragments respectively. The details about building information fragments and $P(Q|ca)$ computation will be discussed in the following subsections.

3.1   Profile Building

A profile is a refined document which describes candidate's knowledge and skills. To construct a profile, evidence of expertise within the corpora should be recognized and collected. Therefore, the method of evidence recognition and collecting is of primary importance to profile building.

In this work, we propose a novel method which uses sliding window extraction and IDF filtering for profile building. Under the intuition that the information around occurrences of a candidate within a certain distance in a document is closely related with him, this method only collects context information around occurrences of a candidate rather than incorporating the whole document. More specifically, a sliding window is used while evidence extraction to prevent irrelevant information being incorporated. The size of window can be fixed or dynamically tuned according to the length of documents.

To further improve the quality of the profile, only the most distinguished words are used while the useless information in candidate context is ignored. In this paper, we adopt IDF which is a measure of the general importance of terms [5] as a criterion to do filtering. Since terms with larger IDF values show more distinction than other terms, only words with IDF values larger than a threshold are selected to build profile of this candidate. The profile of a candidate ca in a document d can be formulated as:

$$c(ca, d) = \{t_i| \, |ind(t_i, d) - ind(ca, d)| \leq \omega, idf(t_i) \geq \tau\} \quad (3)$$

where $c(ca, d)$ denotes the extracted context for candidate ca in document d, $ind(t_i, d)$ is the index of term t_i in document d, $idf(t_i)$ denotes the IDF value of term t_i, ω is the size of the sliding window and τ is the IDF threshold value for filtering. Finally, extracted contexts of a candidate ca are assembled together to form a single profile K_{ca} :

$$K_{ca} = \bigcup_{d_i \in C_{ca}} c(ca, d_i) \tag{4}$$

where C_{ca} is the collection of documents ca occurs. After the profile of a candidate ca is built, the knowledge and skills of ca is modeled as a multinomial probability distribution over a bunch of terms. By using language model, the probability of given topic Q being generated by information fragment I_1 can be estimated as:

$$P(Q|I_1) = \prod_{i}^{|Q|} P(t_i|K_{ca}) \tag{5}$$

Due to data sparsity, Jelinek Mercer smoothing method [20] is used to prevent "zero probability".

$$P(t_i|K_{ca}) = \lambda \frac{tf(t_i, K_{ca})}{|K_{ca}|} + (1 - \lambda) \frac{tf(t_i, C)}{|C|} \tag{6}$$

where $tf(t_i, K_{ca})$ and $tf(t_i, C)$ denote term frequency of t_i in the profile and the repository respectively.

3.2   Weighted Document-Candidate Association

The second information fragment of a candidate, namely I_2, consists of documents which are considered to be related with him. We discuss how to build this information fragment in this subsection.

The "related documents" of a candidate is defined as those documents in which the candidate occurs at least once. After all related documents are found, the next step is to build association between the candidate and his related documents. Instead of treating each document equally, we adopt a weighted association which uses the total occurrences of a candidate in a document to measure the closeness between the candidate and the document. The weight functions are shown in equation 7. There are two logarithm functions: common logarithm function (W2), and natural logarithm function (W3). The Boolean association (W1) which treats associated documents equally is used as our baseline.

$$assoc(ca, d) = \begin{cases} 1, & \text{(W1)} \\ \lg(10 + occur), & \text{(W2)} \\ \ln(e + occur), & \text{(W3)} \end{cases} \tag{7}$$

where $occur$ is the number of occurrences of ca in d.

To estimate the probability of topic Q being generated from I_2, all the documents in I_2 are evaluated at first.Jelinek Mercer Smoothing is employed here to avoid "zero probability" again.

$$P(t|\theta_d) = \lambda \frac{tf(t_i, d)}{|d|} + (1 - \lambda) \frac{tf(t_i, C)}{|C|} \qquad (8)$$

where $tf(t_i, d)$ is the term frequency of t_i in document d, $|d|$ is the word count of d. Then,the probability of topic Q given I_2 can be estimated by combining weighted document-candidate association and document evaluation results:

$$p(Q|I_2) = \prod_i p(t_i|I_2) = \prod_i \{\sum_d P(t_i|\theta_d) * assoc(ca, d)\} \qquad (9)$$

3.3   Similar People

The third information fragment we use is the related people who have similar knowledge structure with a candidate. To find similar people, two different methods can be used. The first method is to measure the similarity of two people by using the similarity between their profiles. The candidates with high similarity scores are considered to be similar people. The second method is to analyze the related documents of each other and select people who share the most same documents as similar people. These two methods make use of pre-built profile subset and related documents respectively. In this paper, the second method is used which can be formulated as:

$$s(ca_1, ca_2) = \frac{|I_{2,ca_1} \cap I_{2,ca_2}|}{|I_{2,ca_1} \cup I_{2,ca_2}|} \qquad (10)$$

where $s(ca_1, ca_2)$ denotes the similarity of ca_1 and ca_2, while $|I_{2,ca_1}|$ and $|I_{2,ca_2}|$ are the sizes of related document sets of ca_1 and ca_2 respectively.

Another issue is to decide the number of related people included in the information fragment for a candidate. Similar people can introduce more useful information in the corpora which may haven't been collected when profile building and related documents collection. On the other hand, it would bring in noisy information if two people are not that similar in knowledge structure, ending up poisoning the scoring of candidates. Therefore, it is important to set an appropriate baseline and find a suitable size for the similar people subset. The similar people information fragment is determined by using:

$$I_3 = \{ca_i|s(ca, ca_i) \geq s_{bl}\}; \qquad (11)$$

where $s(ca, ca_i)$ is similarity between ca and ca_i, s_{bl} denotes threshold set for similar people selection.

The estimation of $P(Q|I_3)$ is based on the scores on I_1 and I_2. When retrieval begins, the scores on I_1 and I_2 of each candidate ca are estimated which results in a raw score for each candidate:

$$P_{raw}(Q|ca) = \prod_{i \in \{1,2\}} P(Q|I_i)^{\psi_i} \qquad (12)$$

where $P_{raw}(Q|ca)$ denotes the raw score estimated for candidate ca. Then, $P(Q|I_3)$ is calculated as:

$$P(Q|I_3) = \sum_{ca_i \in I_3} P_{raw}(Q|ca_i) \tag{13}$$

So far, we have presented the methods of information fragment building and likelihood estimation of the topic with the information fragments. By using equation(3), the probability of topic Q being generated by each candidate can be estimated. Finally, candidates are ranked in a decreasing order of $P(Q|ca)$ and returned to users.

4   Experimental Setup

In this section, we describe the experimental setup, and the evaluation results will be presented in Section 5.

4.1   Test Collection and Topics

The data set we use is CERC (CSIRO Enterprise Research Collection) corpus. The corpus is firstly introduced in TREC 2007 which is used in both the document search task and the enterprise search task. CECR is a crawl of the Australian Commonwealth Scientific and Industrial Research Organization (CSIRO) till March 2007 [3]. The collection consists of 370,715 documents, with a total size 4.2 GB. We took the 50 topics created by TREC 2007 in the Enterprise track. Each topic has a query field and a narrative field. Before using these topics to retrieval, we expanded these query by adding the 3 terms with top IDF values from the narrative field, which is also used in [12].

4.2   Evaluation Metrics

Three measures are used to evaluate the effectiveness of aggregation models: Mean Average Precision (MAP), P@n (Precision at Rank n), and Mean Reciprocal Rank (MRR). MAP is used to measure the precision of expert finding system. In TREC Enterprise Track, MAP is the most important measure [8][19][4]. For people finding systems, users usually only want to see the most relevant people instead of a long list of candidates. Therefore, the precision in the top ranks is of great importance. In this paper,we chose P@5, P@10 (Precision at Rank 5 and 10 respectively) and MRR to measure the precision in the top ranks. MRR is another frequently used measure in information retrieval which is calculated as the mean reciprocal rank of top relevant candidates of each query [18].

4.3   Candidate Recognition

Since there is no candidate list available, we have to build such a list. The method is to firstly extract the email addresses in the corpus which follow the pattern of "Firstname.Lastname@csiro.au" (csiro.au is the domain of CSIRO). We employed regular expressions to parse the documents and got a list of email addresses. By stripping off suffix "@csiro.au" for each email address, we obtained a list of candidates. After careful examination plus analysis on the TREC 2007 qrels file, finally we got 3404 unique names of candidates in total.

The next step is to build an information fragment set for each candidate. To build the set, we have to extract evidences of expertise from the corpus. Candidate recognition is quite important because it is the basis of information fragment building. To recognize appearances of a candidate in the corpus, we used a multiple patterns matching algorithm since there are several different forms for each English name. Take "William Jefferson Clinton" as an example, we used the following seven patterns to match it: "William Jefferson Clinton", "William Clinton", "W. Clinton", "W.J. Clinton", "Clinton, W.J.", "Clinton, W.", "william.clinton@csiro.au". Any pattern matched in a document is recognized as one occurrence of the candidate and marked with a corresponding identifier for further treatment. During pattern matching, there may be ambiguity if people share the same abbreviated name. For example, if there is another candidate named "Walter Clinton", it would be hard to tell whom it stands for given an abbreviation "W. Clinton". In this work, we do not deal with this kind of ambiguity and leave it as future work.

4.4 Association Building

On the basis of candidate recognition, it is not difficult to build the weighted association between candidates and documents. For weighted association building, we calculated total occurrences of a candidate within a document by counting his unique identifier found in this document. Then, weighted association is built by using equation 7.

4.5 Profile Construction

Before profile building, we removed all the HTML markups within each page, turning the document into plain text to avoid markup being extracted as context of the candidates. During sliding window extraction, we used a fix-sized window (100) for simplicity. Sliding window extraction resulted in a number of raw segments which contain evidence of expertise. Then, we used IDF values to filter out relatively unimportant information within each window. Finally, the profile of a candidate was built by piecing its related segments together.

4.6 Similar People Finding

In this paper, similar people are found based on similarity of related documents between candidates. During experiment, we didn't use a baseline to select similar people but only pick top 3 people with highest similarity as similar people for each candidate.

5   Experimental Results

This section gives the experimental results of the aggregation model. We first show the effects of the IDF window based profile building and weighted document-candidate association on the expert finding task. Then the comparison results with previous methods are given. Finally, we show the effects of relevance feedback on the retrieval performance.

5.1   Effectiveness of Profile Building

This section aims to show whether our method of profile building which based on sliding window extraction and IDF filtering delivers benefits to the retrieval results. For comparison, we use the formal candidate model which represents a candidate as a multinomial probability distribution over the vocabulary of terms within his related documents [6] as our BASELINE. Experimental results of these two methods are shown in Figure 1.

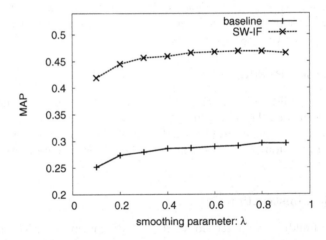

Fig. 1. Effectiveness of differen methods for profile building: Baseline (Formal candidate model), SW-IF (Sliding window extraction and IDF filtering)

The figure shows that our method outperforms baseline method significantly. It is because our profile building method can effectively extract closed related information and filter noise information in extracted windows.

5.2   Effectiveness of Weighted Association

To show the effectiveness of weighted association, we test three forms of association stated in section 3.3. Among them, method W1 which uses boolean association is used as BASELINE. The best result of each method are presented in Table 1. From the table, we can find that weighted association which is based

Table 1. Evaluation results by using different formulations of weighted association: W1 (Boolean), W2 (Common Logarithm), W3 (Natural Logarithm). The smoothing parameter λ of each method is 0.6.

Association	MAP	P@5	P@10	MRR
W1	45.62	25.60	15.80	62.60
W2	48.27	28.00	16.00	64.34
W3	**48.86**	**28.00**	**16.20**	**64.72**

on occurrences of candidates has better performance than uniform distribution which treats documents indiscriminately. It models well the closeness between documents and candidate by taking occurrences into account. For W2 and W3, W3 outperforms W2.

5.3 Comparison with Other Methods

The results of aggregation model compared with the formal candidate model and the formal document model (using candidate-centric association) [6] are shown in Table 2. It can be found that the aggregation model has better performance than formal models for expert finding. Different from other models, the advantage of aggregation model lies in the sufficient evidences of expertise collected from the repository, which express a candidate's knowledge and skills quite well for expert finding task.

Table 2. Results of different expert finding models: Baseline1 (formal candidate model), Baseline2 (formal document model) and AM (aggregation model). The weight assigned to I_1, I_2, I_3 in AM are 0.4, 0.5, 0.1 respectively. The smoothing parameters in I_1, I_2 are both 0.6.

Model	Association	MAP	P@5	P@10	MRR
Baseline1	-	46.90	24.40	14.60	60.93
Baseline2	W2	48.27	**28.00**	16.00	64.34
Baseline2	W3	48.86	**28.00**	**16.20**	64.72
AM	W2	51.48	24.80	15.80	**66.87**
AM	W3	**51.96**	25.60	15.20	66.80

5.4 Relevance Feedback

Relevance Feedback has been proved to be helpful to the performance of information retrieval. The basic idea of relevance feedback is to reformulate the representation of the initial query by taking users' feedback in initial retrieval into account, then perform query again to improve the final result. Pseudo-Relevance Feedback can automatically perform feedback without interaction with users by assuming that top k ranked documents are relevant by default [14]. Evidence such as results published in previous TREC ad hoc tasks, shows that it works

Table 3. Evaluation of effectiveness of relevance feedback: Original (without relevance feedback), RF (with relevance feedback)

query	association	MAP	P@5	P@10	MRR
Original	W2	51.48	24.80	15.80	66.87
Original	W3	51.96	25.60	15.20	66.80
RF	W2	54.22	**26.40**	**16.20**	69.16
RF	W3	**54.73**	26.00	**16.20**	**70.75**

well and produces quite good results. The results of aggregation models with (and without) feedback are shown in Figure 3. For relevance feedback, top 10 terms in top 5 documents returned are selected to revise original query. It can be observed that relevance feedback is also useful for expert finding tasks. When using weighted association $W1$, MAP and MRR witness an increase about 5.30% and 3.40% respectively. And MAP and MRR are increased by 5.33% and 5.91% respectively using association $W2$.

6   Conclusion and Future Work

In this paper, aggregation models were proposed to solve the problem of expert finding in an enterprise repository. Different from previous retrieval models, aggregation models build an information fragment set for each candidate and the probability of being an expert for a given topic can be inferred from the fragment set. More specifically, a novel method based on sliding window extraction and IDF filtering was used to build profiles, and weighted association was used to measure the relationship between candidates and documents. In addition, the information fragment of similar people was also utilized to help identifying authoritative people in a given field. Evaluation results on benchmark data sets demonstrated that aggregation models have better performance than current state-of-the-art models. In the future, we will apply our models on other kinds of entities, such as organizations, products and so on. In addition, we will investigate whether other information fragments are useful for this task.

References

1. http://www.carrozza.com/connex/
2. http://sage.fiu.edu/Mega-Source.htm/
3. Bailey, P., Craswell, N., Soboroff, I., de Vries, A.: The CSIRO Enterprise Search Test Collection. ACM SIGIR Forum, 42–45 (2007)
4. Bailey, P., Craswell, N., de Vries, A., Soboroff, I.: Overview of TREC 2007 Enterprise Track. In: The Sixteenth Text REtrieval Conference Proceedings (2007)
5. Baeza-Yates, R., Ribeiro-Neto, B.: Modern Information Retrieval. Addison Wesley, New Jersey (1999)
6. Balog, K., Azzopardi, L., de Rijke, M.: Formal models for expert finding in enterprise corpora. In: Proceedings of the 29th annual international ACM SIGIR conference, pp. 43–50 (2006)

7. Craswell, N., Hawking, D., Vercoustre, A.M., Wilkins, P.: P@noptic Expert: Searching for Experts Not Just for Documents. In: Ausweb Poster Proceedings (2001)
8. Craswell, N., de Vries, A., Soboroff, I.: Overview of the TREC 2005 Enterprise Track. In: The Fourteenth Text REtrieval Conference Proceedings (2005)
9. Davenport, T.H., Prusak, L.: Working Knowledge: How Organizations Manage What They Know. Harvard Business School Press, Boston (1998)
10. Demartini, G.: Leveraging Semantic Technologies for Enterprise Search. In: The Sixteenth Conference on Information and Knowledge Management (2007)
11. Deng, H., King, I., Lyu, M.R.: Formal Model for Expert Finding on DBLP Bibliography Data. In: The Eighth IEEE International Conference on Data Mining, pp. 163–172 (2008)
12. Duan, H., Zhou, Q., Lu, Z., Jin, O., Bao, S., Cao, Y., Yu, Y.: Research on Enterprise Track of TREC 2007 at SJTU APEX Lab. In: The Sixteenth Text REtrieval Conference Proceedings (2007)
13. Hawking, D.: Challenges in Enterprise Search. In: Proceedings of the 15th Australasian database conference, pp. 15–24 (2004)
14. Lavrenko, V., Croft, W.B.: Relevance-Based Language Models. In: Proceedings of 24th annual international ACM SIGIR conference on Research and development in information retrieval, pp. 120–127 (2001)
15. Mattox, D., Maybury, M., Morey, D.: Enterprise Expert and Knowledge Discovery. In: Proceedings of the 8th International Conference on Human-Computer Interaction, pp. 303–307 (1999)
16. Petkova, D., Croft, W.B.: Hierarchical Language Models for Expert Finding in Enterprise Corpora. In: Proceedings of the 18th IEEE International Conference on Tools with Artificial Intelligence, pp. 599–608 (2006)
17. Serdyukov, P., Aly, R., Hiemstra, D.: University of Twente at the TREC 2008 Enterprise Track: Using the Global Web as an expertise evidence source. In: The Seventeenth Text REtrieval Conference Proceedings (2008)
18. Shah, C., Croft, W.B.: Evaluating High Accuracy Retrieval Techniques. In: Proceeding of the 27th annual international ACM SIGIR conference on Research and development in infromation retrieval, pp. 2–9 (2004)
19. Soboroff, I., de Vries, A., Craswell, N.: Overview of the TREC 2006 Enterprise Track. In: The Fourteenth Text REtrieval Conference Proceedings (2006)
20. Zhai, C., Lafferty, J.: A Study of Smoothing Methods for Language Models Applied to Information Retrieval. ACM Transactionson Information Systems 22(2), 179–214 (2004)
21. Zhu, J.: A Study of the Relationship between Ad hoc Retrieval and Expert Finding in Enterprise Enviroment. In: Proceeding of the 10th ACM workshop on Web information and data management, pp. 25–30 (2008)

Debt Detection in Social Security by Adaptive Sequence Classification*

Shanshan Wu[1], Yanchang Zhao[1], Huaifeng Zhang[2],
Chengqi Zhang[1], Longbing Cao[1], and Hans Bohlscheid[2]

[1] Centre for Quantum Computation and Intelligent Systems (QCIS),
University of Technology, Sydney, Australia
{shanshan,yczhao,chengqi,lbcao}@it.uts.edu.au
[2] Data Mining Section, Payment Reviews Branch, Business Integrity Division,
Centrelink, Australia
{huaifeng.zhang,hans.bohlscheid}@centrelink.gov.au

Abstract. Debt detection is important for improving payment accuracy in social security. Since debt detection from customer transaction data can be generally modelled as a fraud detection problem, a straightforward solution is to extract features from transaction sequences and build a sequence classifier for debts. For long-running debt detections, the patterns in the transaction sequences may exhibit variation from time to time, which makes it imperative to adapt classification to the pattern variation. In this paper, we present a novel adaptive sequence classification framework for debt detection in a social security application. The central technique is to catch up with the pattern variation by boosting discriminative patterns and depressing less discriminative ones according to the latest sequence data.

Keywords: sequence classification, adaptive sequence classification, boosting discriminative patterns.

1 Introduction

In social security, each customer's transactional records form an activity sequence. From a business point of view, these sequences have a close relationship with debt occurrence. Here, a debt indicates an overpayment made by government to a customer who is not entitled to that payment. Debt prevention has emerged as a significant business goal in government departments and agencies. For instance, in Centrelink Australia (http://www.centrelink.gov.au), the Commonwealth Government agency responsible for delivery social security payments and benefits to the Australian community, approximately 63 billion Australian dollars (30% of government expenditure) is distributed each year to 6.4 million customers. Centrelink makes 9.98 million individual entitlement payments and

---

* This work was supported by the Australian Research Council (ARC) Linkage Project LP0775041 and Early Career Researcher Grant from University of Technology, Sydney, Australia.

D. Karagiannis and Z. Jin (Eds.): KSEM 2009, LNAI 5914, pp. 192–203, 2009.

5.2 billion electronic customer transactions annually [1]. In any year, the debt raised by Centrelink is significant, to say the least. In order to achieve high payment accuracy, it is imperative and important for Centrelink to detect and prevent debts based on the customer activity sequence data. In this paper, we present our research work of debt detection in Centrelink using adaptive sequence classification.

For each customer's activity sequence associated with a debt, it is labelled as *debt*. Contrarily, if no debt happens to a customer, the corresponding sequence is labelled as *normal*. Therefore, debt detection can be generally modelled as a sequence classification problem. Sequence classification has been a focused theme in the data mining research community. Since 1990's, along with the development of pattern recognition, data mining and bioinformatics, many sequence classification models have been proposed. Frequent pattern based classification is one of the most popular methodologies and its power was demonstrated by multiple studies [2,3,4].

Most of the conventional frequent pattern based classifications follow two steps. The first step is to mine a complete set of sequential patterns given a minimum support. The second step is to select a number of discriminative patterns to build a classifier. In most cases, mining for a complete set of sequential patterns on a large dataset is extremely time consuming. The discovered huge number of patterns make the pattern selection and classifier building very time consuming too. In fact, the most important consideration in sequence classification is not finding the complete rule set, but discovering the most discriminative patterns. Recently, more attention has been put on the discriminative frequent pattern discovery for effective classification [5,6]. In this paper, we proposed a novel measure, *contribution weight*, to boost the discriminative patterns. Contribution weight is induced by applying the frequent patterns to a set of evaluation data, and expresses the discriminative power of the patterns regarding to the evaluation data. The interestingness measure of frequent patterns are refined by contribution weight, so as to let the discriminative patterns pop up.

Moreover, sequence data represent the evolvement of data sources, and the sequential patterns generalize the trends of sequences. For long running sequence classification issues, even if the sequences come from the same source, the sequential patterns may vary from time to time. Therefore, the classifier built on a sequence dataset in the past may not work well on the current sequence dataset, not to mention future datasets. For example, based on our previous research work in Centrelink Australia, we found out that the classifier built on transaction data generated from Jul. 2007 to Feb. 2008 does not work quite well on the new data generated from Mar. 2008 to Sep. 2008, due to the changes of policies, economic situation and other social influences. Therefore, it is significant to improve the sequence classification to make it adapt to the sequential pattern variation. The most direct way is to rebuilt the classifier with the latest training dataset. However, the training is quite a time-consuming process, and if the pattern variation is not so much, an incremental updates would be much more efficient than rebuilding the classifier. In this paper, we propose an adaptive sequence classification framework to tackle the above problem. The adaptive model adapts the classifier in a timely fashion by adopting the proposed discriminative pattern boosting strategy, so as to catch up with the trends of sequential pattern variation and improve the classification accuracy.

There are three main contributions in this paper. Firstly, we propose a novel method to boost discriminative frequent patterns for sequence classification, which improves the accuracy of classifier. Secondly, we build up an adaptive sequence classification model which upgrades the sequence classification performance on time-varying sequences. Lastly, our strategies are applied to a real-world application, which shows the efficiency and effectiveness of the proposed methods.

The structure of this paper is as follows. Section 2 provides the notation and description to be used in this paper. Section 3 introduces how the discriminative frequent patterns are boosted. Our proposed adaptive sequence classification framework is given in Section 4. The case study is in Section 5, which is followed by the related work in Section 6. Then we conclude the paper in Section 7.

2   Problem Statement

Let S be a sequence database, in which each sequence is an ordered list of *elements*. These elements can be either *simple items* from a fixed set of items, or *itemsets*, that is, non-empty sets of items. The list of elements of a data sequence s is denoted by $< s_1, s_2, \cdots, s_n >$, where s_i is the ith element of s.

Consider two sequences $s =< s_1, s_2, \cdots, s_n >$ and $t =< t_1, t_2, \cdots, t_m >$. We say that s is a subsequence of t if s is a "projection" of t, derived by deleting elements and/or items from t. More formally, s is a subsequence of t if there exist integers $j_1 < j_2 < \ldots < j_n$ such that $s_1 \subseteq t_{j1}$, $s_2 \subseteq t_{j2}$, ..., $s_n \subseteq t_{jn}$. Note that for sequences of simple items the above condition translates to $s_1 = t_{j1}$, $s_2 = t_{j2}$, ..., $s_n = t_{jn}$. A sequence t is said to *contain* another sequence s if s is a subsequence of t, in the form of $s \subseteq t$.

2.1   Frequent Sequential Patterns

The number of sequences in a sequence database S containing sequence s is called the support of s, denoted as $sup(s)$. Given a positive integer min_sup as the support threshold, a sequence s is a frequent sequential pattern in sequence database S if $sup(s) \geq min\_sup$. The sequential pattern mining is to find the complete set of sequential patterns with respect to a given sequence database S and a support threshold min_sup.

2.2   Classifiable Sequential Patterns

Let T be a finite set of *class labels*. A *sequential classifier* is a function

$$\mathcal{F} : \mathcal{S} \to \mathcal{T} \tag{1}$$

In sequence classification, the classifier $\mathcal{F}$ is built on the base of frequent *classifiable sequential patterns* $\mathcal{P}$.

Definition 2.1 (Classifiable Sequential Pattern). *Classifiable Sequential Patterns (CSP) are frequent sequential patterns for the sequential classifier in the form of $p_a \Rightarrow \tau$, where p_a is a frequent pattern in the sequence database $\mathcal{S}$.*

Based on the mined classifiable sequential patterns, a sequential classifier can be formulized as

$$\mathcal{F} : s \xrightarrow{\mathcal{P}} \tau. \tag{2}$$

That is, for each sequence $s \in \mathcal{S}$, $\mathcal{F}$ predicts the target class label of s based on the sequential classifier built with the classifiable sequential pattern set $\mathcal{P}$. Suppose we have a classifiable sequential pattern set $\mathcal{P}$. A sequence instance s is said to be *covered* by a classifiable sequential pattern $p \in \mathcal{P}$ if s contains the antecedent p_a of the classifiable sequential pattern p.

3   Discriminative Frequent Patterns Boosting

Given a dataset, the more samples a pattern can correctly classifies, the more discriminative the pattern is on the dataset. In other words, the more samples a pattern incorrectly classifies, the less discriminative the pattern is on the dataset. To make it more statistically significant, the definitions of *positive contribution ability* and *negative contribution ability* are given as follows.

Definition 3.1 (Positive Contribution Ability). *Given a dataset S, the Positive Contribution Ability (PCA) of pattern P is the proportion of samples that can be correctly classified by P out of all the samples in dataset S.*

Definition 3.2 (Negative Contribution Ability). *Given a dataset S, the Negative Contribution Ability (NCA) of pattern P is the proportion of samples that are incorrectly classified by P out of all the samples in the dataset S.*

For a classifiable sequential pattern P in the form of $p_a \Rightarrow \tau$, PCA of P on S can be denoted as

$$PCA_S(P) = \frac{\|\{s|p_a \subseteq s \land s \in S_\tau\}\|}{\|S\|}, \tag{3}$$

and NCA of pattern P on S can be denoted as

$$NCA_S(P) = \frac{\|\{s|p_a \subseteq s \land s \in S_{\neg\tau}\}\|}{\|S\|}, \tag{4}$$

where S_τ and $S_{\neg\tau}$ represent the subsets of S in which samples are of class τ and are not of class τ, respectively.

Above all, PCA and NCA describe the classification ability of patterns on a given dataset. In order to enhance classification performance, it is intuitive to boost the patterns with higher PCA and lower NCA, while depress those with lower PCA and higher NCA. Thereafter, a measure of *Contribution Weight* is proposed to measure the discriminative power that a pattern contributes to the classification on a dataset.

Definition 3.3 (Contribution Weight). *Given a dataset S, Contribution Weight of a classifiable sequential pattern P is the ratio of Positive Contribution*

Ability $PCA_S(P)$ on S and Negative Contribution Ability $NCA_S(P)$ on S. It can be denoted as

$$CW_S(P) = \frac{PCA_S(P)}{NCA_S(P)} = \frac{\|\{s|p \subseteq s \wedge s \in S_\tau\}\|}{\|\{s|p \subseteq s \wedge s \in S_{\neg\tau}\}\|}. \tag{5}$$

The proposed measure of contribution weight tells the relative discriminative power of a classifiable sequential pattern on a given dataset, which is based on the classification performance of the pattern on the dataset. According to the definition, contribution weight has following characters.

- The greater the value of contribution weight is, the more discriminative a pattern is on a given dataset, and vice versa.
- Contribution weight is a measure with regard to a dataset on which classification performance is evaluated.
- Contribution weight is independent of the algorithm that is used for classifiable sequential pattern mining, and it does not matter which interestingness measure is used for classification.

Therefore, we introduce contribution weight as a factor to boost the discriminative frequent patterns on a certain dataset. The term of *Boosted Interestingness Measure* is defined as follows.

Definition 3.4 (Boosted Interestingness Measure). *For a classifiable sequential pattern P with an interestingness measure R, the corresponding Boosted Interestingness Measure on dataset S is denoted as*

$$R_S^* = R \times CW_S(P). \tag{6}$$

In other words, boosted interestingness measure of a pattern can be regarded as a weighted interestingness measure, and the weight tells how much contribution the corresponding pattern can make to the classification on the given dataset. Patterns that are more discriminative on a given dataset are strongly boosted by higher contribution weights, and vice versa. From this point of view, boosted interestingness measure adjusts the original interestingness measure so as to make it indicating the discriminative ability of classifiable sequential patterns on the given dataset more vividly.

4   Adaptive Sequence Classification Framework

In order to catch up with the pattern variation over time, an adaptive sequence classification framework is introduced in this section. The main idea of the adaptive framework is to include the latest pattern into the classifier with the proposed boosted interestingness measure, so as to improve the classification performance on dataset of near future.

As illustrated in Figure 1, the initial classifiable sequential pattern set CSP_0 is extracted from the dataset DS_0, and then is used to perform prediction/classification on coming dataset DS_1 and get the predicted labels L_1'. Once L_1,

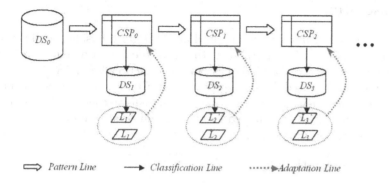

⟹ Pattern Line   ⟶ Classification Line   ·····▶Adaptation Line

Fig. 1. Architecture of Adaptive Sequence Classification

Algorithm 1. Adaptive classification model.

Data: Dataset DS_i and corresponding real labels L_i that are available after
classification/prediction, $i = 0, 1, ...$
Basic classification algorithm$F(F_1$:Classifier construction;F_2:Classifying)

Result: Predicted labels L_i', $i = 1, 2, ...$
Classifiers CSP_i, $i = 0, 1, 2, ...$

1 **begin**
2     $CSP_0 = F_1(DS_0, L_0)$
3     $i = 1$
4   **while** i **do**
5        $L_i' = F_2(DS_i, CSP_{i-1})$
6   *Wait till L_i is available*
7   *Modify CSP_{i-1} with $R^*_{(DS_i, L_i)}$ to get CSP_i*
8        $i = i + 1$

9 **end**

---

the real class labels of dataset DS_1 , is available, interestingness measure of
the classifier CSP_0 could be refined and CSP_0 evolves into CSP_1 with boosted
interestingness measure, which brings the timely trends of patterns in dataset
DS_1 into the classification model. The boosted classifier will be applied to continuously coming dataset for prediction/classification. The procedure goes on
as dataset updates all along, which is generalized in Algorithm 1. The boosted
classifier CSP_i, $i = 1, 2, ...$ not only takes the latest pattern variation into the
classification model, but also tracks the evlovement of the patterns ever since the
initial classifier is built. Therefore, the performance of classification is expected
to outperform that of the initial classifier.

Since the adaptive model is based on boosted interestingness measure, it inherits the properties of boosted interestingness measure congenitally. To be more
precise, it is independent of interestingness measure and classifiable sequence
mining method.

5   Case Study

The proposed algorithm has been applied in a real world business application in Centrelink, Australia. The purpose of the case study is to predict and further prevent debt occurrence based on customer transactional activity data. In this section, the dataset used for debt prediction in Centrelink is described firstly. Then a pre-experiment is given to evaluate the effectiveness of discriminative pattern boosting strategy, followed by the experimental results of adaptive sequence classification framework.

5.1   Data Description

The dataset used for sequence classification is composed of customer activity data and debt data. In Centrelink, every single contact (e.g., a circumstance change) of a customer may trigger a sequence of activities running. As a result, large volumes of activity based transactions are encoded into 3-character "Activity Code" and recorded in activity transactional files. In the original activity transactional table, each activity has 35 attributes, in which 4 attributes are used in the case study. These attributes are "CRN" (Customer Reference Number) of a customer, "Activity Code", "Activity Date" and "Activity Time", as shown in Table 1. We sort the activity data according to "Activity Date" and "Activity Time" to construct the activity sequence. The debt data consist of the "CRN" of the debtor and "Debt Transaction Date". In our case study, only the activities of a customer before the occurrence of his/her debt are kept for the sequence classification.

Table 1. Centrelink Data Sample

CRN	Act_Code	Act_Date	Act_Time
*****002	DOC	20/08/07	14:24:13
*****002	RPT	20/08/07	14:33:55
*****002	DOC	05/09/07	10:13:47
*****002	ADD	06/09/07	13:57:44
*****002	RPR	12/09/07	13:08:27
*****002	ADV	17/09/07	10:10:28
*****002	REA	09/10/07	7:38:48
*****002	DOC	11/10/07	8:34:36
*****002	RCV	11/10/07	9:44:39
*****002	FRV	11/10/07	10:18:46
*****002	AAI	07/02/08	15:11:54

Table 2. Data Windows

Window	Start Date	End Date
W0	02/07/07	31/10/07
W1	01/08/07	30/11/07
W2	01/09/07	31/12/07
W3	01/10/07	31/01/08
W4	01/11/07	29/02/08
W5	01/12/07	31/03/08
W6	01/01/08	30/04/08
W7	01/02/08	31/05/08
W8	01/03/08	30/06/08
W9	01/04/08	31/07/08
W10	01/05/08	31/08/08

5.2   Effectiveness of Boosting Discriminative Patterns

In order to evaluate the effectiveness of discriminative patterns boosting, two groups of experiments are presented in this section. In both groups, we compare the performance of classification which uses discriminative pattern boosting strategy with that does not boost discriminative patterns. In group (a), the activity sequence data generated from Jul. 2007 to Oct. 2007 are used. After data cleaning, there are 6, 920 activity sequences including 210, 457 activity records used. The dataset is randomly divided into the following 3 portions.

- Training data(60%): To generate the initial classifier.
- Evaluation data(20%): To refine classifier.
- Test data(20%): To test the performance of classification.

While in group (b), some data generated in Nov. 2007 is added to the evaluation data and test data, expecting to include some pattern variation.

According to the property of contribution weight, the boosted interestingness measure is independent of basic classification. Therefore, we use the classification algorithm proposed in our previous work [13] to generate the initial classifier on the training dataset. And we use confidence as the base interestingness measure. For classification which uses boosting strategy, the evaluation dataset is used to refine the initial classifier, and the refined classifier is evaluated on the test dataset. While for the classification that does not boost discriminative patterns, we combine training data and evaluation data to generate the initial classifier, and then apply the initial classifier to the test dataset for debt prediction.

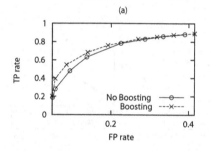 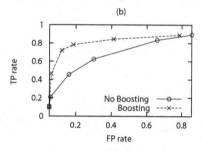

Fig. 2. Effectiveness of Discriminative Patterns Boosting

ROC curve (Receiver Operating Characteristic) is used to plot the fraction of true positives *vs.* the fraction of false positives of each classifier. The best possible prediction method would yield a point in the upper left corner or coordinate (0,1) of the ROC space, representing 100% sensitivity (no false negatives) and 100% specificity(no false positives). Therefore, the more close to the upper left corner the curve is, the better the classification method is. As illustrated in Fig. 2, the boosted classifier outperforms the classifier without boosting in both experiments. In group (a), training data, evaluation data and test data all come from the dataset generated in the same time period. By boosting discriminative patterns with evaluation data, classification power of initial classifier is refined by boosting discriminative patterns and depressing less discriminative patterns, so it outperforms the classification without boosting. As for group (b), since some new data generated in different time period is added to the evaluation data and test data, some pattern variation might be included in the corresponding dataset. In this circumstance, the proposed boosting strategy notices the pattern variation in the updated dataset, refines the interestingness measure of the classifiers with evaluation data, and performs much better in the test data than the classifier without boosting.

In all, the discriminative pattern boosting strategy improves the classification performance, especially when the sequence data evolves with pattern variation.

5.3 Performance of Adaptive Sequence Classification Framework

In this subsection, we will evaluate the adaptive sequence classification framework on the sequence datasets obtained with a sliding window applied on the activity sequence data. After applying sliding window on the sequences generated from Jul. 2007 to Aug. 2008, we get 11 windows listed in Table 2.

Following the framework proposed in Section 4, the classification in our previous work [13] is firstly applied on W_0 and the initial classifier CSP_0 is generated. By discriminative pattern boosting with W_1, CSP_0 is refined to CSP_1 and then is applied to make debt prediction on W_2. Here we still use confidence as the base interestingness measure. The debt prediction performance on W_2 is illustrated in the first graph in Fig. 3. Thereafter, CSP_1 is boosted with sequence data in W_2, and the generated CSP_2 is applied on W_3 to predict debt occurrence. As the procedure goes on continuously, the debt prediction performance on all the

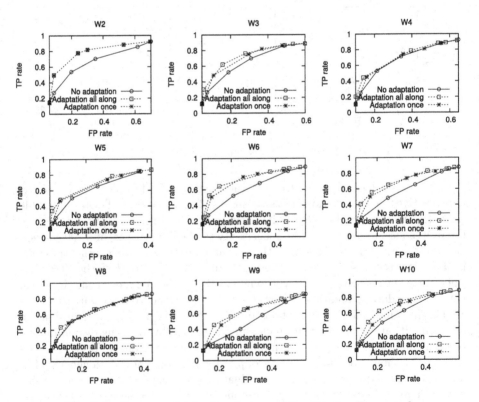

Fig. 3. RoC curves of Adaptive Sequence Classification Framework

following windows are listed in Fig. 3, which is represented by the ROC curves labelled *Adaptation_all_along*. In order to evaluate the performance of adaptive sequence classification framework, debt prediction on each window is also performed with initial classifier CSP_0, whose performance is denoted by the ROC curves labelled *No_adaptation*. According to Fig. 3, we can tell that the proposed adaptive framework outperforms the initial classifier in debt prediction on continuously coming datasets. Since the classifier is continuously updated with the latest data, it catches up the pattern variation in the new dataset and then works well on the debt prediction on the oncoming dataset. Meanwhile, we apply CSP_1, which is boosted once based on initial classifier, to each of the windows and get the performance denoted by the curves labelled *Adaptation_once*. The classifier boosted once still outperforms the initial classifier. While it does not contains the pattern information in the latest datasets, its performance is always worse than that of *Adaptation_all_along* strategy.

Above all, conclusion could be drawn that our proposed adaptive sequence classification framework updates the classifier with new data, includes the sequence pattern variation in the new data, and performs effectively on the continuously arriving data.

6   Related Work

6.1   Sequence Classification

There have been several researchers working towards building sequence classifiers based on frequent sequential patterns. Lesh et al. [2] proposed an algorithm for sequence classification using frequent patterns as features in the classifier. In their algorithm, subsequences are extracted and transformed into sets of features. After feature extraction, general classification algorithms such as Naïve Bayes, SVM or neural network can be used for classification. Their algorithm is the first try on the combination of classification and sequential pattern mining. However, a huge amount of sequential patterns are mined in the sequential mining procedure. Although pruning algorithm is used for the post-processing, there are still a large amount of sequential patterns constructing the feature space. Tseng and Lee [3] proposed an Classify-By-Sequence (CBS) algorithm to combine sequential pattern mining and classification. In their paper, two algorithms, CBS_Class and CBS_All were proposed. In CBS_Class, the database is divided into a number of sub-databases according to the class label of each instance. Then sequential pattern mining is implemented on each sub-database. In CBS_All, conventional sequential pattern mining algorithm is used on the whole dataset. Weighted scoring is used in both algorithms. Exarchos [4] proposed to combine sequential pattern mining and classification followed by an optimization algorithm. The accuracy of their algorithm is higher than that of CBS. However optimization is a very time-consuming procedure.

6.2   Adaptive Classification

Some applications similar to debt detection are financial crime detection, network intrusion detection and spam detection. Bonchi et al. [7] proposed a classification-based methodology for planning audit strategies in fraud detection. The models are constructed by analyzing historical audit data. Then the models are used to plan future audits for the detection of tax evasion. A decision tree algorithm, C5.0, was used in their case study. Although the target problem is similar with ours, the data used are different. What we used are transactional data that record activities related to customers. Because the time order in activities is important for predicting debt occurrences, sequence classifiers instead of decision trees are used in our application. Rosset et al. [8] studied the fraud detection in telecommunications and presented a two-stage system based on C4.5 to find fraud rules. They adapted the C4.5 algorithm for generating rules from bi-level data, i.e., customer data and behaviour-level data. However, the behaviour data they used is the statistics in a short time frame, which is different from sequential patterns in our techniques.

Fawcett and Provost [9] addressed the issue of adaptive fraud detection in telecommunication services. They use a rule-learning program to uncover indicators of fraudulent behaviour from a large database of customer transactions. Then the indicators are used to create a set of monitors, which profile legitimate customer behaviour and indicate anomalies. Xu et al. [10] utilize FP-tree growth algroithm to extract the associations among features from transactions during a certain period in order to profile the user's behavior adaptively. Lu et al. [11] used deviations from the expected Benfords Law distributions as an indicators of anomalous behaviour that are strong indicators of fraud. The adaptive Benfords Law adapts to the incomplete data records, compared with classic Benfords Law. Lee et al. [12] used meta-learning mechanism by combining existing models with new models trained on new intrusion data or new normal data to make intrusion detection models adaptive. Different from transactional fraud detection that attempts to classify a transaction or event as being legal or fraud, our techniques try to predict the likelihood of a customer being fraud based on his/her past activities. It is at customer level instead of transaction level.

7   Conclusion and Future Work

In this paper we proposed a novel adaptive sequence classification framework for long running sequence classification in the circumstance of time-varying sequence patterns. In order to make the classifier catch up with the latest sequence pattern variation, we introduce a discriminative pattern boosting strategy, which boosts discriminative patterns and depresses less discriminative patterns based on the latest sequential data. The proposed methods are tested on a real-world dataset, and the case study shows the effectiveness of the proposed strategy.

Our current adaptive framework refines the classifier round by round, and in each round the adaptation is based on the classifier generated in the last round. Though it tracks the evolvement of sequential patterns, the latest pattern variation is given the same consideration as the previous ones. In our future work, we will study how to put tilted weight to the historical data, which may include more latest sequence pattern characteristics into the classification model.

Acknowledgements. We are grateful to Mr. Peter Newbigin and Mr. Brett D. Clark from Payment Reviews Branch, Business Integrity Division, Centrelink, Australia, for extracting data, providing domain knowledge and helpful suggestions.

References

1. Centrelink. Centrelink annual report 2004-2005. Technical report, Centrelink, Australia (2005)
2. Lesh, N., Zaki, M.J., Ogihara, M.: Mining Features for Sequence Classification. In: Proc.of the fifth ACM SIGKDD International Conference on Knowledge Discovery and Data Mining, San Diego, California, USA, August 1999, pp. 342–346 (1999)
3. Tseng, V.S.M., Lee, C.-H.: CBS: A new classification method by using sequential patterns. In: Proc. of SIAM International Conference on Data Mining (SDM 2005), pp. 596–600 (2005)
4. Exarchos, T.P., Tsipouras, M.G., Papaloukas, C., Fotiadis, D.I.: A two-stage methodology for sequence classification based on sequential pattern mining and optimization. Data and Knowledge Engineering 66(3), 467–487 (2008)
5. Cheng, H., Yan, X., Han, J., Hsu, C.-W.: Discriminative Frequent Pattern Analysis for Effective Classification. In: Proc. of IEEE International Conference on Data Engineering (ICDE 2007), April 2007, pp. 716–725 (2007)
6. Han, J., Cheng, H., Xin, D., Yan, X.: Frequent pattern mining: current status and future directions. Data Mining and Knowledge Discovery 15(1), 55–86 (2007)
7. Bonchi, F., Giannotti, F., Mainetto, G., Pedreschi, D.: A classification-based methodology for planning audit strategies in fraud detection. In: Proc. of the 5th ACM SIGKDD International Conference on Knowledge Discovery and Data Mining, San Diego, CA, USA, August 1999, pp. 175–184 (1999)
8. Rosset, S., Murad, U., Neumann, E., Idan, Y., Pinkas, G.: Discovery of fraud rules for telecommunications - challenges and solutions. In: Proc. of the 5th ACM SIGKDD International Conference on Knowledge Discovery and Data Mining, San Diego, CA, USA, August 1999, pp. 409–413 (1999)
9. Fawcett, T., Provost, F.: Adaptive fraud detection. Data Mining and Knowledge Discovery 1, 291–316 (1997)
10. Xu, J., Sung, A.H., Liu, Q.: Tree based behavior monitoring for adaptive fraud detection. In: Proc. of the 18th International Conference on Pattern Recognition, Washington, DC, USA, pp. 1208–1211 (2006)
11. Lu, F., Boritz, J.E., Covvey, D.: Adaptive fraud detection using Benfords Law. In: Lamontagne, L., Marchand, M. (eds.) Canadian AI 2006. LNCS (LNAI), vol. 4013, pp. 347–358. Springer, Heidelberg (2006)
12. Lee, W., Stolfo, S.J., Mok, K.W.: Adaptive intrusion detection: a data mining approach. Artificial Intelligence Review 14, 533–567 (2000)
13. Zhang, H., Zhao, Y., Cao, L., Zhang, C., Bohlscheid, H.: Customer activity sequence classification for debt prevention in social security. Accepted by Journal of Computer Science and Technology (2009)

Ontology Based Opinion Mining for Movie Reviews

Lili Zhao and Chunping Li

Tsinghua National Laboratory for Information Science and Technology
Key Laboratory for Information System Security, Ministry of Education
School of Software, Tsinghua University, Beijing, China
zhaoll07@mails.tsinghua.edu.cn, cli@tsinghua.edu.cn

Abstract. Ontology itself is an explicitly defined reference model of application domains with the purpose of improving information consistency and knowledge sharing. It describes the semantics of a domain in both human-understandable and computer-processable way. Motivated by its success in the area of Information Extraction (IE), we propose an ontology-based approach for opinion mining. In general, opinion mining is quite context-sensitive, and, at a coarser granularity, quite domain dependent. This paper introduces a fine-grain approach for opinion mining, which uses the ontology structure as an essential part of the feature extraction process, by taking account the relations between concepts. The experiment result shows the benefits of exploiting ontology structure to opinion mining.

Keywords: Ontology, Opinion mining, Sentiment analysis, Feature extraction.

1 Introduction

With the expansion of Web2.0, the number of online reviews grows rapidly. More and more Websites, such as Amazon (http://www.amazon.com) and IMDB (http://www.imdb.com), encourage users write reviews for the products they are interest in. When a person writes a review, he or she probably comments several aspects of the product, moreover, opinions to the aspects are not the same, some aspects receive negative comments, some receive positive ones, and some receive neutral ones. These opinions are useful for both users and manufactures. As a result, the area of opinion mining and sentiment analysis has enjoyed a burst research activity recently.

Opinion mining aims to discover user opinions from their textual statements automatically. It is distinctively different from traditional text mining in that the latter is based on objective topics rather than on subjective perceptions. Specifically, traditional text mining focuses on specific topics (e.g., business, travel) as well as topic shifts in text whereas opinion mining is much more difficult than those for topic mining. This is partially attributed to the fact that topics are represented explicitly with keywords while opinions are expressed with subtlety.

D. Karagiannis and Z. Jin (Eds.): KSEM 2009, LNAI 5914, pp. 204–214, 2009.

Opinion mining requires deeper understanding of language characteristic and textual context[1].

In the past few years, many researchers studied the problem [2][3][4][5][6][7]. Most of the existing works are based on product reviews because a review usually focuses on a specific product and contains little irrelevant information. The main tasks are to discover features that have been commented on and to decide whether the opinions are positive or negative. In this paper, we will focus on movie review. Different from product reviews, movie reviews have following characteristic. The reviews contain users' comments on movie attributes (e.g. screenplay, sound), each attribute may be expressed in different words by different users. For example, "screenplay" may be written by "screen" or "picture" instead. Motivated by the role of ontology in conceptualizing domain-specific information [8], in this work, we attempt to solve the feature identification problem in the domain of movie review by employing ontology structure. Ontology allows us to interpret a movie review at a finer granularity with shared meaning.

The rest of this paper is organized as follows: section 2 discusses related work. Section 3 describes in detail the proposed method. We report in section 4 our experimental results and give our conclusion on this work in section 5.

2   Related Work

In most studies of opinion mining, two main research directions have been given attention, including sentiment classification and feature-level opinion mining. Traditional document classification methods are often employed in sentiment classification, which entire document is classified as positive, negative or neutral. However, these methods are too rough. In most cases, both positive and negative opinions can appear in the same document, for example, "Overall, this movie is good, the story is moving, but the music is bad.". There are some works focused on sentiment classification [1][9][10][11][12][13][14]. They usually manually or half-manually construct a priori knowledge dictionary which contains polarity words. Some classic machine learning methods (Naive Bayes, Maximum Entropy, and SVM) have been experimented in [1], and the authors point out that machine learning methods are more effective than manually labeling. In [14], a model-based approach is proposed, trying to analyze the structure of the document deeply, but there is not any of experiment data. These works are different from ours as we are interested in opinions expressed on each feature.

In the feature level opinion mining [2][3][4][5][15][16][7][17][18] [19], two critic extraction-related sub-tasks were (1) identification features, and (2) extraction of opinions associated with these features. Existing works on identifying object features discuss in reviews often rely on the simple linguistic heuristic that (explicit) features are usually expressed as nouns or noun phrases. In [2], the authors consider product features to be concepts forming certain relationships with the product (for example, for a movie, its music is one of its properties) and try to identify the features connected with the product name through corresponding meronymy discriminators. Note that their approach does not involve opinion mining but simply focuses more on the task of identifying different types of features. In [3], they propose the idea of opinion mining and summarization. They use a lexicon-based method to determine whether the opinion expressed on a product feature is positive or negative. Later, [5] follows the intuition that frequent nouns or noun phrases are likely to be features. They identify frequent features through association mining, and then apply heuristic-guided pruning aimed at removing (a) multi-word candidates in which the words do not appear together in a certain order, and (b) single-word candidates for which subsuming super-strings have been collected (the idea is to concentrate on more specific concepts, so that, "music" is discarded in favor of "background music"). This method is improved in [7] by a more sophisticated method based on holistic lexicon and linguistic rules to identify the orientations of context dependent opinion words. The approach works with many special words, phrases and language constructs which have impacts on opinions based on their linguistic patterns, such as "negation" rules, "but" clauses rules, intra-sentence conjunction rule, etc. In [17], the authors consider three increasingly strict heuristics to select from noun phrases based on POS tag patterns. Both [18] and [19] used conjunction rules to find words from large domain corpora. The conjunction rule basically states that when two opinion words are linked by a conjunction word in a sentence, their opinion orientations are the same.

3   Methodology

3.1   Main Framework

In this work, the main idea is to improve feature level opinion mining by employing ontology. The proposed method integrates of word segmentation, part-of-speech tagging, ontology development, feature identification, and sentiment classification. The framework is illustrated in Fig. 1. Contrast to previous methods, it incorporates the component of ontology.

The functions of main components are as follows:

1. Data preprocessing, including words segmentation and POS tagging. Because many concepts in this field cannot be found in general dictionary, this part is achieved with the help of ontology development, which improves the accuracy of word segmentation.
2. Feature identification. In this part, we integrate ontology to improve the accuracy of feature extraction.
3. Polarity identification. Our approach to polarity measurement, like others, relies on a lexicon of tagged positive and negative sentiment terms which are used to quantify positive/negative sentiment. In this part, SentiWN [20] was used as it provides a readily interpretable positive and negative polarity value for a set of "affective" terms.
4. Sentiment analysis. This part completes two tasks: a) convert the polarity obtained from the 3 step to more precise one by analyzing the context, such as "negation" rules. b) with the help of ontology again, make it possible to compute the polarities of the nodes through the hierarchy relationship.

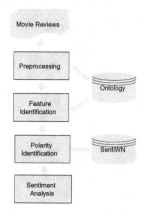

Fig. 1. Overview Framework of the Proposed Method

3.2 Ontology Development

According to the domain dependence, ontology can be divided into four types: generic ontology, domain ontology, task ontology, and application ontology. In the following, we focus on ontology development. The domain ontology is to describe the concepts of special area, including concepts in special area, attributes of the concepts, relationship between concepts, and constraints among the relationships. The target of constructing domain ontology is to define common terminologies in the area, and give the definition of the relationships among the terminologies.

In the domain of opinion mining, ontology is a generic concept. In the proposed method, it is not necessary as much complicated as the structure of OWL. We aim to find the opinion on a-feature-of-movie, or some attributes of feature-of-movie. Therefore, in our movie ontology, the concept is apart into two parts: movie and feature. This paper focuses on movie review, many concepts in this field cannot be found in general dictionary. So during the segmentation, we need domain ontology to complete the segmentation. In this section, we present an iterative approach to construct the ontology (Fig.2).

Our goal is to extract the concepts with a seeds set from movie reviews. The task of ontology construction will be divided into two steps as follows:

- Step 1: Select the relevant sentences including conceptions
- Step 2: Extract the conceptions from those sentences

We choose the sentences following two conditions. The first one is that the sentence contains a conjunction word. The second one is that the sentence contains at least one concept seed. The reason is that the phrases in the conjunction structure have the same characteristics. For example, the sentence "the special effects and sound effect are very good", we already know the "sound effect" is a feature, but do not know "special effects". If applying the conjunction strategy,

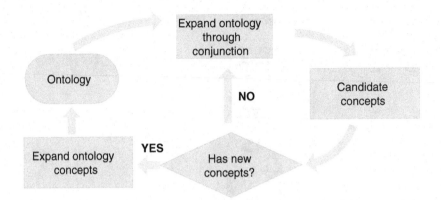

Fig. 2. Ontology Development

we will easily get that "special effects" is also a feature. At the initial stage, we use some manually labeled features as seeds. The whole process will end until no new candidates concept emerges.

The movie model is shown in Fig. 3(a) It is a forest structure which indicates the relationship among movies. Every node has a father node and a son node, moreover, every node has several synonyms which represent the terminologies in the area. For example, the root of the tree is "movie", however, some people often use "film" instead, then "film" is a synonym of "movie", that is why it is a forest. However, Fig. 3(a) does not show the synonyms. In this paper feature is an attribute of movie. The feature model is shown in Fig. 3(b), it is a multi-tree structure which indicates the relationship among features. As well as movie model, every node has a father node and a son node, every node has several synonyms represent the terminologies. Different from movie model, feature model is a multi-tree structure. Because in a movie review, people may say "the movie is good" without "the music is good" but not "the music is good" without "the movie ...". Therefore, the feature model is a multi-tree.

The movie model shows hierarchy relationship of the concepts, the root is "movie". The feature model is similar with movie model except that the root of attribute ontology is "summary". Because the review can focus on only one topic without any attribute, e.g., "the movie is moving" we cannot get any feature information from the sentence.

3.3   Feature Identification

The main idea for feature identification is to use ontology terminologies to extract features. The workflow is shown in Fig. 4.

Starting with a preprocessed movie review, we can identify related sentences which contain the ontology terminologies. From those sentences, we can easily extract the features.

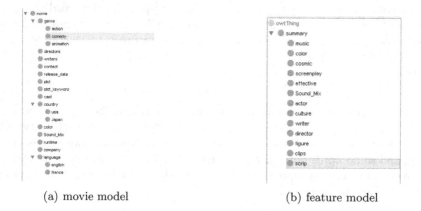

(a) movie model   (b) feature model

Fig. 3. Ontology Development

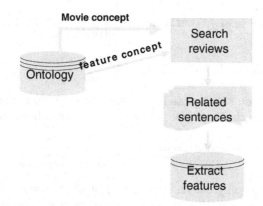

Fig. 4. The interaction between ontology and feature identification

3.4   Polarity Identification

SentiWN [20] introduces a semi-automatic approach to derive a version of Word-Net where word senses are bearing polarities.

Table 1 shows the entry of 'unimaginative' in SentiWN. The first column indicates the word's POS tagger, "a" represents adjective. The numbers under "synset-offset" indicate the word's position in the dictionary. The third and

Table 1. SentiWN: 'unimaginative'

POS	synset-offset	pos	neg	word	sense
a	1775641	0.0	0.0	unimaginative a	2
a	580160	0.125	0.5	unimaginative a	3
a	614153	0.0	0.625	unimaginative a	1

fourth columns with polarity tags (pos, neg) represent the word's polarity strength (the value of maximal strength equals to 1). Column "sense" indicates word's polarity, 1 and 3 represent negative while 2 represents neutral. Since the entry does not give the neutral score directly, we use formula (1) to obtain the neutral score:

$$score(obj) = 1 - score(pos) - score(neg) \qquad (1)$$

So the neutral scores of 'unimaginative' are 1.0, 0.375, 0.375 respectively.

However, we just want to get polarity score. After studying SentiWN, we find that some words have all the three senses, some words have two senses of the three, and some words only have one sense. In order to get one polarity entry of a word or phrase, we propose a computing method as following: If the word or phrase has all the three senses:

$$score('word' = pos) = \frac{\sum_{ws_i \in pos} score_{ws_i} - \sum_{ws_i \in neg} score_{ws_i}}{|word_{ws_i \in pos}|} \qquad (2)$$

$$score('word' = neg) = \frac{\sum_{ws_i \in neg} score_{ws_i} - \sum_{ws_i \in pos} score_{ws_i}}{|word_{ws_i \in neg}|} \qquad (3)$$

$$score('word' = obj) = \frac{\sum_{ws_i \in obj} score_{ws_i}}{|word_{ws_i \in obj}|} \qquad (4)$$

If the word or phrase has two senses of the three, we use two of the above formulas to compute the two senses. The third sense can be calculated as:

$$score('word' = noncompute\_sense) = 1 - \sum_{ws_i \in already\_compute} score_{ws_i} \qquad (5)$$

If the word or phrase only has one sense, we just use the scores in SentiWN.

In above formulas, $score('word')$ denotes the average sense value of word or phrase and ws_i means word sense. $score_{ws_i}$ is the value of SentiWN entry of word sense i while $|word_{wsi}|$ means the cardinal with the word sense i. If the value of $|word_{ws_i}|$ equals to 1 and the value of $\sum_{ws_i} score_{ws_i}$ also equals to 1 then the value of $score('word' = ws_i)$ will be 1 which is not suitable for formula (5). In this case, we use $word_{all\_wordsense}$ instead of $word_{ws_i}$.

In this way, the adjective 'unimaginative' gets a value of 0.33 in neutral, while negatively interpreted it receives 0.5, leaving a 0.17 score to the remaining positive case. Applying this strategy to SentiWN, we have generated an adjective lexicon with prior polarities that blends the numerical values of an adjective's SentiWN entry into three discrete polarity classes.

3.5   Sentiment Analysis

Using the final lists of positive, negative and neutral words or phrases, we analyze the opinion orientation expressed on each feature. Actually, we limit the sentiment analysis on the node of the model, because, some people are not very interested about features, they just want to know the general information about the movie. E.g., maybe people do not understand much terminologies in this area, he just wants to know the overall opinion about one movie, this is obviously, what this user wants to know is a high-level concept but not the information of a node. In this case, we need to use the hierarchy structure to calculate the opinion of high-level concept. Here, we give the formulas (6), (7), (8) to indicate how we get the three opinions on high-level concept through the nodes which are the children of the high level concept.

$$op_{hlc}(neg) = \frac{\sum_{children\_node_{ws_i \in neg}} score(children\_node1_{ws_i \in neg})}{|children\_node_{ws_i \in neg}|} \qquad (6)$$

$$op_{hlc}(pos) = \frac{\sum_{children\_node_{ws_i \in pos}} score(children\_node1_{ws_i \in pos})}{|children\_node_{ws_i \in pos}|} \qquad (7)$$

$$op_{hlc}(neu) = \frac{\sum_{children\_node_{ws_i \in neu}} score(children\_node1_{ws_i \in neu})}{|children\_node_{ws_i \in neu}|} \qquad (8)$$

In above formulas, op_{hlc} represents the opinion on high level concept while $score$ indicates the opinion score with same word sense ws_i on the children nodes. $|children\_node_{ws_i}|$ denotes the cardinal of children nodes with the same opinion. Taking the multi-tree structure of the model into account, this is a process of an iterative calculation.

Also, a set of linguistic rules need to be specially handled as follows:

Negative Words: The negation word or phrase usually reverses the opinion expressed in the sentence. Negation words include such "no", "not", "never". The proposed method detects these words and then modifies the opinion orientation. For example, the sentence, "the music is not good", conforms to the negation word "not", then it is assigned the negative orientation, although "good" is positive.

Conjunction Words: By the limited knowledge in the SentiWN, there are some words that we cannot fully identified. To solve this problem, we use conjunction words to detect the undiscovered words. The underlying intuition is that the conjunction words usually conjoin words of same polarity.

4   Experiment

We now evaluate the proposed method to see how effective it is for opinion mining.The proposed technique has been evaluated the benchmark review data set, which contains 1400 movie review documents randomly selected from IMDB (www.imdb.com) for analyzing. Half of the movie reviews are positive and the other half negative. We randomly select 60 positive review documents and 60 negative reviews to test. For the 120 documents, we manually label the features as shown in Table 2.

In the experiment, we take the labeled features as baseline in contrast to the results from the proposed method which is combined with ontology, which is shown in Table 3.

Table 2. Examples for Labeling Features

review sentence	features
as for acting , it's acceptable , but hardly anything else.	acting
there are some nice flash back sequences and occasionally impressive pacing , but the overall technical aspect of this film is on the ground floor.	flash back sequences, pacing, technical aspect, film
in other words '28 days' doesn't contribute to the moviemaking business on any level.	moviemaking

The result indicates that the accuracy is satisfied, and proves that it is reasonable to compute the polarity score by the proposed method in section 3.4. Through inspection, we find that the factor is the ontology development. As the limited knowledge in this field and a wide variety of literal expressions, some features cannot be detected. Then, we perform the ontology-based method which is

Table 3. Feature Detection Results

Method	Ontology-combined	Baseline	Accuracy
Features	143	186	76.9%

Table 4. Result for Sentiment Analysis

Method	Ontology-based	Baseline	Accuracy
pos	53	60	88.3%
neg	49	60	81.7%

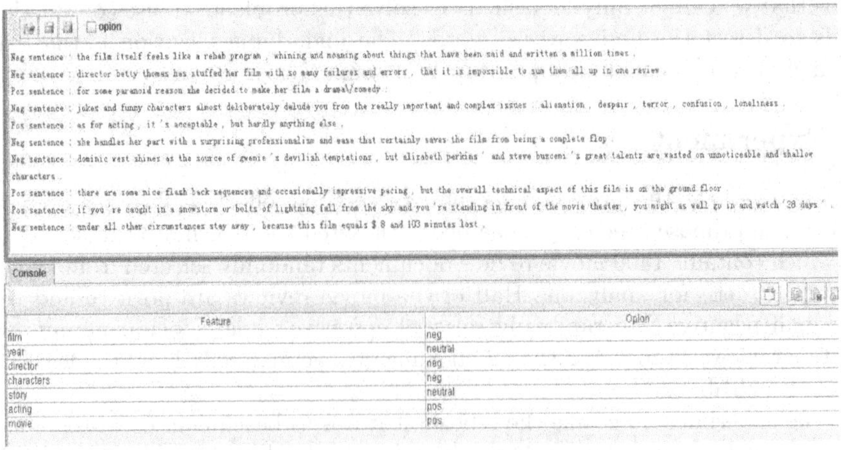

Fig. 5. Results from Sentiment Analysis

proposed in section 3.5 for sentiment analysis. As the selected dataset is already labeled, we use the origin class label (pos and neg) as baseline which is contrast to the result output from the proposed method. The result is shown in Table 4. The achieved accuracy is relatively good.

At last, we investigate each feature in each sentence, and analyze the opinion associated with them. By using the sentiment analysis method mentioned in section 3.5, we can get not only the opinion for the whole review, but also each opinion on each feature. The result is shown in following Fig. 5. The proposed technique can identify the features which are terminologies in the movie model or feature model, and the opinions on them. This result is very encouraging that it indicates ontology can be employed to opinion mining.

5   Conclusion

Ontology is a kind of concept models that could describe system at the level of semantics and knowledge. It aims to access knowledge of a domain in a general way and provides a common understanding for concepts in the domain so as to realize knowledge. Opinion mining is a challenging problem, which is to analyze opinions which appear in a series of texts and given by users towards different features, and determine whether these opinions are positive, negative or neutral and how strong they are. In this paper, we focus on employing ontology to mining movie reviews. Because ontology aims to provide knowledge about specific domains that are understandable by both developers and computers. The experiment is carried out effectively, and the result is good. Therefore, it is rational and effective to employ ontology to opinion mining. This opened up a new way for the opinion mining, especially for the feature extraction and sentiment analysis. As shown above, feature extraction is a difficult task. It may be features that are difficult to accurately define. More work is needed on ontology development. A variety of steps can be taken to extend this work:

1. Develop a more fine-grain ontology. The more fine-grain the ontology is, the more accuracy we can get through feature extraction.
2. Continue to experiment with linguistic rules. In the experiment, we just considered negation words and conjunction words. This is not enough for some complex sentences.

References

1. Pang, B., Lee, L., Vaithyanathan, S.: Thumbs up? sentiment classification using machine learning techniques. In: Proceedings of the Conference on Empirical Methods in Natural Language Processing, EMNLP 2002 (2002)
2. Popescu, A.M., Etzioni, O.: Extracting product features and opinions from reviews. In: Proceedings of the Conference on Empirical Methods in Natural Language Processing, EMNLP 2005 (2005)
3. Hu, M., Liu, B.: Mining and summarizing customer reviews. In: Proceedings of ACM SIGKDD conference, KDD 2004 (2004)

4. Kaji, N., Kitsuregawa, M.: Automatic construction of polarity-tagged corpus from html documents. In: Proceedings of the COLING/ACL on Main conference poster sessions, Association for Computational Linguistics Morristown, NJ, USA, pp. 452–459 (2006)

5. Hu, M., Liu, B.: Mining opinion features in customer reviews. In: Proceedings of AAAI, pp. 755–760 (2004)

6. Carenini, G., Ng, R., Pauls, A.: Interactive multimedia summaries of evaluative text. In: Proceedings of the 11th international conference on Intelligent user interfaces, pp. 124–131. ACM, New York (2006)

7. Ding, X., Liu, B.: The utility of linguistic rules in opinion mining. In: Proceedings of SIGIR 2007 (2007)

8. Gruber, T.R.: A translation approach to portable ontology specifications. Knowledge Acquisition 5, 199–220 (1993)

9. Pang, B.: Seeing stars: Exploiting class relationships for sentiment categorization with respect to rating scales. Ann. Arbor. 100 (2005)

10. Riloff, E., Wiebe, J.: Learning extraction patterns for subjective expressions. In: Proceedings of the Conference on Empirical Methods in Natural Language Processing (EMNLP 2003), pp. 105–112 (2003)

11. Turney, P., et al.: Thumbs up or thumbs down? semantic orientation applied to unsupervised classification of reviews. In: Proceedings of the 40th annual meeting of the Association for Computational Linguistics, pp. 417–424 (2002)

12. Gamon, M., Aue, A., Corston-Oliver, S., Ringger, E.: Pulse: Mining customer opinions from free text. In: Famili, A.F., Kok, J.N., Peña, J.M., Siebes, A., Feelders, A. (eds.) IDA 2005. LNCS, vol. 3646, pp. 121–132. Springer, Heidelberg (2005)

13. Dave, K., Lawrence, S., Pennock, D.: Mining the peanut gallery: Opinion extraction and semantic classification of product reviews. In: Proceedings of the 12th international conference on World Wide Web, pp. 519–528. ACM, New York (2003)

14. Hearst, M.A.: Direction-based text interpretation as an information access refinement, pp. 257–274 (1992)

15. Jacquemin, C.: Spotting and Discovering Terms through Natural Language Processing. MIT Press, Cambridge (2001)

16. Kobayashi, N., Inui, K., Matsumoto, Y.: Collecting evaluative express for opinion extraction. In: Proceedings of the International Joint Conference on Natural Language Processing, IJCNLP (2004)

17. Yi, J., Bunescu, T.N., Niblack, R.W.: Sentiment analyzer: extracting sentiments about a given topic using natural language processing techniques. In: Proceedings of IEEE International Conference on Data Mining, ICDM 2003 (2003)

18. Hatzivassiloglou, V., McKeown, K.: Predicting the semantic orientation of adjectives. In: Proceedings of ACL-EACL 1997 (1997)

19. Kanayama, H., Nasukawa, T.: Fully automatic lexicon expansion for domain-oriented sentiment analysis. In: Proceedings of the Conference on Empirical Methods in Natural Language Processing, EMNLP 2006 (2006)

20. Esuli, A., Sebastiani, F.: Sentiwordnet: A publicly available lexical resource for opinion mining. In: Proceedings of 5th Conference on Language Resources and Evaluation, LREC 2006 (2006)

Concept-Based, Personalized Web Information Gathering: A Survey

Xiaohui Tao and Yuefeng Li

SIT, Queensland University of Technology, Australia
{x.tao,y2.li}@qut.edu.au

Abstract. Web information gathering surfers from the problems of information mismatching and overloading. In an attempt to solve these fundamental problems, many works have proposed to use concept-based techniques to perform personalized information gathering for Web users. These works have significantly improved the performance of Web information gathering systems. In this paper, a survey is conducted on these works. The reviewed scholar report that the concept-based, personalized techniques can gather more useful and meaningful information for Web users. The survey also suggests that improvement is needed for the representation and acquisition of user profiles in personalized Web information gathering.

1 Introduction

Over the last decade, the rapid growth and adoption of the World Wide Web have further exacerbated user need for efficient mechanisms for information and knowledge location, selection and retrieval. Web information covers a wide range of topics and serves a broad spectrum of communities [1]. How to gather useful and meaningful information from the Web, however, becomes challenging to Web users. This challenging issue is referred by many researchers as Web information gathering [23,11].

Given an information needs, Web information gathering aims to acquire useful and meaningful information for users from the Web. The Web information gathering tasks are usually completed by the systems using keyword-based techniques. The keyword-based mechanism searches the Web by finding the documents with the specific terms matched. This mechanism is used by many existing Web search systems, for example, Google and Yahoo!information gathering. Han and Chang [17] pointed out that by using keyword-based search techniques, the Web information gathering systems can access the information quickly; however, the gathered information may possibly contain much useless and meaningless information. This is particularly referred as the fundamental issue in Web information gathering: information mismatching and information overloading [27,28,29,30,71].

In attempting to solve these fundamental problems, many researchers have aimed at gathering personalized Web information for users with better effectiveness and efficiency. These researchers have not only moved information gathering

D. Karagiannis and Z. Jin (Eds.): KSEM 2009, LNAI 5914, pp. 215–228, 2009.

from keyword-based to concept-based, but also take user background knowledge into consideration. In these works, Web user profiles are widely used for user modelling and personalization [22], because they reflect the interest and preferences of users [50]. User profiles are defined by Li and Zhong [30] as the interesting topics underlying user information needs. They are used in Web information gathering to describe user background knowledge, to capture user information needs, and to gather personalized Web information for users [14,17,30,58].

This survey paper attempts to review the development of the concept-based, personalized Web information gathering techniques. The review notes the issues in Web personalization, focusing on Web user profiles and user information needs in personalized Web information gathering. The reviewed scholar reports that the concept-based models utilizing user background knowledge are capable of gathering useful and meaningful information for Web users. However, the representation and acquisition of user profiles need to be improved for the effectiveness of Web information gathering. This survey has contributions to better understanding of existing Web information gathering systems.

The paper is organized as follows. Section 2 reviews the concept-based Web information gathering techniques, including concept representation and extraction. Section 3 presents the survey of personalized Web information gathering, including user profile representation and acquisition. Finally, Section 4 makes the final remarks for the survey.

2   Concept-Based Web Information Gathering

The concept-based information gathering techniques use the semantic concepts extracted from documents and queries. Instead of matching the keyword features representing the documents and queries, the concept-based techniques attempt to compare the semantic concepts of documents to those of given queries. The similarity of documents to queries is determined by the matching level of their semantic concepts. The semantic concept representation and extraction are two typical issues in the concept-based techniques and are discussed in the following sections.

2.1   Semantic Concept Representation

Semantic concepts have various representations. In some models, these concepts are represented by controlled lexicons defined in terminological ontologies, thesauruses, or dictionaries. In some other models, they are represented by subjects in domain ontologies, library classification systems, or categorizations. In some models using data mining techniques for concept extraction, semantic concepts are represented by patterns. The three representations given have different strengthes and weaknesses.

The lexicon-based representation defines the concepts in terms and lexicons that are easily understood by users. WordNet [12] and its variations [3,21] are typical models employing this kind of concept representation. In these models, semantic concepts are represented by the controlled vocabularies defined in terminological ontologies, thesauruses, or dictionaries. Because these are being controlled, they are also easily utilized by the computational systems. However, when extracting terms to represent concepts for information gathering, some noisy terms may also be extracted because of the term ambiguity problem. As a result, the information overloading problem may occur in gathering. Moreover, the lexicon-based representation relies largely on the quality of terminological ontologies, thesaurus, or dictionaries for definitions. However, the manual development of controlled lexicons or vocabularies (like WordNet) is usually costly. The automatic development is efficient, however, in sacrificing the quality of definitions and semantic relation specifications. Consequently, the lexicon-based representation of semantic concepts was reported to be able to improve the information gathering performance in some works [21,35], but to be degrading the performance in other works [59].

Many Web systems rely upon subject-based representation of semantic concepts for information gathering. In this kind of representation, semantic concepts are represented by subjects defined in knowledge bases or taxonomies, including domain ontologies, digital library systems, and online categorizations. Typical information gathering systems utilizing domain ontologies for concept representation include those developed by Lim *et al.* [32], by Navigli [40], and by Velardi *et al.* [60]. Domain ontologies contain expert knowledge: the concepts described and specified in the ontologies are of high quality. However, expert knowledge acquisition is usually costly in both capitalization and computation. Moreover, as discussed previously, the semantic concepts specified in many domain ontologies are structured only in the subsumption manner of *super-class* and *sub-class*, rather than the more specific *is-a*, *part-of*, and *related-to*, the ones developed or used by [14,20] and [71]. Some attempted to describe more specified relations, like [4,51] for *is-a*, [15,44] for *part-of*, and [18] for *related-to* relations only. Tao *et al.* [55,56] made a further progress from these works and portrayed the basic *is-a*, *part-of*, and *related-to* semantic relations in one single computational model for concept representation.

Also used for subject-based concept representation are the library systems, like Dewey Decimal Classification (DDC) used by [20,62], Library of Congress Classification (LCC) and Library of Congress Subject Headings (LCSH) [55,56], and the variants of these systems, such as the "China Library Classification Standard" used by [70] and the Alexandria Digital Library (ADL) used by [61]. These library systems represent the natural growth and distribution of human intellectual work that covers the comprehensive and exhaustive topics of world knowledge [5]. In these systems, the concepts are represented by the subjects that are defined by librarians and linguists manually under a well-controlled process [5]. The concepts are constructed in taxonomic structure, originally designed for information retrieval from libraries. These are beneficial to the information gathering systems. The concepts are linked by semantic relations, such as subsumption like *super-class* and *sub-class* in the DDC and LCC, and *broader*, *used-for*, and *related-to* in the LCSH. However, the information gathering systems using library systems for concept representation largely rely upon the existing knowledge bases. The limitations of the library systems, for example, the focus on the United States more than on other regions by the LCC and LCSH, would be incorporated by the information gathering systems that use them for concept representation.

The online categorizations are also widely relied upon by many information gathering systems for concept representation. The typical online categorizations used for concept representation include the Yahoo! categorization used by [14] and *Open Directory Project*[1] used by [7,41]. In these categorizations, concepts are represented by categorization subjects and organized in a taxonomical structure. However, the nature of categorizations is in the subsumption manner of one containing another (*super-class* and *sub-class*), but not the semantic *is-a, part-of*, and *related-to* relations. Thus, the semantic relations associated with the concepts in such representations are not in adequate details and specific levels. These problems weaken the quality of concept representation and thus the performance of information gathering systems.

Another semantic concept representation in Web information gathering systems is pattern-based representation that uses multiple terms (e.g. phrases) to represent a single semantic concept. Phrases contain more content than any one of their containing terms. Research representing concepts by patterns include Li and Zhong [27,28,29,30,24,31], Wu *et al.* [65,64,63], Zhou *et al.* [73,74], Dou *et al.* [10], and Ruiz-Casado *et al.* [45]. However, pattern-based semantic concept representation poses some drawbacks. The concepts represented by patterns can have only subsumption specified for relations. Usually, the relations existing between patterns are specified by investigation of their containing terms, like [30,63,73]. If more terms are added into a phrase, making the phrase more specific, the phrase becomes a sub-class concept of any concepts represented by the sub-phrases in it. Consequently, no specific semantic concepts like *is-a* and *part-of* can be specified and thus some semantic information may be missed in pattern-based concept representations. Another problem of pattern-based concept representation is caused by the length of patterns. The concepts can be adequately specific for discriminating one from others only if the patterns representing the concepts are long enough. However, if the patterns are too long, the patterns extracted from Web documents would be of low frequency and thus, cannot support the concept-based information gathering systems substantially [63]. Although the pattern-based concept representation poses such drawbacks, it is still one of the major concept representations in information gathering systems.

2.2   Semantic Concept Extraction

The techniques used for concept extraction from text documents include text classification techniques and Web content mining techniques, including association rules mining and pattern mining. These techniques are reviewed and discussed as follows.

Text classification is the process of classifying an incoming stream of documents into categories by using the classifiers learned from the training samples

---

[1] http://www.dmoz.org

[33]. Text classification techniques can be categorized into different groups. Fung *et al.* [13] categorized them into two types: *kernel-based classifiers* and *instance-based classifiers*. Typical kernel-based classifier learning approaches include the *Support Vector Machines* (SVMs) [19] and regression models [47]. These approaches may incorrectly classify many negative samples from an unlabeled set into a positive set, thus causing the problem of information overloading in Web information gathering. Typical instance-based classification approaches include the *K*-Nearest Neighbor (*K*-NN) [9] and its variants, which do not relay upon the statistical distribution of training samples. However, the instance-based approaches are not capable of extracting highly accurate positive samples from the unlabeled set. Other research works, such as [14,42], have a different way of categorizing the classifier learning techniques: *document representations based classifiers*, including SVMs and *K*-NN; and *word probabilities based classifiers*, including Naive Bayesian, decision trees [19] and neural networks used by [69]. These classifier learning techniques have different strengthes and weaknesses, and should be chosen based upon the problems they are attempting to solve.

Text classification techniques are widely used in concept-based Web information gathering systems. Gauch *et al.* [14] described how text classification techniques are used for concept-based Web information gathering. Web users submit a topic associated with some specified concepts. The gathering agents then search for the Web documents that are referred to by the concepts. Sebastiani [47] outlined a list of tasks in Web information gathering to which text classification techniques may contribute: automatic indexing for Boolean information retrieval systems, document organization (particularly in personal organization or structuring of a corporate document base), text filtering, word sense disambiguation, and hierarchical categorization of web pages. Also, as specified by Meretakis *et al.*[38], the Web information gathering areas contributed to by text classification may include sorting emails, filtering junk emails, cataloguing news articles, providing relevance feedback, and reorganizing large document collections. Text classification techniques have been utilized by [36] to classify Web documents into the best matching interest categories, based on their referring semantic concepts.

Text classification techniques utilized for concept-based Web information gathering, however, incorporate some limitations and weaknesses. Glover *et al.* [16] pointed out that the Web information gathering performance substantially relies on the accuracy of predefined categories. If the arbitration of a given category is wrong, the performance is degraded. Another challenging problem, referred to as "cold start", occurs when there is an inadequate number of training samples available to learning classifiers. Also, as pointed out by Han and Chang [17], the concept-based Web information gathering systems rely on an assumption that the content of Web documents is adequate to make descriptions for classification. When the assumption is not true, using text classification techniques alone becomes unreliable for Web information gathering systems. The solution to this problem is to use high quality semantic concepts, as argued by

Han and Chang [17], and to integrate both text classification and Web mining techniques.

Web content mining is an emerging field of applying knowledge discovery technology to Web data. Web content mining discovers knowledge from the content of Web documents, and attempts to understand the semantics of Web data [22,30]. Based on various Web data types, Web content mining can be categorized into Web text mining, Web multimedia data mining (e.g. image, audio, video), and Web structure mining [22]. In this paper, Web information is particularly referred to as the text documents existing on the Web. Thus, the term "Web content mining" here refers to "Web text content mining", the knowledge discovery from the content of Web text documents. Kosala and Blockeel [22] categorized Web content mining techniques into database views and information retrieval views. From the database view, the goal of Web content mining is to model the Web data so that Web information gathering may be performed based on concepts rather than on keywords. From the information retrieval view, the goal is to improve Web information gathering based on either inferred or solicited Web user profiles. With either view, Web content mining contributes significantly to Web information gathering.

Many techniques are utilized in Web content mining, including pattern mining, association rules mining, text classification and clustering, and data generalization and summarization [27,29]. Li and Zhong [27,28,29,30], Wu et al. [64], and Zhou et al. [73,74] represented semantic concepts by maximal patterns, sequential patterns, and closed sequential patterns, and attempted to discover these patterns for semantic concepts extracted from Web documents. Their experiments reported substantial improvements achieved by their proposed models, in comparison with the traditional *Rocchio*, *Dempster-Shafer*, and probabilistic models. Association rules mining extracts meaningful content from Web documents and discovers their underlying knowledge. Existing models using association rules mining include Li and Zhong [26], Li et al. [25], and Yang et al. [68], who used the granule techniques to discover association rules; Xu and Li [67] and Shaw et al. [48], who attempted to discover concise association rules; and Wu et al. [66], who discovered positive and negative association rules. Some works, such as Dou et al. [10], attempted to integrate multiple Web content mining techniques for concept extraction. These works were claimed capable of extracting concepts from Web documents and improving the performance of Web information gathering. However, as pointed out by Li and Zhong [28,29], the existing Web content mining techniques incorporate some limitations. The main problem is that these techniques are incapable of specifying the specific semantic relations (e.g. *is-a* and *part-of*) that exist in the concepts. Their concept extraction needs to be improved for more specific semantic relation specification, considering the fact that the current Web is nowadays moving toward the Semantic Web [2].

3   Personalized Web Information Gathering

Web user profiles are widely used by Web information systems for user modelling and personalization [22]. User profiles reflect the interests of users [50]. In terms of Web information gathering, user profiles are defined by Li and Zhong [30] as the interesting topics underlying user information needs . Hence, user profiles are used in Web information gathering to capture user information needs from the user submitted queries, in order to gather personalized Web information for users [14,17,30,58].

Web user profiles are categorized by Li and Zhong [30] into two types: the *data diagram* and *information diagram* profiles (also called *behavior-based profiles* and *knowledge-based profiles* by [39]). The data diagram profiles are usually acquired by analyzing a database or a set of transactions [14,30,39,52,53]. These kinds of user profiles aim to discover interesting registration data and user profile portfolios. The information diagram profiles are usually acquired by using manual techniques; such as questionnaires and interviews [39,58], or by using information retrieval and machine-learning techniques [14]. They aim to discover interesting topics for Web user information needs.

3.1   User Profile Representation

User profiles have various representations. As defined by [50], user profiles are represented by a previously prepared collection of data reflecting user interests. In many approaches, this "collection of data" refers to a set of terms (or vector space of terms) that can be directly used to expand the queries submitted by users [8,39,58]. These term-based user profiles, however, may cause poor interpretation of user interests to the users, as pointed out by [29,30]. Also, the term-based user profiles suffer from the problems introduced by the keyword-match techniques because many terms are usually ambiguous. Attempting to solve this problem, Li and Zhong [30] represented user profiles by patterns. However, the pattern-based user profiles also suffer from the problems of inadequate semantic relations specification and the dilemma of pattern length and pattern frequency, as discussed previously in Section 2 for pattern-based concept representation.

User profiles can also be represented by personalized ontologies. Tao *et al.* [55,56], Gauch *et al.* [14], Trajkova and Gauch [58], and Sieg *et al.* [52] represented user profiles by a sub-taxonomy of a predefined hierarchy of concepts. The concepts existing in the taxonomy are associated with weights indicating the user-perceived interests in these concepts. This kind of user profiles describes user interests explicitly. The concepts specified in user profiles have clear definitions and extents. They are thus excellent for inferences performed to capture user information needs. However, clearly specifying user interests in ontologies is a difficult task, especially for their semantic relations, such as *is-a* and *part-of*. In these aforementioned works, only Tao *et al.* [55,56] could emphasis these semantic relations in user interest specification.

User profiles can also be represented by a training set of documents, as the user profiles in TREC-11 Filtering Track [43] and the model proposed by Tao *et al.* [54] for acquiring user profiles from the Web. User profiles (the training sets) consist of positive documents that contain user interest topics, and negative documents that contain ambiguous or paradoxical topics. This kind of user profiles describes user interests implicitly, and thus have great flexibility to be used with any concept extraction techniques. The drawback is that noise may be extracted from user profiles as well as meaningful and useful concepts. This may cause an information overloading problem in Web information gathering.

3.2   User Profile Acquisition

When acquiring user profiles, the content, life cycle, and applications need to be considered [46]. Although user interests are approximate and explicit, it was argued by [55,56,30,14] that they can be specified by using ontologies. The life cycle of user profiles refers to the period that the user profiles are valuable for Web information gathering. User profiles can be long-term or short-term. For instance, persistent and ephemeral user profiles were built by Sugiyama *et al.* [53], based on the long term and short term observation of user behavior. Applications are also an important factor requiring consideration in user profile acquisition. User profiles are widely used in not only Web information gathering [55,56,30], but also personalized Web services [17], personalized recommendations [39], automatic Web sites modifications and organization, and marketing research [72]. These factors considered in user profile acquisition also define the utilization of user profiles for their contributing areas and period.

User profile acquisition techniques can be categorized into three groups: the *interviewing*, *non-interviewing*, and *semi-interviewing* techniques. The interviewing user profiles are entirely acquired using manual techniques; such as questionnaires, interviews, and user classified training sets. Trajkova and Gauch [58] argued that user profiles can be acquired explicitly by asking users questions. One typical model using user-interview profiles acquisition techniques is the TREC-11 Filtering Track model [43]. User profiles are represented by training sets in this model, and acquired by users manually. Users read training documents and assign positive or negative judgements to the documents against given topics. Based upon the assumption that users know their interests and preferences exactly, these training documents perfectly reflect users' interests. However, this kind of user profile acquisition mechanism is costly. Web users have to invest a great deal of effort in reading the documents and providing their opinions and judgements. However, it is unlikely that Web users wish to burden themselves with answering questions or reading many training documents in order to elicit profiles [29,30].

The non-interviewing techniques do not involve users directly but ascertain user interests instead. Such user profiles are usually acquired by observing and mining knowledge from user activity and behavior [30,49,53,58]. Typical model is the personalized, ontological user profiles acquired by [56] using a world knowledge base and user local instance repositories. Some other works, like [14,58] and [52], acquire non-interviewing ontological user profiles by using global categorizations such as Yahoo! categorization and Online Directory Project. The machine-learning techniques are utilized to analyze the user-browsed Web documents, and classification techniques are used to classify the documents into the concepts specified in the global categorization. As a result, the user profiles in these models are a sub-taxonomy of the global categorizations. However, because the categorizations used are not well-constructed ontologies, the user profiles acquired in these models cannot describe the specific semantic relations. Instead of classifying interesting documents into the supervised categorizations, Li and Zhong [29,30] used unsupervised methods to discover interesting patterns from the user-browsed Web documents, and illustrated the patterns to represent user profiles in ontologies. The model developed by [34] acquired user profiles adaptively, based on the content study of user queries and online browsing history. In order to acquire user profiles, Chirita *et al.* [6] and Teevan *et al.* [57] extracted user interests from the collection of user desktop information such as text documents, emails, and cached Web pages. Makris *et al.* [37] comprised user profiles by a ranked local set of categories and then utilized Web pages to personalize search results for a user. These non-interviewing techniques, however, have a common limitation of ineffectiveness. Their user profiles usually contain much noise and uncertainties because of the use of automatic acquiring techniques.

With the aim of reducing user involvement and improve effectiveness, the semi-interviewing user profiles are acquired by semi-automated techniques. This kind of user profiles may be deemed as that acquired by the hybrid mechanism of interviewing and non-interviewing techniques. Rather than providing users with documents to read, some approaches annotate the documents first and attempt to seek user feedback for just the annotated concepts. Because annotating documents may generate noisy concepts, global knowledge bases are used by some user profile acquisition approaches. They extract potentially interesting concepts from the knowledge bases and then explicitly ask users for feedback, like the model proposed by [55]. Also, by using a so-called Quickstep topic ontology, Middleton *et al.* [39] acquired user profiles from unobtrusively monitored behavior and explicit relevance feedback. The limitation of semi-interviewing techniques is that they largely rely upon knowledge bases for user background knowledge specification.

4 Remarks

This survey introduced the challenges existing in the current Web information gathering systems, and described how the current works related to the challenges. The scholar reviewed in this survey suggested that the key to gathering meaningful and useful information for Web users is to improve the Web information gathering techniques from keyword-based to concept-based, and from general to personalized. The concept-based systems using user background knowledge were reported capable of gathering useful and meaningful information for Web users. However, research gaps exist for the representation and acquisition of user profiles, in terms of effective user information need capture.

References

1. Antoniou, G., van Harmelen, F.: A Semantic Web Primer. MIT Press, Cambridge (2004)
2. Berners-Lee, T., Hendler, J., Lassila, O.: The semantic Web. Scientific American 5, 29–37 (2001)

3. Budanitsky, A., Hirst, G.: Evaluating WordNet-based measures of lexical semantic relatedness. Computational Linguistics 32(1), 13–47 (2006)
4. Cederberg, S., Widdows, D.: Using lsa and noun coordination information to improve the precision and recall of automatic hyponymy extraction. In: Proceedings of the seventh conference on Natural language learning at HLT-NAACL 2003, pp. 111–118 (2003)
5. Chan, L.M.: Library of Congress Subject Headings: Principle and Application. Libraries Unlimited (2005)
6. Chirita, P.A., Firan, C.S., Nejdl, W.: Personalized query expansion for the Web. In: Proceedings of the 30th annual international ACM SIGIR conference on Research and development in information retrieval, pp. 7–14 (2007)
7. Chirita, P.A., Nejdl, W., Paiu, R., Kohlschütter, C.: Using ODP metadata to personalize search. In: Proceedings of the 28th annual international ACM SIGIR conference on Research and development in information retrieval, pp. 178–185 (2005)
8. Cui, H., Wen, J.-R., Nie, J.-Y., Ma, W.-Y.: Probabilistic query expansion using query logs. In: Proceedings of the 11th international conference on World Wide Web, pp. 325–332 (2002)
9. Dasarathy, B.V. (ed.): Nearest Neighbor (NN) Norms: NN Pattern Classification Techniques. IEEE Computer Society Press, Los Alamitos (1990)
10. Dou, D., Frishkoff, G., Rong, J., Frank, R., Malony, A., Tucker, D.: Development of neuroelectromagnetic ontologies(NEMO): a framework for mining brainwave ontologies. In: Proceedings of the 13th ACM SIGKDD international conference on Knowledge discovery and data mining, pp. 270–279 (2007)
11. Espinasse, B., Fournier, S., Freitas, F.: Agent and ontology based information gathering on restricted web domains with AGATHE. In: Proceedings of the 2008 ACM symposium on Applied computing, pp. 2381–2386 (2008)
12. Fellbaum, C. (ed.): WordNet: An Electronic Lexical Database. MIT Press, Cambridge (1998)
13. Fung, G.P.C., Yu, J.X., Lu, H., Yu, P.S.: Text classification without negative examples revisit. IEEE Transactions on Knowledge and Data Engineering 18(1), 6–20 (2006)
14. Gauch, S., Chaffee, J., Pretschner, A.: Ontology-based personalized search and browsing. Web Intelligence and Agent Systems 1(3-4), 219–234 (2003)
15. Girju, R., Badulescu, A., Moldovan, D.: Automatic discovery of part-whole relations. Comput. Linguist. 32(1), 83–135 (2006)
16. Glover, E.J., Tsioutsiouliklis, K., Lawrence, S., Pennock, D.M., Flake, G.W.: Using Web structure for classifying and describing Web pages. In: WWW 2002: Proceedings of the 11th international conference on World Wide Web, pp. 562–569 (2002)
17. Han, J., Chang, K.-C.: Data mining for Web intelligence. Computer 35(11), 64–70 (2002)
18. Inkpen, D., Hirst, G.: Building and using a lexical knowledge base of near-synonym differences. Computational Linguistics 32(2), 223–262 (2006)
19. Joachims, T.: Text categorization with Support Vector Machines: learning with many relevant features. In: Nédellec, C., Rouveirol, C. (eds.) ECML 1998. LNCS, vol. 1398, pp. 137–142. Springer, Heidelberg (1998)
20. King, J.D., Li, Y., Tao, X., Nayak, R.: Mining World Knowledge for Analysis of Search Engine Content. Web Intelligence and Agent Systems 5(3), 233–253 (2007)
21. Kornilakis, H., Grigoriadou, M., Papanikolaou, K., Gouli, E.: Using WordNet to support interactive concept map construction. In: Proceedings of IEEE International Conference on Advanced Learning Technologies, pp. 600–604 (2004)

22. Kosala, R., Blockeel, H.: Web mining research: A survey. ACM SIGKDD Explorations Newsletter 2(1), 1–15 (2000)

23. Li, Y.: Information fusion for intelligent agent-based information gathering. In: Zhong, N., Yao, Y., Ohsuga, S., Liu, J. (eds.) WI 2001. LNCS (LNAI), vol. 2198, pp. 433–437. Springer, Heidelberg (2001)

24. Li, Y., Wu, S.-T., Tao, X.: Effective pattern taxonomy mining in text documents. In: CIKM 2008: Proceeding of the 17th ACM conference on Information and knowledge management, pp. 1509–1510 (2008)

25. Li, Y., Yang, W., Xu, Y.: Multi-tier granule mining for representations of multidimensional association rules. In: Proceedings of the Sixth IEEE International Conference on Data Mining, pp. 953–958 (2006)

26. Li, Y., Zhong, N.: Interpretations of association rules by granular computing. In: Proceedings of IEEE International Conference on Data Mining, Melbourne, Florida, USA, pp. 593–596 (2003)

27. Li, Y., Zhong, N.: Ontology-based Web mining model. In: Proceedings of the IEEE/WIC International Conference on Web Intelligence, Canada, pp. 96–103 (2003)

28. Li, Y., Zhong, N.: Capturing evolving patterns for ontology-based web mining. In: Proceedings of the 2004 IEEE/WIC/ACM International Conference on Web Intelligence, pp. 256–263 (2004)

29. Li, Y., Zhong, N.: Web Mining Model and its Applications for Information Gathering. Knowledge-Based Systems 17, 207–217 (2004)

30. Li, Y., Zhong, N.: Mining Ontology for Automatically Acquiring Web User Information Needs. IEEE Transactions on Knowledge and Data Engineering 18(4), 554–568 (2006)

31. Li, Y., Zhou, X., Bruza, P., Xu, Y., Lau, R.Y.: A two-stage text mining model for information filtering. In: CIKM 2008: Proceeding of the 17th ACM conference on Information and knowledge management, pp. 1023–1032 (2008)

32. Lim, S.-Y., Song, M.-H., Son, K.-J., Lee, S.-J.: Domain ontology construction based on semantic relation information of terminology. In: 30th Annual Conference of the IEEE Industrial Electronics Society, vol. 3, pp. 2213–2217 (2004)

33. Liu, B., Dai, Y., Li, X., Lee, W., Yu, P.: Building text classifiers using positive and unlabeled examples. In: Proceedings of the Third IEEE International Conference on Data Mining, ICDM 2003, pp. 179–186 (2003)

34. Liu, F., Yu, C., Meng, W.: Personalized web search for improving retrieval effectiveness. IEEE Transactions on Knowledge and Data Engineering 16(1), 28–40 (2004)

35. Liu, S., Liu, F., Yu, C., Meng, W.: An effective approach to document retrieval via utilizing WordNet and recognizing phrases. In: SIGIR 2004: Proceedings of the 27th annual international ACM SIGIR conference on Research and development in information retrieval, pp. 266–272 (2004)

36. Ma, Z., Pant, G., Sheng, O.R.L.: Interest-based personalized search. ACM Transactions on Information Systems (TOIS) 25(1), 5 (2007)

37. Makris, C., Panagis, Y., Sakkopoulos, E., Tsakalidis, A.: Category ranking for personalized search. Data & Knowledge Engineering 60(1), 109–125 (2007)

38. Meretakis, D., Fragoudis, D., Lu, H., Likothanassis, S.: Scalable association-based text classification. In: CIKM 2000: Proceedings of the ninth international conference on Information and knowledge management, pp. 5–11 (2000)

39. Middleton, S.E., Shadbolt, N.R., Roure, D.C.D.: Ontological user profiling in recommender systems. ACM Transactions on Information Systems (TOIS) 22(1), 54–88 (2004)

40. Navigli, R., Velardi, P., Gangemi, A.: Ontology learning and its application to automated terminology translation. IEEE Intelligent Systems 18, 22–31 (2003)

41. Qiu, G., Liu, K., Bu, J., Chen, C., Kang, Z.: Quantify query ambiguity using odp metadata. In: SIGIR 2007: Proceedings of the 30th annual international ACM SIGIR conference on Research and development in information retrieval, pp. 697–698 (2007)

42. Ravindran, D., Gauch, S.: Exploiting hierarchical relationships in conceptual search. In: Proceedings of the 13th ACM international conference on Information and Knowledge Management, pp. 238–239 (2004)

43. Robertson, S.E., Soboroff, I.: The TREC 2002 filtering track report. In: Text REtrieval Conference (2002)

44. Ross, D.A., Zemel, R.S.: Learning parts-based representations of data. The Journal of Machine Learning Research 7, 2369–2397 (2006)

45. Ruiz-Casado, M., Alfonseca, E., Castells, P.: Automatising the learning of lexical patterns: An application to the enrichment of WordNet by extracting semantic relationships from Wikipedia. Data & Knowledge Engineering 61(3), 484–499 (2007)

46. Schuurmans, J., de Ruyter, B., van Vliet, H.: User profiling. In: CHI 2004: CHI 2004 extended abstracts on Human factors in computing systems, pp. 1739–1740 (2004)

47. Sebastiani, F.: Machine learning in automated text categorization. ACM Computing Surveys (CSUR) 34(1), 1–47 (2002)

48. Shaw, G., Xu, Y., Geva, S.: Deriving non-redundant approximate association rules from hierarchical datasets. In: CIKM 2008: Proceeding of the 17th ACM conference on Information and knowledge management, pp. 1451–1452 (2008)

49. Shen, X., Tan, B., Zhai, C.: Implicit user modeling for personalized search. In: CIKM 2005: Proceedings of the 14th ACM international conference on Information and knowledge management, pp. 824–831 (2005)

50. Shepherd, M.A., Lo, A., Phillips, W.J.: A study of the relationship between user profiles and user queries. In: Proceedings of the 8th annual international ACM SIGIR conference on Research and development in information retrieval, pp. 274–281 (1985)

51. Shinzato, K., Torisawa, K.: Extracting hyponyms of prespecified hypernyms from itemizations and headings in web documents. In: COLING 2004: Proceedings of the 20th international conference on Computational Linguistics, Morristown, NJ, USA, p. 938 (2004); Association for Computational Linguistics

52. Sieg, A., Mobasher, B., Burke, R.: Web search personalization with ontological user profiles. In: Proceedings of the sixteenth ACM conference on Conference on information and knowledge management, pp. 525–534 (2007)

53. Sugiyama, K., Hatano, K., Yoshikawa, M.: Adaptive web search based on user profile constructed without any effort from users. In: Proceedings of the 13th international conference on World Wide Web, pp. 675–684 (2004)

54. Tao, X., Li, Y., Zhong, N., Nayak, R.: Automatic Acquiring Training Sets for Web Information Gathering. In: Proceedings of the 2006 IEEE/WIC/ACM International Conference on Web Intelligence, pp. 532–535 (2006)

55. Tao, X., Li, Y., Zhong, N., Nayak, R.: Ontology mining for personalzied web information gathering. In: Proceedings of the 2007 IEEE/WIC/ACM International Conference on Web Intelligence, pp. 351–358 (2007)
56. Tao, X., Li, Y., Zhong, N., Nayak, R.: An ontology-based framework for knowledge retrieval. In: Proceedings of the 2008 IEEE/WIC/ACM International Conference on Web Intelligence, pp. 510–517 (2008)
57. Teevan, J., Dumais, S.T., Horvitz, E.: Personalizing search via automated analysis of interests and activities. In: Proceedings of the 28th annual international ACM SIGIR conference on Research and development in information retrieval, pp. 449–456 (2005)
58. Trajkova, J., Gauch, S.: Improving ontology-based user profiles. In: Proceedings of RIAO 2004, pp. 380–389 (2004)
59. Varelas, G., Voutsakis, E., Raftopoulou, P., Petrakis, E.G., Milios, E.E.: Semantic similarity methods in WordNet and their application to information retrieval on the Web. In: WIDM 2005: Proceedings of the 7th annual ACM international workshop on Web information and data management, pp. 10–16 (2005)
60. Velardi, P., Fabriani, P., Missikoff, M.: Using text processing techniques to automatically enrich a domain ontology. In: FOIS 2001: Proceedings of the international conference on Formal Ontology in Information Systems, pp. 270–284 (2001)
61. Wang, J., Ge, N.: Automatic feature thesaurus enrichment: extracting generic terms from digital gazetteer. In: JCDL 2006: Proceedings of the 6th ACM/IEEE-CS joint conference on Digital libraries, pp. 326–333 (2006)
62. Wang, J., Lee, M.C.: Reconstructing DDC for interactive classification. In: Proceedings of the sixteenth ACM conference on Conference on information and knowledge management, pp. 137–146 (2007)
63. Wu, S.-T.: Knowledge Discovery Using Pattern Taxonomy Model in Text Mining. PhD thesis, Faculty of Information Technology, Queensland University of Technology (2007)
64. Wu, S.-T., Li, Y., Xu, Y.: Deploying approaches for pattern refinement in text mining. In: Proceedings of the Sixth International Conference on Data Mining, pp. 1157–1161 (2006)
65. Wu, S.-T., Li, Y., Xu, Y., Pham, B., Chen, P.: Automatic pattern taxonomy exatraction for web mining. In: Proceedings of IEEE/WIC/ACM International Conference on Web Intelligence, pp. 242–248 (2004)
66. Wu, X., Zhang, C., Zhang, S.: Efficient mining of both positive and negative association rules. ACM Transactions on Information Systems (TOIS) 22(3), 381–405 (2004)
67. Xu, Y., Li, Y.: Generating concise association rules. In: CIKM 2007: Proceedings of the sixteenth ACM conference on Conference on information and knowledge management, pp. 781–790 (2007)
68. Yang, W., Li, Y., Wu, J., Xu, Y.: Granule mining oriented data warehousing model for representations of multidimensional association rules. International Journal of Intelligent Information and Database Systems 2(1), 125–145 (2008)
69. Yu, L., Wang, S., Lai, K.K.: An integrated data preparation scheme for neural network data analysis. IEEE Transactions on Knowledge and Data Engineering 18(2), 217–230 (2006)
70. Yu, Z., Zheng, Z., Gao, S., Guo, J.: Personalized information recommendation in digital library domain based on ontology. In: IEEE International Symposium on Communications and Information Technology, 2005. ISCIT 2005, vol. 2, pp. 1249–1252 (2005)

71. Zhong, N.: Representation and construction of ontologies for Web intelligence. International Journal of Foundation of Computer Science 13(4), 555–570 (2002)
72. Zhong, N.: Toward Web Intelligence. In: Proceedings of 1st International Atlantic Web Intelligence Conference, pp. 1–14 (2003)
73. Zhou, X., Li, Y., Bruza, P., Xu, Y., Lau, R.Y.: Pattern taxonomy mining for information filtering. In: AI 2008: Proceedings of the 21st Australasian Joint Conference on Artificial Intelligence, pp. 416–422 (2008)
74. Zhou, X., Li, Y., Bruza, P., Xu, Y., Lau, R.Y.K.: Two-stage model for information filtering. In: WI-IAT 2008: Proceedings of the 2008 IEEE/WIC/ACM International Conference on Web Intelligence and Intelligent Agent Technology, pp. 685–689 (2008)

The Online Market Observatory:
A Domain Model Approach

Norbert Walchhofer[1], Milan Hronsky[1], and Karl Anton Froeschl[2]

[1] EC3 - e-commerce competence center
Vorlaufstrasse 5/6, 1010 Vienna, Austria
http://www.ec3.at
[2] University of Vienna, Institute of Scientific Computing
Universitaetsstrasse 5, 1010 Vienna, Austria

Abstract. The "Semantic Market Monitoring" (SEMAMO) project aims at the prototypical implementation of a generic online market observatory. SEMAMO is intended to provide a flexible empirical instrument for the continuous collection of data about products and services on offer through WWW portals. Based on a uniform data processing scheme covering all stages from data capture using configurable mediators, through integration of data from multiple sources and persistent storage, up to statistical analyses and reporting functions, SEMAMO delivers a self-contained formal specification framework of market monitoring applications. The formal descriptions of application domains, data integration tasks, and analyses of interest facilitate, by deductive conversions, the arrangement and execution of all internal data and storage structures, observation processes, data transformations, and market analytics, respectively. Specifically, SEMAMO exploits formalised domain structures to adaptively optimise data quality and observation efficiency. The framework is evaluated practically in an application to online tourism.

Keywords: online market monitoring, online information extraction, semantic technologies.

1 Introduction

Because of the gradual spread of the World Wide Web into the business domain during the last few decades [1][2] it seems fairly natural to extend traditional methods of market observation [3] to cover online, or electronic, markets as well. While, with an emphasis on single online portals at a time, Web analytics [4] address many of these questions already, there have been few attempts in gathering online market information simultaneously across a variety of online portals, aiming at the statistical integration of collected observations. As a matter of fact, this endeavour links approaches of "business intelligence" [5] to methods of data integration [6], calling for a smart as possible combination of existing proposals and practices from either field of research. This paper describes a generic instrument of observation reflecting the typical conditions of collecting

D. Karagiannis and Z. Jin (Eds.): KSEM 2009, LNAI 5914, pp. 229–240, 2009.

data about online markets. Assuming a reasonably wide class of empirical problems sharing a common structure, the generic observation model is embedded in a configuration space comprising (i) a group of parameters defining a particular observation setting and (ii) another group detailing the processing flow the observations, once gathered, eventually undergo. The ensuing observation instrument is currently implemented as a research prototype named SEMAMO, for "Semantic Market Monitoring", paying heed to the use of recently so-called *semantic technologies* [7] for the specification of both the observation structure and processing flow of a particular application instance (online tourism markets, this time) of the observatory.

The following sections present first an overview of how to formally characterize the relevant domain structures of an online observation setting, highlighting the generic nature of the devised model using illustrations from online tourism markets [8], before the focus is shifted to the procedural dimension, highlighting the effective utilization of formalized domain relationships in arranging and conducting data pre-processing and data analytical functions of the market observatory: Section 3 sketches a typical task of online observation processing, namely the re-identification of encountered entities as a precondition to linking up successive observations to time series, whereas Section 4 explains the model-driven role of adaptiveness in achieving efficiency of the empirical monitoring process. Section 5, finally, provides a summary and reflects basic assumptions of the SEMAMO approach.

2   The Domain Model

From an economic point of view, a market comprises a multitude of different offer channels, some making use of the Internet in terms of online shops or electronic marketplaces. In order to gain insight about market behaviour, systematic surveillance of *many* such channels making up a market is required. In fact, market transactions depend crucially on the provision of information about qualities and prices of things on offer which in case of online channels, such market-specific *digital* information is both easy to capture and fast to process and disseminate, given appropriate resources for data collection and analysis. The proposed empirical instrument of online market monitoring, SEMAMO, revolves around a persistent integrated storage of incrementally accumulated online market data fed forward to multiple, and possibly not entirely anticipated, analytical uses in terms of statistical aggregates and models ready for subsequent decision making or theoretical investigation. SEMAMO implements a three-tier preformed *input>storage>output* approach, gathering data from (predefined) multiple heterogeneous online sources, storing these data in an internal canonical format persistently, and distributing integrated data to various analytical purposes and uses, most of the time comprising "near real-time" statistical aggregation and reporting. More specifically, all of these processing steps are embedded in a periodically repeated *main control loop* in order to achieve an automatic and efficient as possible mode of operation. SEMAMO represents a particular domain of application in formal terms as an instantiation of its generic configuration space with respect to (i) a *structural* and (ii) a *processing* dimension, using the structural description of an application domain to determine and govern the actual data processing within the main control loop.

2.1 The Structure Model

Any empirical model rests on (i) a sampling population of observation units and (ii) the set of attributes to observe on each population unit [9]. In online market monitoring, well-defined sampling populations can be hardly surveyed directly; rather, they are constructed from a designated set of portals conveying the (quantitative) information of interest: the units of an online population may become accessible through different portals, possibly depending, with at least some of their observable features, on the respective portals. Accordingly, provision has to be made for tracking the appearance of individual population units on different portals. Furthermore, the units of interest may become "clustered" by an intermediating actor, altogether giving rise to (at least) a three-pronged population structure composed of

- a *port* population, designating the portals monitored;
- a *sampler* population, comprising the primary units of observation; linked by
- a *switch* population, the units of which are generally used to represent whole clusters of primary observation (that is, sampler) units made accessible on a subset of portals within a port population.

What is actually observed online, then, always consists of a genuine combination of a port, a switch, and a sampler population, that is, any sampler is accessible at a time in one of the ports, through one of the switches in between. In operative terms, hence, <port,switch,sampler>-unit triplets represent the units of observation proper, entailing yet another emergent *sensor* population of their own. In turn, sensor populations are naturally stratified by ports and switches, opening a combinatorially rich field of comparative statistics. The resulting SEMAMO-Graph is depicted in Fig. 1 (left part), exhibiting a typical (though partial) implementation of the market monitoring schema of the running SEMAMO tourism instance. Note that, for sensor, switch, and sampler populations, the nodes in this schema represent whole populations, respectively, whereas for the port population each node represents a single portal registered.

To further facilitate the formal (schema) definition of populations and population linkages as stated in the SEMAMOGraph, lower-level information structures, such as (structured) value domains, attributes, and the like, are supplied (but not shown in Fig. 1); while these are provided and maintained in supplementary directories and codebooks, *registries* hold the current population frames, that is, the population units encountered during repeated observation processes. A specific registry records information on *sources*, represented by dashed arcs in Fig. 1. Logically, a source delivers data from a port to a defined sensor population, implying generally the configuration, and application, of some mediator [10][11] actually delivering observation data on sensors (cf. Subsection 2.2). The

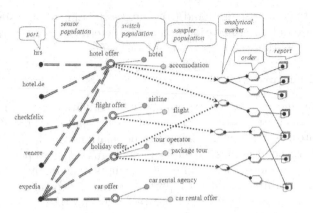

Fig. 1. Tourism Application SEMAMOGraph

SEMAMOGraph, as shown, can be conceived practically also as a direct graphical manipulation interface governing the addition, or update, of schema elements in a context-dependent mode.

Contrary to the perhaps suggestive case illustrated in Fig. 1, switch and sampler populations might be coupled in different ways (by adding further sensor population elements and linking arcs to the SEMAMOGraph) to introduce sensor populations as intended; doing so is reasonable so long as there are real sources contributing data.

Within an individual observation cycle, SEMAMO applies all defined mediators of the established sources to the respective online portals (cf. Section 4). These sources generate, through their attached wrappers (that is, mediators), temporary raw data termed "canonical wrapper records" (CWR, for short) on sensors. More specifically, wrappers implement port-specific query plans covering predefined portions of sensor populations, expressed in terms of domain model structures. Accordingly, the outcome of a wrapper application consists of a set of sensor observations, representing the "state" of these sensors at extraction time. In order to link the "harvested" sensor data to the already registered population units, a re-identification process (cf. Section 3) commences seeking to match CWRs to switch and sampler population units, respectively; in case of success, the harvest (that is, observation) series of a sensor becomes extended by this new observation, otherwise a sensor has been encountered the first time and, hence, population registries have to be updated accordingly. Unit re-identification amounts, of course, to a record linkage problem [12] making use of some distance metrics, again governed by information supplied through the SEMAMO domain model.

From a data model point of view, the interrelation between the different types of data structures representing port, source, switch, and sampler elements revolving around a single sensor population instance could be sketched as shown in Fig. 2. While the top layer of the exhibit indicated the sensor data flow from ports to the staging area pooling the temporary observation CWRs, both master and harvest data areas denote structurally constant SEMAMO storages: obviously, the *master data* area gathers the registries of all populations defined through a SEMAMOGraph, thus depending on a particular SEMAMO application, though reusing same basic structure elements over and over again, whereas the *harvest data* area—holding the collected harvest series for all sensor populations maintained—resorts to a uniform built-in data representation called "canonical harvest records" (CHR, for short).

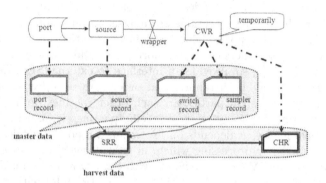

Fig. 2. SEMAMO Registry Structures

Harvest series, built of time-indexed CHRs, are linked to master data through so called "sensor registry records" (SRR, for short) chaining together ports (sources, in fact) with switch and sampler units to individual sensors by means of a simple composite foreign key relationship. Likewise, the SEMAMO domain model establishes the outbound data structures of a SEMAMO application instance. Naturally, the arranged sensor populations of a SEMAMO instance define the interface between storage and analysis areas, respectively, by supplying to the latter the time-dependent harvest series linked to the various available attributes cross-classifying a sensor population. This is to say that, on top of the set of sensor populations in a SEMAMO application, so-called *analytical markets* define the crucial domain elements of the analytical area as depicted in Fig. 1 (right part): each analytical market implements a particular view on the harvest data base of SEMAMO by selecting a targeted subset of gathered sensor (that is, CHR) data.

As a view on harvest data, analytical markets buffer harvest data for subsequent analysis, still referring to individual sensor data so as not to exclude any reasonable type of statistical market analysis. Additionally, for each analytical market arranged, a market pivot table (MPT, for short) is created internally which provides elementary aggregates—or, market indicators—for each of the indicated value dimensions (counts, price, capacity), cross-classified with respect to all market dimensions inherited from the harvest data definition: thus, an MPT in fact represents a basic data warehouse cube [13] in support of all ensuing OLAP. Again, analytical markets are registered in the SEMAMO domain model, guided by information taken from the sensor population registry. With respect to MPTs, also the statistical aggregation functions on value dimensions to be used (e.g., count, average, median, sum of squares, min, max,...) are specified here.

Based on analytical markets, order and report structures are added in a straightforward way; orders simply encode particular types of statistical analyses applied to analytical markets (typically, in a periodical fashion) in order to gain information on market monitoring or decision making levels, triggered by reports defining analytical schedules by grouping sets of analyses composed of tabular and graphical representations of analysis outputs obtained from order processing. Eventually, reports encode certain information interests of decision makers or market researchers (in fact, reports may exhibit a tree-like composite structure) and, in doing so, provide an interface to external SEMAMO service customers. To this end, both reports and orders are recorded in appropriate domain model registries. In particular, each order specifies both (i) the type of analysis alongside with any method-specific parameters, and (ii) its linkage to the analytical market it depends on. Conversely, report schedules determine the due dates for order processing which, in turn, trigger the replenishment of the data buffers (and MPTs) associated with analytical markets. Finally, the SEMAMO domain model provides auxiliary "semantic" information structures, mostly on the level of value domains, such as synonym tables, domain vocabularies/thesauri, or coincidence/correspondence data relating different semantically overlapping classifications or meronomic subdivisions [14] useful in various SEMAMO processing stages.

2.2   The Process Model

The SEMAMO structure model provides formal representations of all the different objects and established object relationships characterizing a particular application, as defined through a SEMAMOGraph instance. In addition to using this formal frame of reference, the actual operation of the online market observatory calls for a minute specification of all processing steps within the generic processing flow. On top, the operation of SEMAMO is controlled by two main processing loops, viz.

- a *harvest cycle* accessing all defined sources attached to Web portals registered for monitoring through the arranged wrappers executing predefined (but adaptively instantiated, depending on different internal harvest parameters; cf. Section 4) query plans;
- an *order cycle* responsible for executing all registered statistical analyses as mandated by report schedules.

Indirectly, both loops interact since the harvest cycle writes periodically to the harvest data base while the order cycle reads from it, although this interaction need not be in sync. As to the job specification, the harvest cycle is by far the more complex one and, contrary to the order cycle, only few processing steps can be composed simply from toolbox functions: in spite of the repetitive nature of Web extraction, data cleansing (or, more generally, so-called ETL processing), and unit re-identification tasks, many transformation functions critically depend on local circumstances and, hence, need careful tuning and testing prior to deployment—activities hardly amenable to automation. Still, there are better defined processing steps such as intermediate schema mappings [15] (notably, the conversion of query plans to parameters controlling wrapper application to portals through binding patterns [16] in Web extraction jobs, or the definition of forward mappings used to integrate sources [17] at sensor population level), for which function libraries can indeed be provided in terms of configurable tool sets.

In the SEMAMO prototype, data integration tasks are located in two places: first, in combining different sources to sensor populations (dashed arcs in Fig. 1) and, secondly and technically much simpler to accomplish, combining different sensor populations to analytical markets (dotted arrows in Fig. 1). While, at both stages, a "global-as-view" integration applies, data requests actually move upstream from reports compiling orders which, implicitly through the analytical markets they refer to, demand the provision of queries to data from some subset of portals. Thus, eventually, queries expressed in terms of analytical market dimensions have to be re-translated into wrapper binding patterns of query plans which harvest portal data to be converted into sensor attributes. In fact, SEMAMO makes use of a weighting scheme reflecting the (sometimes changing) data demands of the order cycle which is then passed on to the harvest cycle.

The least standardized processing stage of SEMAMO concerns cleansing and rectification of portal wrapper data. Generally, wrappers have to be hand-crafted [18] (often supported by visual tools, such as the Lixto Visual Wrapper [19]) and data cleansing operations custom-tailored to the peculiarities and anomalies of individual data sources. However, as data quality is often checked against statistical criteria, many parameters for data validation and editing derive from internal statistical analyses applied to the harvest data base and, thus, are provided in SEMAMO routinely for inclusion in data cleansing functions. The explicit specification of work flows in SEMAMO transformation and schema mapping steps amounts to the (serial) attachment of function modules to SEMAMOGraph arcs much in the same way popular data mining tools [20][21] support GUI-based process specifications. Despite these indispensable customisation tasks, the specification of many core and auxiliary processing steps of SEMAMO can be deduced directly from an application's domain model (cf. Subsection 2.1). This includes ancillary service processes (such as the update of population registries in case of sensors encountered for the first time, or the empirical induction of coincidences between terms of different classification hierarchies, the continuous update of data validation statistics, etc.) as well as more fundamental building blocks of the main harvest cycle of SEMAMO such as the periodic computation of the adaptive harvest schedule actually governing wrapper application (cf. Sec. 4).

3   Unit Re-identification

Careful maintenance of population registries is a salient prerequisite of good data quality [22]. In SEMAMO, the reference to population units is established through the observation process: by scanning online portals, units are registered and, in case of repeated observation, have to be re-identified properly in order to assign time-dependent observations correctly to (re-)identified population units. For the sake of distinction, let *canonical unit* denote the rectified representation of a real entity encountered in an online observation process whereas a *harvest unit* (represented in a CHR; cf. Subsection 2.1) refers to an observation not yet assigned to a canonical unit. Contrary to canonical units bearing unique registry entries by definition, harvest units may feature ambiguous representations, possibly depending on the source they originate from. Figuring out this assignment constitutes the critical task of unit re-identification. The SEMAMO domain model defines type and structure of population units used in an application. In particular, monitored portals contribute switch and sampler units to the population registries such that, in general, portal sub-populations overlap. Accordingly, population registries are induced by sifting out matching harvest units and maintaining the unique ones as canonical population units in a rectified representation (cf. Fig. 2). Apparently, this matching rests on probabilistic decisions, with a re-identification process composed of two stages, viz. (i) *pair-wise record matching* and (ii) a *"link analysis"* chaining pair-wise matching instances to the identified canonical unit. However, contrary to stage (i), stage (ii) does not depend on the domain model of an application and, hence, is not dealt with any further in what follows. Pair-wise record matching in SEMAMO is a combinatorial process based on some population schema $S(A_i, \ldots, A_k)$ comprising the $k > 0$ attributes $A_i, \ldots, A_k$ defined in a SEMAMO domain model. The extension of schema S is the relation $R(A_i, \ldots, A_k) = \{r(a_1, \ldots, a_k) | a_j \in \mathrm{dom}(A_j), 1 \leq j \leq k\}$. Let $r(B)$ denote the projection of a tuple $r \in R(A_i, \ldots, A_k)$ over a non-empty attribute subset $B \subseteq \{A_i, \ldots, A_k\}$. Furthermore, let $r' \equiv r''$ express the fact that $r', r'' \in R(A_i, \ldots, A_k)$ represent the very same canonical unit (equivalence) whether or not $r' = r''$ (syntactical equality). To facilitate unit re-identification, SEMAMO resorts to two kinds of semantic rules. First, value ambiguities are resolved using edit rules covering the case of "near functional" attribute dependencies: for some $J \subseteq \{A_i, \ldots, A_k\}$ provided that $r'(J) = r''(J)$ for a record pair $r', r'' \in R(A_i, \ldots, A_k)$, $r' \equiv r''$ can be assumed. Edit rules are justified on statistical grounds that J is chosen such that the entailed rate of false positive matches remains very low (although, strictly speaking, $r' \equiv r''$ is logically not warranted even in case $r' = r''$ because of undetected homonymy). For instance, letting the attributes $A_1 =$'hotel name', $A_2 =$'street address', $A_3 =$'ZIP code', $A_4 =$'city name', and $A_5 =$'country code' in a hotel population schema, it is fairly safe to expect that $J = \{A_1, A_2, A_3, A_5\}$ already match a pair of hotel harvest records even if city names in the records compared differ: $r'(\{A_4\}) \neq r''(\{A_4\})$. Viewed another way, given such an edit rule, the disparate values of attributes in set $M = \{A_1, \ldots, A_k\} - J$ represent *synonyms* relative to the matching values of J (known synonyms can also be tried, of course, as a replacement of originally non-matching attributes in J). Quite frequently, a set $E \subseteq R(A_1, \ldots, A_k)$ such that, for $r', r'' \in E$, $r'(J_E) = r''(J_E)$ for some non-empty $J_E \subseteq \{A_1, \ldots, A_k\}$ may not justify to conclude $r' \equiv r''$ from statistical evidence. In those cases, *discriminant rules* are applied to decide whether or not two records match. To this end, letting $M_E = \{B_1, \ldots, B_m\} = \{A_1, \ldots, A_k\} - J_E$, a proximity vector function

$$\boldsymbol{p_E} = (p_1, \ldots, p_m) : \times_{j=1}^{m} (B_j \times B_j) \to [0, 1]^m \tag{1}$$

is arranged, composed of component proximity functions p_j depending on the data type of attribute $B_j \in M_E$ (for instance, string comparison metrics [23] for alphanumeric attributes). Each classifier $D_E(r'(M_E), r''(M_E))$ has to be trained, of course, using a representative sample of application-specific data, but may be as simple as a linear (perceptron-type) normalised separator criterion $\boldsymbol{p_E} \cdot \hat{\boldsymbol{w_E}}$ with estimated weights $\hat{\boldsymbol{w_E}} > \boldsymbol{0}$, yielding acceptable results in the tourism application case, that is, in detecting matches between hotel units harvested from four different online hotel booking portals. Albeit expressed in numerical terms, the trained classifiers encode salient knowledge about the application domain.

4 Adaptive Online Observation

A pivotal function of any empirical instrument consists in generating observations. In case of online markets, observations originate from taking "snapshots" of Web portals using mediators wrapping portal content or, more specifically, capturing attribute values of defined population units in a systematic fashion. Moreover, monitoring online markets entails the repeated observation of population units in order to capture market dynamics appropriately, suggesting a cyclic organisation of the observation process (cf. Subsection 2.2). Taking observations online is subject to quite specific "experimental" conditions, invalidating many commonly presupposed sampling conditions:

- the conceived online populations can rarely be enumerated; in other words, well-defined sampling frames and, thus, selection probabilities are lacking;
- selecting individual population units at random within portals may not be feasible; rather, because of technical access properties, bulks of population units can be sampled only, which may impede representativeness and coverage of the sample data gathered;
- the number of observations taken within a period of time happens to be bounded because, otherwise, the virtually unrestricted gathering of observations for purely observational purposes interferes with the normal business mode of portal operation [24], possibly provoking access denials;
- observation attempts simply may fail (that is, resulting in unit or item nonresponse), introducing a further kind of non-sampling error into the data collection process.

To embrace a large as possible class of monitoring scenarios, SEMAMO implements a fairly generic observation model which attaches a harvest series of successive measurements of a quantitative variable (price, in general) to a single sensor (unit of observation as defined in Subsection 2.1). Thus, essentially, each sensor keeps track of the prices for a product or service on offer (sampler unit) accessible online through a portal and, inside the portal, commercially through a vendor or market intermediary (switch unit). The set of sensors in an observed population is broken down multi-dimensionally in terms of recorded offer features as defined by virtue of the SEMAMO application domain model, in addition to portals and switch units providing further stratifications. Through repeated observation, harvest series per sensor are generated incrementally and stored persistently in the SEMAMO harvest data base for later use in statistical analyses. The dynamics of the observed quantitative variables may be modelled in different ways. In the tourism application, for instance, price trajectories are conceived as a step function, that is, prices are assumed to change discretely once in a while, after an average (but unknown) interval of time. This way the temporal behaviour of a sensor mirrors the updating of online content in terms of a change rate model [25][26]. Generally speaking, for sampling purposes the sensors' change rates are estimated from observations recorded in the harvest data base, taking into account also failed observation attempts, and converted into adaptive harvest weights based on a heuristic which maps (price) change dynamics to observation frequencies [27]: basically, this is to say that the sampling probability (or, frequency) of a sensor is made dependent adaptively on the evident variability of previous measurements. In each harvest cycle, by grouping sensors according to hitherto observed change dynamics, a stratified *harvest schedule* is derived which, then, is mapped onto available query plans for execution through the pre-configured portal wrappers. In a sense, schedule strata segment markets reflecting differences in observed dynamics. In general, query plans intersect with schedule strata to varying degrees, so that, in order to strike an optimal balance between the scheduled sample of sensors determined for observation and the tolerated access quotas of the portals monitored, the selection of query plans (resource allocation) in a harvest schedule is accomplished by applying a meta-heuristic [28].

The derivation of adaptive harvest schedules is built into the main harvest cycle of SEMAMO. This feedback harvesting scheme implements a self-calibrating model of online market monitoring, determined and governed through a range of parameters provided through the SEMAMO domain model, such as (i) the subject-matter defining both sensor and constituent populations, (ii) the quantitative variables of interest, (iii) the choice of a model underlying the dynamics of observed quantitative variables, (iv) the query plans available to schedule implementation, (v) a set of numerical parameters and threshold values controlling the heuristics involved, and (vi) access quotas and cost functions per portal relevant for schedule optimisation. Hence, by arranging (or changing) the domain model of an application as outlined in Section 2, this observation instrument is attuned, though indirectly, to the respective experimental set-up of a market monitoring application, provided that the required data sources (wrapper configurations) have been arranged appropriately.

5 Conclusions

Amongst other effects, the "digital turn" of commerce tremendously increases the pace of all business processes, bearing competitive forces and responses. In particular, online markets offer the access to a wealth of business-related information on market sizes, market dynamics, and competitive market behaviour at large. Certainly, early trend anticipation and quick reaction to market changes become even more of crucial importance to business success once major fractions of markets have gone online. In some sectors, such as the tourism industry, this situation is no doubt established already. Besides the individual market participant's view, of course, a surveillance of online markets is of considerable macro-economic interest as well, be it, say, for estimating the growth in sales volumes, or with respect to assessing concentration tendencies possibly calling for antitrust or other regulatory measures.

All of this suggests the use of powerful and flexible means of online market monitoring, exploiting the abundance of publicly available data jointly mirroring an important part of today's economic reality. Presuming a general structure of empirical approaches to online market dynamics, the presented SEMAMO approach prototypically implements a fairly generic strategy of online market monitoring, defining the empirical process in principle, and leaving the detailed specification of the various structural and procedural parameters of the instrument to the formal-symbolic representation of each application domain from which the internal gears and operations of the instrument are deduced mechanically as far as possible. This way, the use of the instrument itself is economised in terms of a specialised toolbox facilitating "online market statistics on demand" across a wide range of application domains. It remains to be seen, mainly through practical experimenting, whether the SEMAMO approach in fact meets its expectations and, in particular, if the concepts developed may extend to a still more general view of a "Web Observatory".

Acknowledgments. The authors gratefully mention the valuable discussions with Wilfried Grossmann, Marcus Hudec, Wolfgang Jank, Patrick Mair, and Kurt Hornik on statistical issues of SEMAMO. The project is supported by grant Fit-IT "Semantic Systems and Services"' No. 815.135 of the Austrian Research Promotion Agency (FFG).

References

1. Shaw, M., Blanning, R., Strader, T., Whinston, A. (eds.): Handbook on Electronic Commerce. Springer, Heidelberg (1999)
2. Romm, C.T., Sudweeks, F. (eds.): Doing Business Electronically - A Global Perspective of Electronic Commerce. Springer, London (1998)
3. Kotler, P., Armstrong, G.: Principles of Marketing. Pearson, Prentice Hall, New Jersey (2007)
4. Dhyani, D., Ng, W.K., Bhowmick, S.S.: A Survey of Web Metrics. ACM Computing Surveys 34(4), 469–503 (2002)

5. Fayyad, U., Piatetsky-Shapiro, G., Smyth, P., Uthurusamy, R. (eds.): Advances in Knowledge Discovery and Data Mining. AAAI/MIT Press, Menlo Park (1996)
6. Lenzerini, M.: Data Integration: A Theoretical Perspective. In: PODS 2002, pp. 233–246 (2002)
7. Davies, J., Studer, R., Warren, P.: Semantic Web Technologies: Trends and Research in Ontology-based Systems. Wiley, New York (2006)
8. Werthner, H., Klein, S.: Information Technology and Tourism. In: A Challenging Relationship. Springer, Wien (2008)
9. Levy, P.S., Lemenshow, S.: Sampling of Populations: Methods and Applications. Wiley, New York (1999)
10. Wiederhold, G., Genesereth, M.: The Conceptual Basis for Mediation Services. IEEE Expert 12(5), 38–47 (1997)
11. Baumgartner, R., Frölich, O., Gottlob, G.: The Lixto Systems Applications in Business Intelligence and Semantic Web. In: Franconi, E., Kifer, M., May, W. (eds.) ESWC 2007. LNCS, vol. 4519, pp. 16–26. Springer, Heidelberg (2007)
12. Fellegi, I., Sunter, A.: A Theory for Record Linkage. JASA 64(328), 1183–1210 (1969)
13. Kimball, R., Ross, M.: The Data Warehouse Toolkit: The Complete Guide to Dimensional Modeling. Wiley, New York (2001)
14. Visser, U.: Intelligent Information Integration for the Semantic Web. LNCS (LNAI), vol. 3159. Springer, Heidelberg (2004)
15. Batini, C., Lenzerini, M., Navate, S.B.: A Comparative Analysis of Methodologies for Database Schema Integration. ACM Computing Surveys 18(4), 323–364 (1986)
16. Rajaraman, A., Sagiv, Y., Ullman, J.D.: Answering Queries Using Templates with Binding Patterns. In: PODS 1995, pp. 105–112 (1995)
17. Halevy, A.Y.: Answering Queries Using Views: A Survey. VLDB Journal 10(4), 270–294 (2001)
18. Chidlovskii, B., Borghoff, U., Chevalier, P.: Towards Sophisticated Wrapping of Web-based Information Repositories. In: RIAO 1997, Montreal, pp. 123–155 (1997)
19. LiXto Visual Wrapper, http://www.lixto.com/li/liview/action/display/frmLiID/12/ (last accessed June 24, 2009)
20. SAS Enterprise Miner, http://support.sas.com/documentation/onlinedoc/miner/index.html (last accessed June 24, 2009)
21. SPSS Modeler (previously: Clementine), http://www.spss.com/software/modeling/modeler/ (last accessed June 24, 2009)
22. Herzog, T.N., Scheuren, F.J., Winkler, W.E.: Data Quality and Record Linkage Techniques. Springer, New York (2007)
23. Winkler, W.E.: String Comparator Metrics and Enhanced Decision Rules in the Fellegi-Sunter Model of Record Linkage. ASA Proc. of the Sect. on Survey Res. Methods, 354–359 (1990)
24. Hess, C., Ostrom, E.: Ideas, Artifacts, and Facilities: Information as a Common-pool Resource. Law and Contemporary Problems 66, 111–145 (2003)
25. Cho, J., Garcia-Molina, H.: Estimating frequency of change. ACM Transactions on Internet Technology 3(3), 256–290 (2003)
26. Grimes, C., Ford, D.: Estimation of web page change rates. In: JSM, pp. 3968–3973. Denver, Co. (2008)
27. Walchhofer, N., Froeschl, K.A., Hronsky, M., Hornik, K.: Adaptive Web Harvesting in Online Market Monitoring. Submitted to Journal for Advanced Data Analysis and Classification (2009)
28. Kellerer, H., Pferschy, H.U., Pisinger, D.: Knapsack Problems. Springer, Heidelberg (1983)

Data Driven Rank Ordering and Its Application to Financial Portfolio Construction

Maria Dobrska*, Hui Wang, and William Blackburn

University of Ulster at Jordanstown
BT37 0QB United Kingdom
dobrska-m@email.ulster.ac.uk,
{h.wang,wt.blackburn}@ulster.ac.uk

Abstract. Data driven rank ordering is a the search for optimal ordering in data based on rules inherent in historical data. This is a challenging problem gaining increasing attention in the machine learning community. We apply our methodology based on pairwise preferences derived from historical data to financial portfolio construction and updating. We use share price data from the FTSE100 between 2003 and 2007. It turned out the portfolio of shares constructed and updated this way produced significant outperformance compared to two benchmarks - firstly a portfolio constructed by a buy and hold strategy and secondly a portfolio created and updated based on neural network predictions.

Keywords: ranking, learning to order, pairwise preference, portfolio construction.

1 Introduction

Rankings are very intuitive sources of information, therefore they are utilised for making decisions on an everyday basis. They also provide potentially more powerful information about data than, for example, class membership of data.

Ranking is also a subject of study in many areas of research. It is so fundamental for decision making that it has been extensively studied in economics, psychology, operations research, and other human-centered disciplines, usually under the name of preference [11]. Ranking has in recent years found applications in many areas of artificial intelligence, including recommender systems, e-commerce, multi-agent systems, planning and scheduling, and intelligent financial data analysis [11].

There are two approaches towards the problem of learning to rank order — based on pairwise preferences or on performance measures for each single instance. Many researchers agree that rankings based on pairwise preferences are more powerful and informative [14].

In existing studies, it is usually assumed that the pairwise preferences are provided by some ranking experts. However in some cases such preferences are not

---

* The first author is a PhD student.

D. Karagiannis and Z. Jin (Eds.): KSEM 2009, LNAI 5914, pp. 241–252, 2009.

directly available, so we have to learn their structure elsewhere. The importance of preference learning has been highlighted in [10,11].

In this paper we consider the problem of how to mine such preferences from the financial data and apply them in the financial portfolio construction and updating system. We introduce a preference function and propose a method of optimization of its parameters. Such a training scheme for the preference function utilizes existing rankings present in the historical data. Our hypothesis is that such a trained function has a generalization power and can be employed to rank unseen instances.

We evaluated our methodology for financial portfolio management, and compared it with a neural network based approach and also with a standard portfolio evaluation method (buy and hold strategy). In experiments conducted on data from the London Stock Exchange our hypothesis has been supported, as our portfolio construction and trading scheme based on trained preference rankings has yielded encouraging results.

This paper is organized as follows: in Section 2 we discuss existing work on the problem of rank ordering, learning to order and its applications. In Section 3 we describe our methodology for rank ordering based on a trained preference function. In Section 4 we discuss existing methods for portfolio construction and stock market trading and apply our methodology in this area. Section 5 provides information on our experimental results. Conclusions and future work are discussed in Section 6.

2   Rank Ordering — Background

Early research work addressing the subject of rank ordering based on preference judgements was reported by Hans Buhlmann and Peter Huber. In [4] the problem of choosing the best ranking for a given preference matrix is discussed. Since the instances to be rank ordered are participants in a tournament whose scores are known, the availability of preference judgements can be assumed as they are based on the score of each game in the tournament.

William Cohen extensively studied the problem of rank ordering based on pairwise comparisons. In [7] he and his co-workers addressed these issues with the use of preference judgements which provide information about priority between each two given instances. In this study the availability of "primitive preference functions" for a set of instances to be rank ordered is assumed, i.e. the presence of some "ranking experts". The preference training involves the search for the optimal weighting of such experts, or more precisely their judgements regarding the preferences among instances.

The construction of the ranking from the set of pairwise preference judgements requires formalization of the notion of agreement between a ranking and the preference function. In [7] the agreement is measured as a sum of values of the preference function for those pairs of instances for which the preference function and the ordering agree, i.e. if the preference function indicates that a should be ranked ahead of b, the ordering reflects this preference. Finding the ordering whose agreement with preference function exceeds some rational number is NP-complete. This is why a "greedy algorithm" for the search of quasi-optimal ordering is introduced. The greedy algorithm finds an ordering whose agreement with a given function is not smaller than half the value of the agreement of the optimal ordering.

An improvement of the greedy algorithm can be found in [8]. The greedy algorithm treats the instances as nodes of a graph and weights of directed edges between those nodes are derived from preference judgements. Improvement of the algorithm involves dividing the initial graph into subgraphs referred to as strongly connected components. The ordering for each of the subgraphs is conducted with the use of the greedy algorithm.

According to [14] pairwise preferences are very useful information for ranking creation, as they are simple and potentially sharper than other approaches. The search for optimal ranking based on pairwise preferences involves maximization of the probability of full agreement between an ordering and pairwise preferences. Pairwise preferences are based on the estimated probability of one instance outperforming another. The application of this theory to the ranking of financial portfolios will be discussed in Section 4.1.

In [9] the problem of ranking of labels for given instances is addressed. The main point is to design a methodology which ranks the labels (from a given fixed set). It is then compared with classification methodology, where only one label is associated with each instance. The problem of rank ordering is approached from the pairwise preference angle. For each pair of class labels a separate model is created, based on training instances for which preferences among labels are known. The models are trained with the use of the decision tree learner C4.5.

In [13] a methodology based on pairwise preferences for ordinal regression is proposed. Ordinal regression is a problem which lies between multi-class classification and regression — there is a fixed number of rank labels which are to be associated with each instance, and there is an ordering among these labels, i.e. misclassifying an instance into an adjacent class is less harmful than into a remote one. The proposed framework based on Support Vector Machine classification methodology is (unlike statistical models) distribution independent. For a pair of instances a preference is obtained via large margin classifier, i.e. one class contains pairs whose first member is preferred over the second one and the other class contains pairs where the second instance should be ranked ahead of the first one.

3   Learning to Rank Order with Preference Function Optimization

Many researchers agree that rankings based on preference judgements are potentially more powerful and sharper than those based on a performance measure of each single instance to be rank ordered [14]. However in real world applications, e.g when ranking of stocks is considered, expert opinions about pairwise preferences may not be available.

3.1   Preference Functions for Rank Ordering

The lack of ready-to-use preference judgements can, however, be overcome with the use of a preference function. Such a function can be obtained if some training data are available. In this work we consider a simple type of function. The optimization process aims at finding an optimal vector of the function's parameters.

Let $\mathbf{P} = \{S_1, S_2, \cdots, S_n\}$ be a set to be rank ordered, where S_i is an element in an m-dimensional Euclidean space defined by m attributes. We write $S_i = \langle s_{i_1}, s_{i_2}, \ldots, s_{i_m} \rangle$, where $s_{i_j} \in \mathbb{R}$. We define a preference function as $Pref$: $\mathbf{P} \times \mathbf{P} \to \mathbb{R}$ such that

- if $Pref(S_i, S_j) > 0$, S_i is ranked ahead of S_j;
- if $Pref(S_i, S_j) < 0$, S_j is ranked ahead of S_i;
- if $Pref(S_i, S_j) = 0$, there is no preference between S_i and S_j;
- $\mathrm{sgn}(Pref(S_i, S_j)) = -\mathrm{sgn}(Pref(S_j, S_i))$ or more strictly: $Pref(S_i, S_j) = -Pref(S_j, S_i)$.

In this work we consider the following family of functions: $Pref_W$: $\mathbb{R}^m \times \mathbb{R}^m \to \mathbb{R}$,

$$Pref_W(S_i, S_j) = \sum_{k=1}^{m} w_{<i,j>_k}(s_{i_k} - s_{j_k}), \tag{1}$$

where $w_{<i,j>_k}$ is the weight of attribute k for the instance pair $< S_i, S_j >$. The higher value of the function, the more S_i is preferred over S_j. Other types of preference functions are also feasible, however the choice of preference function is not the subject of this particular work.

3.2   Optimization of the Preference Function

In (1) the set of attribute weights is unknown. An optimization process is employed to search for attribute weights which maximize the agreement between the preference function based ranking and the actual ranking.

Let $L = \{l_1, l_2, \ldots, l_n\}$ be a set of the *actual* (numerical) rank labels of instances in $\mathbf{P}$, where l_i is the rank label of S_i, i.e., l_i represents the position of instance S_i in the ranking. In terms of preference $l_i < l_j$ indicates that the i-th instance is ranked ahead of the j-th one in the actual ranking. The actual preference function $F : \mathbf{P} \times \mathbf{P} \to \mathbb{R}$ can be derived from the actual ranking as follows:

$$F(S_i, S_j) = l_j - l_i.$$

The optimal preference function $Pref^* : \mathbf{P} \times \mathbf{P} \to \mathbb{R}$ satisfies:

$$\forall_{i,j \leq n} \ (l_i < l_j \Rightarrow Pref^*(S_i, S_j) > 0)$$

and

$$\forall_{i,j \leq n} (l_i = l_j \Rightarrow Pref^*(S_i, S_j) = 0)$$

The aim of the optimization process is to minimise the cumulative error for all pairs of instances from the set $\mathbf{P}$, which is defined as follows:

$$E_W = \sum_{i=1}^{n-1} \sum_{j=i+1}^{n} |F(S_i, S_j) - Pref_W(S_i, S_j)| \tag{2}$$

There are several techniques which can be employed to minimise the error E_W. Here we use gradient descent – a technique for finding a local minimum taking steps in the opposite direction to the gradient vector. By substituting (1) into (2) we have:

$$E_W = \sum_{i=1}^{n-1} \sum_{j=i+1}^{n} |F(S_i, S_j) - \sum_{k=1}^{m} w_{<i,j>_k}(s_{i_k} - s_{j_k})|.$$

To apply gradient descent, partial derivatives of the error function are required:

1. If $F(S_i, S_j) < Pref(S_i, S_j)$, then $\dfrac{\partial E_W}{\partial w_{<i,j>_k}} = -s_{i_k} + s_{j_k}$;

2. If $F(S_i, S_j) > Pref(S_i, S_j)$, then $\dfrac{\partial E_W}{\partial w_{<i,j>_k}} = s_{i_k} - s_{j_k}$;

3. If $F(S_i, S_j) = Pref(S_i, S_j)$, then E_W is not differentiable.

Note that E_W is not differentiable when there is no disagreement between the function $Pref_W$ and the actual preference function F. It is then reasonable to assume that in this case $\frac{\partial E_W}{\partial w_{<i,j>_k}} = 0$.

In the gradient descent optimization process the weights are updated iteratively:

$$w_{<i,j>_k}^{(t)} = w_{<i,j>_k}^{(t-1)} - \gamma \frac{\partial E_W^{(t-1)}}{\partial w_{<i,j>_k}}, \tag{3}$$

where $\gamma > 0$ is a (small) learning factor and $t > 0$ is an iteration number. The process stops after a fixed number of rounds, or when the value of the error function E_W is smaller than a predefined threshold, or when the value of E_W does not change in the consecutive iteration. The choice of $\gamma > 0$ and number of rounds or threshold value should be based on empirical studies for given data and may differ from one domain to another. We denote the preference function obtained this way by $Pref_{W^*}$.

Once the preference function is obtained, it can be used to construct a ranking function for prediction. In this work we used the greedy algorithm from [8] to obtain total order from pairwise preferences.

4   Financial Forecasting and Portfolio Management

The rank ordering based on a trained preference function can be applied to various domains. In this work we concentrate on financial applications and use the methodology for stock portfolio construction and updating.

Changes in stock prices reflect a mixture of different factors associated with investors. These include knowledge, fears, optimism or greed of market players [17]. It should be noted that investors' behavior is often repeatable which leads to an assumption that historical patterns and relationships can be employed to predict future market conditions. Such forecasting cannot be performed with the use of a single indicator, however a reasonable set of market indicators should provide some clues about future behavior of stock prices [17].

4.1 Background

Various trading strategies are proposed in the literature. They employ different indicators associated with price movements to support the creation of profitable portfolios.

In [14] ordering by pairwise comparisons is applied to rank financial portfolios. Estimates of probabilities of pairwise dominances (or preferences) derive from parameters associated with portfolios' performances. The experiments are conducted for 8 mutual funds and market indices. All 8! possible orderings are evaluated in searching for the one which optimally fits the pairwise dominance probability estimates. A serious disadvantage of this method is the poor time efficiency. Even a small increase in the number of instances (portfolios) to be rank ordered drastically affects the time required to search for the optimal one.

Artificial neural networks (ANN) are often employed in financial data mining. In [1] a decision support system combing ANN with the genetic algorithm is proposed for dealing on Tokyo Stock Exchange. Neural networks are utilized to predict the highest increase rate and lowest decrease rate of the TOPIX (Tokyo Stock Exchange Prices Index) within the next four weeks. Based on these predictions, a decision support system – optimized with the genetic algorithm which generates buy/sell/hold signals for traders – is introduced. Also in [15] feedforward and recurrent ANN are employed in a search for non-linear patterns present in foreign exchange data. In [5] Cao and Tay report that a support vector machine (SVM) utilized for financial time series forecasting outperformed back-propagation neural networks and yielded results comparable to radial basis function neural networks, and that SVM have good generalization performance.

In [12] a hidden Markov model (HMM) is employed in order to predict the next day closing prices for (airline) stocks. Estimation of HMM parameters is performed with the historical data. According to the experimental results in [12] the use of HMM in stock price prediction yields results comparable to those obtained with the use of certain artificial neural networks. The advantage of HMM lies in its ability to explain the models which are analyzed.

Generating buy/sell signals by identifying real turning points in stock prices is discussed in [2]. The system based a on simplified Markov model is trained with the use of past data to answer the question as to whether a candidate turning point is a real trend reversal or noise-generated fluctuation. The model predictors are based on technical indicators.

In [16] Andrew W. Lao *et al.* argues that technical analysis can provide meaningful information from market prices. The approach using smoothing techniques (e.g. non-parametric kernel regression) extracts non-linear patterns from data. An algorithm which detects technical patterns by construction of kernel estimators is also introduced in [16].

None of the above methodologies and techniques directly deals with the problem of portfolio construction. The question "which entities should be invested in or sold at a particular point in time?" is critical in investment decision making. Little work can be found on portfolio updating strategies and data-driven comparative studies on the assets available in the market.

4.2   Learning to Rank Order for Portfolio Construction

In this section we will analyse the problem of choosing stocks for a portfolio of a predefined size so that it yields a target level of profits in the near future. According to our findings discussed in Section 5 the methodology discussed in Section 3 can lead to outperformance of the market in a significant proportion of cases.

Suppose we have n stocks. The ith stock on trading day t is represented by $S_i(t) = \langle s_{i_1}(t), s_{i_2}(t), \cdots, s_{i_m}(t) \rangle$, where $s_{i_j}(t)$ is a technical indicator of the stock on the t-th trading day. Then $\mathbf{P}(t) = \langle S_1(t), S_2(t), \cdots, S_n(t) \rangle$ represents the states of all n stocks on day t. Our aim is to, having seen $\mathbf{P}(0), \mathbf{P}(1), \cdots, \mathbf{P}(t)$, predict the ranking of these stocks with respect to their future return potential on day $t + r$. If we get a good ranking, trading decisions can then be made on the basis of this ranking.

More specifically, let us define the preference function for portfolio construction as in (1):

$$Pref_W(S_i(t), S_j(t)) = \sum_{k=1}^{m} w_{<i,j>_k}(s_{i_k}(t) - s_{j_k}(t)), \tag{4}$$

The actual rank labels $\langle l_1(t), l_2(t), \cdots, l_n(t) \rangle$ can be obtained from the returns generated by stocks between day t and $t + r$, where $r \in \mathbb{N}$ is a return period length. In other words, the greater the return generated by the i-th stock during a given time period, the higher the position in the ranking, i.e. the smaller the rank label $l_i(t)$.

The aim of a weight optimization process as in (3) is to reproduce the actual ranking (which is based on future returns) with the use of a preference function without utilizing information about future price movements.

Our hypothesis is that weightings trained on historical data can be employed in the future for the creation of rankings supporting trading strategies. Top-ranked instances (stocks) can be taken into consideration in decision-making about changing stocks within the portfolio.

4.3   Exemplary Technical Indicators

In this work we used five well known technical indicators:

1. $s_{i_1}(t)$ — Relative Strength Index (RSI),

$$RSI = 100 - \frac{100}{1 + RS}, \; (RS = \text{relative strength})$$

$$RS = \frac{\text{average gain in previous } x \text{ trading days}}{\text{average loss in previous } x \text{ trading days}}$$

2. $s_{i_2}(t)$ — Money Flow Index (MFI),

$$MFI = 100 - \frac{100}{1 + MR}, \; (MR = \text{money ratio})$$

$$MR = \frac{\text{positive money flow}}{\text{negative money flow}},$$

$$\text{money flow} = \frac{\text{high} + \text{low} + \text{close}}{3} \times \text{volume}$$

3. $s_{i_3}(t)$ — Fast Stochastic Oscillator (%K)

$$\%K = \frac{\text{close} - \text{lowest in past } x \text{ days}}{\text{highest in past } x \text{ days} - \text{lowest in past } x \text{ days}}$$

4. $s_{i_4}(t)$ — Bollinger Band (%b)

$$\%b = \frac{\text{close} - \text{lowerBB}}{\text{upper BB} - \text{lower BB}}$$

5. $s_{i_5}(t)$ — Bollinger Band Width (BBW),

$$BBW = \frac{\text{upper BB} - \text{lower BB}}{\text{middle BB}}$$

Detailed discussion on technical indicators can be found in [3,17,18].

5   Experimental Results

In this section we present initial results of applying our preference training methodology to financial portfolio construction.

5.1   Experimental Settings

In our experiments we utilized the data from the London Stock Exchange. We used a set of fifty three stocks in FTSE100 index from 1 January 2003 until 31 December 2007 for which the full data was available at [19]. Five technical indicators described in the previous section were calculated from the raw data, which consist of daily open, high, low and close prices along with trading volume.

The data has been divided into twenty five 3-year long periods with the use of a sliding window of a month-size (with the first period starting on 1st January 2003 and ending on 31 December 2005, the second starting on 1st February 2003 ending on 31 January 2006, etc. and the last one starting on 1st January 2005 and ending on 31 December 2007). Each of such 3-year long periods has been

Table 1. Predicted portfolio against buy and hold strategy

Updating frequency	Portfolio size	Mean gain (%)	Outperformance count
1	3	18.7	14
1	5	15.8	17
1	7	15.1	18
1	10	11.5	14
3	3	18.6	16
3	5	16	19
3	7	17.9	21
3	10	14.1	23
5	3	10.4	15
5	5	9.9	14
5	7	12.4	18
5	10	9.5	21
10	3	13.9	17
10	5	11.3	18
10	7	11.1	19
10	10	7.6	18

Table 2. Predicted portfolio against portfolio based on ANN prediction

Updating frequency	Portfolio size	Mean gain (%)	Outperformance count
1	3	28	19
1	5	20.5	16
1	7	17.7	20
1	10	5.5	17
3	3	28.5	20
3	5	25.1	23
3	7	21.2	23
3	10	16.1	23
5	3	14.4	14
5	5	11	15
5	7	12.1	20
5	10	6.8	18
10	3	19.4	17
10	5	14.9	17
10	7	11.6	19
10	10	5.8	17

split into two subsets – training $\mathbf{P}_{tr_k}$ (consisting of the first 2 years) and testing $\mathbf{P}_{test_k}$ (consisting of the last year), $k \leq 25$.

The first period has been used only for estimation of gradient descent parameters – learning factor γ and number of rounds q have been set up to: $\gamma = 0.1$ and

$q = 7$, respectively. The past period length for calculating technical indicators as described in Section 4.3 has been set to $x = 8$.

For each trading day t from training set $\mathbf{P}_{tr_k}$ the optimization as described in Sections 3 and 4.2 has been performed to find an optimal weight vector W_t. Actual rankings in the training process were based on 1-, 3-, 5- and 10-day returns (in separate experiments). The final trained weight vector $W*$ was defined as a mean of the optimized vectors for all training trading days. The vector W was then utilized in the testing phase.

The training performance was evaluated with the use of test data from $\mathbf{P}_{test_k}$. The portfolios were constructed based on top ranks yielded by ranking based on the trained preference function. Experiments were conducted for portfolios consisting of 3, 5, 7 and 10 top stocks, where equal amount of money was invested in each of the stocks in the portfolio. Each portfolio was updated with the frequency equal to the return period length applied in the training, i.e. if training was conducted based on 3-day returns, the testing portfolio was updated every 3 trading days.

Our methodology was compared with the performance of the buy and hold strategy and portfolios based on predictions generated by ANN [6]. The ANN were trained to predict future returns based on the five technical indicators discussed in Section 4.3. Data was divided into experimental periods and training and testing subsets were determined as in the experiments involving preference function training.

5.2 Results

It should be noted that experiments involving each of the 3-year long periods with different return period lengths were conducted independently. Table 1 summarizes the experimental results for different portfolio sizes and return period lengths. The column "Mean gain (%)" contains the average percentage **gain of our method over a buy and hold (B&H) strategy** (where equal amount of money is invested in each of the fifty three stocks at the beginning of the testing period and the related holdings are sold at the end). The column "Outperformance count" contains information about the number of cases (out of all 24 experimental sets) when our method outperformed the buy and hold strategy.

Table 2 contains equivalent information from a comparison between performance of portfolios based on our methodology and portfolios based on ANN predictions (outperformance of our method over ANN). Note that the results do not take into account trading costs (buy/sell spread, purchase stamp duty, commission).

From Tables 1 and 2 it follows that our methodology applied for twenty four experimental periods outperforms a buy and hold strategy and portfolios based on ANN predictions. The best results are yielded when the portfolio is updated every trading day, however such high trading frequency would involve higher trading costs.

The results show that our method for portfolio construction based on rankings with a trained preference function yields significant outperformance for those experimental periods which exhibited overall strong upward price movements. However the results were less significant over other periods. This issue is to be addressed in the future.

6 Conclusions and Future Work

We discussed our novel approach towards the financial portfolio construction and updating problem. It is based on a trained preference function which provides information about pairwise preferences in the data, i.e. which stock should be ranked ahead of the other one on a particular trading day. Such a preference function can be very useful when there is conflicting or no expert opinion about pairwise preferences among instances to be rank ordered. However such a function has to be carefully chosen to reflect the actual preferences. This is why we employed optimization of the preference function based on training (historical) data for which actual rankings are known.

The preference function was trained with the use of historical data containing the information about actual rankings which have been derived from the returns generated by each stock to be rank ordered. The experiments performed on data from the London Stock Exchange are encouraging regarding both the methodology and the financial application. With various portfolio sizes and updating frequencies our method of portfolio construction outperformed the buy and hold strategy and ANN-based portfolios in the majority of experimental cases.

This work is based on a fixed structure of preference function where only a vector of weights may be updated. In the future the issue of using different preference functions and optimization techniques will be addressed. Deterioration of performance in a falling market should also be analyzed. Applications involving non-time series data can also be addressed.

References

1. Baba, N., Inoue, N., Yan, Y.: Utilization of soft computing techniques for constructing reliable decision support systems for dealing stocks. In: Proceedings of the 2002 International Joint Conference on Neural Networks, 2002. IJCNN 2002., pp. 2150–2155 (2002)
2. Bao, D., Yang, Z.: Intelligent stock trading system by turning point confirming and probabilistic reasoning. Expert Syst. Appl. 34, 620–627 (2008)
3. Bollinger Bands, http://www.bollingerbands.com
4. Buhlmann, H., Huber, P.: Pairwise Comparison and Ranking in Tournaments. The Annals of Mathematical Statistics 34, 501–510 (1963)
5. Cao, L., Tay, F.: Support vector machine with adaptive parameters in financial time series forecasting. IEEE Transactions on Neural Networks 14, 1506–1518 (2003)
6. SPSS Clementine, http://www.spss.com/clementine
7. Cohen, W., Schapire, R., Singer, Y.: Learning to order things. In: NIPS 1997: Proceedings of the 1997 conference on Advances in neural information processing systems, pp. 451–457. MIT Press, Cambridge (1998)

8. Cohen, W., Schapire, R., Singer, Y.: Learning to order things. Journal of Artificial Intelligence Research 10, 243–270 (1999)
9. Fürnkranz, J., Hüllermeier, E.: Pairwise Preference Learning and Ranking. In: Lavrač, N., Gamberger, D., Todorovski, L., Blockeel, H. (eds.) ECML 2003. LNCS (LNAI), vol. 2837, pp. 145–156. Springer, Heidelberg (2003)
10. Furnkranz, J., Hullermeier, E.: Preference Learning. Künstliche Intelligenz 19, 60–61 (2005)
11. Goldsmith, J., Junker, U.: Preference handling for artificial intelligence. AI Magazine 29, 9–12 (2008)
12. Hassan, M., Nath, B.: StockMarket Forecasting Using Hidden Markov Model: A New Approach. In: ISDA 2005: Proceedings of the 5th International Conference on Intelligent Systems Design and Applications, pp. 192–196. IEEE Computer Society, Los Alamitos (2005)
13. Herbrich, R., Graepel, T., Obermayer, K.: Support Vector Learning for Ordinal Regression. In: International Conference on Artificial Neural Networks, pp. 97–102 (1999)
14. Hochberg, Y., Rabinovitch, R.: Ranking by Pairwise Comparisons with Special Reference to Ordering portfolios. American Journal of Mathematical and Management Sciences 20 (2000)
15. Chung-Ming, K., Tung, L.: Forecasting Exchange Rates Using Feedforward and Recurrent Neural Networks. Journal of Applied Econometrics 10, 347–364 (1995)
16. Lo, A., Mamaysky, H., Wang, J.: Foundations of Technical Analysis: Computational Algorithms, Statistical Inference, and Empirical Implementation. The Journal of Finance. LV (2000)
17. Pring, M.: Technical Analysis Explained. McGraw-Hill, Inc., New York (1991)
18. StockCharts.com, http://stockcharts.com/school/doku.php?id=chart_school
19. Yahoo! Finance, http://uk.finance.yahoo.com

Prioritizing Non-functional Concerns in MAMIE Methodology

Hakim Bendjenna[1,2], Mohamed Amroune[1], Nacer-eddine Zarour[1], and Pierre-jean Charrel[2]

[1] LIRE laboratory, Computer Science Department, Mentouri University, Constantine, Algeria
[2] Toulouse University and Institut de Recherche en Informatique de Toulouse, Toulouse, France
hakim.bendjenna@etu.univ-tlse2.fr, medamroune@gmail.com, nasro-zarour@umc.edu.dz, charrel@univ-tlse2.fr

Abstract. The increasing globalization of markets and companies demands more and more investigations of distributed requirements engineering. Requirements elicitation is organized as one of the most critical activities of the requirements engineering process. It is a difficult task enough when done locally, but it is even more difficult in a distributed environment due to cultural, language and time zone boundaries. In our previous research we have proposed a methodology to elicit requirements for an inter-company co-operative information system. In this methodology the analyst must specify a priority value for each non-functional concern that may be considered in the inter-company co-operation process. To do, s/he must take in consideration several viewpoints of concerned stakeholders. In order to help the analyst to better accomplish this task, we propose in the present paper a process based on Grunig and Hunt model to identify concerned stakeholders and Michell et al model to prioritize them. A mathematical function proposed to return the final non-functional concern priority constitutes a cornerstone of this process. Preliminary results suggest that this process is of valuable help to analysts during requirements elicitation.

Keywords: Requirements elicitation, Non-functional concern, Priority, Co-operative information system.

1 Introduction

Developing requirements for an Inter-Company co-operative Information System (ICIS) requires continuous and effective coordination of tasks, resources, and people. Thus, the challenge of requirements engineering is compounded by the various boundaries inherent in distributed projects, be they geographical, temporal, organizational, or cultural. Several researchers have started to examine the intricate interplay of these boundaries and their effects [5, 9]. Most of the effort has been devoted to elicitation and facilitation techniques. The need for improved collaborative tools has also been discussed [7, 10], whereas little research has been done on how to elicit requirements for an ICIS [5, 7].

D. Karagiannis and Z. Jin (Eds.): KSEM 2009, LNAI 5914, pp. 253–262, 2009.

We early propose in [2] a methodology named MAMIE (for MAcro to MIcro level requirements Elicitation) to elicit requirements for an inter-company co-operative information system. MAMIE aims at analyzing an inter-company cooperative information system by means of UML based so-called *co-operation use cases* and encapsulates the requirements into so-called *viewpoints* of different stakeholders.

In MAMIE, a functional concern F.C represents a primary business goal, and a non functional concern NF.C is related to a specific F.C and represents a quality or a property of this F.C (e.g. time, cost). The analyst assigns a rank of priority (from 1-very low- to 5-very high-) for each NF.C that may be considered in the inter-company cooperation process. In the environment of a ICIS, assigning a rank of priority for each NF.C is not a simple task: several stakeholder may have some interest for the same NF.C and thus affect a specific value of priority to it. The different values may be different, even oppose to each other, e.g. very low for one stakeholder and very important for another). According to [1,4,11,12,14], the priority of a non-functional requirement is to be assigned by stakeholders to reflect business criticality, importance to the customers or users, urgency, importance for the product architecture, or fit to release theme.

In the present paper, we enhance MAMIE methodology to give response to the following question: how an analyst assigns a rank of priority to a specific non-functional concern by considering all viewpoints of concerned stakeholders? The assignment process takes into consideration the priority of each stakeholder, following Mitchell et al. model [13]. Grunig and Hunt model [9] is also used to identify stakeholders before prioritize them.

The rest of the paper is organized as follows: section 2 presents an overview of MAMIE methodology. The proposed process is detailed in section 3. Finally, section 4 draws some conclusions and highlights some future works.

2   MAMIE Macro to Micro Level Requirements Elicitation Method: An Overview

MAMIE is a method which aims to guide a system analyst in order to elicit requirements for an ICIS. It is composed of two phases (figure 1) which specify respectively the macro and micro levels co-operation.

Phase 1 consists to elicit and specify co-operative activities, constraints of cooperation between companies, and the relevant non-functional concerns of the future system (the macro level co-operation). Essential requirements are issued from interviews, questionnaire, documents study or other elicitation techniques.

Phase 1 is composed of 3 steps:

In step 1.1, the analyst elicits and models *co-operation use cases* and relations between them: we define a co-operation use case as a use case attached to one of the companies of the ICIS which relates to at least one other use case attached to another company. All use cases included in a *co-operation use case* are considered also as *cooperation use cases*. In this step of the analysis, the analyst assigns each company to its corresponding *co-operative use case*. For each of these use cases, the analyst describes, in a general way, what the related company do:

- in order to accomplish delegated goals (normal scenario);
- when it cannot normally perform delegated goals (exceptional scenarios).

In step 1.2, for each *co-operation use case* which is not included in another use case, the analyst uses the sequence diagram to elicit and model requirements which describe interaction and constraints of co-operation between companies.

In step 1.3, the analyst identifies and specifies all the NF.Cs which must be taken into account, their relationships, and priorities within the use cases defined at step 1.1. NF.Cs can be identified using several approaches such as those used to identify goals. We have taken the main ideas from Chung [4], in particular NF.C catalogue. The analyst with the other stakeholders could identify which of these entries are applicable in this case.

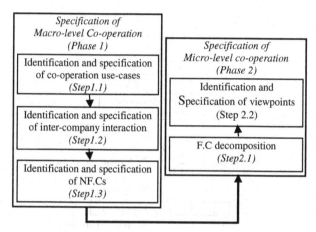

Fig. 1. MAMIE: the different steps

The question in phase 2 is: *ow to get necessary requirements to set up goals, constraints and NF.Cs identified in phase 1?* The answer to this question describes the micro level of inter-company co-operation.

Phase 2 is composed of 2 steps:

Step 2.1 starts by decomposing F.Cs corresponding to the goals of *co-operation use cases* which are not included in other *co-operation use cases*, identified in Step 2.1).

Step 2.2 consists to identify and specify viewpoints which contain requirements representing response to the main question of this phase.

At the end of the whole process, the specification of the future system is organized in terms of a set of ordered viewpoints which contain requirements issued from initial business goals. MAMIE proposes templates for *co-operation use cases*, NF.Cs and viewpoints identified respectively in step 1.1, step 1.3 and step 2.2. The "MAMIE Tool" prototype implements partially MAMIE method (see [4] for details).

3 The Proposed Priority to Non-functional Concerns

3.1 Overview

The proposed process aims to assign a rank of priority to a specific non-functional concern on the basis of the concerned stakeholders' priorities. A simple view of this process is depicted in figure 2, where:

- NF.C denotes the current non-functional concern;
- S_i represents the i-th stakeholder ;
- ps_i is the i-th stakeholder's priority using Mitchell et al. prioritization model;
- $pNfc_i$ denotes the priority given by the i-th stakeholder to NF.C, $pNfc_i$ $\in$ {1:very low; 2:low; 3:medium; 4:high; 5:very high};
- PNfc denotes the NF.C priority resulting from the proposed prioritization function , $PNfc \in [1,5]$;
- i=1..n, where n is the number of the identified stakeholders.

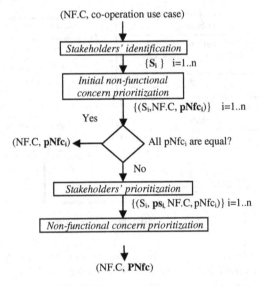

(NF.C, co-operation use case)

Stakeholders' identification

{S_i } i=1..n

Initial non-functional concern prioritization

{(S_i,NF.C, **pNfc**$_i$)}   i=1..n

Yes

(NF.C, **pNfc**$_i$) ◄———   All pNfc$_i$ are equal?

No

Stakeholders' prioritization

{(S_i, **ps**$_i$, NF.C, pNfc$_i$)} i=1..n

Non-functional concern prioritization

(NF.C, **PNfc**)

Fig. 2. The proposed process

The process is partitioned vertically in four main parts:

- Stakeholders' identification: this process starts by identifying stakeholders who are concerned with the *co-operation use case* given as parameter. Grunig and Hunt model [8] is used to identify Stakeholders (see section 3.2). The result of this step is a set of stakeholders.
- InitialNF.C prioritization: in this step, each identified stakeholder assigns a specific rank of priority to the current NF.C. If all stakeholders give the same priority to this NF.C, the process ends up and the NF.C takes this value of priority.
- Stakeholders' prioritization: this step aims to prioritize stakeholders identified in the first step according to their roles in the company. In order to prioritize them we adopt Mitchell et al. [13] model (see section 3.3).
- NF.Cs prioritization function: by considering the result of the previous step, the last step of this process returns a rank of priority for the given NF.C (see section 3.4).

3.2 Stakeholders Identification

The model developed by Grunig and Hunt [8] identifies a stakeholder according to the kind of linkage - relationship - to her/his organization: enabling linkage, functional linkage, diffused linkage, and normative linkage.

- The enabling linkage identifies stakeholders who have some control and authority over the organization, such as board of directors, governmental legislators and regulators, etc. These stakeholders enable an organization to have resources and autonomy to operate. When enabling relationships falter, the resources can be withdrawn and the autonomy of the organization restricted.
- The functional linkage is essential to the function of the organization, and is divided into input functions that provide labor and resources to create products or services (such as employees and suppliers) and output functions that consume the products or services (such as consumers and retailers).
- The normative linkage relates too associations or groups with which the organization has a common interest. Stakeholders in the normative linkage share similar values, goals or problems and often include competitors that belong to industrial or professional associations.
- The diffused linkage is the most difficult to identify because it includes stakeholders who do not have frequent interaction with the organization, but become involved based on the actions of the organization. These are the publics that often arise in times of a crisis. Diffused linkage includes the media, the community, activists, and other special interest groups.

Going through the linkage model should help the organization to identify all its stakeholders.

3.3 Stakeholders Prioritization

In [12], Mitchell et al. developed a model that includes the attributes of power, legitimacy and urgency (cf. figure 3). Stakeholders have power when they can influence other parties to make decisions the party would not have otherwise made. Mitchell et al. relied on Etzioni's [6] categorization of power: coercive power, based on the physical resources of force, violence, or restraint; utilitarian power, based on material or financial resources; and normative power, based on symbolic resources.

Legitimacy is determined by whether the stakeholder has a legal, moral, or presumed claim that can influence the organization's behavior, direction, process or outcome. Stakeholders are risk-bearers who have "invested some form of capital, human or financial, something of value, in a firm." Mitchell et al. used the notion of risk to narrow stakeholders with a legitimate claim. The combination of power and legitimacy is authority.

Urgency exists under two conditions: "(1) when a relationship or claim is of a time sensitive nature and (2) when that relationship or claim is important or critical to the stakeholder". Urgency, then, requires organizations to respond to stakeholder claims in a timely fashion. Urgency alone may not predict the priority of a stakeholder, especially if the other two attributes are missing.

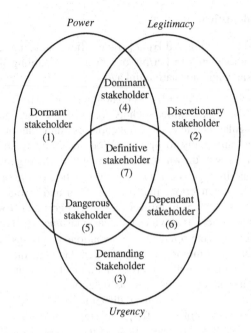

Fig. 3. Stakeholders prioritization [12]

Mitchell et al. [12] use the combination of the three attributes to develop a prioritization strategy. Accordingly, latent stakeholders possess only one of the attributes; expectant stakeholders possess two attributes, and definitive stakeholders possess all three attributes. If individuals or groups do not possess any of the attributes, they are not considered as stakeholders.

- The latent stakeholders have lower salience to an organization because they only have one attribute. They are identified as dormant, discretionary, and demanding.
  - The dormant stakeholder has power but no legitimacy or urgency in its claim. Therefore its power remains unused.
  - Discretionary stakeholders possess legitimacy, but no power to influence and no urgency in the claim, and therefore are reliant on the good will of the organization rather than through any other pressure.
  - The demanding stakeholder has urgency, but no legitimacy or power.

These groups could be bothersome, but not dangerous.

- Expectant stakeholders possess two attributes and are organized into dominant, dependent, and dangerous stakeholders.
  - Dominant stakeholders have power and legitimacy, and because they can act on their claims, they receive much of management's attention.
  - Dependent stakeholders have legitimacy and urgency. Organizations should be socially responsible to stakeholders that have a legitimate and urgent claim, and who depend on the organization to address and resolve the claim.

The inclusion of a dependent relationship is important because it recognizes that stakeholder priority is not limited to influence over the organization.
- Dangerous stakeholders have urgency and power, but lack legitimacy.
These stakeholders generally use formal channels to affect change, but may they become violent or coercive to achieve their claims. Social activist groups sometimes engage in forms of protests, boycotts, and (in extreme cases) damage to property and lives.

- The stakeholders who have all three attributes are definitive stakeholders and have the highest priority.

An important tenet of this model is that each attribute is variable. In other words, any group can acquire (or lose) power, legitimacy, or urgency depending on the situation.

3.4 Non-functional Concern Prioritization

This section presents the pivotal step of the priority assignment process. We describe how an analyst gives a rank of priority of a specific NF.C on the basis of stakeholders` priorities deduced from Mitchell et al. model and their evaluation of a specific NF.C.

Let's assume that σ denotes the proposed prioritization function. The function σ will be:

$$\sigma \left(\begin{array}{c} (S_1, ps_1, NF.\,C, pNfc_1), (S_2, ps_2, NF.\,C, pNfc_2), ... \\ (S_n, ps_n, NF.\,C, pNfc_n) \end{array} \right) \rightarrow (NF.\,C, PNfc) \tag{1}$$

Where:

$$PNfc = \Sigma(ps_i * pNfc_i) \Big/ \Sigma ps_i \tag{2}$$

Before giving an example for using this function, let us seen the responses (R1, R2 and R3) for three important questions (Q1.Q2 and Q3) related to the equation (2):

Q1- P is it always less than or equal to 5?
Q2- What is the result of this function if all the stakeholders give the same value of priority?
Q3- What is the result of this function if all the stakeholders have the same priority?
R1- $P \leq 5$, this means that:

$$\Sigma(ps_i * pNfc_i) \Big/ \Sigma ps_i \leq 5 \tag{3}$$

$$\sum ps_i * pNfc_i) \leq 5 * \sum ps_i \tag{4}$$

We have $1 \leq pn_i \leq 5$, so (4) is always true and thus $P \leq 5$.

R2- If all stakeholders give the same value of priority (let be pNfc be this value) for a specific NF.C. The prioritization function must return (pNfc) as result.

In this case, we can define the function σ as:

$$\sigma \begin{pmatrix} (S_1, ps_1, NF.\,C, pNfc), (S_2, ps_2, NF.\,C, pNfc), \ldots \\ (S_n, ps_n, NF.\,C, pNfc) \end{pmatrix} \rightarrow (NF.\,C, PNfc) \tag{5}$$

Where:

- $W = \dfrac{(\sum ps_i * pNfc)}{\sum ps_i} = pNfc$ (6)

- $i = 1..n$

R3- If all the stakeholders have the same priority ps for example. The function σ may be defined as:

$$\sigma \begin{pmatrix} (S_1, ps, NF.\,C, pNfc_1), (S_2, ps, NF.\,C, pNfc_2), \ldots \\ (S_n, ps, NF.\,C, pNfc_n) \end{pmatrix}$$

$\rightarrow (NF.\,C, PNfc)$

Where:

- $W = \sum(ps * pNfc_i) / \sum ps_i$

    $= (ps * \sum pNfc_i) / (n * ps)$

    $= \sum pNfc / n$ (8)

- $i = 1..n$

The equation (8) represents the average of the importance value given by all stakeholders, which represents a logical result if all identified stakeholders have the same priority.

Let us now considering a simple example where an analyst has identified two stakeholders: a government regulator (S_1) and an employee (S_2). According to Grunig and Hunt model S_1 has an enabling linkage with the organization while S_2 has a functional linkage. S_1 estimates that *security* non functional concern has a high priority (4), however it has a very low priority according to S_2 (1). The given priority values are different, which brings us to consider the stakeholders' priorities.

S_1 has the three attributes together (power, legitimacy and urgency) and thus classified as a definitive stakeholder (priority=7), on the other hand S_2 has at most two attributes (power and urgency) which make her/him a dangerous stakeholder (priority=5) according to Mitchell et al .model.

So, the result of the function σ will be:

$$\sigma\,((S_1, \mathbf{7}, security, 4), (S_2, \mathbf{5}, security, 1)) \rightarrow (security, P) \tag{9}$$

Where:

$P = (7*4+5*1)/(7+5)$

$= 33/12$

$= 2.7$

We could remark that the value of W returned by σ function in this example is nearest to the NF.C priority value attributed by the stakeholder S_1 having the highest priority.

4 Conclusion

The process presented in this paper aims to help an analyst to prioritize non-functional concern that could be considered in an inter-company co-operation process. We consider this process as a complementary to MAMIE methodology proposed in our previous research to elicit requirements for an inter-company co-operation information system.

The model of Grunig and Hunt is used in the first step of this process to identify stakeholders concerned with the current situation. However, Mitchell et al. model is used to prioritize stakeholders when the identified stakeholders have different points of view on the priority of the current non-functional concern. At this stage, we have proposed a mathematical function having as parameters a set of quadruples. Each one of them encapsulates essentially the priority value of the corresponding stakeholder and the weight giving by her/him to the current NF.C. This function provides a decision support for analysts, making possible to considering various stakeholders' evaluation about a specific NF.C.

Planned extension to this work includes an investigation of the following research question: *how a stakeholder could effectively evaluate the priority or the importance of a given non-functional concern according to her/his role in the company?* This means that s/he must knows all the parameters that influence on this decision, but also how they affect it?

References

1. Azar, J., Smith, R.K., Cordes, D.: Value Oriented Prioritization. IEEE Software (January 2006)
2. Bendjenna, H., Zarour, N., Charrel, P.J.: MAMIE: A Methodology to Elicit Requirements for an inter-company cooperative information system. In: Innovation on Software Engineering conference (ISE 2008) IEEE conference, Vienne, Austria, pp. 290–295 (2008), ISBN 13: 978-0-7695-3514-2
3. Berander, P., Andrews, A.: Requirements Prioritization. In: Aurum, A., Wohlin, C. (eds.) Engineering and Managing Software Requirements, pp. 69–94. Springer, Heidelberg (2005)
4. Chung, L., Nixon, B., Yu, E., Mylopoulos, J.: Non-Functional Requirements in Software Engineering. Kluwer Academic Publishers, Dordrecht (2000)
5. Damian, D.E., Zowghi, D.: Requirements Engineering Challenges in Multi-site Software Development Organization. Requirements Engineering Journal 8, 149–160 (2003)
6. Etzioni, A.: Modern Organizations. Prentice-Hall, Englewood Cliffs (1964)
7. Evaristo, R., Waston-Manheim, M.B., Audy, J.: E-Collaboration in édistributed Requirements Determination. International journal of e-Collaboration 1, 40–55 (2005)
8. Gruning, J., Hunt, T.: Managing public relations. Holt, Rinehart, and Winston, p. 141 (1984)
9. Hanish, J.: Understanding the culturaland Socil Impacts on Requirements Engineering Processes- Identifying some problem challenging Virtual Team interaction with clients. In: The European éconference on Information Sytems, Bled. Slovenia (2001)

10. Kazman, R., Asundi, J., Klein, M.: Quantifying the Cost and Benefits of Architectural Decisions. In: Proc. Int. Conf. Software Eng., pp. 297–306 (2001)
11. Lehtola, L., Kauppinen, M., Kujala, S.: Requirements Prioritization Challenges in Practice. In: Proc. of 5th Intl. Conf. on Product Focused Software Process Improvement, PROFES (2004), Kansai Science City, Japan, pp. 497–508 (2004)
12. Mitchell, R.K., Agle, B.R., Wood, D.J.: Toward a theory of stakeholder identification and salience: defining the principal of who and what really counts. Academy of Management Review 22(4), 853–887 (1997)
13. Ryan, A.: An Approach to Quantitative Non-Functional Requirements in Software Development. In: Proc. the 34th Annual Government Electronics and Information Association Conference (2000)
14. Wolf, T., Dutoit, A.H.: Supporting Traceability in Distributed Software Development Projects. In: Workshop on distributed Software development, Paris (2005)

Verifying Software Requirements Based on Answer Set Programming

Kedian Mu[1], Qi Zhang[1,2], and Zhi Jin[3]

[1] School of Mathematical Sciences
Peking University, Beijing 100871, P.R. China
[2] Microsoft Server and Tools Business China
Building 5, 555 Dong Chuan Rd, Shanghai 200241, P.R. China
[3] Key Laboratory of High Confidence Software Technologies, Ministry of Education,
School of Electronics Engineering and Computer Science
Peking University, Beijing 100871, P.R. China

Abstract. It is widely recognized that most software project failures result from problems about software requirements. Early verification of requirements can facilitate many problems associated with the software developments. The requirements testing is useful to clarify problematical information during the requirements stage. However, for any complex and sizeable system, the development of requirements typically involves different stakeholders with different concerns. Then the requirements specifications are increasingly developed in a distributed fashion. This makes requirements testing rather difficult. The main contribution of this paper is to present an answer set programming-based logical approach to testing requirements specifications. Informally, for an individual requirements test case, we consider the computation of the output of the system-to-be in requirements testing as a problem of answer set programming. In particular, the expected responses of the requirements test case is viewed as an intended solution to this problem. Based on the requirements and the input of the requirements test case, we design a logic program whose answer sets correspond to solutions of the problem. Then the testing is performed by an answer set solver. Finally, we identify the disagreement between the answer sets and the intended solution to detect the defects in software requirements.

1 Introduction

The success of a software system project depends on to a large extent the quality of requirements of the system. It has been widely recognized that most software project failures result from shortcomings of requirements. Errors being made during the requirements stage account for 40 to 60 percent of all the defects found in a software project [1,2]. Most practitioners identify poor requirements, incorrect specifications, and ineffective requirements management as major sources of problems in the development process [3,4]. To make matters worse, when defects leak into the subsequent phases in the software development life cycle, it is

D. Karagiannis and Z. Jin (Eds.): KSEM 2009, LNAI 5914, pp. 263–274, 2009.

extremely expensive to detect and correct them [5,6]. Consequently, early verification of requirements presents an opportunity to reduce the costs and increase the quality of software system.

Generally, the requirements documents review is the most widely used technique to verify requirements. During the stage of requirements review, analysts have to carefully read and examine specifications to detect and correct defects in requirements before they leak through the subsequent development phases. However, people find it is very difficult to imagine or simulate the behavior of the system-to-be in a certain environment by only reading written statements of requirements in many cases, and extremely difficult to reveal those inexplicit requirements problems such as incompleteness and ambiguity.

To address this, a thought of testing requirements has been presented to verify requirements [7,8]. Informally speaking, a process of testing requirements is a simulation of running of the system-to-be in an environment given by a conceptual test case. To a large extent, requirements testing is like a black box testing. The gist of this simulation is to imagine or "compute" the output of the system-to-be in the environment. Similar to software testing, the disagreement between the output of the system-to-be and the expected response of the conceptual test case is viewed as a trigger to identify defects in requirements. Then design of conceptual test cases, computation of output, identification of defects may be considered as major activities of requirements testing.

There are at least three reasons to support requirements testing. Firstly, it is feasible to derive the conceptual functional test cases from the use cases and functional requirements at early development stage, and then the conceptual test cases could be further developed into the software test cases during software testing stage [9,10,11]. In this sense, the conceptual test cases could be viewed as useful prototypes of the software test cases. Secondly, conceptual test cases may assist developers to discern the behavior of the system-to-be in a certain environment. It is useful to clarify many problematical requirements [8]. Finally, the later in the system development life cycle a defect is detected, the more costly it is to amend the final product [12]. That is, requirements testing can detect and fix defects at cheaper cost than software testing. For the simplicity of discussion in this paper, we use the term *requirements test case* to denote the conceptual test case derived from functional requirements and use cases.

Unfortunately, there is still no systematic tool to support the requirements testing. At present, the requirements testing is often performed manually by tracking the related requirements in specifications [7]. On a small software project, it may not be difficult to test requirements by stakeholder' cooperation. But it seems to be awkward to simulate the behavior of the system-to-be for larger or contentious projects. Although the requirements testing is a black box testing and we just focus on the output of the system in a given environment, it is rather difficult to design the simulation to compute the output for any complex system. Additionally, the development of any complex system invariably involves many different stakeholders. The requirements specifications are increasingly developed in a distributed fashion. Each stakeholder is just familiar with a special aspect of the system-to-be. Thus, to simulate the behavior of the system-to-be, it is necessary to involve many practitioners in requirements testing. It may result in extra cost of verification, in particular, for geographically separated teams. Moreover, too many participants often impair efficiency of requirements testing. Actually, in many projects, stakeholders already faced with tight schedules are reluctant to take on additional simulation.

As a solution to this problem, we present an approach to requirements testing based on Answer Set Programming in this paper. Answering set programming is a novel logic programming paradigm [13,14,15] for solving combinational search problems appearing in knowledge representation and reasoning. The following characteristics of answer set programming motivate us to choose it as paradigm for requirements testing:

(a) **It is a kind of declarative programming languages.** In procedural programming languages such as C++, the programmer tells the computer how to reach a solution via a sequence of instructions. Procedural programming is well suited for problems in which the inputs are well specified and for which clear algorithms can be carried out. In contrast, in a declarative programming such as answer set programming, the programmer describes what the computer should do rather than how to do and then leverage an answer set solver to find the solution. In this sense, answer set programming provides a promising way to facilitate the difficulties in computing the output of the system-to-be in requirements testing. On the other hand, since declarative programming language requires only what a solution should look like, even the requirements analyst can easily code programs after a short training. Furthermore, the declarative nature renders the flexibility when facing poorly conditioned inputs and frequently changing user requirements.

(b) **It provides a framework for formalizing non-monotonic reasoning.** Frequent requirements changes make non-monotonicity necessary to reasoning about requirements [12,16]. Answer set programming is a non-monotonic framework that represents defeasible inference. Answer set solver draw conclusions tentatively, reserving the right to retract them if further facts are appended to the fact base while rules are remained unchanged. Thus, we consider that answer set programming is more suitable and scalable when representing and reasoning requirements.

The main contribution of this paper is to present a novel approach to using answer set programming as a logic tool for requirements testing. Informally speaking, for an individual requirements test case, we consider the computation of the output of the system-to-be in requirements testing as a problem in answer set programming. In particular, the expected responses of the requirements test case is viewed as an intended solution to this problem. Then we constitute a logic program to find solutions to the problem based on requirements and the input of the requirements test case. Finally, we identify the disagreement between the answer sets and intended solution to detect the defects in requirements. The rest of the paper is organized as follows: Section 2 gives a brief overview of answer set programming. Section 3 presents an approach to requirements testing based on answer set programming. Conclusions are given in section 4.

2   Preliminaries: Answer Set Programming

Answer set programming emerged in the late 1990s as a novel logic programming paradigm [13,14,15] for solving combinational search problems appearing in knowledge representation and reasoning. To solve a problem, the basic idea of answer set programming is to design a logic program whose models or answer sets correspond to solutions of the problem, and then use an answer set solver such as Smodels [17], DLV [18] and ASSAT [19] to find the answer sets.

We first define the syntax of literals, rules and logic programs. A *literal* is a formula of the form A or $\neg A$, where A is an atom. A *rule* is an expression of the form

$$L_0 \leftarrow L_1, \ldots L_m, not\ L_{m+1}, \ldots, not\ L_n. \tag{1}$$

where $m \geq 0, n \geq 0$, L_i is a literal and *not* is the negation as failure symbol. The intuitive reading of a rule is that L_0 is true if $L_1, \ldots L_m$ are true and the program gives no evidence that $L_{m+1}, \ldots, L_n$ are true. The left side of $\leftarrow$ is called the *head* of a rule, while the right side is called its *body*. Usually, a rule is called a *fact* when body is empty indicating the head literal holds unconditionally and $\leftarrow$ can be omitted. A rule is called a *constraint* if head is empty which means that it is not acceptable whenever the body holds. A *logic program* is a finite set of rules.

The semantics of a logic program in this paper coincides with the answer set semantics [20]. Let Π be a logic program and X a set of literals. The *reduct of Π relative to X* is the program obtained from Π by

- deleting each rule such that $\{L_{m+1}, \ldots, L_n\} \cap X \neq \emptyset$ and
- replacing each remaining rule $L_0 \leftarrow L_1, \ldots L_m, not\ L_{m+1}, \ldots, not\ L_n.$ by $L_0 \leftarrow L_1, \ldots L_m.$

This program will be denoted by Π^X. We say that X is an *answer set* for Π if X is minimal among the sets of literals that satisfy Π^X [20]. Note that if no answer set can be calculated, i.e., the answer set is $\emptyset$, the program must be inconsistent.

Variables are supported by the language to turn answer set programming into a usable representational and computational tool. Commonly, we consider rules with variables that are taken as abbreviations for all ground instances over a finite set of constants. Usually variables are capitalized while atoms are not. These are illustrated by the following example.

Example 1. Consider the well-known graph 3-colorability problem. Given a graph shown in Figure 1, how to decide whether there exists a coloring of nodes using only red, green and blue in which no two nodes connected by an arc have the same color?

Generally, we should provide a concrete algorithm to find a solution. But in answer set programming, we just rewrite the problem in the language of logic program and then call an existing answer set solver to find solutions.

The graph in Figure 1 can be encoded in the following facts:

$$node(a).\ node(b).\ node(c).\ node(d).$$
$$arc(a,b).\ arc(b,c).\ arc(c,d).\ arc(d,a). \tag{2}$$

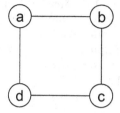

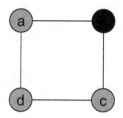

Fig. 1. Graph   **Fig. 2.** Coloring

Here $node(a)$ represents that a is a node in the graph, and $arc(a, b)$ represents that there is an arc between nodes a and b in the graph. The other facts could also be interpreted in the same way.

The specification of a coloring using only reg, green and blue in which no two nodes connected by an arc have the same color can be encoded in the following rules:

$$color(X, red) \leftarrow node(X),\ not\ color(X, green),\ not\ color(X, blue).$$
$$color(X, green) \leftarrow node(X),\ not\ color(X, red),\ not\ color(X, blue).$$
$$color(X, blue) \leftarrow node(X),\ not\ color(X, red),\ not\ color(X, green). \tag{3}$$
$$\leftarrow arc(X, Y),\ color(X, C),\ color(Y, C).$$

Here $color(X, C)$ represents that a node X has a color C. X and Y are variables standing for unspecified nodes of the graph, and C is a variable standing for an arbitrary color in $\{red, green, blue\}$. The first three rules express that every node should have exactly one color among red, $green$ and $blue$. The last constraint prohibits the same coloring of two neighboring nodes.

When program (2) + (3) is submitted to an answer set solver, 18 answer sets are calculated as coloring schemes. One solution is shown below:

$$color(a, green).\ color(b, blue).\ color(c, green).\ color(d, red). \tag{4}$$

This solution is illustrated in Figure 2.

Note that we do not provide any concrete algorithm to find a coloring of nodes satisfying the conditions in answer set programming. We just describe only what a solution should look like in (3). It embodies the nature of declarative programming.

3   Requirements Testing Based on Answer Set Programming

The requirements testing consists of several clearly defined activities, including designing the requirements test cases, testing, analysis (or defects detection), as well as evaluation. As a scenario description of application, a requirements test case is a document that contains the objective, the test condition, the input, the action, and the expected response. The input and the expected response are its main body. The input describes some relevant facts to model the scenario. The expected response provides a criterion to determine whether some feature of an application works correctly or not. For the simplicity of discussion below, a requirements test case consists of an input set and a set of expected responses. Moreover, as mentioned above, we consider development of requirements test cases, computation of output of the system-to-be, and identification of defects as three activities of requirements testing in this paper.

Now we transform the requirements testing into an issue of answer set programming. From the perspective of answer set programming, for a given requirements test case, we re-state the requirements testing as follows.

(1) consider the computation of the output of the system-to-be in requirements testing as a problem I in answer set programming such that the set of expected responses will be its intended answer set.
(2) encode I as a logic program P, such that solutions of I are represented by models of P. As a declarative programming, inspectors or analysts can just describe what the system-to-be should do rather than how to do. As such, we design the logic program P by the following steps:
  – rewriting the requirements in the *rules* of P;
  – rewriting the input of the requirements test cases in the *facts* of P.
(3) submit the program P to an answer set solver to calculate the answer sets of I.
(4) identify the defects in requirements based on the comparison between the answer sets of I in (3) and the intended answer set.

3.1   Representing Requirements and Requirements Test Cases

The requirements may be encoded into the rules to a very large extent in answer set programming in a natural way. In answer set programming, a requirements collection, denoted S, may be considered as a set of rules.

We use a tuple $\langle I, E \rangle$ to denote a requirements test case, where I and E are the input set and the set of expected response, respectively. Each member of I is an individual fact to model the scenario. Then it can be represented by a fact in answer set programming. For example, *"John is a valid user"* can be encoded in the fact *valid_user(john)*. Thus I is a set of facts. On the other hand, E is a set of expected output of the system-to-be, then it is an intended answer set of problem I. So logic program P consists of S and I. The requirements testing is in essence to determine whether S ∪ I ⊢ E is true.

Consider the following example.

Example 2. Suppose that a library manager gives the following demands for developing a library management system.

  – *If an application for borrowing a book is given by a valid user, then the application is acceptable.*
  – *If an application is acceptable and the book is available, then the user should get the book.*

- *If an application for borrowing a book is given by a invalid user, then the application is unacceptable.*
- *If an application is unacceptable, then the user can not get the book.*

We may encode these requirements in the following set of *rules*, denoted S:

$$valid\_application(X,Y) \leftarrow borrow(X,Y), \; valid\_user(X).$$
$$get(X,Y) \leftarrow valid\_application(X,Y), \; available(Y).$$
$$\neg valid\_application(X,Y) \leftarrow borrow(X,Y), \; \neg valid\_user(X).$$
$$\neg get(X,Y) \leftarrow \neg valid\_application(X,Y).$$

Furthermore, suppose that we want to test the requirements above. We derive a requirements test case from the related requirements as follows:

- **Input**
  - *John is a valid user for the library.*
  - *John applies to borrow the book of "Model Theory" from the library.*
  - *The book of "Model Theory" is available.*
  - *Alice is not a valid user for the library.*
  - *Alice applies to borrow the book of "Machine Learning" from the library.*
  - *The book of "Machine Learning" is available.*
- **Expected Responses**:
  - *John gets the book of "Model Theory".*
  - *Alice does not gets the book of "Machine Learning" .*
  - *Alice is reminded to register as a valid user.*

From the requirements test case, clearly, the problem I is that what should result from the applications of *John* and *Alice*. Then we may use $< I, E >$ to describe the requirements test case, where

$$I = \{ \; valid\_user(john).$$
$$borrow(john, modelTheory).$$
$$available(modelTheory).$$
$$\neg valid\_user(alice).$$
$$borrow(alice, machineLearning).$$
$$available(machineLearning).\} \quad \text{and}$$
$$E = \{ \; get(john, modelTheory).$$
$$\neg get(alice, machineLearning).$$
$$remind(alice, register).\}.$$

Then the logic program P contains the rules in S and facts in I.

3.2 Requirements Testing and Analysis

As mentioned above, the process of testing can be viewed as a process of computing models of P using an answer set solver. If there is no particular demand, inspectors or analysts may use some existing answer set solvers such as Smodels [17], DLV [18] and ASSAT [19] to compute the models of P. If there exists an answer set E_P of the program P such that $E \subseteq E_P$, we consider that the result of requirements testing corresponds with the expected responses.

Example 3. Consider the example above again. We can get an answer set of P as follows:

$$E_P = \{get(john, modelTheory).\ \neg get(alice, machineLearning).\}.$$

However, $E_P \subset E$, we don't think that the result of requirements testing corresponds with our expectation. As shown later, it will lead to identification of requirements defects.

The main goal of the requirements testing is to identify requirements defects. As mentioned earlier, the identification of requirements defects is always triggered by the disagreement between E_P and the expected responses E. However, it is not easy to find a unifying method to identify all kinds of the requirements defects as these defects may caused by lots of different factors. Here we provide a set of strategies for identifying some typical defects.

(1) If there exists an answer set E_P such that $E \subseteq E_P$, we consider that the result of requirements testing corresponds with the expected responses. Generally, there is no evident defect in the requirements that can be detected by this testing. Further, if we perform testing for some subsets of the requirements S based on the requirements test case, we may detect redundancy of S.

    (a) **Identification of Redundancy.** Redundancy is undesirable since it may result from an incorrect modeling or merging of several sources. For several reasons, the problem of redundancy of a requirements specification is noteworthy [21]. At first, removing redundant requirements leads to a simplification of requirements specification. This may be attractive to any project with resource limitations. Every redundancy will consume the resources. Then under limited resources such as a short timeline, developers need to avoid the requirements that incur a high cost but provide relatively low value. Moreover, simplifying requirements specification leads to a new specification of the same system that is easier to understand for all the stakeholders, as the process of removing redundancy can improve the development team's shared understanding. Thirdly, the simplification of requirements specification may improve the quality of the system-to-be, as a large amount of redundancy may affect the running efficiency of the system-to-be. Based on our requirements testing framework, the strategy for identifying redundancy could be:

        – Let S^* be a subset of S and P^* the logic program of S^*. If there exists an answer set E_P^* of P^* such that $E \subseteq E_P^*$, then $S - S^*$ may be considered as a *possible* redundancy relative to the requirements test case. Essentially, if E can be derived from a subset of S, then there is possibly unnecessary requirements in S. Of course, it needs a further judgement to determine whether $S - S^*$ is *really* redundant.

(2) If there is no answer set E_P such that $E \subseteq E_P$, then there are some defects in the union of the requirements S and the input I.

    (a) **Identification of Inconsistency.** It seems to be inevitable to confront inconsistency in the development involving many stakeholders. Many studies [22,23,24,25,26] have paid attention on inconsistency management. However, the term of inconsistency has different definitions in requirements engineering [27]. Most logic-based research such as [24,27,12] concentrated on a particular kind of inconsistency, logical contradiction: any situation in which some fact α and its negation $\neg\alpha$ can be simultaneously derived from the same requirements collection. In this paper, we shall be also concerned with the logical contradiction. As mentioned above, $E_P = \emptyset$ if and only if the program P is inconsistent. Consequently,

$$E_P = \emptyset \text{ implies that } S \cup I \text{ is inconsistent.}$$

(b) **Identification of Incompleteness.** Intuitively, inconsistency and redundancy result from "too much" information about some situation. However, it is also common to have "too little" information about some other situations in practice. The initial requirements statements for the proposed system is often incomplete and possibly vague [27]. Based on our requirements testing framework, the strategy for identifying incompleteness could be:
   - For each answer set E_P of the program P, if $E \cap E_P \subset E$, then S is incomplete relative to the requirements test case. That is, if there are some expected responses that can not be derived from the requirements, then the requirements associated with these expected responses are missing.

Now we give the examples to illustrate how to test requirements with answer set programming.

Example 4. Consider the setting mentioned in *Example* 2 and *Example* 3. Clearly, $E \cap E_P \subset E$. The expected response $remind(alice, register)$ does not appear in E_P. Actually, there is no answer set of P containing $remind(alice, register)$, since $remind$ does not appear in program P. It signifies that the *incompleteness* of the requirements collection S is identified by this requirements testing. We may consider the requirements associated with $remind(alice, register)$ are missing in the requirements collection S.

Example 5. Suppose that the second library manager gives the following demands about valid users.

   - *If a user has registered correctly, then the user is valid.*
   - *If a user has not registered, then the user is not valid.*
   - *All the library managers are valid users.*
   - *Library managers need not register.*

Now we encode these requirements in the following rules:

$$valid\_user(X) \leftarrow register(X). \quad \neg valid\_user(X) \leftarrow \neg register(X).$$
$$valid\_user(X) \leftarrow manager(X). \quad \neg register(X) \leftarrow manager(X).$$

Further, we design a requirements test case as follows:

$$I = \{ register(tom). \quad manager(jordan).\} \text{ and}$$
$$E = \{ valid\_user(tom). \quad valid\_user(jordan).\}.$$

For this setting, the answer set of the program P is $\emptyset$. Then we can say that the requirements given by the second manager is *inconsistent*.

Furthermore, suppose that the second manager changes the last statement as follows: *Library managers should register.* Then the corresponding rule is

$register(X) \leftarrow manager(X)$. Using the same requirements test case to test the modified requirements. The corresponding program P_1 is

$$valid\_user(X) \leftarrow register(X).$$
$$\neg valid\_user(X) \leftarrow \neg register(X).$$
$$valid\_user(X) \leftarrow manager(X).$$
$$register(X) \leftarrow manager(X).$$
$$register(tom).$$
$$manager(jordan).$$

We get an answer set

$$E_{P_1} = \{valid\_user(tom). \quad valid\_user(jordan).\}.$$

It corresponds with E. Further, if we delete the following rule

$$register(X) \leftarrow manager(X).$$

from the program P_1, then for the rest of the program, denoted P_2,

$$\{valid\_user(tom). \quad valid\_user(jordan).\}$$

is also an answer set of P_2. So, the last demand provided by the second manager is possibly *redundant*.

4   Related Work and Conclusion

We have presented an approach of using answer set programming to requirements testing. The main contribution of this paper is to make an attempt at transforming the requirements testing into an issue of answer set programming. It facilitates the difficulties in computing the output of the system-to-be in requirements testing. Given a requirements test case and a requirements collection, we encode the union of the requirements collection and the input of the test case into a logic program P. By leveraging an existing answer set solver, we may find the answer sets of P as the system outputs. By using the answer set programming language, the inspectors or requirements analysts can easily describe what the system-to-be would look like due to its declarative nature. This approach also shows its flexibility when facing poorly conditioned inputs and frequently changing user requirements. Furthermore, we propose some strategies for identifying several typical defects including inconsistency, redundancy, and incompleteness based on disagreement between answer sets of P and the expected responses.

At present, logic-based techniques have received much attention in requirements engineering. Most logic-based studies such as [12,24,25] focus on inconsistency identification and management. In contrast, the gist of the logic-based requirements testing presented in this paper is to identify more general defects. The declarative nature of answer set programming and available answer set solvers facilitate the difficulties of requirements testing. It provides a basis for automated requirements testing.

On the other hand, similar to most logic-based techniques, the automated requirements testing must provide a bridge from natural language to logic formulas. However, a prototype system [12] that translates software requirements in natural language into logic formulas has been announced recently. An integration of such prototype systems and our work may bring up an automated requirements testing which will be within our future work.

Acknowledgements

This work was partly supported by the National Natural Science Foundation of China under Grant No. 60703061, the National Key Research and Development Program of China under Grant No. 2009CB320701, the Key Project of National Natural Science Foundation of China under Grant No. 90818026, the National Natural Science Fund for Distinguished Young Scholars of China under Grant No.60625204, and the NSFC and the British Royal Society China-UK Joint Project.

References

1. Davis, M.: Software Requirements: Objects, Functions, and States. Prentice Hall, Englewood Cliffs (1993)
2. Leffingwell, D.: Calculating the return on investment from more effective requirements management. American Programmer 10, 13–16 (1997)
3. CHAOS: Software Development Report by the Standish Group (1995), http://www.standishgroup.com/chaos.html
4. Ibanez, M.: European user survey analysis. Tech. rep. ESI report TR95104. European Software Institute, Zamudio, Spain (1996), www.esi.es
5. Davis, A., Jordan, K., Nakajima, T.: Elements underlying the specification of requirements. Ann. Softw. Eng. 3, 63–100 (1997)
6. Leffingwell, D., Widrig, D.: Managing Software Requirements: A Use Case Approach. Addison-Wesley, Boston (2003)
7. Wiegers, K.: Software Requirements, 2nd edn. Microsoft Press, Redmond (2003)
8. Beizer, B.: Software Testing Techniques, 2nd edn. Van Nostrand Reinhold, New York (1990)
9. Ambler, S.: Reduce development costs with use-case scenario testing. Software Development 3, 52–61 (1995)
10. Collard, R.: Test design. Software Testing and Quality Engineering 1, 30–37 (1999)
11. Hsia, P., Kung, D., Sell, C.: Software requirements and acceptance testing. Annals of software Engineering 3, 291–317 (1997)
12. Gervasi, V., Zowghi, D.: Reasoning about inconsistencies in natural language requirements. ACM Transaction on Software Engineering and Methodologies 14, 277–330 (2005)
13. Lifschitz, V.: Answer set planning. In: International Conference on Logic Programming, pp. 23–37. Massachusetts Institute of Technology (1999)

14. Niemela, I.: Logic programs with stable model semantics as a constraint programming paradigm. Annals of Mathematics and Artificial Intelligence 25, 72–79 (1998)
15. Marek, W., Truszczynski, M.: Stable models and an alternative logic programming paradigm. In: The Logic Programming Paradigm: a 25-Year Perspective, pp. 375–398. Springer, Heidelberg (1999)
16. Mu, K., Jin, Z., Lu, R.: Inconsistency-based strategy for clarifying vague software requirements. In: Zhang, S., Jarvis, R.A. (eds.) AI 2005. LNCS (LNAI), vol. 3809, pp. 39–48. Springer, Heidelberg (2005)
17. Niemela, I., Simons, P., Syrjanen, T.: Smodels: A system for answer set programming. CoRR cs.AI/0003033 (2000)
18. Leone, N., Pfeifer, G.: The dlv system for knowledge representation and reasoning. ACM Transactions on Computational Logic 7, 499–562 (2006)
19. Lin, F., Zhao, Y.: Assat: Computing answer sets of a logic program by sat solver. Artificial Intelligence 157, 115–137 (2004)
20. Lifschitz, V.: Foundations of logic programming. In: Principles of Knowledge Representation, pp. 69–127. CSLI Publications, Stanford (1996)
21. Mu, K., Jin, Z., Lu, R., Peng, Y.: Handling non-canonical software requirements based on annotated predicate calculus. Knowledge and Information Systems 11, 85–104 (2007)
22. Spanoudakis, G., Finkelstein, A.: Reconciling requirements: a method for managing interference, inconsistency and conflict. Annals of Software Engineering 3, 433–457 (1997)
23. Spanoudakis, G., Zisman, A.: Inconsistency management in software engineering: Survey and open research issues. In: Chang, S.K. (ed.) Handbook of Software Engineering and Knowledge Engineering, pp. 329–380. World Scientific Publishing Co., Singapore (2001)
24. Hunter, A., Nuseibeh, B.: Managing inconsistent specification. ACM Transactions on Software Engineering and Methodology 7, 335–367 (1998)
25. Easterbrook, S., Chechik, M.: A framework for multi-valued reasoning over inconsistent viewpoints. In: Proceedings of International Conference on Software Engineering (ICSE 2001), Toronto, Canada, pp. 411–420 (2001)
26. Chechik, M., Devereux, B., Easterbrook, S.: Efficient multiple-valued model-checking using lattice representations. In: Proceedings of the International Conference on Concurrency Theory, Aalborg, Denmark, pp. 21–24 (2001)
27. Zowghi, D., Gervasi, V.: On the interplay between consistency, completeness, and correctness in requirements evolution. Information and Software Technology 45, 993–1009 (2003)

Blending the Sketched Use Case Scenario with License Agreements Using Semantics

Muhammad Asfand-e-yar[1], Amin Anjomshoaa[1], Edgar R. Weippl[2], and A Min Tjoa[1]

[1] Vienna University of Technology Wien, Favoritenstraße 9-11 / 188-1, Wien 1040, Austria
{asfandeyar,anjomshoaa,amin}@ifs.tuwien.ac.at
[2] Secure Business Austria Wien, Favoritenstraße 16, Wien 1040, Austria
eweippl@securityresearch.at

Abstract. Software end-users need to sign licenses to seal an agreement with the product providers. Habitually, users agree with the license (i.e. terms and conditions) without fully understanding the agreement. To address this issue, an ontological model is developed that formulates the user requirements formally. This paper, introduces this ontological model that includes an abstract license ontology that contains the common features found in different license agreements. The abstract license ontology is then extended to a few real world license agreements. The resulting model can be used for different purposes such as querying the appropriate licenses for a specific requirement or checking the license terms and conditions with user requirements.

1 Introduction

In general, installing software requires the end-user's agreement to the software's terms and conditions. Software agreements describe issues such as software's functionality, restrictions on its use, and the maximum number of copies that can be made. Agreements may also specify the number of users that can work in a connected environment, state any relevant laws, rules pertaining to the distribution of the software, modifications that can be made, how to download updates, the payment procedure, and much more. The concepts contained in software agreements are often difficult for the end-user to understand. The terms and concepts are different in every software agreement. Usually, users do not read through such lengthy license agreements. Therefore, they usually bypass the whole agreement and click the "accept" button underneath without reading any of the details, such as the software's capacities and restrictions. Nonetheless, clicking the "accept" button, without any deeper understanding of the agreement might lead to a user's inadvertent implication in an illegal act. For example, a user might be violating the law by making "illegal" copies, installing software in a restricted jurisdiction or distributing copies without paying a share to the owner. Legal violations are not the only problem. By not reading the agreement, the end-user is unable to benefit from all of his/her rights with regard to the software. Some agreements allow for free updates, while others only allow updates by payment. In some agreements it says whether or not an end-user receives a replacement in case of damage. Some agreements stipulate a trial period after which,

D. Karagiannis and Z. Jin (Eds.): KSEM 2009, LNAI 5914, pp. 275–284, 2009.

either the software stops functioning or the users is required to make payment. Software agreements apply penalties; if terms and conditions are violated then users are subject to the law.

To address these issues, a common understanding of machine-interpretable knowledge is required, that would provide a license repository and make it easier for users to select an appropriate product. For this purpose, we propose a semantic model for licensing that makes it possible to answer user queries in conjunction with the user's specific requirements.

To develop a semantic model in OWL, as initial step we selected 3 license agreements: Mozilla [1], Apache [2], and Adobe Photoshop [3]. In order to extract the requirements, we studied each of these licenses thoroughly. Subsequently, a user scenario was defined. On the basis of this user scenario, rules were delineated. Finally, the Semantic model was created in Protégé [7] from extracted requirements and defined rules.

We discuss related work in section 2 and conflicts of license agreement with local law in section 3. An overview of the proposed solution is given in section 4. The user scenario, the ontology model, and its realization in Protégé are discussed in sections 5 and 6 respectively. Section 7 focuses on the applied rules to the model and the resulting inferences. Finally, the conclusion and future work are discussed in section 8 and section 9 respectively.

2   Related Work

In the domain of computing literature, studies have been conducted for developing a semantic web model to fulfill user requirements. Özgür and his team developed a Semantic Web and service-driven information gathering environment for mobile platforms (SMOP). In SMOP a user requests a Geographical Positioning System (GPS) data from his/her mobile and receives results on his/her mobile from GPS. A semantic matching service (SMS) and map service (MS) is used to match the request of user with GPS. A server in XML format is used to translate the user requests to SMS and MS and XML parser is used to translate results for the user [8]. In the KAoS tool, a hierarchy of ontology concepts is used to build policies. In the tool, a user-defined policy is confirmed when the policy is automatically planned and executed in semantically described workflows. The KAoS tool, permits for listing, managing, solving conflicts and enforcing the policies in specific contexts (for example, a user context) [9].

The work in Semantic Digital Rights Management for Controlled P2P RDF Metadata Diffusion is carried out to develop a legally binding agreement between a user and a provider. A small ontology is proposed to implement this handshake mechanism. A four step exchange process is used in which, 1) a user makes a request to a provider, 2) on the basis of this request a proposal is made, 3) the proposal is then sent to the user, 4) and finally the user signs the proposal and sends it back to the provider. A secure channel is created between user and provider. After creation of the secure channel, the product and related information are transferred between user and provider [10]. The work defined above is similar to our proposed approach; however the user requirements are considered as a canonical reference. A user lists his/her requirements and these requirements are fulfilled by selecting the appropriate license agreements using ontology description of these agreements.

3 Conflicts of License Agreement with Local Law

As discussed in Section 1, a user needs to agree to the terms and conditions of the license agreement. In addition to terms and conditions, the license agreement also includes information about the software product such as limitation of product (i.e. where to use and how to purchase), updates (allowed freely or should be purchased), transfer, distribution, backup copies, etc. In cases where the jurisdiction law conflicts with the license agreement, different conditions are applied according to license agreements as described below.

In the license agreements the conditions of selling, distributing and purchasing are also defined with the product information. For example Microsoft [4] and Adobe [3] do not allow selling, distributing or purchasing their product in Syria, Sudan, Cuba, Iran and North Korea. Android's [5] agreement is a legally binding agreement between user and Google. It states that the sale is void when the license agreement conflicts with the local law. While according to Adobe [3] and Cisco [6] when there is a conflict then international law is applicable. In some cases of conflicts, for example the disclaimer of liabilities cannot be applied in the law conflict area or jurisdiction [6]. Termination of agreement betides (i.e. applied) whenever license agreement conflicts with jurisdiction law [4]. Open source agreement depends on the license of modified modules; therefore a developer can apply his/her own conditions on conflicts of agreements [1] [2].

4 Overview

The work presented in this paper consists of modeling user profiles and license ontologies. A user profile includes information about user requirements that are extracted from a use-case scenario. A web interface may facilitate creation of such user profile for the novice users who are not familiar with user modeling concepts and terminologies.

The licensing ontology is developed on basis of some use case scenarios in context of license agreement. The license ontology consists of an abstract ontological model and real world license ontological models that extend the abstract model and describe a specific license agreement. The abstract ontological model is developed from common extracted information of different license agreements. The abstract ontological model will be then extended by real world license models in order to complete the meaning of license agreements in the license ontological model.

The functionality of entire process, i.e. user profile and licensing ontology, depends on the search and comparison of user requirements with license ontological models. The functionality is completed when rules (i.e. description logic) are defined. The rules depend on both user profile and license ontology and facilitate the search and comparison of the user requirements in license ontology. Results are obtained by applying queries on ontological models. These queries are built on the basis of defined rules. The query results are then sent back to the user for final approval. The whole scenario is depicted in Figure 1.

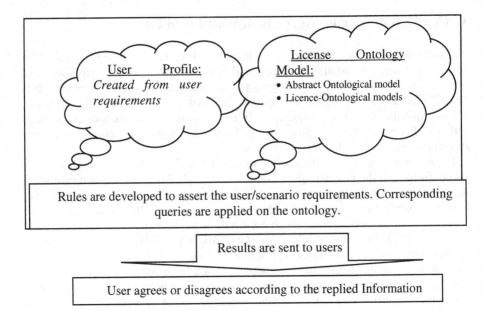

Rules are developed to assert the user/scenario requirements. Corresponding queries are applied on the ontology.

Results are sent to users

User agrees or disagrees according to the replied Information

Fig. 1. A general model to describe the work flow

5 Use Case Scenario

To develop a Semantic Web model a use case scenario is presented that identify user requirements and consequently rules are defined. The defined rules not only evaluate ontological models but also validate the Semantic Web model. The use case scenario is defined below:

A university needs a software product for its students. The requirements to select software product are categorized as follows:

1. Online payment will be made.
2. Product can be downloaded or received by mail.
3. If the product is received damaged, a replacement is required.
4. Source code is required for the development of additional components or changes could be applied to the existing source code.
5. The product development will take place on Solaris systems only.
6. Backup copies are necessary in case of error in development period.
7. The university logo will be placed on the modified version.
8. Modified versions will be available for everyone to download from the university's web site.

These eight rules are used to develop a formal description of user requirements (i.e. user profile). To validate the proposed approach, three widely used licenses namely Mozilla, Apache and Adobe Photoshop license agreements, were selected that potentially match the use case scenario.

6 Ontology Model and Its Realization in Protégé

An ontological approach is proposed to address the issues related to user requirements. The user requirements are extracted from within license agreements. To

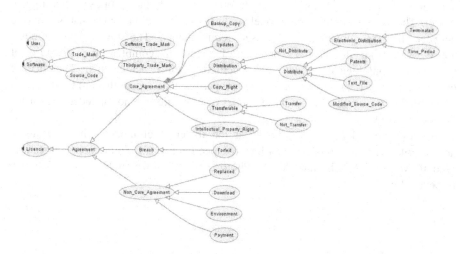

Fig. 2. Class diagram of abstract model

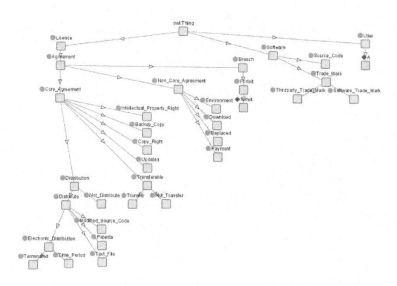

Fig. 3. Abstract ontology model

construct the ontological model, the license agreements are studied and the common features are extracted from the license agreements as discussed in Section 5. These common features were consolidated to construct the abstract ontological model. Figures 2 and 3 show the abstract ontological model.

The Semantic model contains an abstract ontological model on a higher level, and license ontological models on lower levels of the proposed ontological model. The abstract ontological model is designed to facilitate information sharing between license agreements and user requirements. The lower level models are derived from the abstract ontological model to complete the structure of the license ontological model for specified license agreements. The license ontological model has extra classes or properties that cover the requirements of a specific license agreement but these classes or properties are not included in abstract license ontology.

Instances are added in the classes at the lower level model. These instances are described in each model according to specific license requirements. Instances used in Mozilla, Apache and Adobe Photoshop are shown in Figures 4, 5 and 6 respectively:

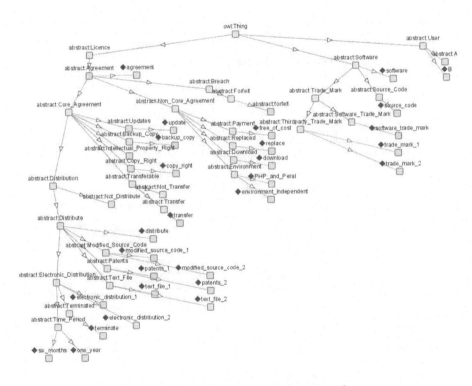

Fig. 4. Mozilla ontology, imported abstract ontology

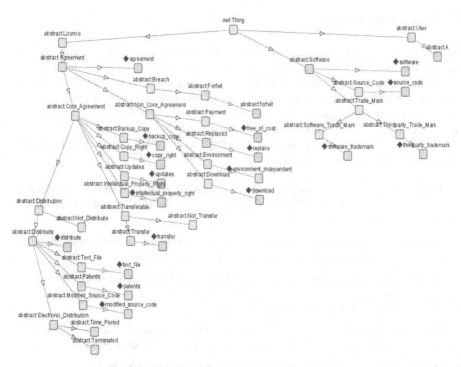

Fig. 5. Apache ontology, imported abstract ontology

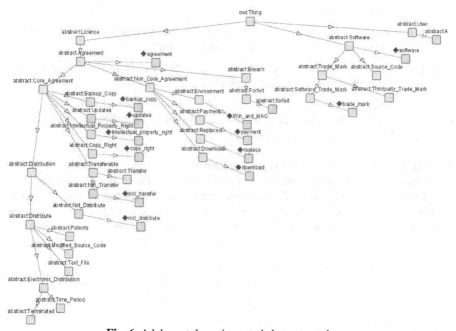

Fig. 6. Adobe ontology, imported abstract ontology

7 Rules and Results

In this section, the eight rules will be introduced that are built according to the previously discussed user requirements in Section 5. Each rule is translated into SPARQL query [11] and the queries deliver the compliance of the user requirements with the license agreement. The overall steps of this process has been already shown in Figure 1, where the user profile interacts with license ontological model using the defined rules, and finally results are sent back to the end-user.

The rules that are formulated as SPARQL queries for the user requirements defined in Section 5 are as follows:

Rule 1: (?user profile:wantsToPay abstract:online)(?product abstract:hasa ?license) (?license abstract:Online ?Payment) → (?user abstract:canPayOnlineFor ?source)

Rule 2: (?user profile:onlinepayment abstract:download)(?product abstract:can_be ?receive)→(?user abstract:online_download ?receive)

Rule 3: (?user profile:replace_damange abstract:if_damage)(?product abstract:if_received_damage ?return) → (?user abstract:if_received_damage ?return)

Rule 4: (?user profile:modification abstract:allows)(?product abstract:provides ?modification) → (?user abstract:can_perform ?modification)

Rule 5: (?user profile:development abstract:environment) (?software abstract:works_on ?environment)→(?user abstract:develop_in ?environment)

Rule 6: (?user profile:requires_backup abstract:backup)(?product abstract:allows ?backup) → (?user abstract:can_make ?backup)

Rule 7: (?user profile:places_logo abstract:logo)(?license abstract:allows_personal ?logo) → (?user abstract:can_provide ?logo)

Rule 8: (?user abstract:available ?source_code)(?product abstract:provides ?source_code) (?source_code abstract:available_on ?website) → (?user abstract:sourcecode_on ?website)

According to these rules, all 3 investigated license ontologies (Figures 4, 5, and 6) fulfill the user requirements with the exception of Abode Photoshop license that does not satisfy some requirements. For example, the Adobe Photoshop license agreement neither allows the customer to transfer and distribute the product nor does it permits to distribute its source code. This has been shown in Figures 8 and 9.

Fig. 8. Rejecting transfer and distribution of product

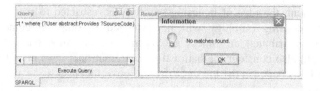

Fig. 9. Not providing source code of product

Mozilla and Apache fulfill user requirements described in Section 5. The results for Mozilla and Apache are obtained using the rules discussed in this section. After receiving all the positive results according to the user requirements, user can select either both license agreements or the license agreement of interest separately.

8 Conclusion

The approach focuses on user requirements and ontology design, and their interaction with one another. A scenario is defined that consequently define the rules. On the basis of these rules an abstract ontology is developed. The license agreement ontology imports the abstract ontology that helps to retrieve results by applying the identified rules.

The main idea behind the proposed approach is to provide an abstract ontology that can be easily extended by other companies' to make the machine-readable form of their licensing agreements. This abstract ontology is a common ontology that covers most of the required concepts of different license agreement. A company needs only to extend this ontology by adding its missing requirements (that will be minimal in most cases) and complete the ontology by providing instances based on its license agreement. Users will query for their requirements on the license agreement repository and related products will be displayed to them. The users can easily check their requirements by clicking on a button without having to read the lengthy license agreements and without having to know the product's licensing details.

The proposed system is developed in Protégé. License agreements are implemented as OWL-DL ontology [12].

9 Future Work

The license ontology is currently elaborated according to user requirements. The presented idea has been implemented as proof of concepts and the ontological models and query in those ontological models are tested. The next phase is to implement the relevant web services and license repository to make the search convenient for end-users. It is important to note that searching of an appropriate license for a user is one aspect of whole idea and the concept should be extended to support creating license ontology for a software developer. Our aim is to develop such a system that not only provide searching facility for a user but also provide a platform for software developers to create license ontology for their product.

At the moment, the idea of licensing ontology is restricted to software products; however the licensing ontology will be extended to other agreements between geographically distributed customers and companies. In such cases the licensing ontology should also consider the national and international laws in order to identify and solve conflicts between the laws of different governments.

References

1. Mozilla Public License Version 5.0 (2007), http://support.mozilla.com/en-US/kb/Terms+of+Service (last visited September 2009)
2. Apache License, version 2.0 (January 2004), http://www.apache.org/licenses/ (last visited September 2009)
3. Adobe software License Agreements, http://www.adobe.com/products/eulas/pdfs/Gen_WWCombined-combined-20080623_1026.pdf (last visited September 2009)
4. Master End-User License Agreement, MSDN, the Microsoft developer network subscription, http://msdnaa.oit.umass.edu/Neula.asp (last visited September 2009)
5. Terms and Conditions – Andorid Software Development Kit License Agreement, http://developer.android.com/sdk/terms.html (last visited September 2009)
6. End-User License Agreement (Product & Services) – Cisco Systems, http://www.cisco.com/en/US/docs/general/warranty/English/EU1KEN_.html (last visited September 2009)
7. Protégé 3.4 (March 16, 2009), http://protege.stanford.edu/ (last visited September 2009)
8. Gümüs, Ö., Kardas, G., Dikenelli, O., Erdur, R.C., Önal, A.: SMOP: A Semantic Web and Service Driven Information Gathering Environment for Mobile Platforms. In: Ontologies, Databases and Applications of Semantics (ODBASE) 2006 International Conference, pp. 927–940 (2006)
9. Uszok, A., Bradshaw, J.M., Johnson, M., Jeffers, R.: KAoS Policy Management for Semantic Web Services. Institute for Human and Machine Cognition, IEEE Intelligent Systems archive 19(4), 32–41 (2004)
10. Diffusion, M., García, R., Tummarello, G.: Semantic Digital Rights Management for Controlled P2P RDF. In: 2nd Semantic Web Policy Workshop, SWPW 2006 (2006)
11. SPARQL Protocol for RDF; W3C Recommendation (January 15, 2008), http://www.w3.org/TR/rdf-sparql-protocol/
12. OWL Profile, http://www.w3.org/2007/OWL/wiki/Profile_Explanations (last visited September 2009)

A Comparative Analysis for Detecting Seismic Anomalies in Data Sequences of Outgoing Longwave Radiation

Yaxin Bi[1], Shengli Wu[1], Pan Xiong[2] and Xuhui Shen[2]

[1] School of Computing and Mathematics, University of Ulster,
Co. Antrim, BT37 0QB, United Kingdom
{y.bi,s.wu}@ulster.ac.uk
[2] Institute of Earthquake Science, China Earthquake Administration,
Beijing, 100036, China
xiong.pan@gmail.com, shenxh@seis.ac.cn

Abstract. In this paper we propose to use wavelet transformations as a data mining tool to detect seismic anomalies within data sequences of outgoing longwave radiation (OLR). The distinguishing feature of our method is that we calculate the wavelet maxima curves that propagate from coarser to finer scales in the defined grids over time and then identify strong singularities from the maxima lines distributing on the grids by only accounting for the characteristics of continuity in both time and space. The identified singularities are further examined by the Holder exponent to determine whether the identified singularities can be regarded as potential precursors prior to earthquakes. This method has been applied to analyze OLR data associated with an earthquake recently occurred in Wenchuan of China. Combining with the tectonic explanation of spatial and temporal continuity of the abnormal phenomena, the analyzing results reveal that the identified singularities could be viewed as the seismic anomalies prior to the Wuchan earthquake.

1 Introduction

By studying remote sensing data, researchers have found various abnormal activities on the earth, atmosphere and ionosphere prior to large earthquakes, which are reflected in anomalous thermal infrared (TIR) signals [1], outgoing longwave radiation [2] and surface latent heat flux (SLHF) [3, 4] and anomalous variations of the total electron content (TEC) [5, 6] prior to the earthquake events. The latest advancements in lithosphere – atmospheric – ionospheric models provide a possible explanation to the origin of these phenomena [5, 6] and also permit us to investigate into the spatial and temporal variability of remote sensing data before and during earthquakes.

Several studies have been carried out to analyze thermal infrared anomalies appearing in the area of earthquake preparation a few days before the seismic shock [7, 8]. These include an analytical comparison on a single image of pre (vs. post) earthquake satellite TIR imagery [8]; a multispectral thermal infrared component analysis on the Moderate Resolution Imaging Spectroradiometer (MODIS) on Terra

D. Karagiannis and Z. Jin (Eds.): KSEM 2009, LNAI 5914, pp. 285–296, 2009.

and Aqua satellites using Land Surface Temperature (LST) [9]; investigation into OLR data to discover anomalous variations prior to a number of medium to large earthquakes [2]. More recently a new data mining method has been developed on the basis of wavelet analysis to detect anomalous SLHF maxima peaks associated with four coastal earthquakes [3, 4].

OLR is the thermal radiation flux emerging on top of the atmosphere and connected to the earth–atmosphere system. It is often affected by cloud and surface temperature. Due to OLR resulting from infrared band telemetry, not only OLR data is continuous and measurable, but also it is sensitive to the sea surface layer and the ground temperature change. It can be therefore viewed as a very useful source for monitoring the symptoms of some natural disasters linking to "hot" origin of phenomena, like earthquakes.

Precisely detecting seismic precursors within OLR data prior to earthquakes is important to sufficiently avail of OLR resources to monitor varying conditions of active faults beneath the earth and to identify the potential earthquake zones. It is appreciated that advanced data mining methods could offer a potential solution to detecting abnormal events embedded in OLR data. However a challenge facing data mining research is to properly and rapidly digest massive volumes of OLR data in order to detect abnormal events.

In this paper we propose to use wavelet transformations as a data mining tool to detect seismic anomalies within OLR data, which is built on two real continuous Daubechies Wavelet and Gaussian Derivative Wavelet. The proposed method first calculates the wavelet maxima that propagate from coarser to finer scales over the defined grids and then identifies strong singularities from the maxima lines distributing on the grids by only accounting for the characteristics of continuity in both time and space. The identified singularities are further examined by the Holder exponent to determine whether the identified singularities can be regarded as potential precursors prior to earthquakes. In this context, the continuity of time means that the singularities appear in the same period or with a short delay of each other, while the continuity of space means that the detected singularities distribute over a two-dimensional space in accordance with a precise geometrical constraint conforming to the geological settings of the region.

The proposed method has been applied to analyze the OLR data associated with the earthquake recently occurred in Wenchuan of China. Combining with the tectonic explanation of spatial and temporal continuity of the abnormal phenomena, the analyzing results have shown the identified singularities could be referred to as the seismic anomalies prior to the Wuchan earthquake.

2 Overview of Wavelet Transforms

Over the past decade wavelets transforms have been formalized into a rigorous mathematical frame work and have found numerous applications in diverse areas such as numerical analysis, signal and image processing, and others. Wavelet analysis is not only capable of revealing characteristics of data sequences, like trends, breakdown points, discontinuities in data streams, and self-similarity, but also it has the ability to provide a different view for time series data or data sequences in a compact way of compression or de-noising without appreciable degradation [9].

Which of wavelets will be used analyze OLR data really depends the nature of OLR data and what we require in detecting seismic anomalies. We have undertaken an empirical analysis on several wavelet methods and selected two for our study. The first method is one of the Daubechies Wavelets, called db1, and the second is the Gaussian Derivative Wavelets called gaus3. Both of these methods employ one dimensional continuous wavelet transformations. We use these two methods to analyze the continuity of modulus maximum in both time and space and to detect singularities within the OLR data sequences.

2.1 Wavelet Transformation

Wavelets transforms are categorized into discrete wavelet transform (DWT) and continuous wavelet transform (CWT) [10]. The idea of DWT is to separate an input sequence into low and high frequency parts and to do that recursively in different scales. The results of DWT over data sequences give wavelet coefficients which encode the averages. CWT is defined as the sum over all time of the data sequence multiplied by scaled, shifted versions of the wavelet function ψ :

$$Wf(s, p) \quad = \frac{1}{\sqrt{s}} \int_{-\infty}^{\infty} f(t) \; \psi \, (\frac{t-p}{s}) dt \qquad (1)$$

where s is a scale, p is a position and $1 / \sqrt{s}$ is a energy normalization factor for the different scales. The results of CWT are wavelet coefficients in terms of Wf, which are a function of scale and position. The key difference between the two transforms is the way of operating on the set of scales and positions in calculating wavelet coefficients. CWT operates at every scale, from that of the original data up to the maximum scale, whereas DWT chooses scales and positions based on powers of two, the so called dyadic scales and positions. Thus there is a need to trade off between detailed analysis and efficiency in order to choose either CWT or DWT. In this study we mainly focus on CWT that is more suitable for analyzing data drift within data sequences.

The resulting coefficients Wf of a time domain measurements provide a compact way to reveal characteristics of data sequences, such as breakdown points and discontinuities in data sequences, implying concept changes or abnormal activities within the sequences. These coefficients can be alternatively expressed in a two-dimensional time scale map as shown in Fig. 1, which consists of three parts. The first one is original OLR measurements – a data stream. The second part is the visualization of the resulting coefficients Wf, where x-axis represents the duration of data streams, y-axis represents scales of the CWT transform and entry values are coefficient magnitudes of different scales at time points. Within the map large magnitude components in the terms of modulus maxima will present at time points where the maximum change in the data sequence has occurred or the data sequence at these time points is discontinuous. By connecting all these modulus maxima points across the scales, we can obtain a number of modulus maxima curves as shown in the third part of Fig. 1. The colors in map legend indicate the degrees of maxima magnitudes from large to small.

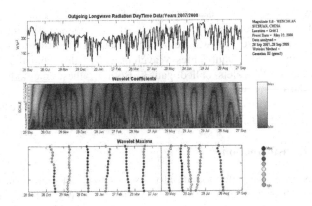

Fig. 1. A time scale plot of coefficients Wf

Mathematically, if we assume ψ in Eq. (1) have zero mean, then the integral measures the variation of p in a neighborhood of time of size proportional to the scale factor s > 0. (s_0, p_0) is defined to be a modulus maximum if $| Wf(s0, p0)|$ is a local maximum with the following condition [9]:

$$\frac{\partial |Wf(s_0, p_0)|}{\partial p} = 0 \qquad (2)$$

This condition means $Wf(s_0, p)$ strictly increases to the left of p_0 or strictly decreases to the right of p_0 in the neighborhood of time point. For each p_0, we can connect all $Wf(s, p_0)$ values across all the scales to construct a group of connected curves in a time scale plot. These curves are referred to as maxima lines. It appears that the group of such maxima lines provides a particularly useful representation of the entire CWT.

2.2 Calculating Singularities by the Holder Exponent

An alternative way of extracting maxima lines from CWT coefficients is to examine the decay of the wavelet modulus across the scales for each time point. Points where large or cusp changes occur in the data sequence, called singularities, will have large coefficients at all the different scales, thus having little decay. Elements such as noise, however, only produce large coefficients in the finer scales but will not persist to the coarser scales, therefore they would show more decay across all of the scales. Such decay of the data at a given time point can be measured by means of the Holder exponent.

The Holder exponent, also known as the Lipschitz exponent [9], is a tool that provides information about the regularity within data sequences. In essence, the regularity determines what order a function is differentiable. For instance, if a data sequence is expressed by $f(t)$, it is differentiable at $t=0$, then it has a Holder exponent of 1. If $f(t)$ is discontinuous but bounded in the neighbourhood of $t=0$, such as a step function, then the Holder exponent is 0. Therefore measuring the regularity of the data in the sense of time can be used to detect singularities. Through examining the decay of large coefficients at all the different scales, the point-wise Holder regularity of the data can be determined. The Holder regularity is defined as follows.

Assume that a signal $f(t)$ can be approximated locally at t_0 by a polynomial of the form [9]

$$f(t) = c_0 + c_1(t - t_0) + \cdots + c_n(t - t_0)^n + C|t - t_0|^\alpha$$
$$= P_n(t - t_0) + C|t - t_0|^\alpha \tag{3}$$

where P_n is a polynomial of order n and C is a coefficient. The exponent α can be thought of as the residual that remains after fitting a polynomial of order n to a data sequence, or as part of the data sequence that does not fit into an $n+1$ term approximation [9]. The local regularity of a function at t_0 can then be characterised by this 'Holder' exponent:

$$|f(t) - P_n(t - t_0)| \leq C|t - t_0|^\alpha \tag{4}$$

A higher value of α indicates a better regularity or a smoother function. In order to detect singularities, a transform is required such that omits the polynomial part of the data sequence. Transformation of Eq. (4) using a wavelet with at least n vanishing moments then provides a method for estimating the values of the Holder exponent in time.

$$|Wf(s, p)| \leq C * s^\alpha \tag{5}$$

Applying the log to each side of Eq. (5) and omitting the constant $\log(C)$, we have Eq. (6), where α is the slope, the decay of the wavelet modulus across its scales.

$$\alpha = \frac{\log(|Wf(s, p)|)}{\log(s)} \tag{6}$$

To perform the Holder exponent analysis over OLR data, we define three studying areas and divide each of them into a set of grids. The analyzing results on the different grids are combined into a $n \times m$ matrix, where rows n represent to the scales at the grids in which the wavelet analysis has been performed, columns m correspond to time, and the entry values of the matrix are either maxima values or zero if none has been detected at this particular time point. As a result, maxima lines constructed in each grid can be connected together in a grid graph, which may be continuous in both space and time of the grid path. The estimation of the Holder exponent will be performed on each grid, the resulting exponents will be arranged in a single plot, which can be directly compared with maxima lines in the gird graph.

2.3 Experimental Procedure

In this study, we select the Wuchan earthquake for evaluating the proposed method. The Wuchan earthquake occurred on 12th May 2008 with a 8.0 magnitude. The location of its epicenter is at 30.986°N, 103.364°E, and its depth is 19 km. The earthquake was followed by a series of smaller aftershocks.

The experimental data is OLR observed by National Oceanic and Atmosphere Administration (NOAA) satellites [11]. These OLR data have been recorded twice-daily by the several polar-orbiting satellites for more than eight years, making up time series data sequences over the different periods of time along with the spatial coverage of the entire earth. The original OLR data were processed by the interpolation technique to minimize the distance in either space or time. The detailed interpolation technique has been reported in [12]. For our study we downloaded twice-daily means from the NOAA centre. Their spatial coverage is 1×1 degree of latitude by longitude covering the area of 90°N – 90°S and 0°E – 357.5°E, and the time range is from 3rd September 2006 to 28th September 2008.

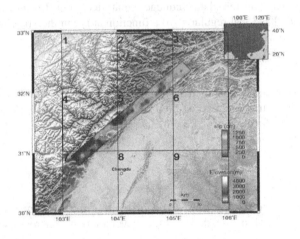

Fig. 2. Grids defined for the Wenchuan

For our experiments, we first studied the characteristics of tectonic background, continental boundaries and active faults along with historical seismic activities in several regions of China and selected three experimental areas, namely Regions 1 and 2, and the Wenchuan earthquake region (main region). The main region covers the earthquake. Region 1 is adjacent to an active fault line and Region 2 is far from the fault zones and there are no historical earthquakes recorded. Each of the regions is divided into a set of grids as shown in Fig. 2. The location of the main region is from 30°N, 103°E to 33°N, 106°E, Region 1 is from 28°N, 105°E to 31°N, 108°E, and Region 2 is from 44°N, 113°E to 47°N, 116°E.

Applying db1 and gaus3 to analyze OLR data in each grid, we can obtain sets of wavelet maxima values, and then we take all sets of maxima values in the nine grids and rearrange them onto one figure as shown in Fig. 3. In the figure the x-axis represents time in day units, and the y-axis represents the grids in a sequential order. The magnitudes of maxima represent the strength of singularities, where the larger the magnitude, the higher the degree of seismic anomalies. The figure heading lists the earthquake name, the period of selected data, data type, region of data, grid path and the wavelet method used. The red line indicates the day when the earthquake occurred.

Examining Fig. 3, several maxima curves with large magnitudes can be identified, they appear before and after the Wuchan earthquake and on the day when the earthquake occurred. Interestingly some of them are continuous across the grids. These maxima curves could be viewed as singularities, which could be caused by the large energy flux before the earthquake, by the release of a large amount of energy when the earthquake occurred and by many aftershocks of the earthquake. Therefore these singularities could be regarded as seismic anomalies, which are highlighted with the dashed red ovals.

3 Results and Discussion

The first analyzing results over the three regions are illustrated in Figs. 4, 5 and 6, respectively. These results were produced by gaus3. The red lines indicate the day when the Wenchuan earthquake occurred.

In Fig. 4 several continuous singularities are identified, some of them are around the Wenchuan earthquake. These singularities may be caused by the large amount of energy generated by the Wenchuan earthquake. Compared with Fig. 5, the maxima curves are more disorder, but one continuous singularity can be clearly observed. Looking at Fig. 6, the maxima lines are completely disorder.

The distribution of the singularities in Fig. 5 is similar to those in Fig. 4. However in Fig. 4 the maxima lines are more continuous with larger magnitudes, and a clear singularity appears on the day when the earthquake occurred. Although a similar distribution appears in Fig. 6, the maxima lines are discontinuous and the magnitudes of the maxima are also smaller. Considering the factors of geographic region and tectonic background of the earthquake, we could conclude that the singularities from the wavelet maxima curves of the Wenchuan region are more informative than those in the other two regions. In particular, the maxima lines in Region 2 are completely

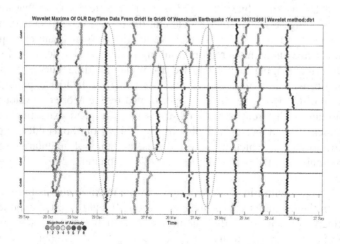

Fig. 3. The curves of wavelet maxima

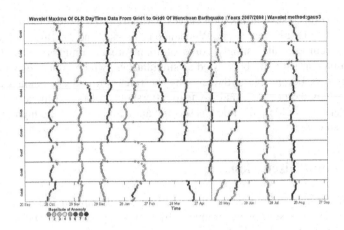

Fig. 4. Maxima curves of the Wenchuan region

disorder, this might be due to the fact that this region is stable and there are almost no earthquakes in past decades. The singularities can also be discovered in Region 1 on the day when the Wenchuan earthquake occurred, these could be a result of that Region 1 is close to the active fault line and the epicenter of the Wenchuan earthquake.

In order to investigate the periodicity of the occurrence of singularities, we selected OLR data in a different year based on the same grids, i.e. from 28th September, 2006 to 28th September, 2007, and carry out a further experiment using db1. The result is illustrated in Figs 3 and 7, respectively.

Comparing Figs 3 and 7, we can see that Fig. 3 has more continuous maxima lines than Fig. 7 does, and the maxima curves in Fig. 7 is discontinuous and disorder. Importantly the singularities identified in Fig. 3 are not present in Fig. 7 and the distribution of the maxima lines in Fig. 3 is entirely different from that depicted in Fig. 7. These findings make us to conclude that the occurrence of the maxima curves in Fig. 3 is not periodic. Thus we believe that the Wenchuan earthquake had significant influence on the main region, resulting in the singularities as in Fig. 3.

To validate the singularities identified from the maxima curves as illustrated Figs. 4, 5 and 6, we carry out a further analysis using the Holder exponent method. The estimation of the Holder exponent is performed by the Morlet wavelet transform for OLR data and a different number of scales 64 is used in extracting the Holder exponent values. Fig. 8 presents a plot of the values of the Holder exponent α, corresponding to Fig. 4, which provide the measurements of singularity strength of the Wf coefficients over the period of one year (27th September 2007 – 28th September 2008).

Comparing Fig. 5 with Fig. 9, we can find that the maxima curves in Fig. 5 are not correlated with the exponent values presented in Fig. 9, but the small exponent values in Fig. 9 clearly present from the middle of May to the end of June 2008. This phenomenon is consistent with that as shown in Fig. 8. The similar findings can be also found from Figs. 6 and 9.

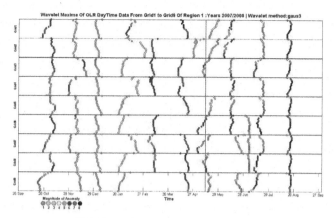

Fig. 5. Maxima curves of the Region 1

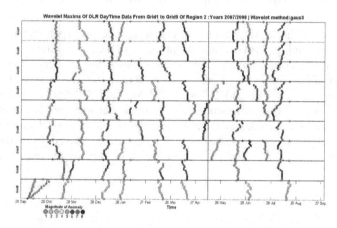

Fig. 6. Wavelet maxima analysis curves of the Region 2

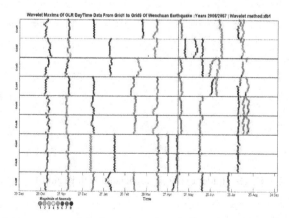

Fig. 7. Maxima curves of the Wenchuan from 28th Sep, 2006 to 28th Sep, 2007

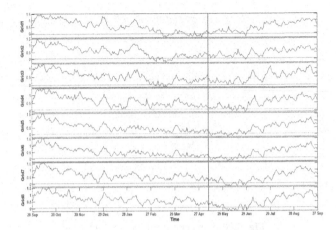

Fig. 8. Holder exponent of the Wenchuan region

Examination of the maxima curves in Fig. 4 and the corresponding α values that are less than zero in Fig. 8 reveals that the presence of the singularities in grids 1, 2, 3, 5 and 6 is basically corresponding to the small α values. Some singularities associated with the maxima curves in Fig. 4 are clearly visible in this plot at some time points, such as in between 27th Feb and 28th Mar, 29th Mar and 27th Apr, 29th May and 29th Jun, etc., but lines of high magnitude coefficients in between 29th Dec and 28th Jan, and on the day of the earthquake are not clearly visible across all these grids 1-6. These findings could not confirm that the singularities derived from the maxima curves in between 29th Dec and 28th Jan in Fig. 4 are directly related to the seismic anomalies.

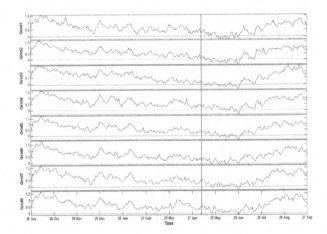

Fig. 9. Holder exponent associated with Region 1

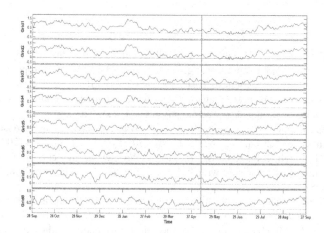

Fig. 10. Holder exponent associated with Region 2

4 Conclusion

This paper presents a comparative analysis on the selected OLR data associated with the Wenchuan earthquake and investigates how the OLR singularities discovered are related to the earthquakes. The wavelet modulus maxima and Holder exponent discussed in this study can be used new data mining techniques to detect concept drifting or changes within data streams. The comparative experiments and analyses on time and space of the Wenchuan earthquakes conclude that the prominent OLR singularities as seismic anomalies could be discovered by incorporating wavelet maxima curves with irregular singularities obtained by the Holder exponent technique.

Theoretically, some concept changes or abnormal activities found in singularities are not necessarily associated with a certain maxima pattern. The more regular the values of singularities, the larger the value of α. Holder exponent detects mainly the singular behavior of data sequences. Because there is no relationship between the strength of singularity and the reflection coefficient of a geological boundary, it is generally considered that Holder exponent is independent of seismic amplitude or high noise values. High amplitudes do not mean high degrees of singularities or very abrupt bursts [9]. This distinguishes Holder exponent from many other types of seismic attributes.

Acknowledgements

This work is supported by the project of "Data Mining with Multiple Parameters Constraint for Earthquake Prediction" (founded by the Ministry of Science and Technology of China, Grant No.:2008BAC35B05). The authors would like to acknowledge NOAA for making OLR data available for various research communities.

References

1. Carreno, E., Capote, R., Yague, A., et al.: Observations of thermal anomaly associated to seismic activity from remote sensing. In: General Assembly of European Seismology Commission, Portugal, September 10-15, pp. 265–269 (2001)
2. Ouzounov, D., Liu, D., et al.: Outgoing long wave radiation variability from IR satellite data prior to major earthquakes. Tectonophysics 431, 211–220 (2007)
3. Cervone, G., Kafatos, M., Napoletani, D., et al.: Wavelet Maxima Curves Associated with Two Recent Greek Earthquakes. Nat. Hazards Earth Syst. Sci. 4, 359–374 (2004)
4. Cervone, G., Singh, R.P., Kafatos, M., Yu, C.: Wavelet maxima curves of surface latent heat flux anomalies associated with Indian earthquakes. Nat. Hazards Earth Syst. Sci. 5, 87–99 (2005)
5. Pulinets, S.A., Boyarchuk, K.A.: Ionospheric Precursors of Earthquakes, p. 316. Springer, Heidelberg (2004)
6. Hayakawa, M., Molchanov, O.A.: Seismo Electromagnetics: Lithosphere-Atmosphere-Ionosphere Coupling, TERRAPUB, Tokyo, p. 477 (2002)
7. Tronin, A.A.: Satellite thermal survey application for earthquake prediction. In: Hayakawa, M. (ed.) Atmospheric and Ionospheric Phenomena Associated with Earthquakes. TERRAPUB, Tokyo, pp. 717–746 (1999)
8. Tronin, A.A., Hayakawa, M., Molchanov, O.A.: Thermal IR satellite data application for earthquake research in Japan and China. J. Geodyn. 33, 519–534 (2004)
9. Ouzounov, D., Freund, F.: Mid-infrared emission prior to strong earthquakes analyzed by remote sensing data. Advances in Space Research 33(3), 268–273 (2004)
10. Paul, S.: Addison Illustrated Wavelet Transform Handbook, Introductory Theory And Applications in Science, Engineering, Medicine And Finance. Institute of Physics Publishing (2002)
11. NCAR and NOAA: ftp ftp.cpc.ncep.noaa.gov; cd precip/noaa* for OLR directories (2008)
12. Liebmann, B., Smith, C.A.: Description of a Complete (Interpolated) Outgoing Longwave Radiation Dataset. Bulletin of the American Meteorological Society 77, 1275–1277 (1996)

On Optimization of Predictions in Ontology-Driven Situation Awareness*

Norbert Baumgartner[1], Wolfgang Gottesheim[2], Stefan Mitsch[2],
Werner Retschitzegger[3], and Wieland Schwinger[2]

[1] Team Communication Technology Mgt. Ltd., Goethegasse 3, 1010 Vienna, Austria
[2] Johannes Kepler University Linz, Altenbergerstr. 69, 4040 Linz, Austria
[3] University of Vienna, Dr.-Karl-Lueger-Ring 1, 1010 Vienna, Austria

Abstract. Systems supporting situation awareness in large-scale control systems, such as, e. g., encountered in the domain of road traffic management, pursue the vision of allowing human operators prevent critical situations. Recently, approaches have been proposed, which express situations, their constituting objects, and the relations in-between (e. g., road works causing a traffic jam), by means of domain-independent ontologies, allowing automatic prediction of future situations on basis of relation derivation. The resulting vast search space, however, could lead to unacceptable runtime performance and limited expressiveness of predictions. In this paper, we argue that both issues can be remedied by taking inherent characteristics of objects into account. For this, an ontology is proposed together with optimization rules, allowing to exploit such characteristics for optimizing predictions. A case study in the domain of road traffic management reveals that search space can be substantially reduced for many real-world situation evolutions, and thereby demonstrates the applicability of our approach.

1 Introduction

Situation awareness. Situation awareness is gaining more and more importance as a way to help human operators cope with *information overload* in large-scale control systems, as, e. g., encountered in the domain of road traffic management. In this respect, a major vision of situation-aware systems is to support human operators in anticipating possible future situations in order to pro-actively prevent critical situations by taking appropriate actions. As first steps towards making this vision come true, situation-aware systems aggregate information about physical *objects* (e. g., road works) and *relations* among them (e. g., causes) into relevant *situations* (e. g., road works cause a traffic jam).

Situation evolution and prediction. Situations evolve continuously, resulting from alterations of real-world objects over time (e. g., road works cause a traffic jam which grows, then shrinks, and finally dissolves). Such evolutions are

---

* This work has been funded by the Austrian Federal Ministry of Transport, Innovation and Technology (BMVIT) under grant FIT-IT 819577.

D. Karagiannis and Z. Jin (Eds.): KSEM 2009, LNAI 5914, pp. 297–309, 2009.

also dealt with in the fields of context prediction [1] and time series analysis [2] to predict future developments. However, techniques proposed in these fields are often based on quantitative data and make use of learning from historic data to achieve domain-independence. Therefore, they are only able to detect situations that have already occurred in the past, and, hence, are not applicable for predicting critical situations before they occur for the first time [1]. In situation awareness, however, such critical situations often endanger life and are, besides that, not observable in sufficient quantity to obtain meaningful training data for machine learning (e. g., a wrong-way driver rushing into a traffic jam). In our previous work [3],[4],[5],[6], we therefore proposed—in accordance with Llinas et al. [7]—to pursue a different approach using domain-independent ontologies describing qualitative facts for achieving situation awareness, as well as techniques thereupon predicting future situations without relying on historic data. For this, we described situation evolution in terms of *transitions* of the relations contributing to a situation. Based on this, prediction of whether or not a current situation can evolve into a critical situation, is achieved by computing all possible paths of such transitions between the current and the critical situation.

Optimization potentials for prediction. The number of such paths, however, depends on the contributing objects and the relations defined among them, resulting in a vast search space of possible transitions between two situations [5]. This holds the risk of inducing unacceptable runtime performance and limiting the expressiveness of predictions. We argue that both issues can be remedied by taking into account inherent characteristics of involved objects, describing in detail which relations among them are actually possible, and which transitions between these relations may occur. Up to now, such knowledge is often incorporated only implicitly, like, e. g., in terms of subsumption rules for assessing situations [8]. In this paper, we propose a *domain-independent ontology* for modeling *object evolution characteristics* describing how a real-world object anchored in time and space can change, and based on this, *optimization rules* for reducing prediction search space.

Structure of the paper. In Section 2, we summarize our previous work on situation awareness and exemplify the potentials of our approach in the domain of road traffic management. In Section 3, an ontology together with optimization rules is proposed for representing and exploiting object evolution characteristics. Next, we evaluate the applicability of such characteristics on basis of a case study in Section 4. Finally, we provide an overview of related work in Section 5, before we conclude the paper in Section 6 by indicating further prospects of our work.

2 Motivating Example

Road traffic management systems, responsible for, e. g., ensuring safe driving conditions, are a typical application domain for situation awareness. In this section, examples from the domain of road traffic management are used to summarize our previous work on situation awareness and to illustrate the potentials of incorporating object evolutions characteristics into prediction of situations.

Situation awareness in road traffic management. In principle, human operators of road traffic management systems observe highways for critical situations like, e. g., a wrong-way driver heading towards a traffic jam, in order to resolve them by taking appropriate actions. In our previous work [4] we introduced a framework for building situation-aware systems on basis of a domain-independent ontology. This framework was used to build a prototype [6] supporting human operators and, thereby, the feasibility of our approach to situation awareness was shown. In the framework's ontology, objects and relations among them, which are derived from object properties, are aggregated into situations. Information on such objects and relations (and thereby on traffic situations) is obtained from various sources such as, e. g., from traffic flow monitoring systems and drivers reporting traffic information. Changes of object properties over time, like, e. g., movement on a road, reported by such sources cause changes in the relations between objects, leading in turn to *situation evolution*.

Spatial and temporal relations. For describing such relations between objects, we discern families of *spatial and temporal relations*, with each family modeling a certain real-world aspect [3]. These comprise *mereotopological reasoning about regions*, describing, e. g., whether a traffic jam occurs in a tunnel, *positional and orientational reasoning about points*, expressing, e. g., that an accident happened in front of a traffic jam, as well as *size* and *distance*. Temporal relation families allow us to express that, e. g., an accident occurred shortly before a traffic jam. For describing these relation families, we base upon well-known calculi further detailed in Section 3.

Predicting future situations. By employing these relation calculi, prediction of situations provides the basis for early detection of possibly emerging critical situations. Let us suppose that human operators of a road traffic management system want to be informed of a critical situation "Wrong-way driver in the area of road works", as depicted in Fig. 1. In order to pro-actively take actions such as, e. g., issue warnings to motorists, human operators want to be informed already if such a situation is possibly emerging, which is indicated by the initial situation "Wrong-way driver very close to road works". In Fig. 1, we illustrate the combination of relations from appropriate relation families—in this case, mereotopology, distance, orientation, and size—which have to be valid between objects in order to constitute the initial situation. All other situations in this example would be formalized with similar combinations, but details are left out for brevity. For predicting whether or not this situation can evolve into the critical situation, we derive all possible paths of transitions between these two situations (Fig. 1 shows an exemplary subset thereof). A transition between two such situations occurs, if a transition is possible in at least one of the contributing relation families. By that, we can predict to reach the situation depicted in step 1 with one transition describing a change in the distance between the wrong-way driver and road works. With one further step—one transition in our relation family describing mereotopology and one transition in the family describing spatial distance—we predict to reach the critical situation. Due to relation families

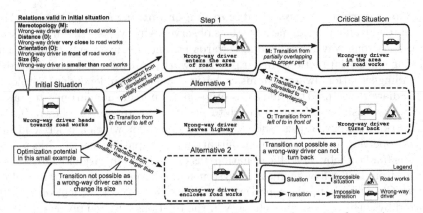

Fig. 1. Exemplary prediction of situations towards a critical situation

being defined independently from concrete objects, alternative paths via different situations are predicted, as depicted in Fig. 1. However, not all of these predicted situations and paths can actually occur in real-world.

Considering object evolution. The approach we present in this paper aims to eliminate such impossible future situations and transitions between them by incorporating *object evolution characteristics* into the prediction process. Prerequisite for this is to remove impossible relations and transitions from relation families. For example, knowledge that a wrong-way driver can only change its position, but obviously not its size, allows us to rule out a predicted future situation, in which the wrong-way driver encloses the area of road works, as depicted in Fig. 1. Another impossible situation occurs, if our information sources report that the wrong-way driver turned left at the depicted junction (i. e., leaving the highway), instead of following step 1. Object evolution characteristics enable us to conclude that no further evolution towards the critical situation is possible, as road works can not change their position, and a wrong-way driver can not turn back towards the road works. Hence, in Fig. 1 we can exclude the situation "Wrong-way driver turns back" from the possible paths towards the critical situation, which allows us to take back issued warnings. Without object evolution characteristics, we would not be able to detect this circumstance, and would still predict that the critical situation is about to emerge.

Summarizing, by using object evolution characteristics, we are able to reduce the number of relevant relations and transitions in a relation family as a pre-requisite for removing impossible situations and transitions between them. Thereby, the number of possible paths between two situations, i. e., prediction search space, is reduced.

3   Object Evolution Characteristics

Object evolution characteristics, as laid out in the previous section, bear potential for increasing performance during prediction of future situations, as well as for increasing prediction expressiveness. In Fig. 2, the principle idea of our approach is depicted, by showing the interaction between the different constituents. An evolution ontology defines the vocabulary for describing object evolution characteristics, which is exploited by optimization rules to optimize relation families. On basis of such optimized relation families, a prediction algorithm—presented in our previous work [5]—generates optimized possible evolutions between situations. In the following, these constituents are described in detail.

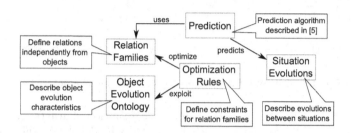

Fig. 2. Overview of our approach

Object evolution ontology. Situation evolution, as described above, depends on the ability of objects to change over time. Considering object evolution characteristics in more detail, one can naturally differentiate between *situation-independent* ones describing inherent characteristics of objects that do not affect or depend on other objects (e. g., road works can not change their size), and *situation-dependent* ones defining how an object evolves in relation with other objects (e. g., a traffic jam caused by an accident in front of it can only dissolve after the accident has been cleared). Although the potential benefit of situation-dependent object evolution characteristics are likely to be higher than those of situation-independent ones, they incur a fundamental increase in modeling complexity, as object evolution characteristics would have to be modeled individually for every situation. Therefore, in the object evolution ontology proposed in the following, we focus on situation-independent object evolution characteristics, and later show in Section 4 that, thereby, already substantial optimizations can be achieved with reasonable modeling effort.

We differentiate in the ontology depicted in Fig. 3 between *non-changing objects* and *changing objects* being able to change at least in some of their properties, thus contributing to situation evolution. As objects in situation awareness are anchored in time and space, corresponding types describe an object's typical lifespan and size, as well as an object's particular kind of supported spatial change (*scaling*—changes size, *translational motion*—changes position, and *rotation*[1]—changes orientation). A complementary characteristic—boundary permeability—describes, whether or not objects are able to share the same region with other objects (e. g., a tunnel has a permeable boundary, meaning that other

---

[1] In the field of computer graphics, a fourth change—reflection—is known, but this is of little practical benefit for describing real-world objects in situation awareness.

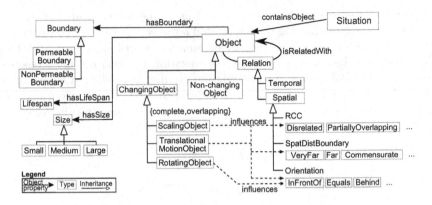

Fig. 3. Ontology for describing object evolution characteristics

objects can be inside the tunnel). The object evolution characteristics of concrete objects (e. g., a wrong-way driver) are then modeled by inheriting from the appropriate types of the evolution ontology (cf. Table 2 for an exemplary tabular representation).

Relation families. For formalizing the relation families introduced in Section 2, we use well-known calculi from the field of spatio-temporal reasoning: the Region Connection Calculus (RCC, cf. [9]) describes mereotopology, Spatial Distance of Boundaries (SpatDistBoundary, cf. [10]) formalizes spatial distance, and the Oriented Points Relation Algebra (OPRA, cf. [11]) describes orientation. Each relation family is described with a directed graph representing relations between objects as nodes and possible transitions in-between as edges (called *conceptual neighborhood graph*—CNG [12]), as depicted in Figure 4[2].

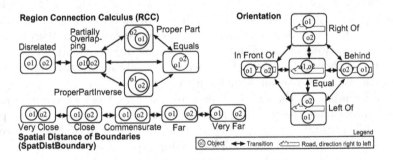

Fig. 4. Conceptual neighborhood graphs of relation families

---

[2] Note that in this paper we use a rather basic notion of an orientation calculus describing relative orientation between two objects on their centroids in order to increase comprehensibility. In contrast, OPRA would result in a very large CNG.

Optimization rules. On basis of the ontology introduced above, optimizations on the level of single relation families are enabled using optimization rules detailed in the following. Relation families describing mereotopology, distance, and orientation are influenced by types representing object evolution characteristics, and thereby drive the formalization of optimization rules. In particular, mereotopological relations are influenced by characteristics describing whether and how objects support translational motion or scaling, and whether an object's boundaries are permeable. For example, a car having a non-permeable boundary can be a proper part of a tunnel having a permeable boundary, but the inverse relation is not possible. Relations describing positions and distances are influenced by translational motion and scaling only. For example, road works and accidents can not move, inducing that the spatial distance between them will not change throughout their lifespan. Finally, relations describing orientation are influenced by rotation, translational motion, and scaling.

Optimization rules—on basis of the object evolution characteristics—define conditions for removing relations and transitions from the relation family's CNG. These optimization rules are formalized in terms of a simplified syntax based on the Semantic Web Rule Language (SWRL, [13]). In this syntax, an object o being a member of the type T is expressed as T(o), whereas two objects o1 and o2 taking part in the relation R are expressed as R(o1,o2). Table 1 lists optimization rules for the relations and transitions of the three relation families RCC, SpatDistBoundary, and Orientation together with their effect on relations and transitions. In general, it can be seen that, concerning RCC, optimization rules reduce the number of possible relations and transitions. We pick out the relation ProperPart of RCC to exemplify the meaning of these optimization rules: ProperPart(o1,o2) between the objects o1 and o2 is removed from RCC and, therefore, not taken into account during prediction of situations, if o2 does not have a permeable boundary and if o2 is smaller than o1. All other rules can be interpreted analogously. With similar rules on combined relations of different relation families, we are able to optimize complex situation types. The

Table 1. Optimization rules for three relation families

Optimization Rule	Optimization Effect
RCC	
Not optimizable	Disrelated is not removable
1 $NonPermeableBoundary(o2)$	Removes PartiallyOverlapping(o1,o2)
2 $NonPermeableBoundary(o2) \lor IsSmaller(o2, o1)$	Removes ProperPart(o1,o2)
3 $NonPermeableBoundary(o1) \lor IsSmaller(o1, o2)$	Removes ProperPartInverse(o1,o2))
4 $NonPermeableBoundary(o1) \lor NonPermeableBoundary(o2) \lor IsLarger(o1, o2) \lor IsLarger(o2, o1)$	Removes Equals(o1,o2)
5 $\neg(ScalingObject(o1) \lor ScalingObject(o2) \lor TranslationalMotionObject(o1) \lor TranslationalMotionObject(o2))$	Removes all transitions
SpatDistBoundary	
Not optimizable	Relations are not removable
6 $\neg(ScalingObject(o1) \lor ScalingObject(o2) \lor TranslationalMotionObject(o1) \lor TranslationalMotionObject(o2))$	Removes all transitions
Orientation	
Not optimizable	Relations are not removable
7 $\neg(ScalingObject(o1) \lor ScalingObject(o2) \lor TranslationalMotionObject(o1) \lor TranslationalMotionObject(o2))$	Removes InFrontOf $\leftrightarrow$ Equal and removes Equal $\leftrightarrow$ Behind
8 $\neg((RotatingObject(o1) \land (ScalingObject(o1) \lor TranslationalMotionObject(o1))) \lor (RotatingObject(o2) \land (ScalingObject(o2) \lor TranslationalMotionObject(o2))))$	Removes all other transitions

optimizations achievable for different object combinations and transitions between situations are shown in the next section.

4   Road Traffic Management Case Study

In this section, we employ a case study in the domain of road traffic management to discuss the potentials of situation-independent object evolution characteristics for optimizing prediction of future situations. For this, we base upon road traffic objects with their evolution characteristics, as well as situations defined in collaboration with the Austrian highways agency ASFINAG[3] (cf. [14] for a detailed overview of the more than 100 road traffic objects, and 16 critical situations considered in our prototype). In Table 2, we use the evolution ontology introduced above to describe evolution characteristics of a small subset of these objects, which was selected to show each spatial characteristic at least once.

Table 2. Evolution characteristics of real-world objects

Object	Size	Translational-MotionObject	Scaling-Object	Rotating-Object	Boundary
RoadWorks	Large	-	-	-	PermeableBoundary
Accident	Medium	-	-	-	NonpermeableBoundary
WrongWayDriver	Small	✓	-	✓	NonpermeableBoundary
TrafficJam	Large	✓	✓	✓	PermeableBoundary

Optimization of a single relation family's relations and transitions. To start with, we evaluate the effects of these object evolution characteristics on relations and transitions of single relation families. Fig. 5 shows the optimizations for the RCC relation family achieved by applying the optimization rules introduced above to the objects Accident and RoadWorks described in Table 2. Naturally, these objects with few evolution characteristics are best suited to reduce the CNGs of relation families. For example, as neither road works nor accidents can change their position or size, optimization rule 5 removes all transitions from RCC. This means that, once a particular relation between two objects is detected, no further evolution can occur. Additionally, the relations ProperPartInverse and Equals are removed by the optimization rules 3 and 4, because an accident's non-permeable boundary does not allow other objects to enter the region occupied by the accident. These optimizations reduce the search space, because less relations and/or transitions need to be taken into account during prediction of situations. At the same time, prediction expressiveness is increased, because only those relations and transitions being actually relevant for the involved objects are used for predicting situations. For example, on basis of such an optimized CNG, as depicted in Fig. 5, we no longer would predict that an accident can become ProperPartInverse of road works. Similar optimizations are also possible between wrong-way drivers and traffic jams. If objects are capable to change in many ways, like, e.g., traffic jams do, such optimizations are, as already mentioned, not possible.

---

[3] www.asfinag.at

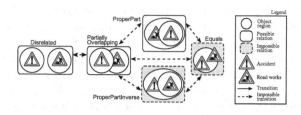

Fig. 5. Optimized RCC relation family

Optimization of situations and transitions between two situations. As described before, relation families optimized in such a way reduce the number of situations and transitions towards a critical situation. In Fig. 6, a wrong-way driver is close to a traffic jam, indicating a possibly emerging critical situation, which is reached when the wrong-way driver rushes into the traffic jam. In general, intermediary situations and transitions between these two situations are predicted if, as mentioned above, a transition is possible in at least one of the contributing relation families. Without taking object evolution characteristics into account, prediction would result in a large number of such intermediary situations and transitions. From these, first of all three situations with their corresponding transitions can be ruled out employing the optimized CNGs, because they rely on impossible relations. Additionally, another situation is not possible, because its combination of relations demands object evolution characteristics not being supported by the wrong-way driver: the situation `PartiallyOverlapping & Close` describes that a wrong-way driver and a traffic jam partially overlap (i. e., a wrong-way driver is partly outside and partly inside a traffic jam), while at the same time their boundaries are close to each other. This is not possible, because the size of a wrong-way driver only allows `VeryClose` when overlapping another object. Such an optimization rule is not listed in Table 1, because it is actually situation-dependent. Overall, the optimizations reduce the number of possible situations between the initial and the critical situation from nine to five, and the transitions from 19 to four. If additionally removing non-reachable situations, a further reduction to three situations, and three transitions is possible.

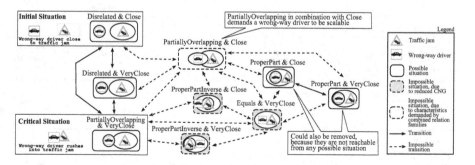

Fig. 6. Situation evolution graph with optimized transitions

Optimization on basis of semantics of combined relation families. However, in this example we still find transitions, which are actually not possible in a real-world scenario. These impossible transitions are characterized by implicit semantics resulting from combined relations. For example, we predict to reach `PartiallyOverlapping` & `VeryClose` with a single transition from `Disrelated` & `Close`, because it only requires one transition in RCC and one transition in SpatDistBoundary. But two real-world objects would first become `Disrelated` & `VeryClose`, before they can become `PartiallyOverlapping`. Removing such transitions will be subject of our ongoing work (cf. Section 6).

Summary of optimization results. In Table 3, we summarize the achievable optimizations for ten out of the 16 situations defined with ASFINAG. For formally describing the situations, we use the syntax introduced for our optimization rules in Section 3. We then compare the number of situations and transitions predicted between an initial and a critical situation with and without considering object evolution characteristics. In general, it can be seen that considering object evolution characteristics—unless the evolution paths are already optimal, as it is the case in the fourth example—leads to a substantial reduction of both situations and transitions between them by more than 50%. Particularly interesting is the last example, which—caused by the initial situation not containing any relation—leads to a large number of possible situations and transitions, because potentially every situation comprising a combination of relations from the families `RCC` and `Orientation` leads to the critical

Table 3. Summary of achievable optimizations

Situation description and formalization	Optimization
Initial situation: wrong-way driver close to traffic jam $WrongWayDriver(o1) \wedge TrafficJam(o2) \wedge Disrelated(o1, o2) \wedge Close(o1, o2)$ **Critical situation**: wrong-way driver rushes into traffic jam $WrongWayDriver(o1) \wedge TrafficJam(o2) \wedge PartiallyOverlapping(o1, o2) \wedge VeryClose(o1, o2)$	Without: 18 situations, 7 transitions; With: 3 situations, 4 transitions (scale 0–20). ■ Number of situations ■ Number of transitions
Initial situation: poor driving conditions near the border area of traffic jam $PoorDrivingConditions(o1) \wedge TrafficJam(o2) \wedge Disrelated(o1, o2) \wedge Commensurate(o1, o2)$ **Critical situation**: poor driving conditions in the area of traffic jam $PoorDrivingConditions(o1) \wedge TrafficJam(o2) \wedge PartiallyOverlapping(o1, o2) \wedge VeryClose(o1, o2)$	Without: 24 situations, 9 transitions; With: 4 situations, 5 transitions (scale 0–30). ■ Number of situations ■ Number of transitions
Initial situation: poor driving conditions cause an accident $PoorDrivingConditions(o1) \wedge Accident(o2) \wedge PartiallyOverlapping(o1, o2) \wedge VeryClose(o1, o2)$ **Critical situation**: poor driving conditions move away from the accident $PoorDrivingConditions(o1) \wedge Accident(o2) \wedge Disrelated(o1, o2) \wedge Commensurate(o1, o2)$	Without: 63 situations, 13 transitions; With: 4 situations, 15 transitions (scale 0–100). ■ Number of situations ■ Number of transitions
Initial situation: traffic jam potentially grows together with another traffic jam $TrafficJam(o1) \wedge TrafficJam(o2) \wedge Disrelated(o1, o2) \wedge Close(o1, o2)$ **Critical situation**: traffic jams are grown together $TrafficJam(o1) \wedge TrafficJam(o2) \wedge Disrelated(o1, o2) \wedge VeryClose(o1, o2)$	Without: 1 situation, 1 transition; With: 1 situation, 1 transition (scale 0–1). ■ Number of situations ■ Number of transitions
Initial situation: road works become active $RoadWorks(o1)$ **Critical situation**: road works cause abnormal traffic $RoadWorks(o1) \wedge AbnormalTraffic(o2) \wedge PartiallyOverlapping(o1, o2) \wedge InFrontOf(o1, o2)$	Without: 97 situations, 19 transitions; With: 4 situations, 6 transitions (scale 0–100). ■ Number of situations ■ Number of transitions

situation. In such settings, object evolution characteristics are particularly helpful to reduce the search space: if taking into account that road works are stationary and that a traffic jam can only be caused by road works being in front of it, all situations not containing InFrontOf are invalid.

5 Related Work

In this section, we present related research on prediction techniques in various domains, as well as more closely related work on the optimization and combination of calculi stemming from the area of qualitative spatial reasoning.

Prediction in other domains. Predicting future situations based on current and historic data is a task relevant in numerous domains. Various techniques in the field of time-series analysis, often being based on quantitative data, exist to, e. g., forecast future developments of stocks [2]. Efforts applying these approaches to the domain of road traffic management exist [15], but naturally they do not focus on predicting situations, but instead aim to predict traffic flow in terms of vehicle throughput. Solutions relying on alternative techniques, but still being based on quantitative data, like, e. g., neural networks [16] or bayesian belief networks [17], in a similar manner only predict traffic flow. In [18], a neural network-based solution to predict traffic accidents based on recognizing patterns in vehicle tracking data is presented. Compared to our work, this approach is tailored to the specific problem of accident prediction, whereas we aim for prediction of arbitrary situations from qualitative data in a domain-independent manner. Research in the area of context awareness postulates object evolution characteristics [19] and proposes to predict a situation's future on its historic attribute values. However, similar to the work on context history presented in [1], such an approach is unable to predict events that did not occur in the past.

Optimization of relation calculi. Dylla et al. [20] analyze a single calculus (OPRA) for optimization by manually eliminating transitions between relations being impossible due to characteristics like, e. g., locomotion. In contrast to this work, we propose domain-independent concepts for describing such characteristics and apply these concepts to automatically optimize different calculi. Moreover, we describe optimizations for combined calculi, which are frequently used in situation awareness to define situations. Possible combination methods, as well as comparisons of their reasoning performance are presented in Wölfl et al. [21]. The authors differentiate between orthogonal combinations expressing new calculi by combining existing formalisms without semantic interdependencies, and non-orthogonal combinations consisting of calculi with semantic interdependencies. In situation awareness, we typically encounter non-orthogonal combinations, which should best be combined using so-called tight integration techniques, as Wölfl et al. suggest. Such tight integrations are more expressive—i. e., contain more relations—than loose ones, which might due to their algorithmic design already rule out important relations. However, this advantage has its downside in prediction performance, as a large number of relations induces a large number of predictable situations and transitions. With our approach, we argue that prediction performance of tightly integrated calculi can be improved by removing impossible relations not fitting the characteristics of involved objects.

6   Future Work

We aim for further extending our object evolution ontology with additional object evolution characteristics, and in particular, with types for describing object evolution patterns. Such patterns promise a finer-grained representation of an object's changes over time (e. g., a traffic jam first grows, then begins to shrink, and finally dissolves), facilitating prediction of future situations with knowledge about complex object behavior. By recognizing these patterns we envision to provide operators of systems supporting situation awareness with better recommendations for actions to be taken. Second, our optimization rules are to be extended with situation-dependent rules, and with rules exploiting semantics of combined relation families, which are currently only implicitly available in ontology-driven situation awareness. Finally, we want to measure the effects on runtime performance resulting from removing relations and transitions of single relation families, as well as from removing situations and transitions between them from predicted situation evolutions. For this, we will extend our prototype and base on real-world test data provided by ASFINAG.

References

1. Mayrhofer, R., Radi, H., Ferscha, A.: Recognizing and predicting context by learning from user behavior. Radiomatics: Journal of Communication Engineering, special issue on Advances in Mobile Multimedia 1(1), 30–42 (2004)
2. Cheng, C.H., Teoh, H.J., Chen, T.L.: Forecasting stock price index using fuzzy time-series based on rough set. 4th Intl. Conf. on Fuzzy Systems and Knowledge Discovery 3, 336–340 (2007)
3. Baumgartner, N., Retschitzegger, W.: Towards a situation awareness framework based on primitive relations. In: Proc. of the IEEE Conf. on Information, Decision, and Control, Adelaide, Australia, pp. 291–295. IEEE, Los Alamitos (2007)
4. Baumgartner, N., Retschitzegger, W., Schwinger, W.: Lost in time, space, and meaning—an ontology-based approach to road traffic situation awareness. In: Proc. of the 3rd Workshop on Context Awareness for Proactive Systems, Guildford, UK (2007)
5. Baumgartner, N., Retschitzegger, W., Schwinger, W., Kotsis, G., Schwietering, C.: Of situations and their neighbors—Evolution and Similarity in Ontology-Based Approaches to Situation Awareness. In: Kokinov, B., Richardson, D.C., Roth-Berghofer, T.R., Vieu, L. (eds.) CONTEXT 2007. LNCS (LNAI), vol. 4635, pp. 29–42. Springer, Heidelberg (2007)
6. Baumgartner, N., Retschitzegger, W., Schwinger, W.: A software architecture for ontology-driven situation awareness. In: Proc. of the 23rd Annual ACM Symposium on Applied Computing, Fortaleza, Ceara, Brazil, pp. 2326–2330. ACM, New York (2008)

7. Llinas, J., Bowman, C., Rogova, G., Steinberg, A.: Revisiting the JDL data fusion model II. In: Proc. of the 7th Intl. Conf. on Information Fusion, Stockholm, Sweden, pp. 1218–1230 (2004)
8. Kokar, M.M., Matheusb, C.J., Baclawski, K.: Ontology-based situation awareness. International Journal of Information Fusion 10(1), 83–98 (2009)
9. Cohn, A.G., Bennett, B., Gooday, J.M., Gotts, N.: RCC: A calculus for region based qualitative spatial reasoning. GeoInformatica 1, 275–316 (1997)
10. Hernández, D., Clementini, E., Felice, P.D.: Qualitative distances. In: Kuhn, W., Frank, A.U. (eds.) COSIT 1995. LNCS, vol. 988, pp. 45–57. Springer, Heidelberg (1995)
11. Moratz, R., Dylla, F., Frommberger, L.: A relative orientation algebra with adjustable granularity. In: Proc. of the Workshop on Agents in Real-Time and Dynamic Environments (2005)
12. Freksa, C.: Temporal reasoning based on semi-intervals. Artificial Intelligence 54(1), 199–227 (1992)
13. Horrocks, I., Patel-Schneider, P.F., Boley, H., Tabet, S., Grosof, B., Dean, M.: SWRL: A semantic web rule language combining OWL and RuleML (2004), http://www.w3.org/Submission/SWRL
14. Baumgartner, N.: BeAware! — An Ontology-Driven Framework for Situation Awareness Applications. PhD thesis, Johannes Kepler University Linz (2008)
15. Sabry, M., Abd-El-Latif, H., Yousef, S., Badra, N.: A time-series forecasting of average daily traffic volume. Australian Journal of Basic and Applied Sciences 1(4), 386–394 (2007)
16. Dia, H.: An object-oriented neural network approach to short-term traffic forecasting. European Journal of Operational Research 131(2), 253–261 (2001)
17. Sun, S., Zhang, C., Yu, G.: A bayesian network approach to traffic flow forecasting. IEEE Transactions on Intelligent Transportation Systems 7(1), 124–132 (2006)
18. Hu, W., Xiao, X., Xiea, D., Tan, T.: Traffic accident prediction using vehicle tracking and trajectory analysis. In: Proc. of the 6th Intl. Conf. on Intelligent Transportation Systems (2003)
19. Padovitz, A., Loke, S.W., Zaslavsky, A., Burg, B.: Towards a general approach for reasoning about context, situations and uncertainty in ubiquitous sensing: Putting geometrical intuitions to work. In: Proc. of the 2nd Intl. Symposium on Ubiquitous Computing Systems, Tokyo, Japan (2004)
20. Dylla, F., Wallgrün, J.O.: Qualitative spatial reasoning with conceptual neighborhoods for agent control. Journal of Intelligent Robotics Systems 48(1), 55–78 (2007)
21. Wölfl, S., Westphal, M.: On combinations of binary qualitative constraint calculi. In: Proc. of the 21st Intl. Joint Conference on Artificial Intelligence, Pasadena, CA, USA, pp. 967–972 (2009)

On Undecidability of Cyclic Scheduling Problems

Grzegorz Bocewicz[1], Robert Wójcik[2], and Zbigniew Banaszak[3]

[1] Koszalin University of Technology, Department of Computer Science and Management,
Śniadeckich 2, 75-453 Koszalin, Poland
bocewicz@ie.tu.koszalin.pl
[2] Wrocław University of Technology, Institute of Computer Engineering, Control and Robotics,
Wybrzeże Wyspiańskiego 27, 50-370 Wrocław, Poland
robert.wojcik@pwr.wroc.pl
[3] Warsaw University of Technology, Department of Business Informatics, Narbutta 85,
02-524 Warsaw, Poland
Z.Banaszak@wz.pw.edu.pl

Abstract. Cyclic scheduling concerns both kinds of problems relevant to the deductive and inductive ways of reasoning. The first class of problems concentrates on rules aimed at resources assignment so as to minimize a given objective function, e.g. the cycle time, the flow time of a job. In turn, the second class focuses on a system structure designed so as to guarantee that the assumed qualitative and/or quantitative measures of objective functions can be achieved. The third class of problems can however be seen as integration of the previous ones, i.e. treating design and scheduling or design and planning, simultaneously. The complexity of these problems stems from the fact that system configuration must be determined for the purpose of processes scheduling, yet scheduling must be done to devise the system configuration. In that context, the contribution provides discussion of some Diophantine problems solubility issues, taking into account the cyclic scheduling perspective.

Keywords: Diophantine problem, cyclic scheduling, timetabling, multi-criteria optimization.

1 Introduction

The way an enterprise's production capacity is utilized decides about its competitiveness. In that context studies aimed at designing of decision support systems (DSS) dedicated to discrete processes scheduling, and especially cyclic scheduling, are of primary importance.

A cyclic scheduling problem is a scheduling problem in which some set of activities is to be repeated an indefinite number of times, and it is desired that the sequence be repeating. Cyclic scheduling problems arise in different application domains (such as manufacturing, time-sharing of processors in embedded systems, digital signal processing, and in compilers for scheduling loop operations for parallel or pipelined architectures) as well as service domains (covering such areas as workforce scheduling (e.g., shift scheduling, crew scheduling), timetabling (e.g., train timetabling,

D. Karagiannis and Z. Jin (Eds.): KSEM 2009, LNAI 5914, pp. 310–321, 2009.

aircraft routing and scheduling), and reservations (e.g., reservations with or without slack, assigning classes to rooms) [4], [5], [7], [14], [15]. The scheduling problems considered belong to the class of NP-hard ones and are usually formulated in terms of decision problems, i.e. as searching for an answer whether a solution possessing the assumed features exists or not [10].

More formally, a decision problem can be seen as a question stated in some formal system with a yes-or-no answer, depending on the values of some input parameters. The decision problems fall into two categories: decidable and non decidable problems [12], [13]. In this context, the primary question is to what kind of above-mentioned problems the real-life cyclic scheduling ones belong.

A decision problem is called decidable or effectively solvable if an algorithm exists which terminates after a finite amount of time and correctly decides whether or not a given number belongs to the set. A classic example of a decidable decision problem is the set of prime numbers. It is possible to effectively decide whether a given natural number is the prime one, by testing every possible nontrivial factor. A set which is not computable is called non-computable or non-decidable (undecidable). The relevant illustration provides the problem of deciding whether a Diophantine equation (multivariable polynomial equation) has a solution in integers.

In this paper, we provide illustrative examples proving the Diophantine nature of timetabling originated cyclic scheduling models. In that context, the paper's objective is to provide the methodology employing the reverse approach while aimed at DSS designing. The rest of the paper is organized as follows: Section 2 describes the case of manufacturing processes cyclic scheduling. The concept of Diophantine problem is then recalled in Section 3. In Section 4, a case of timetabling like scheduling problem is investigated. Conclusions are presented in Section 5.

2 Systems of Concurrent Cyclic Processes

Consider the flexible manufacturing cell composed of four industrial robots, two machine tools, and the input and output buffers shown in Fig. 1. Robots transporting work pieces from input buffer or machine tool to another machine tool or output buffer serve for handling work pieces – two kinds of work pieces processed along two different production routes. Production routes can be seen as repetitively executed cyclic processes supporting the flow of work pieces along different production routes.

In general case, the cells considered can be organized in much more complex flexible manufacturing systems (see Fig. 2). In such structures, different cyclic scheduling problems can be considered. The typical problem concerns routing of work pieces, in which a single robot has to make multiple tours with different frequencies. The objective is to find a minimal makespan schedule in which the robots repeat their handlings with assumed frequencies.

A system of repetitive manufacturing processes consists of a set of process-sharing common resources while following a distributed mutual exclusion protocol (see Fig. 3). Each process P_i, $(i=1,2,...,n)$, representing one product processing, executes periodically a sequence of operations using resources defined by $Z_i = (R_{i1}, R_{i2}, ..., R_{il(i)})$, where $l(i)$ denotes a length of production route. The operation times are given by the sequence $ZT_i = (r_{i1}, r_{i2}, ..., r_{il(i)})$, where $r_{i1}, r_{i2}, ..., r_{il(i)} \in N$ are defined in uniform time units (N – set of natural numbers). For instance the system shown in Fig.1

consists of six resources and five processes. The resources R_3, R_4, are shared resources, since each one is used by at least two processes, and the resources R_1, R_2, R_5, R_6 are non-shared because each one is exclusively used by only one process. The processes P_1, P_2, P_3, P_4, P_5 are executing operations using resources given by the sequences: $Z_1 = (R_1, R_3)$, $Z_2 = (R_2, R_3)$, $Z_3 = (R_3, R_4)$, $Z_4 = (R_4, R_5)$, $Z_5 = (R_4, R_6)$.

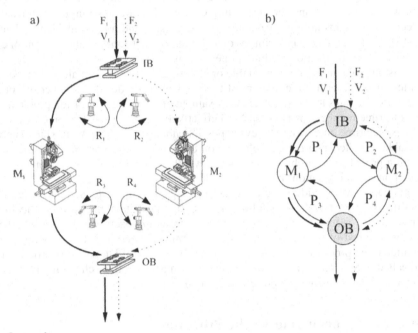

Legend:
R_i – the i-th robot; M_j – the j-th machine tool; IB, OB – the input and output buffers
P_i – the i-th cyclic work piece transportation/handling process,
F_i – the i-th production route,
V_i – the batch processed along the i-th - production route,

➞ - the cyclic process flow direction,
    - the i-th production route.

Fig. 1. Robotic cell: a) the cell layout, b) the model of cyclic processes

Different problems can be stated for the considered class of systems. For example, the following questions can be formulated [1], [2]: Does such an initial process allocation exists that leads to a steady state in which no process waits to access the common shared resources? What set of priority dispatching rules assigned to shared resources guarantee, if any, the same rate of resources utilization?

In general case, the questions may concern the qualitative features of system behavior such as deadlock and/or conflict avoidance [8]. For example they may be aimed at satisfaction of conditions, which guarantees system's repetitiveness for a given initial state and/or allocation of dispatching rules.

a)   b)

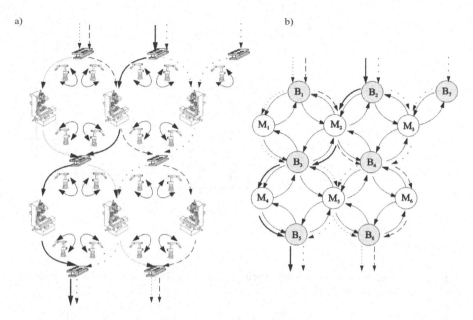

Fig. 2. Structure of the Flexible Manufacturing System: a) the fractal-like layout structure,
b) the graph model of the fractal-like layout structure

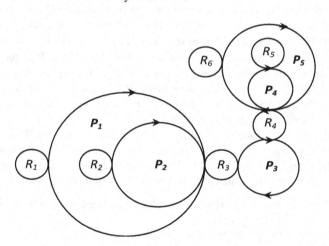

Fig. 3. Repetitive concurrent processes

In order to illustrate the Diophantine character of the model under study, let us
consider the system of concurrently flowing cyclic processes shown in Fig. 4. At the
initial state (see Fig. 4 a) the steady state cyclic system behaviour, illustrated by
Gantt's chart (see Fig. 4 b), is characterized by periodicity $T=5$ (obtained under assumption $t_{C1}=t_{E1}=t_{F1}=t_{C2}=t_{B2}=t_{D2}=t_{D3}=t_{A3}=t_{E3}=1$).

Note that besides initial process allocation (see the sequence $S_0 = (F,C,A)$), the cyclic
steady state behavior depends on routing direction as well as priority rules determining the

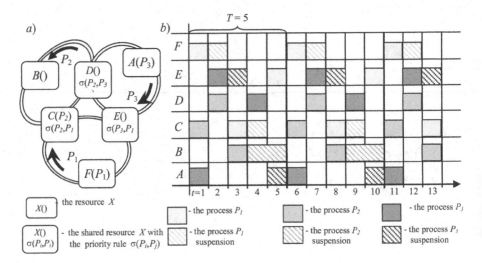

Fig. 4. System of concurrently flowing cyclic processes: a) initial state; b) Gantt's chart encompassing the cyclic steady state

order in which processes make their access to the common shared resources (for instance, in case of resource D, $\sigma_D = \sigma(P_2,P_3)$ – the priority rule determines the order in which processes can access to the common shared resources D, i.e. at first to the process P_2, then to the process P_3, then P_2 and once again to P_3, and so on). For instance, changing the priority rules into $\sigma_C = \sigma(P_2,P_1)$, $\sigma_D = \sigma(P_2,P_3)$, $\sigma_E = \sigma(P_3,P_1)$ is guaranteeing the cyclic steady state behavior, and if $\sigma_C = \sigma(P_2,P_1)$, $\sigma_D = \sigma(P_3,P_2)$, $\sigma_E = \sigma(P_1,P_3)$, then the resultant state is a deadlock state. In turn, for the initial state of process allocation $S_0 = (F,B,A)$ (i.e., the process P_1 is allocated to the resource F, the process P_2 to the resource B, and the process P_3 to A), the resultant steady state is characterized with the periodicity $T = 6$ (see Fig. 5).

In general case, the periodicities of the cyclic steady states as well as corresponding sets of dispatching rules can be calculated from the linear Diophantine equation of the following form: $3y + 3x = T + z$. The formulae considered have been obtained through the following transformations, following the assumptions below:

- the initial state and set of dispatching rules guarantee that an admissible solution exists (i.e. cyclic steady state),
- the structure of the graph model is consistent.

Consider the following set of equations (1):

$$\text{i) } x \cdot (t_{C1} + t_{E1} + t_{F1}) + y \cdot t_{C2} + z \cdot t_{E3} = T$$
$$\text{ii) } y \cdot (t_{C2} + t_{B2} + t_{D2}) + x \cdot t_{C1} + z \cdot t_{D3} = T \qquad (1)$$
$$\text{iii) } z \cdot (t_{D3} + t_{A3} + t_{E3}) + x \cdot t_{E1} + y \cdot t_{D2} = T$$

where:

t_{ij} – the execution time of the operations executed on the i-th resource along the j-th process,

T – periodicity of the system of concurrently executed cyclic processes.

Subtracting equation iii) from equation ii) the resulting equation has the form:

$$y \cdot t_{C2} + y \cdot t_{B2} + x \cdot t_{C1} - z \cdot t_{A3} - z \cdot t_{E3} - x \cdot t_{E1} = 0$$

and after adding it to equation i), the resultant formulae has the form:

$$y \cdot (2 \cdot t_{C2} + t_{B2}) + x \cdot (2 \cdot t_{C1} + t_{F1}) - z \cdot t_{A3} = T.$$

Consequently, the obtained equation:

$$y \cdot N + x \cdot M = T + z \cdot K$$

is Diophantine equation, where $N = M = 3$ and $K = 1$ under the following assumption $t_{C1} = t_{E1} = t_{F1} = t_{C2} = t_{B2} = t_{D2} = t_{D3} = t_{A3} = t_{E3} = 1$, i.e. it takes the following form: $3y + 3x = T + z$. First three solutions of this equation are shown in Table 1.

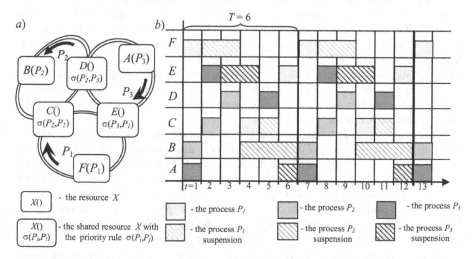

Fig. 5. System of concurrently flowing cyclic processes: a) initial state; b) Gantt's chart encompassing the cyclic steady state

Table 1. First three solutions of the Diophantine equation: $3y + 3x = T + z$

T	x	y	z
5	1	1	1
6	1	2	3
10	2	2	2

3 Diophantine Problems

A Diophantine equation (named in honour of the 3rd-century Greek mathematician Diophantus of Alexandria) is an indeterminate polynomial equation that allows the variables to be integers only, i.e. an equation involving only sums, products, and powers in which all the constants are integers and the only solutions of interest are integers [12], [13]. For example, $3x + 7y = 1$ or $x^2 - y^2 = z^3$, where x, y, and z are integers (see Pythagoras' theorem or Fermat's last theorem). Apart from polynomial, the exponential Diophantine equations where an additional variable or variables occur as exponents, can be considered [9], e.g., Ramanujan-Nagell equation, $2^n - 7 = x^2$.

Therefore, the Diophantine problems have fewer equations than unknown variables and involve finding integers, which satisfy all equations. The following questions belong to the most frequently asked: Are there any solutions? Are there any solutions beyond some that are easily found by inspection? Are there finitely or infinitely many solutions? Can all solutions be found, in theory? Can one in practice compute a full list of solutions?

The solvability of all Diophantine problems proposed in 1900 by David Hilbert and known as Hilbert's 10th problem, has been settled negatively [6], [9]. It has been proved that no general algorithm exists for determining whether a Diophantine equation is soluble [15]. Such an algorithm does exist for the solution of first-order equations, in general case, however, Diophantine problems are unsolvable.

In that context, Diophantine equations fall into three classes: those with no solutions, those with only finitely many solutions, and those with infinitely many solutions. For example, the equation $6x - 9y = 29$ has no solutions, but the equation $6x - 9y = 30$, which upon division by 3 reduces to $2x - 3y = 10$, has infinitely many. In case of $x^2 + y^2 = z^2$, infinite set of solutions (3, 4, 5), (5, 12, 13), (7, 24, 25), (9, 40, 41),..., exists, however, in case of $x^3 + y^3 = z^3$ the relevant set is empty. In turn, $x = 20$, $y = 10$ is a solution, and so is $x = 20 + 3t$, $y = 10 + 2t$ for every integer t, positive, negative, or zero. This is called a one-parameter family of solutions, with t being the arbitrary integer parameter.

In order to analyze polynomial Diophantine equations, the few general approaches, i.e. based on *Hasse principle* and *infinite descent* method, are usually employed. In case of exponential Diophantine equations, however, a similar general theory is not yet available so ad hoc or trial and error methods are used so far. Consider a given set of Diophantine equations modelling the system under study. Assuming the set of variables and Diophantine equations (polynomial or exponential) encompass the system's structure while the set of solutions following given features of system functioning as system's behavior, one can ask the following questions:

Does a control procedure exist enabling to guarantee an assumed system behavior subject to system's structure constraints?

Does a system's structure exist such that an assumed system behavior can be achieved?

Therefore, taking into account non decidability of Diophantine problems one can easily realize that not all behaviors can be obtained under constraints imposed by system's structure. The similar observation concerns the system's behavior that can be achieved in systems possessing specific structural characteristics. That means, the exhaustive searching for assumed control in the system at hand can be replaced by designing of a system aimed at the behavior requested.

4 Timetabling

As already mentioned, timetabling understood as planning classes can be seen as an example of cyclic scheduling problem. The case considered below, belonging to a class of multi-criteria decision problems, can also be modeled as a Diophantine problem.

Consider the set of lectures $P = \{A_1, B_1, A_2, B_2\}$. Lectures A_1 and A_2 belong to group A, and lectures B_1 and B_2 belong to group B. Given are two class teams E_1 and

E_2. To team E_1, 5 units of time are assigned regarding lectures for group A (including A_1 and A_2) and 4 units of time for group B (including B_1 and B_2). To team E_2, 8 units of time are assigned regarding lectures for group A and 6 units of time for group B. Moreover, there are periods limiting class duration in the respective teams. So, in case of team E_1, the period limiting lecture A_1 is $t_{1,A1} = 2$; in case of B_1, $t_{1,B1} = 2$; in case of A_2, $t_{1,A2} = 3$; and B_2, $t_{1,B2} = 2$; the relevant duration times are included in the sequence $T_1 = (2,2,3,2)$. In case of team E_2, the relevant duration times are denoted similarly and are included in the sequence $T_2 = (4,2,4,4)$. That means:

$$t_{1,A1}+t_{1,A2} = 5; \quad t_{1,B1}+t_{1,B2} = 4; \quad t_{2,A1}+t_{2,A2} = 8; \quad t_{2,B1}+t_{2,B2} = 6. \tag{2}$$

Classes A_1 and B_1 are conducted by the lecturer L_1, while A_2 and B_2 by lecturer L_2. The timetable sought should be free of gaps both for teams as well as for lecturers. In other words, we are looking for permutations Q_1, Q_2 encompassing gap-free timetables:

$$
\begin{aligned}
Q_1 &= (q_{1,1}, q_{1,2}, q_{1,3}, q_{1,4}); \quad q_{1,1} \neq q_{1,2} \neq q_{1,3} \neq q_{1,4}; \quad q_{1,1}, q_{1,2}, q_{1,3}, q_{1,4} \in P; \\
Q_2 &= (q_{2,1}, q_{2,2}, q_{2,3}, q_{2,4}); \quad q_{2,1} \neq q_{2,2} \neq q_{2,3} \neq q_{2,4}; \quad q_{2,1}, q_{2,2}, q_{2,3}, q_{2,4} \in P.
\end{aligned}
\tag{3}
$$

Permutations Q_1, Q_2 do not allow the cases where the same classes A_1, A_2, B_1, B_2 are simultaneously conducted for different teams. That means, in case of A_1, for instance, the terms of this class beginning, i.e. $x_{1,A1}, x_{2,A1}$, in different teams, i.e., E_1, E_2 have to follow the formulae (4):

$$(x_{1,A1} + t_{1,A1} \leq x_{2,A1}) \vee (x_{2,A1} + t_{2,A1} \leq x_{1,A1}). \tag{4}$$

Therefore, course A_1 conducted for E_1 will either precede or succeed its repetition for E_2. Similar formulae can be considered for B_1, A_2, B_2:

$$
\begin{aligned}
(x_{1,B1} + t_{1,B1} \leq x_{2,B1}) &\vee (x_{2,B1} + t_{2,B1} \leq x_{1,B1}), \\
(x_{1,A2} + t_{1,A2} \leq x_{2,A2}) &\vee (x_{2,A2} + t_{2,A2} \leq x_{1,A2}), \\
(x_{1,B2} + t_{1,B2} \leq x_{2,B2}) &\vee (x_{2,B2} + t_{2,B2} \leq x_{1,B2}).
\end{aligned}
\tag{5}
$$

Moreover, analogously similar conditions can be formulated for lecturers L_1, L_2. That is, in case of L_1, for instance, the terms of his/her duties beginning $x_{1,A1}, x_{2,B1}, x_{2,A1}, x_{1,B1}$ and concerning classes A_1 and B_1, have to follow formulae (6):

$$
\begin{aligned}
(x_{1,A1} + t_{1,A1} \leq x_{2,B1}) &\vee (x_{2,B1} + t_{2,B1} \leq x_{1,A1}), \\
(x_{2,A1} + t_{2,A1} \leq x_{1,B1}) &\vee (x_{1,B1} + t_{1,B1} \leq x_{2,A1}).
\end{aligned}
\tag{6}
$$

Analogously, similar conditions can be formulated for lecturer L_2. So, assuming that lecturer L_2 conducts the courses A_2, B_2, the relevant formulae are as follows:

$$
\begin{aligned}
(x_{1,A2} + t_{1,A2} \leq x_{2,B2}) &\vee (x_{2,B2} + t_{2,B2} \leq x_{1,A2}), \\
(x_{2,A2} + t_{2,A2} \leq x_{1,B2}) &\vee (x_{1,B2} + t_{1,B2} \leq x_{2,A2}).
\end{aligned}
\tag{7}
$$

In particular, the guarantee the timetable sought will gap-free requires that the beginnings $x_{1,A1}, x_{1,B1}, x_{1,A2}, x_{1,B2}, x_{2,A1}, x_{2,B1}, x_{2,A2}$ of lectures A_1, A_2, B_1, B_2 have to follow lecture order constrains assumed by permutations Q_1 and Q_2. That means the following equations have to hold:

$$
\begin{aligned}
x_{1,q_{1,1}} \cdot x_{2,q_{2,1}} &= 0, \\
x_{1,q_{1,i}} &= x_{1,q_{1,i-1}} + t_{1,q_{1,i-1}} \quad \text{for } i = 2,3,4, \\
x_{2,q_{2,i}} &= x_{2,q_{2,i-1}} + t_{2,q_{2,i-1}} \quad \text{for } i = 2,3,4,
\end{aligned}
\tag{8}
$$

where:

$x_{1,q_{1,1}}$ – denotes a term of class beginning for team E_1 related to element $q_{1,1} \in P$ defined in permutation $Q_1(3)$,

$t_{1,q_{1,1}}$ – denotes a class duration time for team E_1 related to element $q_{1,1} \in P$ defined in permutation $Q_1(3)$.

The above conditions guarantee the timetable for E_1, E_2 is gap-free. Similar equations have to hold for lecturers L_1, L_2. Therefore, in case of lecturer L_1, the gap- free timetable requires the following equation holds:

$$
\begin{aligned}
t_{1,A1} + t_{2,B1} &+ t_{2,A1} + t_{1,B1} = \\
= \max \big\{ &\big|x_{1,A1} - t_{2,B1} - x_{2,B1}\big|, \big|x_{1,A1} - t_{2,A1} - x_{2,A1}\big|, \big|x_{1,A1} - t_{1,B1} - x_{1,B1}\big|, \quad (9)\\
&\big|x_{2,B1} - t_{2,A1} - x_{2,A1}\big|, \big|x_{2,B1} - t_{1,B1} - x_{1,B1}\big|, \big|x_{2,A1} - t_{1,B1} - x_{1,B1}\big|\big\},
\end{aligned}
$$

and for lecturer L_2:

$$
\begin{aligned}
t_{1,A2} + t_{2,B2} &+ t_{2,A2} + t_{1,B2} = \\
= \max \big\{ &\big|x_{1,A2} - t_{2,B2} - x_{2,B2}\big|, \big|x_{1,A2} - t_{2,A2} - x_{2,A2}\big|, \big|x_{1,A2} - t_{1,B2} - x_{1,B2}\big|, \quad (10)\\
&\big|x_{2,B2} - t_{2,A2} - x_{2,A2}\big|, \big|x_{2,B2} - t_{1,B2} - x_{1,B2}\big|, \big|x_{2,A2} - t_{1,B2} - x_{1,B2}\big|\big\}.
\end{aligned}
$$

Finally, the following equations have to hold in order to guarantee the gap-free timetable:

$$
\begin{cases}
t_{1,A1} + t_{1,A2} = 5 \\
t_{1,B1} + t_{1,B2} = 4 \\
t_{2,A1} + t_{2,A2} = 8 \\
t_{2,B1} + t_{2,B2} = 6 \\
E\big((x_{1,A1} + t_{1,A1} \leq x_{2,A1}) \vee (x_{2,A1} + t_{2,A1} \leq x_{1,A1})\big) = 1 \\
E\big((x_{1,B1} + t_{1,B1} \leq x_{2,B1}) \vee (x_{2,B1} + t_{2,B1} \leq x_{1,B1})\big) = 1 \\
E\big((x_{1,A2} + t_{1,A2} \leq x_{2,A2}) \vee (x_{2,A2} + t_{2,A2} \leq x_{1,A2})\big) = 1 \\
E\big((x_{1,B2} + t_{1,B2} \leq x_{2,B2}) \vee (x_{2,B2} + t_{2,B2} \leq x_{1,B2})\big) = 1 \\
E\big((x_{1,A1} + t_{1,A1} \leq x_{2,B1}) \vee (x_{2,B1} + t_{2,B1} \leq x_{1,A1})\big) = 1 \\
E\big((x_{2,A1} + t_{2,A1} \leq x_{1,B1}) \vee (x_{1,B1} + t_{1,B1} \leq x_{2,A1})\big) = 1 \\
E\big((x_{1,A2} + t_{1,A2} \leq x_{2,B2}) \vee (x_{2,B2} + t_{2,B2} \leq x_{1,A2})\big) = 1 \\
E\big((x_{2,A2} + t_{2,A2} \leq x_{1,B2}) \vee (x_{1,B2} + t_{1,B2} \leq x_{2,A2})\big) = 1 \\
x_{1,q_{1,1}} \cdot x_{2,q_{2,1}} = 0 \\
x_{1,q_{1,i}} = x_{1,q_{1,i-1}} + t_{1,q_{1,i-1}} \quad \text{for } i = 2,3,4 \\
x_{2,q_{2,i}} = x_{2,q_{2,i-1}} + t_{2,q_{2,i-1}} \quad \text{for } i = 2,3,4 \\
t_{1,A1} + t_{2,B1} + t_{2,A1} + t_{1,B1} = \max\{R_1\} \\
t_{1,A2} + t_{2,B2} + t_{2,A2} + t_{1,B2} = \max\{R_2\}
\end{cases}
\qquad (11)
$$

where:

$$
\begin{aligned}
R_1 = \big\{ &\big|x_{1,A1} - t_{2,B1} - x_{2,B1}\big|, \big|x_{1,A1} - t_{2,A1} - x_{2,A1}\big|, \big|x_{1,A1} - t_{1,B1} - x_{1,B1}\big|, \\
&\big|x_{2,B1} - t_{2,A1} - x_{2,A1}\big|, \big|x_{2,B1} - t_{1,B1} - x_{1,B1}\big|, \big|x_{2,A1} - t_{1,B1} - x_{1,B1}\big|\big\}, \\
R_2 = \big\{ &\big|x_{1,A2} - t_{2,B2} - x_{2,B2}\big|, \big|x_{1,A2} - t_{2,A2} - x_{2,A2}\big|, \big|x_{1,A2} - t_{1,B2} - x_{1,B2}\big|, \\
&\big|x_{2,B2} - t_{2,A2} - x_{2,A2}\big|, \big|x_{2,B2} - t_{1,B2} - x_{1,B2}\big|, \big|x_{2,A2} - t_{1,B2} - x_{1,B2}\big|\big\}.
\end{aligned}
$$

Of course, the above delivered set of equations can be seen as typical set of nonlinear Diophantine equations. So, the question is whether it is decidable or not. Consequently, assuming that classes A_1, B_1 are conducted by lecturer L_1, while classes A_2, B_2 conducted by L_2, the sought timetable should guarantee time windows free schedule, both, for all teams and all lectures. The Oz Mozart [11] implementation of that problem results in response: Lack of solutions.

It follows, from the exhaustive search of all possible solutions, that all available schedules possess either at least one free time window, in class of teams, or in class of lectures.

Lack of feasible solutions still provides an opportunity to formulate so called reverse scheduling problem, i.e. the problem where besides classes of orders searched there are their periods guaranteeing free time window schedules for both teams of students and lecturers. For illustration, let us consider the case assuming the periods of particular classes are unknown while the total sum of time units, devoted to the classes, remains the same as in the previous case, i.e.:

$$t_{1,A1}+t_{1,A2} = 5, \text{ (5 time units for group } A \text{ of team } E_1);$$
$$t_{1,B1}+t_{1,B2} = 4, \text{ (4 time units for group } A \text{ of team } E_1);$$
$$t_{2,A1}+t_{2,A2} = 8, \text{ (8 time units for group } A \text{ of team } E_2); \qquad (12)$$
$$t_{2,B1}+t_{2,B2} = 6, \text{ (6 time units for group } A \text{ of team } E_2).$$

The considered question concerns: does such set of periods included in sequences T_1, T_2 as well as permutations of classes exist guaranteeing that the schedules of student teams and lecturers do not contain free time windows? The Oz Mozart implementation of this problem results in the following: the periods of classes corresponding to the teams E_1, E_2 are the following ones $T_1 = (3,3,2,1)$, $T_2 = (4,4,4,2)$. In case of team E_1, the sequence of classes is as follows (A_1, B_1, A_2, B_2), while in case of E_2, the sequence of classes is (A_2, B_2, A_1, B_1). The solution obtained is illustrated in Fig. 6.

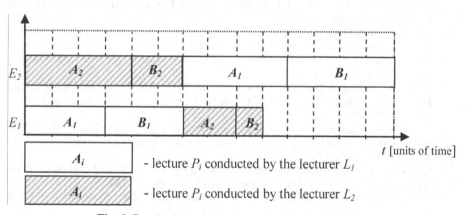

Fig. 6. Gantt's chart without lecturer's free time window

It should be noted that the solution obtained, i.e. $T_1 = (3,3,2,1)$, $T_2 = (4,4,4,2)$ can be seen as a set of sufficient conditions guaranteeing required solution of the two criteria decision problem. Also the relationship between structure of scheduling problem (i.e. number of teams, classes, periods, and so on) and its possible solutions encompassing the required system behavior (encompassing demanded features, e.g. lack of free time windows) can be easily observed. Consequently, it becomes quite obvious that not all such behavior can be achieved in any system structure.

Therefore, the reverse problem formulation allows one to search for sufficient conditions guaranteeing the required demands (we are looking for) hold. Due to the abductive paradigm assumed, the conditions, i.e. values of variables, are sought for arbitrarily chosen subsets of decision variables. Effectiveness of such approach has been examined on an projects portfolio prototyping case [1].

5 Concluding Remarks

The discussion presented explains why some timetabling like problems have an non decidable character. Diophantine models of sequential scheduling problems impose necessity for development of sufficient conditions guaranteeing the solvability of problems considered. Since Diophantine equations can be treated as a set of constraints, the constraint programming [1], [2], [3] i.e. languages descriptive in their character, can be directly implemented.

The alternative approach expresses itself in the possibility of reverse formulation of a timetabling problem. In both cases, i.e. in case of sufficient conditions development and strait, and reverse timetabling solution, the constraint programming environment seems to be well suited.

References

1. Bach, I., Bocewicz, G., Banaszak, Z.: Constraint Programming Approach to Time-Window and Multi-Resource-Constrained Projects Portfolio Prototyping. In: Nguyen, N.T., Borzemski, L., Grzech, A., Ali, M. (eds.) IEA/AIE 2008. LNCS (LNAI), vol. 5027, pp. 767–776. Springer, Heidelberg (2008)
2. Bocewicz, G., Banaszak, Z., Wójcik, R.: Design of Admissible Schedules for AGV Systems with Constraints: a Logic-Algebraic Approach. In: Nguyen, N.T., Grzech, A., Howlett, R.J., Jain, L.C. (eds.) KES-AMSTA 2007. LNCS (LNAI), vol. 4496, pp. 578–587. Springer, Heidelberg (2007)
3. Barták, R.: Incomplete Depth-First Search Techniques: A Short Survey. In: Figwer, J. (ed.) Proceedings of the 6th Workshop on Constraint Programming for Decision and Control, pp. 7–14 (2004)
4. Birger, R., El-Houssaine, A., Wout, D.: Cyclic Scheduling of Multiple Tours with Multiple Frequencies for a Single Vehicle. International Journal of Logistics Systems and Management 5(3/4), 214–227 (2009)
5. Cai, X., Li, K.N.: A Genetic Algorithm for Scheduling Staff of Mixed Skills under Multi-criteria. European Journal of Operational Research 125, 359–369 (2000)
6. Davis, M.: Hilbert's Tenth Problem is Unsolvable. Amer. Math. Monthly 80, 233–269 (1973)
7. Ernst, A.T., Jiang, H., Krishnamoorthy, M., Owens, B., Sier, D.: An Annotated Bibliography of Personnel Scheduling and Rostering. Annals of Operations Research 127, 21–144 (2009)
8. Gaujal, B., Jafari, M., Baykal-Gursoy, M., Alpan, G.: Allocation Sequences of Two Processes Sharing a Resource. IEEE Trans. on Robotics and Automation 11(5), 353–748 (1995)

9. Jones, J.P., Matiyasevich, Y.V.: Exponential Diophantine Representation of Recursively Enumerable Sets. In: Proceedings of the Herbrand Symposium, Marseilles, pp. 159–177. North-Holland, Amsterdam (1982)
10. Pinedo, M.L.: Planning and Scheduling in Manufacturing and Services. Springer, New York (2005)
11. Schutle, H., Smolka, G., Wurtz, J.: Finite Domain Constraint Programming in Oz. German Research Center for Artificial Intelligence, Germany, D-66123 Saarbrucken (1998)
12. Sprindzuk, V.G.: Classical Diophantine Equations. Lecture Notes in Mathematics, vol. 1559. Springer, Berlin (1993)
13. Smart, N.P.: The Algorithmic Resolution of Diophantine Equations. In: London Mathematical Society Student Text, vol. 41. Cambridge University Press, Cambridge (1998)
14. Heo, S.-K., Lee, K.-H., Lee, H.-K., Lee, I.-B., Park, J.H.: A New Algorithm for Cyclic Scheduling and Design of Multipurpose Batch Plants. Ind. Eng. Chem. Res. 42(4), 836–846 (2003)
15. Tan, B., Karabati, S.: Stochastic Cyclic Scheduling Problem in Synchronous Assembly and Production Lines. Journal of the Operational Research Society 49(11), 1173–1187 (1998)

Combination of Two KM Strategies* by Web 2.0

Quoc Trung Pham[1] and Yoshinori Hara[2]

[1] Graduate School of Economics, [2] Graduate School of Management
Kyoto University, Yoshida-Honmachi, Sakyo-ku, Kyoto 606-8501, Japan
`pham.trung@kt4.ecs.kyoto-u.ac.jp`, `hara@gsm.kyoto-u.ac.jp`

Abstract. The previous approaches for KM focused on either of two main strategies: Codification or Personalization (which focused on explicit or tacit knowledge, respectively). Those approaches lead to serious problems, such as: difficulty for integration, inconvenience in knowledge accessibility, limitation of participation, disjunction of knowledge... The approach for solving those problems of this paper is to set up a new integrated platform for combining both strategies. That platform is based on web 2.0, a set of advanced technologies and a new approach to knowledge management. Using web 2.0 for knowledge management, both kinds of knowledge can be supported at the same time and knowledge creating cycle is also facilitated. In this paper, a KMS based on web 2.0 is specified and a demo system is implemented for testing this solution, which shows many advantages in comparison with previous KMS.

Keywords: KMS, KM Strategy, Explicit, Tacit, Web 2.0, Social Network.

1 Introduction

Nowadays, in the knowledge age, knowledge management is considered the best strategy to improve the performance and to strengthen the competitive capability of any enterprise. However, the implementation of KMS in practice is not so successful because of many reasons. The most important reason is that the enterprises have to choose to follow either of the two main KM strategies: Codification and Personalization. Recently, there are some efforts to balance those two strategies, but they meet another obstacle: lack of a suitable platform for the combination.

According to Hanssen et al. [7], there are two main strategies for KM:

♦ Codification: to systematize and store information that constitutes the company's knowledge, and to make this available to people in the company.
♦ Personalization: to support the flow of information in a company by having a centralized store of information about knowledge sources, like a "yellow pages" of who knows what in a company.

Codification strategy focus on explicit knowledge, while Personalization focus on tacit knowledge. Following just one strategy is proven to be not successful in reality, but combining those two strategies at a properly rate is considered a good solution.

---

* Two main KM strategies: Personalization and Codification.

D. Karagiannis and Z. Jin (Eds.): KSEM 2009, LNAI 5914, pp. 322–334, 2009.

According to Hanssen et al., most of companies that use knowledge effectively pursue one strategy predominantly (80%) and use the second strategy to support the first (20%). This showed the imbalance in utilizing two kinds of knowledge in problem solving and decision making, which required both of tacit and explicit knowledge. Can both kind of knowledge be supported equally at the same time? This can be done only if there is a suitable technology for KMS, which can support both kinds of knowledge. Using a suitable platform for knowledge management, both KM strategies can be followed at the same time. This solution ensures the success of KMS implementation and improves the overall performance of the enterprise.

Recently, Web 2.0, which is famous for social networking, is not only ideal platform for creating cooperative knowledge, but also for sharing and communicating between users. It could be used to create a web-based KMS, where explicit knowledge could be searched from a codified knowledge-base and available experts could be found and communicated with to find solutions for a problem. Such platform can satisfy requirements of both strategies.

Therefore, the purpose of this paper is exploring the ability to use web 2.0 in combining two KM strategies and suggesting a solution for KMS implementation based on web 2.0. This paper is organized as follows: (2) Problems of previous knowledge management approaches; (3) Approach for solving those problems; (4) Apply web 2.0 for building a KMS; (5) Demonstration and evaluation of the KMS based on web 2.0; (6) Conclusion and implications.

2 Problems of Previous Knowledge Management Approaches

2.1 Typical Techniques/Approaches of Knowledge Management

Traditional techniques and approaches for KM could be shown as follows:

Table 1. KM processes and the potential role of IT [1]

KM processes	Knowledge Creation	Knowledge Storage/ Retrieval	Knowledge Transfer	Knowledge Application
Supporting IT	Data mining Learning tools	E-bulletin boards Knowledge repository Databases	E-bulletin boards Knowledge directories	Expert systems Workflow systems
Platform technologies	Groupware and communication technologies Intranets/ Web 1.0			

Table 2. Previous Typical Knowledge Management approaches [4]

System	Description	Strategy
Expert Finder	Expert finder is a tool that can be used to identify and contact the relevant expert in problem solving process. It is usually build on a database in which the expert's profile including a description of his knowledge domain stored.	Personalization
Virtual Teamwork	Virtual teamwork refers to work conducted by a team consisting of geographically dispersed members usually enabled through IT support.	

Table 2. (*continued*)

Lessons Learned Database	Lessons learned refers to a concept that deals with the documentation and preservation of positive as well as negative experiences which occurred in practical situation in problem solving process in order to ensure their reutilization respectively their future prevention.	Codification
Case Based Reasoning	Case based reasoning (CBR) is a concept describing a problem solving process based upon the solutions of previous similar problems. For that reason a database is created with past cases including respective solution. Those cases could be searched & adapted later for similar situations.	
Virtual/ Augmented Reality	Virtual reality (VR) refers to an environment simulated by computers as a general term whereas augmented reality describes the use of see-through displays to enable computer generated images to be overlaid on top of the user's view of the real world. This technology can be used to support or assist human actors in certain kinds of knowledge processes through visualization of e.g. instructions.	

2.2 Significant Problems of Previous KM Techniques/Approaches

Problems of previous techniques for KM could be as follows:

A. *Diversity data sources*: previous techniques use independent data sources (database, data store, data warehouse), which is hardly to be converted or transfer through various phases of knowledge processing. This leads to the difficulty in sharing knowledge throughout the enterprise.

B. *Difficulty for integration*: previous techniques are not suitable for integrating many kinds of business application services, which is required in building an effective KMS. It is difficult to integrate many services with various purposes into the same environment without vast changes in structure and data.

Problems of previous approaches for KM could be as follows:

C. *Inconvenience in knowledge accessibility*: knowledge accessibility is not convenient to all of employees. For example, Expert Finder lets employee access to knowledge through experts, who are not always available; Virtual Teamwork through cooperating of group's member in problem solving; Case Base and Lesson Learned through database of previous situations; Virtual Reality let people access to knowledge for training only.

D. *Limitation of participant*: previous approaches do not allow broadly participation of employees in knowledge accumulation process. For example, above approaches accumulate knowledge from a few experts, employees, who are selected for problem solving. Moreover, some democratic mechanism for getting public feedback, such as: voting, rating…, is not used frequently.

E. *Disjunction of knowledge*: Expert finder and Virtual Teamwork follow Personalization strategy, which is only focus on tacit knowledge located inside experts' head; while other approaches follow Codification strategy, which is only concentrate on explicit knowledge. In fact, tacit knowledge and explicit knowledge are both required in problem solving or decision making.

3 Approach for Solving Those Problems

3.1 Necessity of Combination and Integration for Solving Those Problems

The main cause of those problems is in the strategic aspect of those approaches. There exists a trade-off strategy of focusing on either tacit knowledge or explicit knowledge. It is shown in the division between two main KM strategies: Personalization and Codification. Personalization strategy aims to handle tacit knowledge by locating someone who knows, point the knowledge seeker to that person, and encourage them to interact. In contrast, Codification strategy focuses on explicit knowledge by capturing and codifying knowledge for reusing.

Although in some cases or in a short period, tacit knowledge seems to be more important than explicit knowledge or vice versa, but in whole enterprise context or in a long period, a combination of both kind of knowledge will increase the effectiveness of business works and bring back long-term benefits. Therefore, tacit knowledge and explicit knowledge should be considered equally in KM strategy of any enterprise. According to Choi et al. [3], integrating explicit-oriented with tacit-oriented KM strategies is found to result in better performance of the enterprise.

Besides, according to Prof. Nonaka [8], knowledge creation is a spiraling process of interactions between explicit and tacit knowledge. Based on Nonaka's SECI model, two main strategies of knowledge management are realized to focus on two different parts of knowledge creating cycle. Codification focused on Externalization and Combination, whereas Personalization focused on Internalization and Socialization. This separation is the main reason for unsuccessful KMS implementation because a complete knowledge creating cycle is not established. Therefore, the combination of two main strategies of KM is very necessary in building a successful KMS.

However, if enabling techniques for KMS could not be integrated, the flow of information, knowledge between processes and between people could not be established. Therefore, it will prevent the interaction and combination between tacit and explicit knowledge. In order to make this combination success in practice, a suitable platform for integrating data, information, knowledge and applications is very important.

Moreover, the vast change of technologies for KM today requires a unique platform for integrating various types of information, tools, and applications in building a stable KMS. In the globalization, a need for transferring knowledge throughout the world will increase, so that a suitable technical platform for integration of many kinds of information and knowledge is a requisite for the success of any KM strategy.

3.2 Approach for Solving Those Problems

From above analysis, there are two main causes for those problems: disjunction of knowledge and disintegration of technology. In order to solve those problems, a platform is needed to integrate various data sources/ applications and to support both kinds of knowledge. In this paper, web 2.0 is selected as a solution because:

-   It can be used for integrating data, information and applications.
-   It encourages users' participation and contribution.
-   It is suitable for sharing both kinds of knowledge (tacit and explicit).

Besides, according to a paper of O'Donovan et al. [9] about the key success factors in KMS implementation, three main factors of IT which have most effects on the success of KMS are: Ease of use, Security and Openness, and User involvement. These requirements can be satisfied easily by using technical platform based on web 2.0.

According to Smith [10], the three anchor points for Web 2.0 are:

- ◆ *Technology and architecture*: including the infrastructure of the Web and the concept of Web platforms. Examples of specific technologies include Ajax, representational state transfer (REST) and Really Simple Syndication (RSS).
- ◆ *Community and social*: looks at the dynamics around social networks, communities and other personal contents publish/share models, wikis and other collaborative content models.
- ◆ *Business and process*: Web services-enabled business models and mashup[1]/ remix applications. Examples: advertising, subscription models, software as a service (SaaS), and long-tail economics. A well-known specific example is connecting a rental-housing Web site with Google Maps to create a more useful service that automatically shows the location of each rental listing.

The first two anchor points show that web 2.0 can be used as a platform for integration and a social network for attracting users' participation. The last anchor point shows that web 2.0 can be used as a business model, but how it can support the combination of tacit and explicit knowledge is not known. In order to prove that web 2.0 can support both kinds of knowledge, the following case studies about current outstanding web 2.0 services are used. Most of those services can support both of Codification strategy (capture and reuse explicit knowledge) and Personalization strategy (handle tacit knowledge by communication and collaboration).

Table 3. Web 2.0 as a platform for supporting Codification and Personalization

Service	Codification supporting	Personalization supporting
Wikipedia (wikipedia.org)	- This online encyclopedia is the work of the crowd. - It is the biggest encyclopedia in the world. - Anyone can edit any item. - Any work can be rolled back. - The knowledge quality is controlled by a set of rules - The success of this service is on user-participation.	- This environment could be used for creating a list of experts in each subject - Discussion is allowed for participators to exchange ideas
Delicious (del.icio.us)	- Bookmarks for all useful resources on the internet. - A live storage of cooperative knowledge. - Easy for searching a topic because of folksonomy. - Easy to add, edit a bookmark and share with other people.	- Bookmarks to experts based on their professional. - Run add-in applications for communicating with people in the community.
YouTube (youtube.com)	- This video library is a very useful resource on web. - Number of uploaded videos increases day by day. - Video is uploaded, viewed and downloaded easily. - Videos are classified into categories and channels. - Quality of videos based on rate, view and comment. - Quantity and quality of video attract more or less user.	- Users can subscribe for any channel that he/ she likes. - Commenting and rating are ways for communicating or discussing between users. - By the time, users with good channel will attract more people. They could be experts in a subject.

---

[1] Mashup - a Web site or Web application that combines content from more than one source.

Table 3. (*continued*)

GoogleMap (maps.google.com)	- A useful tool for searching any location in the world map. - Users can create their own map, mark, take note and draw on that map... - Map could be view in many forms, and could be attached with pictures, descriptions... - This service could be combined with other services, such as: real estate, traffic...	- User's map can be shared with friends as directions. - Especially useful in finding an expert in real world. - Expert network could be created with links to their homepage, their knowledge or online presentation to communicate with.
FaceBook (facebook.com)	- Collection of profiles, pictures, notes, friends, videos, events, links and so on. - Allow add-in applications to be run and shared. - Easy to keep up with any change from friends by RSS, e-mail. - Support for trace of log in, comment, message...	- Ideal environment for friend making and socializing. - Support mailing and chat system for communication between users. - Easy to search for friends based on their profiles.
YahooAnswer (answers.yahoo.com)	- Questions and answers are accumulated through the time by user participation. - Quality of answers based on rating and pointing system. - Questions are classified into categories for searching later. - Anyone can answer any question in allowed period.	- User can ask for more detail of a question by giving many answers. - Rating is the feedback information for an answer. - User with high rank and good answers will get more appreciate from other users in community.
Blogspot (blogspot.com)	- Allow bloggers to write anything they like. - Easy to embed picture, audio and video. - Accept comment for each blog entry or whole blog. - Tag system help searching easier. - Bookmarks are also used to save useful links.	- Store information about blogger for friend searching. - Easy to follow a blogger, make friend and contact directly.

However, web 2.0 services are different from a KMS in following aspects:

Table 4. Web 2.0 services vs. Knowledge management system

	Web 2.0 services	Knowledge management system
Function	Independent services	Integrated functions/ processes
Focus	Individual content	Organizational knowledge
Purpose	General purpose	Specific purpose for KM/ problem solving
Participants	Public	Staffs, Experts, Partners
Data quality	Unverified and low quality	Verified and high quality
Motivation	Rich content, Large community	Ease of use, Effectiveness

4   Apply Web 2.0 for Building a KMS

4.1   Solutions for Problems in (2.2) Based on Web 2.0

Based on above analysis, web 2.0 is realized to be able to play an important role in solving problems of KMS implementation by providing an integrated environment for both KM strategies. But web 2.0 only cannot satisfy requirements of a KMS, so that, a new KMS based on web 2.0 is needed to solve problems of KMS implementation by taking advantages of web 2.0. In fact, using web 2.0, all problems of KMS implementation in (2.2) can be solved completely as shown in the following table:

Table 5. Web 2.0 as a solution for KMS implementation problems

Problem	Solution
A. Diversity data sources	XML technology, core of web 2.0, could be a solution for this problem. All kind of data source now can be converted and transferred easily on web 2.0 through XML technology. Besides, semantic web also make data become more understandable and sharable between many systems and users.
B. Difficulty for integration	Web 2.0 platform allows many software applications to be integrated easily by using IDE (integrated development environment) and a lot of open source toolkits. With mash-up environment, user can do many things, such as: mail check, instant chat, view stock quotes, view temperature or read newspaper... on the same web screen.
C. Inconvenience in knowledge accessibility	With web 2.0, employees through out the enterprise can access to knowledge source from anywhere and at anytime to use it in problem solving as well as get benefits from learning general knowledge. They can also contribute their own knowledge to open problems. Knowledge can be classified in tags system and displayed in visual forms.
D. Limitation of participant	Web 2.0 requires the participation of many people. The more participation, the more valued knowledge could be gained. Wiki environment allows people to read and write about anything as well as discuss and rate on other people's writings. In this environment, everyone could be a potential expert.
E. Disjunction of knowledge	Web 2.0 environment is very suitable for the combination of both strategies of knowledge management. It focuses on both tacit knowledge (communicating, socializing, cooperating...) and explicit knowledge (stored experiences, cases, guides...).

4.2 Combining Two Main KM Strategies Based on Web 2.0

Not only solving problems of previous KM approaches, web 2.0 can also help facilitating knowledge creating cycle as in following concept model:

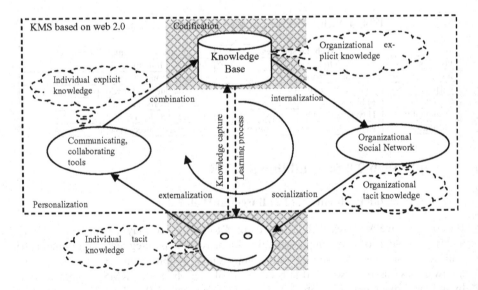

Fig. 1. Concept model of a KMS combining both KM strategies by web 2.0

In this model, web 2.0 can contribute to KMS implementation in two aspects:

- Create knowledge base for knowledge capturing and sharing (supports Codification strategy by storing explicit knowledge for searching and reusing).
- Create environment for collaboration, communication and social networking (supports Personalization strategy by allowing employees to access tacit knowledge of other experts or employees).

From above figure, a combination strategy is suggested based on web 2.0, in which both tacit knowledge and explicit knowledge are supported at a balanced rate. Besides, through communicating tools and organizational social networks, a complete knowledge cycle is established: individual tacit knowledge -> individual explicit knowledge -> organizational explicit knowledge -> organizational tacit knowledge.

An example of knowledge creating cycle is facilitated through this system:

- Leaders' intentions for operational improvement or new problems that need to be solved are published on organizational social network (notice board, blog's stories, online and offline meetings...) => this socialization process converts organizational tacit knowledge into individual tacit knowledge.
- Employees' response to those intentions or problems by asking for more information, telling their ideas or comments, submitting relating analysis, report... => this externalization process converts individual tacit knowledge into individual explicit knowledge.
- Through communication and collaboration tools, those explicit knowledge are classified, arranged, and stored in a logical order for referring, discussing and choosing final solution for initial intentions or problems => this combination process turns individual explicit knowledge to org. explicit knowledge.
- Using organizational explicit knowledge of previous cases in knowledge base as lessons for other divisions throughout the enterprise (through assessment, award system, blog's stories...) could inspire new intentions or find new similar problems to be solved based on previous cases => this internalization process converts org. explicit knowledge into org. tacit knowledge.

Moreover, individual tacit knowledge (of experts/ employees) can be codified and stored in knowledge base directly through knowledge capture process, and employees can also get benefits from the knowledge base through searching and self-learning process. This helps enrich the organizational knowledge base by importing previous lessons learnt from experts/ employees in addition to using knowledge cycle. Besides, this also helps to train new employees by using the knowledge base.

4.4 Framework for a KMS Based on Web 2.0

According to above analysis, a framework for applying web 2.0 in building up a knowledge management system (KMS_W2) is proposed. This framework includes general architecture, knowledge source, people, and processes as follows:

- ➤ *Architecture:* KMS_W2 is an integrated environment based on web 2.0 platform. The architecture of KMS based on web 2.0 is shown in figure below.

- ➤ *Knowledge Source:* a combination of operational database, data-warehouse, document base, content base, knowledge base and expert directory. Knowledge base is used to store codified knowledge which includes problem, cause, related discussion, final solution and hyperlinks to references. The quality of knowledge-base is ensured by knowledge officers and through voting mechanism of system. Moreover, to access tacit knowledge, an updated list of experts with their profiles, experiences and contact information is collected and stored in expert directory.

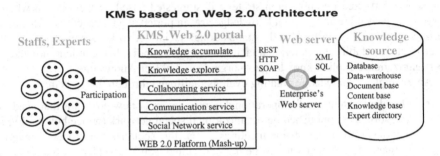

Fig. 2. Architecture of KMS based on web 2.0

➤ *People and Processes*: With this KMS_W2 portal, employees can log on and do their daily works with online applications, communicate with inside/ outside partners, searching for solutions related to their problems, cooperating for problem solving, sharing ideas, voting for group decisions and learning new knowledge from a public knowledge base. Moreover, external users, such as: customers, partners, experts... are also encouraged to use this system. The following use-case diagram shows processes and people of KMS_W2:

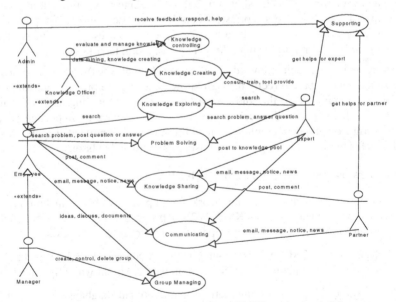

Fig. 3. Use-case diagram of proposed KMS based on web 2.0

The following table shows functional capabilities of KMS_W2:

Table 6. Functional capabilities of some Web 2.0 services and KMS based on Web 2.0

Functions	Wiki pedia	Delici ous	You Tube	Google Map	Yahoo Answer	Face book	Blog spot	KMS_ Web2
Profile					X	X	X	X
Friend making			X			X	X	X
Chat						X		X
Comment	X		X	X	X		X	X
Rating/ Marking		X	X		X			X
Mail			X			X		X
Notice board			X		X	X	X	X
Group managing	X					X		X
Discussion	X					X		X
Blog							X	X
Wiki	X							X
Questioning	X			X	X			X
Upload multimedia			X					X
Channel/ Subscribe			X					
Bookmark	X	X		X			X	X
Map/ GIS				X				
RSS		X					X	X
Mash-up				X		X	X	X
Tagging		X					X	X
TPS/ OAS/ MIS								X
DSS/ OLAP								X
Data Mining								X

5 Demonstration and Evaluation of the KMS Based on Web 2.0

In order to understand how this system works, a demo system is implemented based on an open source code of web 2.0 (from www.elgg.org). The environment for testing server includes: Windows Vista OS., Apache web server, My SQL DBMS and PHP programming language. Requirement for client environment is networked PC with a Web browser (such as: Internet Explorer or Mozillar Firefox). Knowledge source is mainly stored in My SQL database, but some kinds of knowledge object, such as: file, video, audio… are stored in a local folder or a link to embedded object on the internet. Testing system is an intranet website, which can be accessed through the following URL: http://dev.oq.la/elgg1.5/ (See appendix for details).

From this prototype system, suggested KMS and its functions, which are implemented (black) or not implemented (blue, italic), are shown as follow:

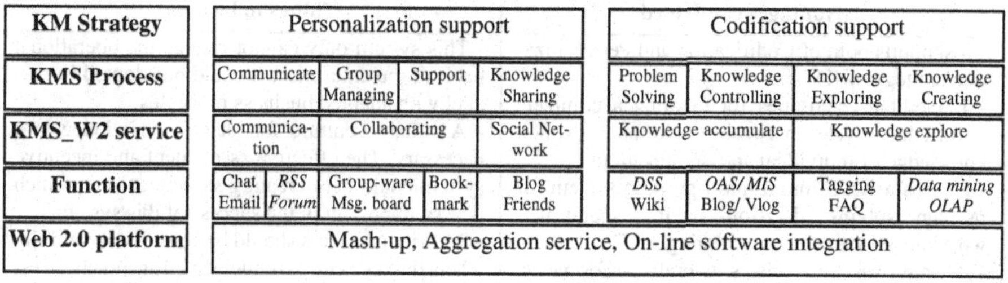

KM Strategy	Personalization support				Codification support			
KMS Process	Communicate	Group Managing	Support	Knowledge Sharing	Problem Solving	Knowledge Controlling	Knowledge Exploring	Knowledge Creating
KMS_W2 service	Communica- tion	Collaborating		Social Net- work	Knowledge accumulate		Knowledge explore	
Function	Chat Email / RSS Forum	Group-ware Msg. board	Book- mark	Blog Friends	DSS Wiki	OAS/ MIS Blog/ Vlog	Tagging FAQ	Data mining OLAP
Web 2.0 platform	Mash-up, Aggregation service, On-line software integration							

Fig. 4. Demo of KMS based on web 2.0 for supporting both KM strategies

In this system, the following knowledge management processes are also supported:

♦ *Knowledge accumulation*: this process is strongly supported by blog, wiki…
♦ *Knowledge evaluation*: this process is supported by voting, commenting…
♦ *Knowledge sharing*: knowledge is shared by organizational social network.
♦ *Knowledge utilization*: tag, chat, group… for searching and using knowledge.

Using this system, the interaction between employees and experts is encouraged through communication and collaboration. By the time, tacit knowledge will be captured more and more. This is an automatic knowledge extraction method. The consistence of knowledge is ensured by the voting mechanism with various weights for employees, experts and knowledge officers. However, users have the right to possess and to share their knowledge by setting "what will be shared with whom" to protect their privacy. In the enterprise context, 'public' sharing level is encouraged by many policies, such as: awards, compliments, incentives based on contribution…

In general, this system is better than previous solutions in the following aspects:

Table 7. Significant benefits of KMS based on Web 2.0

Benefits	Reasons
This system can be deployed quickly and easily with the support of open source code or renting services.	Without web 2.0, integration of just a function required a lot of time and cost. With web 2.0 platform, it is as simple as installing a software.
Users play active role in developing their own working environment. Functional extensibility is allowed according to users' need. This characteristic will attract users for participation and contribution.	With a little training, users can easily add new available tools to their screen or manage their knowledge and chat with friends in their networks. They are free in doing their jobs here.
It is convenient for employees as well as managers to access explicit and tacit knowledge through social network and knowledge base of this system.	Tacit knowledge can be converted to explicit knowledge in many ways and stored in knowledge base for reusing by others in a later time.
This system decreases communication and collaboration cost, therefore increases the effectiveness of any project taken place inside the enterprise.	It is easy to communicate with other friends in one's networks. Group service makes it easy for discussing, sharing and collaborating.
Using this system, the managers can easily monitor their human resource through analyzing the activities' log. They can also make suitable policies or decisions for improving the company performance.	Activities' log helps managers in monitoring their employees. Organizational social network can be used to convey ideas, to give advice… Admin tools help them to change system policy.

In summary, KMS based on web 2.0 has the following advantages:

Table 8. Advantages of KMS based on Web 2.0 and things to be done

Advantages confirmed	Things to be done
- It supports both of Codification and Personalization strategy of KM. - It facilitates activities for creating a complete knowledge cycle between tacit and explicit knowledge of individual and organization. - It turns a KMS into a multi-purpose system for problem solving, decision making, social networking and knowledge exploring. - It improves enterprise's overall performance, brings benefits to all members by facilitating their jobs and broadening their knowledge.	- This system only cannot change the operational habit of the enterprise. It could be adopted in parallel with adapted business processes. - A suitable culture for sharing knowledge is a necessary. The effective assessment and incentive system could help creating sharing culture, which is very important in the success of this system. - Security problems should be considered carefully when this system is implemented in practice, especially when outside partners are allowed to join or renting services are used.

6 Conclusion

In general, implementing a KMS effectively requires a proper strategy for KM. Most of the previous KMS were not successful because they focused on either of two main strategies for KM (Personalization or Codification). The division between tacit and explicit knowledge of two strategies is the main cause of problems in KMS implementation. This paper found that web 2.0 can be used as a platform for combining two main KM strategies in utilizing both kinds of knowledge.

The combination of two strategies for KM by web 2.0 helps to facilitate the knowledge creating cycle inside the enterprise and to provide an open environment for integration, which can be used for many purposes, such as: problem solving, decision making, social networking and knowledge exploring. This combination strategy ensures the effectiveness of KMS and improves the performance of the enterprise.

Moreover, a new KMS based on web 2.0 is suggested in detail and a demo system is tested in practice. This demo shows many advantages in comparison with previous systems. It also creates the creative environment and ensures the democracy in the enterprise, where knowledge is created, accessed and evaluated easily by everyone.

Besides many advantages, there are also some things to be implemented of a real KMS based on web 2.0. Some implications for the future works are as follows: analyzing cultural and organizational problems in deploying such system in the enterprise in practice; more detailed specification and evaluation of the system, especially focusing on the security aspect.

References

1. Alavi, M., Leidner, D.E.: Knowledge Management And Knowledge Management Systems: Conceptual Foundations And Research Issues. MISQ Review (2001)
2. Bontis, N.: Managing organizational knowledge by diagnosing intellectual capital: training and advancing the state of the field. IGI Publishing (2001)
3. Choi, B., Lee, H.: An empirical investigation of KM styles and their effect on corporate performance. Information & Management 40(5) (2003)
4. Derballa, V., Pousttchi, K.: Extending Knowledge Management to Mobile Workplaces. ACM, New York (2004)
5. Fisher, G., Ostwald, J.: Knowledge Management: Problems, Promises, Realities and Challenges. IEEE, Los Alamitos (2001)
6. Jennex, M.E., Olfman, L.: Assessing KM Success. IJKM (2005)
7. Hansen, M.T., Nohria, N., Tierney, T.: What is your strategy for managing knowledge? Harvard Business Review (1999)
8. Nonaka, I., Konno, N.: The Concept of 'Ba': Building Foundation for Knowledge Creation. California Management Review 40(3) (1998)
9. O'Donovan, F., Heavin, C., Butler, T.: Toward a model for understanding the key factors in KMS implementation. LSE (2006)
10. Smith, D.M.: Web 2.0: structuring the Discussion, Gartner Research (2006)

Appendix

Dashboard of demo (http://dev.oq.la/elgg1.5/) - After logging in, employees can see the Dashboard, where most of useful tools are shown. All tools can also be called from menu Tools. Besides, employees can configure their environment display, notification method or privacy policy to their contents by using menu Settings. Menu Administration is shown only for admin to monitor, configure and make change to functions, users, policies and displays for the whole system. Search box is used for searching in knowledge base.

Profile page - This page allows users to update their own profile (personal information, hobbies, skills…), their knowledge (ideas, writings, messages, videos…) and their friendship (friends, groups, activities…). Those things can be shared with their friends by changing privacy setting to public/ limited mode.

Measuring KM Success and KM Service Quality with KnowMetrix – First Experiences from a Case Study in a Software Company

Franz Lehner

University of Passau
Innstrasse 43, D-94032 Passau, Germany
franz.lehner@uni-passau.de

Abstract. It is commonly accepted that knowledge management (KM) and knowledge transfer are critical success factors for competing enterprises. What is not yet known sufficiently is how to measure the status of KM in order to develop concrete and effective actions to improve KM and knowledge transfer. What is lacking is a methodological base to support the development of such diagnostic tools. This paper sets out to fill this gap. Firstly, an overview of the approaches applied so far is provided. From this, a concrete instrument was developed for the use of success factor analysis in a stepwise, methodical manner. The results from a first case study showed that the transfer of the general success factor analysis technique to the specific context of KM is possible. The method is called KnowMetrix and is presented together with first experiences of its application in a software company.

Keywords: KM success, KM success factors, success measurement, KnowMetrix, KM service quality.

1 Introduction

Knowledge management (KM) has become an accepted discipline within the field of Information Systems, is now well anchored in the operations of many enterprises. However, the results of KM are so far not always considered in a positive light and even researchers in the field have become aware of unrealistic expectations over the last decade. Now at last this seems to be replaced by a more realistic view of what the benefits – and costs – of KM can be. Still, attempts to find concrete proof for the benefits of KM often meet with considerable difficulties because the results of KM activities are not always explicitly observable and/or directly measurable.

An overview of the approaches applied and the empirical research undertaken so far in the evaluation of the results associated with, and achieved through, KM is provided in this paper. Because the studies reviewed have not yet delivered entirely satisfactory explanations the already well known method of success factor analysis has been investigated for its usefulness in KM and with regard to what would be required to transfer the method to the specific requirements of KM. From these foundations a concrete concept was derived for the use of success factor analysis in a stepwise,

D. Karagiannis and Z. Jin (Eds.): KSEM 2009, LNAI 5914, pp. 335–346, 2009.

methodical manner in the context of KM activities and the institutional structures for KM. From the literature on positive and negative factors ('accelerators' and 'barriers') for KM in enterprises a preliminary list of partially standardised indicators for the evaluation success for KM activities was produced and adapted for use as a prototype instrument.

2 Measurement of Success in KM and Measurement Methods

The first question to be asked is whether and how KM makes a contribution to the success of a firm and how KM service quality can be measured [3]. have shown with examples that KM and knowledge transfer indeed have a measurable effect on the success of an enterprise – better KM processes reduce the number of errors and mistakes and increase the number of successfully concluded operational processes. The inherent time savings have an indirect, but nevertheless significant, effect on the cost structure of companies.

However, there are four areas where measurements could add value to the assessment of the KM activities of the firm [10]:

- the knowledge (base) itself;
- KM processes;
- projects in KM and knowledge transfer;
- information systems in the KM sphere.

Diagnosing the state of art of knowledge management in an enterprise has long been in the focus of research and practice, often in connection with the evaluation of its intellectual capital. The latter is not considered any further, because the focus here is on the diagnosis of the knowledge management function. There is a large number of instruments to perform such a diagnosis, but these are still not satisfactory. A preliminary study conducted at the University of Passau, investigated instruments for a KM diagnosis in a comprehensive literature review (for full details see [12].

An instrument that concentrates on the visualisation of existing knowledge materials is the Knowledge Management Assessment Tool KMAT [14] and the Knowledge Framework [8]. An instrument that uses a framework from the information systems discipline is the Success Factors of Knowledge Management Systems [13], based on the information systems success theorem of DeLone and McLean [5]. The Knowledge Process Quality Model [16] as well as the Knowledge Management Maturity Model [6] focus on the way knowledge is acquired, shared and managed. Similarly, a number of studies used the Balanced Scorecard (BSC) approach to depict the current status of KM activities. Examples are given in [9], [1], [17]. Building on the principles of the BCS DeGooijer [4] establishes a Knowledge Management Performance Framework that adds behavioural science elements to the instrument. Concentrating on the economical corollaries of KM the Process-Oriented Performance Framework (PPM) uses systems analysis to derive a systematic schema of indicators for the effectiveness of KM. Similarly, "Wissensbilanz A2006" attempts to construct a visual 'balance sheet' of the status of KM and knowledge transfer in the enterprise [2].

A somewhat different perspective is taken with models that are based on critical success factor and multi criteria analysis elements. Often based on the SECI model

[15] these assessment tools, however, give only a snapshot at a specific time and reflecting a specific context.

None of the instruments fulfils all the necessary requirements. Although some of them are already on a high level of development, the following critique must be stated:

- The instruments are mostly used only for analysis and do not provide guidelines or strategies to reinforce identified strengths or remove identified weaknesses.
- Few of the instruments have been empirically validated and in terms of the peculiarities of smaller companies (and SW business is dominated by SMEs) only the last two are explicitly designed for smaller companies.

Many instruments have been proposed for the evaluation of the success of KM and knowledge transfer efforts. As one can see from the state of the art in diagnosing knowledge management there are still drawbacks with the existing instruments. These are addressed by the instrument proposed in this paper. We concentrate our review on the models and frameworks that have a direct relevance to the analysis of success factors.

3 KnowMetrix - Adapting Success Factor Analysis for KM

Critical success factors are areas of top priority for the executive management in a specific firm – they are specific for the individual firm at a specific time. Of special interest for the area of KM and knowledge transfer is the work that went on to develop the correct strategic focus for the information technology efforts of the firm, pioneered by the studies of Rockart and Martin in the USA [11]. The idea of analysing the importance and the impact of particular factors of a specific enterprise function can be adapted to the field of knowledge management.

The object of the success factor analysis for knowledge management is the knowledge management function within an enterprise while the objective of the success factor analysis is the measurement of the contribution of certain factors to management function's success, by identifying strengths as well as weaknesses of knowledge management and utilising the analysis to deduce actions to extend strengths and to dismantle weaknesses.

The success factor analysis must be based on a validated list of success factors. For the application of the success factor analysis in the field of IT management Rockart has empirically elaborated a number of success factors ordered in the four types of service, communication, IS human resources, and repositioning the IS function [11]. For the field of knowledge management there are also investigations of success factors (see e.g [7]). However, the investigations are very heterogeneous in their understanding of knowledge management and also in the research design. The investigation of Helm et al. is the most elaborate so far and unites 39 investigations and presents knowledge management success factors within the four dimensions of personnel, structure, culture, and knowledge management activities. The result of Helm et al. was used as a basis because the factors presented in their work are empirically tested.

The application of the success factor analysis is structured as follows, whereby the process is adapted to the application in the field of knowledge management without changes:

- Selection and adaption of success factors,
- Determination of the participants,
- Create questionnaire,
- Conduct the analysis,
- Analysis and presentation of the results, and
- Interpretation of the results.

Knowledge management will be understood as a combination of different tasks from known and established functions within an enterprise (e.g. management itself, personal management, IT management) under a perspective of dealing with knowledge. There are numerous definitions for success in knowledge management. Often success is related to the contribution of knowledge management to an enterprises' success, although it is hard to determine this contribution. Here, it appears to be appropriate to define knowledge management success by the satisfactory accomplishment of the knowledge management function. The success factors analysis could be a suitable way to assess KM success within a company and is named KnowMetrix within that context.

4 Application of KnowMetrix in a Software Company

Knowledge supply and knowledge services are essential success factors in the software business. For this reason Infoserv (anonymized name of the company) was selected as a first case to practically test KnowMetrix. Different knowledge management measures were already implemented by the company selected and different software systems exist that support the transfer of knowledge. The company was founded in 1997 and develops software-solutions for the management of product information (PIM) as well as the output channels online, print and stationary point of sale (POS). The software regulates different functions for the administration and presentation of the product, price, catalogue etc. as well as the interaction with the sales partners, suppliers, wholesale clients and retail clients. Thereby the automation of communication, sales and support processes is being supported. The great flexibility and modular layout of the software makes adjustments possible to the individual needs and processes of the small companies to the big companies.

More then 150 world wide operating entrepreneurs and businesses use the companies Software on more then 1500 web sites. Among the clients are leading sales- and production companies e.g. Zavvi, TUI, Lufthansa, Puma, Ulla Popken, Nokia, Reebok, Grundfos, Demag, Sika, Hagemeyer, Phonak and Bechtle. The software company is organized in eight departments:

- Research and Development
- Product Management
- Professional Services
- Presales
- Sales

- Training
- Marketing
- IT-Administration

The software company employs altogether about 90 staff members, about 60 of them in Munich. Apart from the head quarters, the company has further branches in the United Kingdom, the Netherlands, Switzerland, Austria, Sweden, Poland and the USA. For the staff member survey about Knowledge Management, not only staff members of the main office were polled but also of the branches in London (5 employees), Poland (20 employees), Zurich (2 employees) und Stockholm (2 employees). These foreign branches operate in the area of Professional Services. In the London, Zurich and Stockholm branches, staff members also are operating in Sales. In addition, an extensive network of sales and technology partners exist in all of Europe and the USA.

The central knowledge information system is named K-net. Thereby, it is a company intern Wiki, which is divided in different areas such as e.g. projects, documentations, Poland etc. Other then these, programme Subversion and Jina are used. Subversion supports the version administration of data and schedules. Jina is used for administration and problem treatment. Furthermore, lately, different knowledge management measures have been realized. To be able to find company specialist in different knowledge areas, an expert register was compiled (Yellow Pages). Furthermore a partner news letter as well as an intern newsletter were established. The partner newsletter addresses itself to the implementation partners and deals with technical problems, solutions and current developments. The internal newsletter covers only technical problems. Moreover, K-nets´ homepage was optimised in the area of documentation for different user roles. Next to the electronical exchange possibilities, open, personal communications is encouraged by the board of directors.

At Infoserv, different locations and time pressure during projects present the greatest obstacles for knowledge management. Project documentation suffers the most from this time pressure. Consequently, the goals of 2009 are the securing of knowledge transfer between Munich and especially Poland, as well as the improvement of information flow out of projects (Documentations, Lessons Learnt, Best Practices). The explicit facilitation and encouragement through the directors and managers are vital advantages for knowledge management at Infoserv.

The procedure of the empirical survey and the adjustments necessary were predetermined in KnowMetrix. In the conventional procedure, the strategic and operative goals of the company are established in phase 1, as well as the identification of the knowledge needs, size of influence, dependence, systems, areas of responsibility and Knowledge Clusters. Phase 2 includes the identification of critical success factors for knowledge management, which will be concretised thought indicators in Phase 3.

The assignments of the three phases in the present study were conducted in the following way: first of all, relevant information about the company and the specific situation of knowledge management has been collected. Based on the information attained, suitable success factors were identified, indicators were derived and a questionnaire was developed. Subsequently, the questionnaire was reviewed and adjusted with the responsible Knowledge Manager of Infoserv Ltd. The main emphasis was here in the information polled in the general part (e.g. departments, locations). To be able to involve as many Infoserv Ltd. branches as possible, a German and English version of the questionnaire was provided.

Phase four consisted of the selection of the participants, the carrying out of the survey and the analysis of the data. The participants of the poll were determined by the investigated company according to the differential characteristics asked for in the general part. In the case described here staff members of the departments "Research & Development", "Professional Services", "Presales" and "Product management" were polled. Furthermore, a difference was made between the head quarters and the other branches. In this way, the staff members survey covered the production part of the company.

The survey was carried out online because of the different international branches of Infoserv Ltd. To this means, the online-questionnaire system „Limesurvey" was used. In case individual staff members would decline to use the online questioner resp. in the case of technical problems with Limesurvey, a form was established, which could be filled out electronically. The survey took place between 02.21.2009 and 03.06.2009. The survey was send per Email to about 65 staff members and 42 took part. The return rate was 65%. 2,24% of the questions were not answered. The choice of appropriate success factors and indicators was based upon previous work by Lehner et al. [12].

5   Analysis of KM Success at the Software Company

In Part I of the questionnaire, general information is collected to categorize the organisation of the company. As this part of the questionnaire was mandatory, all the information about the participants is complete. With ca. 67% of surveyed participants (28 persons) most of them working in the head quarter. Ca. 33% of the participants (14 persons) are active in other surveyed branches (London, Zurich, Stockholm). Ca. 52% of the participants (22 persons) work in the Department of Professional Services, ca. 29% in the Department of Research and Development (12 persons) and ca. 10% in the Departments of Presales (4 persons) and Product management (4 persons) respectively.

Dividing up the participants according to the amount of time they have worked for the company, shows that the half of the polled participants were employed less than 2 years (21 persons), which means that there are relatively many new employees. Ca. 26% of the participants were employed since 2-4 years (11 persons), ca. 10% since 5-7 years (4 persons) und ca. 14% since 8-10 years (6 persons) by Infoserv Ltd. The average employment in the foreign branches was clearly lower with about 1. 65 years as in the headquarters with about 3. 54 years.

General Satisfaction with KM services at the software company
Besides the general questions, the participants were asked in Part I of the survey, to give a general evaluation about the subject of knowledge management in Infoserv Ltd. To this effect, they were given a 7 steps point scale. An evaluation was possible starting at value 1 "not satisfied" to value 7 "very satisfied". As this part was also mandatory, these questions were answered by all participants. The average of general satisfaction reached 4.69. This corresponds with the average of ca. 67% of the best value 7.

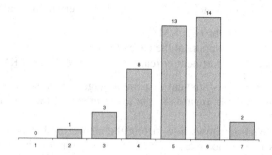

Fig. 1. Overall satisfaction with KM services

In observing, the general satisfaction according to the location shows that there is hardly any difference between the locations, actually the foreign branches (London, Zurich, and Stockholm) are only 0.07 lower than the headquarters.

Observations of the satisfaction according to the department they work in, shows that the Department of Research and Development and Product management prove to have higher satisfaction then the Department of Professional Services and Presales. Research and Development points to a satisfaction of ca. 75% of the best value 7, Product Management of ca. 71%, Professional Services of ca. 64% and Presales of ca. 54%. In fact, 3 out of 5 possible Presales staff members were polled, 60% in this branch as well as one foreign employee, nevertheless based on such a small survey, a conclusive assessment is not possible. On the contrary, the 4 participants in Product Management represented the complete team. The group of staff members that have worked 5-7 years in the company stand out in the analysis of general satisfaction according to the time of employment. It is with a rate of satisfaction of ca. 46%, a distinctly less satisfied group then others. As only 4 participants belong to this group (ca. 10% of the spot check) these results are not statistically significant.

Analysis of the performance-assessment
In this section, the performance-assessments of the survey participants concerning the 23 examined indicators will be considered. The rating average of these indicators amounts to 5.03. This is equivalent to about 72% of the best value 7. After a closer observation, one notices that six of the factors have a very low yield of less than 4.5:

- clear accountability and responsibility for KM activities resp. knowledge and information relaying
- possibilities and time for the documentation and the acquirement of knowledge
- knowledge and experience exchange during official meetings
- existence of standardised knowledge processes
- knowledge transfer between locations
- knowledge transfer between project partners

On the other hand three indicators stand out through a very high yield of more than 6:

- noticeable success through KM
- interdepartmental support in the case of problems
- improvement of department or team performance through KM

The standard deviation is a statistical dispersion gage, which sheds light on the unity of the survey participants in regards to the assessments of the performance-status of the indicator.

The standard deviation shows how strong the answering behaviour of participants resp. indicators varies. The average standard deviation of the performance-value amounts to 1.28. It appears that the above identified as indicators with insufficient yield, except for "Clear accountability and responsibility for KM resp. knowledge and information relaying" show an above average standard deviation. And the other hand, other indicators such as e.g. "Willingness to share information and knowledge", "Possibilities and time to share experiences with co-workers" and "Up to date ness of the electronically memorised information" stand out.

Analysis of Importance-Assessment

In this section, the importance-values of the surveyed participants are going to be considered towards the 23 questioned indicators. These values indicate the importance or the priority of an indicator. The average of a importance-value of an indicator amounts to 6.02. This is equivalent to ca. 86% of the best value 7. The comparison with the average of the performance-value of 5.03 shows that a clearer and better situation is wished for by the survey participants. The higher the respective value, the more important the indicator is assessed by the participants.

Altogether one can determine that the answering behaviour of the participants resp. the indicator is not as varied as in the evaluation of the performance-situation. The average standard deviation is lower with 0.99 as the answers of the participants to the importance of the indicator lays mostly at a higher level.

Importance-Performance-Comparison

In focus with KnowMetrix stands the so called performance difference, which means the difference between the should and performance-assessment. It sheds light on the necessity of improvement measures. The greater the difference between should and performance-assessments the higher the need for measures of improvement.

The average of the variances of all of the indicators was of 0.99, which means that the 23 indicators are evaluated as 1 measure more important then they are fulfilled presently. The following indicators show an especially high difference of more then 1.6:

- clear accountability and responsibility for KM activities resp. knowledge and information relaying
- possibilities and time for the documentation and the acquirement of knowledge
- existence of standardised knowledge processes
- knowledge transfer between locations
- knowledge transfer with project partners

In comparing these five indicators with those identified as less effective, one notices that all are identical except for "Knowledge and experience exchange during official meetings". This indicator was omitted since next to its present effectiveness, it was not considered of great importance. These five indicators should be kept in focus as improvement measures. An especially high performance difference is noticed in both the indicators "Knowledge transfer between locations" and "Knowledge transfer with project partners".

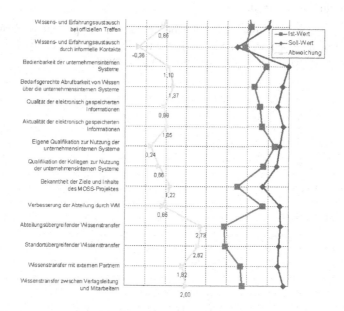

Fig. 2. Comparing importance and performance values of the indicators

On the other hand some indicators have a very low difference with less then 0.2 thus having a high correlation of is and importance values:

- noticeable success through KM
- interdepartmental support in the case of problems
- knowledge and experience exchange through informal contact
- qualification for the use of the company intern system
- improvement of department resp. team performance through KM
- direct methods of communication for the solving of complex tasks

The four-fielded-Matrix has two dimensions. On the coordinate, the dimension performance (Performance-State) is represented, on the ordinate, the dimension priority (Importance-state). The vertical boundaries between the fields constitute the average performance (5.03), the horizontal boundaries are defined thought the average of Priorities (6.02.). So a classification was made of the indicators in 4 fields, out of which arose different recommendations for improvement:

- In the right upper field are the indicators, of which their present performance and importance are considered to be relatively high .They prove to have a low performance difference and there is no immediate need to carry out any measures of improvement.

- The indicators in the left bottom field were all under averaged in importance and the present performance is also considered to be under averaged. Here also, no immediate need to carry out any measures of improvement.
- Indicators in the bottom right field indicate a relatively low importance and a relatively high performance. Here a reduction of investments in these factors could be examined.

As relatively important assessed indicators, of which the present performance is as relatively low assessed are to be found in the top left field. Especially these indicators should be examined regarding measures of improvement.

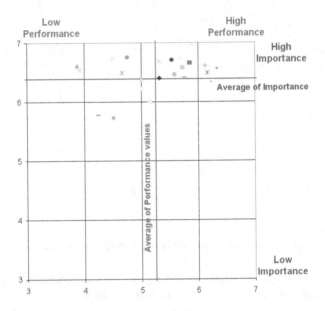

Fig. 3. Contrasting importance and performance values in a matrix

Analysis of success values
As already mentioned, the determined data could also be evaluated according to success. Success values describe the ability of a measuring indicator to contribute to the success of knowledge management in the examined company.

The average success rate amounts to 4.94. It shows that the analysis according to success values leads to the same conclusions as the importance-performance-comparison. The following five indicators are again noticed to be negative:

- clear accountability and responsibility for KM activities resp. knowledge and information relaying
- possibilities and time for documentation and acquirement of knowledge
- existence of standardised knowledge processes
- knowledge transfer between locations
- knowledge transfer with project partners

Next to the success value per indicator, we could consider the success value per surveyed participant. This will be compared with the general satisfaction noted in part I of the questionnaire. Although the average of success through all staff members amounts to 4.94 and the general satisfaction to 4.69 being of very little difference, the difference per staff member could be much larger.

The comparison of general satisfaction (4.69) and success value (4.94) show that the staff members of Infoserv Ltd assess the success of knowledge management as to have a negative difference of only 0.25. This difference can also be an indication of unclearness resp. knowledge management. Another reason could be the individual importance of a participant resp of a few indicators.

6   Experience and Outlook

In summary it may be stated that for academia as well as for practitioners there is sufficient knowledge about barriers and success factors for KM. However, there was up to now no practical and easy-to-use method to measure and diagnose the KM situation. According to the different levels of success factors there is already research in progress (see [12]). The initial list of indicators is based on currently known barriers and success factors as well as indicators for KM processes. Barriers to KM and, particularly, to knowledge transfer, are viewed as equally important criteria for the success of KM initiatives and projects. However, newer findings indicate that both seem to be complementing each other: success does not rely solely on the presence of required success factors but also needs the absence of barriers to manifest itself. Future research is needed to delimit clearly the barriers from the success factors, especially in situations where they are not just negatively coupled, i. e. barriers that are not just the absence of success factors but are independent factors in their own right. In addition to their identification, the respective direction and strength of such factors on the success of KM initiatives need to be assessed.

The results of a first case study applying KnowMetrix showed that the transfer of the general success factor analysis technique to the specific context of KM is possible. As further projects showed in the meantime the KnowMetrix approach can be applied to other companies in the same way as described here and help to uncover weaknesses in KM. Furthermore, the instrument was well accepted by the pilot participants. The results can be seen as a first step in the development of an 'industry-strength' assessment tool for the utility of KM, specifically tailored to the individual needs of an enterprise. More work, however, is necessary in a further, more operational definition of what constitutes 'success' in KM and what indicators would be best to assess it.

References

1. Bodrow, W., Bergmann, P.: Wissensbewertung in Unternehmen. Berlin (2003)
2. Brandner, A., et al.: Assess – Wissensbilanz A2006 – Leitfaden für Klein- und Mittelbetriebe (2006), http://assess.daa.at/download.asp?id=156
3. Davenport, T.H., Prusak, L.: Working Knowledge. In: How Organizations Manage What They Know. Harvard Business School Press, Boston (1998)

4. DeGooijer, J.: Designing a knowledge management performance framework. Journal of Knowledge Management 4(4), 303–310 (2000)
5. DeLone, W.H., McLean, E.R.: Information Systems Success: The Quest for the Depend Variable. Information Research 3(1), 259–274 (1992)
6. Ehms, K., Langen, M.: Ganzheitliche Entwicklung von Wissensmanagement mit KMMM (2000), http://www.wissensmanagement.net/online/archiv/2000/08_0900/Wissensmanagement.shtml
7. Helm, R., et al.: Systematisierung der Erfolgsfaktoren von Wissensmanagement auf Basis der bisherigen empirischen Forschung. ZfB 77(2), 211–241 (2007)
8. Jordan, J., Jones, P.: Assessing your Company's Knowledge Management Style. Long range planning 30(3), 392–398 (1997)
9. Kaps, G.: Erfolgsmessung im Wissensmanagement unter Anwendung von Balanced Scorecards. In: Nohr, H. (ed.) Arbeitspap. Wissensmanagement, Stuttgart, vol. 2 (2001)
10. Lehner, F.: Wissensmanagement. Grundlagen, Methoden und technische Unterstützung, 3rd edn., Hanser, München (2009)
11. Lehner, F.: Success Factor Analysis as an Instrument for Information Management. Journal of Computer Information Systems 33(3), 58–66 (1993)
12. Lehner, F., et al.: Erfolgsbeurteilung des Wissensmanagements. In: Diagnose und Bewertung der Wissensmanagementaktivitäten auf der Grundlage der Erfolgsfaktorenanalyse, Research Report, 2nd edn., No. W-24-07, University of Passau (2008)
13. Maier, R., Hädrich, T.: Modell für die Erfolgsmessung von Wissensmanagementsystemen. Wirtschaftsinformatik 43/5, 497–509 (2001)
14. North, K.: Wissensorientierte Unternehmensführung. Wertschöpfung durch Wissen, 3rd edn. Gabler, Wiesbaden (2002)
15. Nonaka, I., Takeuchi, H.: The Knowledge-Creating Company. In: How Japanese Companies Create the Dynamics of Innovation, New York (1995)
16. Paulzen, O., Perc, P.: A Maturity Model for Quality Improvement in Knowledge Management. In: Wenn, A., et al. (eds.) Proceedings of the 13th Australasian Conference on Information Systems (ACIS 2002), pp. 243–253 (2002)
17. Wu, A.: The integration between Balanced Scorecard and intellectual capital. Journal of Intellectual Capital 6(2), 267–284 (2005)

Knowledge Creation Spaces: The Power of Project Teams

Andrew J. Sense

Associate Professor, School of Management and Marketing, Faculty of Commerce,
University of Wollongong
Wollongong, NSW, 2522, Australia
asense@uow.edu.au

Abstract. This paper examines how and why project teams may be conceived as highly creative and generative knowledge creation spaces – as opposed to traditionally and primarily being conceived as only temporal task focused entities where knowledge is simply exchanged. In this qualitative analysis which draws on situated learning theory, project teams offer significant and yet generally underexploited personal knowledge growth opportunities for participants. The arguments presented in this paper also support a notion that project teams can serve as important action locations for organizations seeking to unlock employee learning potential and harness their creativity, and thereby contribute to the development of a learning culture. Through its challenge to conventional perspectives about project teams and their limitations, this paper provides a conceptual grounding from which to pragmatically and productively approach an important and most often neglected project phenomenon of social learning.

Keywords: Knowledge Creation, Learning Spaces, Project Teams, Situated Knowledge, Organizational and Individual Learning Capability, Group Learning, Work Based Learning.

1 Introduction

Organizational managers engage different structural arrangements in attempting to adaptively and successfully respond to dynamic and unstable competitive markets and the human, political and technological challenges that organizations confront [1, 2, 3, 4, 5]. For example, these may include different operational structures such as networked organizations, self-managing work teams and temporal organizational arrangements such as projects and project teams. In particular, and extending well beyond the traditional disciplines of engineering and construction, projects and project teams are increasingly used to accomplish a diverse and often complex set of technological and/or cultural changes that may otherwise be much less obtainable through the actions of the permanent organization [6, 7, 8]. This is particularly the case where achieving a relatively rapid speed of completion and a high quality of outcome is considered extremely desirable e.g. in new product development or information technology projects. Furthermore, cultural or organizational change programs frequently engage the project form as a key structural means to progress their change efforts and to actively engage participant stakeholders in the change processes. In

D. Karagiannis and Z. Jin (Eds.): KSEM 2009, LNAI 5914, pp. 347–357, 2009.
© Springer-Verlag Berlin Heidelberg 2009

these ways, projects and project teams become the vehicles to deliver discrete physical project outcomes and can also be seen or utilized as cultural change engines – as has often been demonstrated in socio-technical change projects and total quality management projects across a broad spectrum of industries [9].

Accompanying this increased utilization and deployment of the project form to provide solutions to difficult organizational challenges is a far more intricate set of social relationships and performance measures or expectations between projects and their host organizations than that normally suggested in traditional interpretations of projects [6, 10]. Traditional views can limit our interpretations of a project for example, to one being entirely rational and socially separated from the 'normal' organization and which focuses only on cost, specification and time performance outcomes. On the contrary however, projects are intimately, socially and contextually entwined within and often beyond their host organizations and frequently pursue goals (declared or implicit) that go well beyond traditional tangible measures e.g. personal learning and knowledge development, micro-political gains or organizational cultural changes.

As one might intuitively deduce and observe in practice, in pursuit of such a range of declared and undeclared goals or objectives, project teams necessarily create knowledge and also acquire new knowledge from multiple sources. Consequently, and in accordance with those views espoused by Sense [9], Smith and Dodds [11], Arthur, DeFillippi and Jones [12] and Morris [13], one might also presume that projects are rife with significant personal and organizational learning opportunities. Therefore, learning, and in particular, situated learning (alternatively referred to as learning-on-the-job in this paper) is actually a very dominant and practically oriented issue in the processes of projects and underpins the quality of project outcomes and the learning and knowledge development of participants. It seems rather intriguing then that this phenomenon of learning-on-the-job within this context till relatively recently in the project management field (See for example Sense [9]) has attracted so little scholarly and practitioner attention. In the traditional model of projects and project teams, learning-on-the-job has more generally been viewed as simply an inherent, amorphous and opportunistic part of the project management process [14]. In this limited perspective, the generative learning-on-the-job potential of project teams is not fully nor systematically acknowledged, exploited or developed. Such a condition corresponds to a notion of project teams being underutilized generative 'learning spaces'. Arguably, an important and seminal contribution towards achieving a deeper understanding and an ability to nurture such a learning space, involves an examination of how one can conceive project teams and their purposes. Accordingly, the intention of this paper is to conceptually examine how one might consider a project team as a creative and generative learning or knowledge creation space. By establishing this foundation, one is better ontologically informed to further investigate or to develop project team learning capacity and capability within these temporal contexts. The arguments forwarded in this paper are firmly anchored in an epistemology of social constructionism and draw on situated learning theory, and reflect the social and practical dimensions of the project environment.

To advance this discussion further, the following section outlines the traditional conceptions of projects and project teams. This discussion provides a useful and comparative conceptual backdrop to the main arguments posited within this paper.

Thereafter, in the next section, the theoretical learning framework informing those arguments is outlined. The main arguments concerning a view of a project team as a generative learning or knowledge creation space and the potential implications of that conceptualization in the practice world of participants are articulated in the final discussion section of this paper.

2 A Conventional Conception of Project Teams

In most traditional definitions of projects the characteristic emphasis is on: separation and temporality i.e. a separate process involving a finite time for task completion; the uniqueness or specificity of objective and of the project activity in comparison to on-going functional or departmental operations in organizations, and; the flexible, rational management or coordination of many interrelated activities and the resources of the project, by the people associated with the project [9]. Building on these character-istic features the traditional view of project teams and those managing them is one of, implementers i.e. to make things happen [15]. In more sophisticated terms, project teams are traditionally conceived to be a group of people who construct a perceived common identity (mobilized around a project objective) so that they can work to-gether using a set of values or behaviours (derived from past custom and practice) to deliver a project's management system and the project objectives [16, 17, 18]. These descriptions convey a traditional or foundation perspective of a project team as some type of separate social form that maintains a rational output focus on achieving par-ticular and tangible project outcome/s. This ostensibly limited perspective is also somewhat representative of the positivist foundation of the project management field. One can observe that this traditional and restricted perspective fails to explicitly ac-knowledge or incorporate learning or knowledge creation as an attribute. It is also reasonable to suggest that this foundation perspective is also highly inadequate or inappropriate in being able to take account of the messy, dynamic, social and contex-tual character of learning [19] in projects [9]. A traditional perspective of a project team therefore tends to ignore or overlook learning and its potential that is arguably, inherent in every project situation. The identification and adoption of an alternate but complementary perspective of project teams conceived as learning or knowledge creation spaces may provide one way to address that oversight and may also stimulate interest in pursuing personal and organizational learning through projects.

3 Theoretical Framework

The theoretical framework that informs the arguments presented in this paper is a situated theory of learning. In intimate alignment with a social constructivist view of learning, situated learning theory presumes that most learning occurs in culturally embedded ways within communities of practice [20, 21, 22, 23, 24, 25]. Conse-quently, the focus of situated learning theory is concerned with learning as social par-ticipation within these communities of [19, 23, 26]. Situated learning actually evolves (explicitly or implicitly) through the learning processes of observation, dialogue, sto-rytelling and conversations between people as they participate and interact within a practice, and can be referred to in more pragmatic terms as learning-on-the-job.

Therein, participants mutually develop their technical and social competencies and negotiate and construct their identities and common meanings around situations and objects within their evolving practices [20, 21, 23, 24, 27, 28, 29, 30, 31]. Hence, knowledge can be considered a direct result of [and conjoined to] the practices and the mediating socio-cultural conditions of the context, and participation and interaction can be seen as critical for learning and knowledge development. This 'participative community' feature of situated learning theory aligns closely with Brown and Duguid's [32] advocacy of information being embedded in social relationships and institutions and Nonaka and Takeuchi's [33] contention that a key characteristic of knowledge creation is that teams (or communities) of people play a central role in the knowledge creation process (through their interaction, sensemaking [34], dialogue and discussion activities).

The significant themes of situated learning theory particularly relevant in a project team milieu include:

* Knowledge and learning reside within a practice and project knowledge is therefore a direct result of the project practices and the mediating socio-cultural conditions of the project space
* The participation and interaction of people within a project practice is crucial for learning and knowledge competency development
* Individual project participants successfully contribute to and develop a team practice via their collective sensemaking [34] activities
* Participants' construct their project identities through developing both their social and technical competencies within a team [9, 14].

These significant themes of situated learning theory also resonate either explicitly or implicitly in the workplace vocational learning literature [See for example 35, 36, 37]. Embracing this perspective on learning clearly suggests that the project workplace practices serve as the foundation for learning and knowledge development within a project. The 'reality of learning' within a project team can thus be considered constructed, maintained and reproduced through meaningful human practices within a project social context [38]. Indeed, a project team in particular may be regarded as a 'hotspot' for situated learning activity.

4 Considering Project Teams as Generative Learning and Knowledge Creation Spaces

A learning space is a space for interactions between individuals, between individuals and their environment, and between individuals and information [39]. Being of particular importance for generating new learning, Nonaka and Konno [40] and Kolb and Kolb [41] point out that tacit knowledge is embedded in these spaces, and that it is acquired through a person's own experience or reflections on the experiences of others, as they share their feelings, experiences and thoughts while in the learning space. A project team by design establishes a project team's boundaries [which can be quite permeable and malleable according to the project type, process phase or changing situational contexts]. This formality of declaring or acknowledging a temporal and bounded social grouping that is discrete from the normal organization effectively establishes a practice space which can reduce moral and social risks for participants, who in turn will increase their willingness to experiment in their communicative actions [42] – thereby promoting situated learning activity through their interactions and conversations while on-the-job. As such, project teams are shape-givers which provide a space for people to grow [39, 43, 44] through sharing and creating new knowledge [40].

In addition to defining a bounded arena for situated learning (learning-on-the-job), the constituency of a project team is also significant in considering this as a generative learning space. Project teams play key roles in knowledge creation processes because they serve as a place where the commingling of various and diverse learning spaces, which people are part of or participate within, occurs [40]. That being, participants have multiple memberships of other diverse learning spaces and these multiple experiences come together by default or by design in a project setting. A good example of this condition is when participants from various functional departments or indeed separate organizations form a project team. This notion is analogous to a conceptualization of a project team posited by Sense [9, 45], where he utilized situated learning theory and communities of practice to describe a project team from a situated learning perspective. In essence, he considers project teams as a temporary amalgam of individual participants' many different communities of practice and that substantial learning opportunities and new and unique practices and learning emerge at the interfaces of these communities of practice because participants are exposed to new perspectives and challenges [46]. These project team participants remain intimately connected to broader social networks of power i.e. the participants involved in the learning space are still subject to previous, or external to the space social/political/cultural conditions [47]. Therefore, as articulated by Sense [9, 45], participants' learning trajectories are generally more outward bound towards their memberships of their other external communities of practice e.g. their 'normal' department for work. This condition places a very strong emphasis on better understanding and encouraging participatory practices [48] within the project team space so that ideally, participants learning trajectories become more inbound and collectively focused on the project space learning opportunities. Nonetheless, primarily due to the abutment of participants' communities of practice within a project setting, project teams represent significant "opportunity structures" [22, p.50] for both tacit and explicit knowledge to be exposed and exchanged and new knowledge created. Thus, the assembly of diverse and often quite normally unconnected people around a project objective can act as an initiating base and as a stimulus driving creativity in problem solving and in new ideas generation.

The details of how the membership of a team are to work effectively together, what processes they are to develop and use, and how they might actually source and create relevant knowledge within a setting are most often not formally decreed by external sources. Instead, intentionally or unintentionally, they are more the result of less formal and creative social constructivist and sensemaking [34] processes within a setting. In these ways, a project team actually represents an unfettered opportunity to self-organize a space to generate new knowledge - provided the project social environment is well understood and can be facilitative of participant's on-the-job learning activities.

Moreover, project teams usually provide a myriad of formal, quasi-formal and informal situations for learning to develop (e.g. formal team meetings, frequent opportunistic casual conversations and socializing over lunch) in what is typically an energized environment where participants can be both working and not working but in a learning sense are still highly productive [47]. In addition to facilitating structured knowledge transfer, formal project team meeting activities can be instrumental in initiating and supporting participants' critical and communal reflection activities and in further developing their informal working relationships. That is, the formal meetings create the recurrent base opportunities for participants to meet, share knowledge through personal exchanges and narratives and to jointly develop creative solutions to project problems – which in turn better equips them to pursue the building of learning relationships with colleagues in an informal manner during and beyond a project's life cycle [49]. Here also, as in other contexts, such informal, spontaneous discussions and conversations are very important to learning at work, since it is not only what is being learnt that is important but also the learning capability that is being developed in the process of learning. That being, the immediate product of learning is the ability to learn more, or, participants learn-how-to-learn [39]. The multiple, recurrent and often intense social interactions in project teams, be they formal or informal, are therefore opportunities for people to inquire, reflect and interpret and make sense of their experiences [50] and having these opportunities for both personal reflection and reflection through interaction with other participants is an important part of any team learning and development process [51].

The quality of this formal and informal learning-on-the-job in project teams is facilitated (or alternatively impeded) through a combination of physical attributes e.g. a shared office with co-located participants, and mental and emotional attributes within the space e.g. the provision of conditions where participants can genuinely share concerns or experiences and listen to others [40]. If these attributes are constructively addressed, a project team can serve as a space where people have multiple opportunities to openly explore issues with others with some perceived degree of psychological safety – relative to their normal organizational arrangements. The achievement of such a condition (which involves trust and effective patterns of social communication between participants [52]) in virtual project teams (i.e. project teams having members geographically and emotionally dispersed relative to each other and connected primarily through electronic technologies) may be particularly problematic. However, in more common project team arrangements, creating this perception of psychological safety can range for example, from the project team space simply encouraging participants to express their own experiences while facing and embracing differences, to, supporting peoples' actions and reflection activities which allow them to take charge of their own learning and to develop their expertise in an area of their interest and aspirations [41]. This ideally involves participants having control of their project team space (i.e. they are generally free to make decisions about their actions without the influences of external others impeding those processes) and the support to freely explore and communally or individually reflect on issues involved in the project or in the organization – including the managerial ones [42, 53]). Learning spaces which are free from fear and allow people to express their views openly, are a critical condition for ideal speech situations that facilitate people learning about the situations they are in, and about their relationships with each other [42]. Consequently, achieving high levels of psychological safety for participants is beneficial in helping to create a receptive space for learning and conversations between people to develop [51] and for nurturing participants' passion for learning within their projects [9].

Furthermore, perhaps due predominantly to the teaming of the eclectic membership most often seen in a project team, participants more often than not accept that no one person's view is entirely authoritative, or that individual participants have a claim to a privileged perspective on specific project actions, and consequently, ideas are often hotly contested between them. These characteristics are also indicative of a project team operating as a learning space [53]. In more established organizational structures, there are normalising pressures which work against creating such a learning space and these can include for example, traditional hierarchical arrangements and power structures causing people to defer to experts or authority figures to provide answers to issues rather than seek out diverse opinions/ideas from many others [42]. Such hierarchy dependence [9] established through the power or authority differentials between people can also inhibit their will and capacity to communicate freely their representations of experiences and associated emotions – which means, their accounts may become sanitized and censored and incomplete, which shuts down opportunities for people to learn [42]. Contrastingly, in a project team, all participants can demonstrate a greater freedom and propensity to think and to explore and to explicitly challenge issues (such as the degrees of managerial control they are subject to), and it is where they can more readily embrace differences and diversity and multiple meanings or realities [54]. Whilst pragmatically not divorced from the influences of organizational power arrangements and individual authority levels, the project form does however open up the opportunity to minimize, negate or to explore the negative learning effects of some power differentials observed and experienced between people in more permanent organizational settings. Accordingly, the uncustomary and transient power relationships experienced in a project team may also facilitate more 'open' exchanges and critical reflection practices on difficult project issues e.g. the participants' awareness of the distribution of expertise knowledge in the team and how best to utilize it for the benefit of the project and their learning and knowledge development. This then tends to encourage (by default or by design) partnership and imagination which helps develop trust between participants, which further fuels their conversation processes [51] and the building of their cross-functional learning relationships [9].

Practitioners may hardly recognize a project team as such a significant, socially complex and dynamic learning situation and indeed, treat any opportunity to attend to learning-on-the-job within this context with banality or ambivalence. As such, operational project structures within a team more often and predominantly reflect only the pursuit of tangible and rational project goals and exclude systematic and deliberate attention to any learning goals [9]. Even so, and as argued in this paper, project teams have the potential to be highly generative learning spaces because they are structurally and operationally differentiated to standard organizational forms. That is, their form may help participants cut across or dilute what Coopey [55] refers to as the fragmented, competitive and introspected social milieu of normal organizational life, so that they can learn and can learn to trust, build relationships and to grow - thereby build the human and social capital [56] of a project team. As also alluded to in this paper, this consideration of project teams as generative learning and knowledge creation spaces also signals a clear movement towards project learning being a more conspicuous, deliberate and systematic social activity by participants and not simply considered as a random, opportunistic and coincidental act that is grounded in one's project workplace experience. For practitioners, such conceptual shifts concerning project teams and their actions are quite profound and will challenge contemporary project practice.

On a different level, the implications of considering project teams as generative learning and knowledge creation spaces also supports a notion that projects can be viewed as learning episodes [12] or 'learning experiments' [57]. This proposition too, is not easily or readily embraced from a traditional project management perspective because it emphasizes divergent themes or objectives of searching, experimentation, discovery, innovation and risk at the project level [12]. Risk in particular is something that project stakeholders usually tend to want to minimize or avoid. Moreover, when engaging such divergent themes, projects may be considered even more so than usual as highly political events, because such a learning and knowledge focus may prompt challenges and changes to conventional project processes, project outcomes and even affect individual stakeholder identities within their host organizations. With such potentially significant impacts these challenges may prove difficult for practitioners to confront and to pragmatically engage with. Therefore, any initial actions to develop projects as learning episodes and project teams as learning spaces should be pursued prudently. Such development in itself can be viewed as an important learning exercise, whereby individual participants, a project team, a project and a host organization can ultimately realize significant knowledge benefits.

5 Conclusion

Beyond any myopic traditional perspective of project teams as only short-term and task-focused entities, they actually represent significant [and under-utilized] sites for situated learning and creativity to flourish, and the benefits to be realized through that, likely extend well beyond each project episode. However, the formative conditions of project teams in particular, often present very difficult circumstances for participant learning to readily germinate while they are interacting on-the-job. That being, participants may enter a project situation with immature attitudes towards learning in projects and they may not have previously interacted and formed any direct working and social relationships [58]. Compounding this condition further may be project time pressures which do not always facilitate a caring exchange of ideas between participants [59]. This situation resembles what Eskerod and Skriver [60] refer to as, lonely cowboys operating within their separate knowledge silos. In such a circumstance, and in recognition of the potential project and organizational benefits to be accrued, it would seem that organizations and project teams have a joint responsibility to arrange stimulant conditions that deliberately and actively encourage the learning exchanges and knowledge sharing interactions between participants – thereby, both accessing and contributing to the situated learning potential inherent in a project team space.

As explored in this paper, a seminal antecedent condition for such actions to occur however, involves a re-conceptualization of project teams as substantial social learning spaces. This constitutes an ontological base from which to forthrightly pursue productive, practical, locally relevant, and participant focused approaches towards learning-on-the-job within project teams. As such, learning (and knowledge creation) within a project team may become a more prominent and a more organized project action that energizes a team into becoming an increasingly creative unit. Given project participants are intimately connected to or embedded in complex webs of relationships external to a project team, any improvements realized in the learning attitudes and skills of participants within specific project episodes, may over time be readily disseminated into other operational and project settings – thereby, further contributing to organizational creativity and knowledge development.

References

1. Boud, D., Garrick, J.: Understandings of workplace learning. In: Boud, D., Garrick, J. (eds.) Understanding Learning at Work, pp. 1–12. Routledge, Great Britain (1999)
2. Hedberg, B.: How Organizations learn and unlearn. In: Nystrom, P.C., Starbuck, W.H. (eds.) Handbook of Organizational Design: Adapting Organizations to Their Environments, vol. 1, pp. 3–27. Oxford University Press, Oxford (1981)
3. Kezsbom, D., Edward, K.: The New Dynamic Project Management: Winning Through the Competitive Advantage, 2nd edn. John Wiley and Sons Inc., USA (2001)
4. Leonard-Barton, D.: The factory as a learning laboratory. Sloan Management Review, Fall, 23–38 (1992)
5. Leonard-Barton, D.: Wellsprings of Knowledge: Building and Sustaining the Sources of Innovation. Harvard Business School Press, USA (1995)
6. Antoni, M., Sense, A.J.: Learning within and across projects: A comparison of frames. Journal of Work, Human Environment and Nordic Ergonomics 2, 84–93 (2001)
7. Ayas, K., Zeniuk, N.: Project based learning: Building communities of reflective practitioners. Management Learning 32(1), 61–76 (2001)
8. Lundin, R.A., Hartman, F.: Pervasiveness of projects in business. In: Lundin, R.A., Hartman, F. (eds.) Projects as Business Constituents and Guiding Motives, pp. 1–10. Kluwer Academic Publishers, USA (2000)
9. Sense, A.J.: Cultivating Learning Within Projects. Palgrave Macmillan, England (2007a)
10. Lundin, R.A., Midler, C.: Evolution of project as empirical trend and theoretical focus. In: Lundin, R.A., Midler, C. (eds.) Projects as Arenas for Renewal and Learning Processes, pp. 1–9. Kluwer Academic Publishers, USA (1998)
11. Smith, B., Dodds, B.: Developing Managers Through Project Based Learning. Gower Publishing Limited, England (1997)
12. Arthur, M.B., DeFillippi, R.J., Jones, C.: Project based learning as the interplay of career and company non-financial capital. Management Learning 32(1), 99–117 (2001)
13. Morris, P.W.G.: Managing project management knowledge for organizational effectiveness. In: Proceedings of PMI® Research Conference, Seattle, USA, July 2002, pp. 77–87 (2002)
14. Sense, A.J.: Re-characterizing projects and project teams through a social learning lens. In: European Group for Organizational Studies (EGOS) conference, Amsterdam, Netherlands (July 2008)
15. Frame, J.D.: The New Project Management: Tools for an Age of Rapid Change, Corporate Reengineering, and Other Business Realities. Jossey-Bass Publishers, San Francisco (1994)
16. Block, R.: The Politics of Projects. Yourdon Press, New York (1983)
17. Cleland, D.I.: Project Management: Strategic Design and Implementation, 3rd edn. McGraw-Hill Companies Inc., USA (1999)
18. Turner, J.R.: The Handbook of Project-based Management, 2nd edn. McGraw-Hill Publishing Company, UK (1999)

19. Senge, P.M., Scharmer, O.: Community action research: Learning as a community of practitioners, consultants and researchers. In: Reason, P., Bradbury, H. (eds.) Handbook of Action Research: Participative Inquiry and Practice, pp. 238–249. Sage Publications Ltd., London (2001)

20. Brown, J.S., Duguid, P.: Organizational learning and communities of practice: Towards a unified view of working learning and innovation. Organization Science 2(1), 40–57 (1991)

21. Lave, J., Wenger, E.: Situated Learning: Legitimate Peripheral Participation. Cambridge University Press, Cambridge (1991)

22. Saint-Onge, H., Wallace, D.: Leveraging Communities of Practice for Strategic Advantage. Butterworth-Heinemann, Amsterdam (2002)

23. Wenger, E.: Communities of Practice: Learning, Meaning and Identity. Cambridge University Press, USA (1998)

24. Wenger, E., McDermott, R., Snyder, W.M.: Cultivating Communities of Practice: A Guide to Managing Knowledge. Harvard Business School Press, USA (2002)

25. Wenger, E., Snyder, W.: Communities of practice: The organizational frontier. Harvard Business Review, 139–145 (January-February 2000)

26. Park, P.: People, knowledge, and change in participatory research. Management Learning 32(2), 141–157 (1999)

27. Cook, S.D.N., Yanow, D.: Culture and organizational learning. Journal of Management Inquiry 2(4), 373–390 (1993)

28. Dixon, N.: The Organizational Learning Cycle: How We Can Learn Collectively, 2nd edn. Gower, England (1999)

29. Gherardi, S.: Learning as problem-driven or learning in the face of mystery?*" Organization Studies 20(1), 101–124 (1999)

30. Gherardi, S., Nicolini, D.: To transfer is to transform: The circulation of safety knowledge. Organization 7(2), 329–348 (2000)

31. Hildreth, P., Kimble, C., Wright, P.: Communities of practice in the distributed international environment. Journal of Knowledge Management 4(1), 27–38 (2000)

32. Brown, J.S., Duguid, P.: The Social Life of Information. Harvard Business School Press, Boston (2000)

33. Nonaka, I., Takeuchi, H.: The Knowledge Creating Company: How Japanese Companies Create the Dynamics of Innovation. Oxford University Press, USA (1995)

34. Weick, K.E.: Sensemaking in Organizations. Sage Publications Inc., USA (1995)

35. Billett, S.: Guided learning at work. Journal of Workplace Learning: Employee Counselling Today 12(7), 272–285 (2000)

36. Marsick, V.J.: New paradigms for learning in the workplace. In: Marsick, V. (ed.) Learning in the Workplace, pp. 11–30. Croom Helm, USA (1987)

37. Solomon, N.: Culture and difference in workplace learning. In: Boud, D., Garrick, J. (eds.) Understanding Learning at Work, pp. 119–131. Routledge, Great Britain (1999)

38. Greenwood, J.: Action research and action researchers: Some introductory considerations. Contemporary Nurse 3(2), 84–92 (1994)

39. Shani, A.B., Docherty, P.: Learning by Design: Building Sustainable Organizations. Blackwell Publishing Ltd., Oxford (2003)

40. Nonaka, I., Konno, N.: The concept of 'Ba': Building a foundation for knowledge creation. California Management Review 40(3), 40–54 (1998)

41. Kolb, D.A., Kolb, A.Y.: Learning styles and learning spaces: Enhancing experiential learning in higher education. Academy of Management Learning and Education 4(2), 193–212 (2005)

42. Coopey, J., Burgoyne, J.: Politics and organizational learning. Journal of Management Studies 37(6), 869–885 (2000)

43. Kolb, D.A., Baker, A.C., Jensen, P.J.: Conversation as experiential learning. In: Baker, A.C., Jensen, P.J., Kolb, D.A. (eds.) Conversational Learning: An Experiential Approach to Knowledge Creation, pp. 51–66. Quorum books, USA (2002)
44. Phillips, A.: Creating space in the learning company. In: Burgoyne, J., Pedler, M., Boydell, T. (eds.) Towards the Learning Company: Concepts and Practices, pp. 98–109. McGraw-Hill Book Company Europe, Great Britain (1994)
45. Sense, A.J.: Learning generators: Project teams re-conceptualized. Project Management Journal 34(3), 4–12 (2003)
46. Wenger, E.: Communities of practice and social learning systems. In: Nicolini, D., Gherardi, S., Yanow, D. (eds.) Knowing in Organizations: A Practice-based Approach, pp. 76–99. M.E. Sharpe Inc., USA (2003)
47. Solomon, N., Boud, D., Rooney, D.: The in-between: Exposing everyday learning at work. International Journal of Lifelong Education 25(1), 3–13 (2006)
48. Billett, S.: Workplace participatory practices: Conceptualising workplaces as learning environments. Journal of Workplace Learning 16(6), 312–324 (2004)
49. Davenport, T.H., Prusak, L.: Working Knowledge: How Organizations Manage What They Know. Harvard Business School Press, USA (1998)
50. Seibert, K.W., Daudelin, M.W.: The Role of Reflection in Managerial Learning: Theory, Research, and Practice. Quorum Books, USA (1999)
51. Baker, A.C.: Receptive spaces for conversational learning. In: Baker, A.C., Jensen, P.J., Kolb, D.A. (eds.) Conversational Learning: An Experiential Approach to Knowledge Creation, pp. 101–123. Quorum books, USA (2002)
52. Gray, C.F., Larson, E.W.: Project Management: The Managerial Process, 2nd edn. Irwin/McGraw-Hill, Boston (2003)
53. Fulop, L., Rifkin, W.D.: Representing fear in learning in organizations. Management Learning 28(1), 45–63 (1997)
54. Rifkin, W., Fulop, L.: A review and case study on learning organizations. The Learning Organization 4(4), 135–148 (1997)
55. Coopey, J.: Learning to trust and trusting to learn: A role for radical theatre. Management Learning 29(3), 365–382 (1998)
56. Boxall, P., Purcell, J.: Strategy and Human Resource Management, 2nd edn. Palgrave Macmillan, China (2008)
57. Björkegren, C.: Learning for the Next Project: Bearers and Barriers in Knowledge Transfer within an Organization, Licentiate thesis No. 32 from the International Graduate School of Management and Industrial Engineering. Institute of Technology, Linköpings Universitet, Linköping, Sweden (1999)
58. Sense, A.J.: Structuring the project environment for learning. International Journal of Project Management, Special issue on the European Academy of Management (EURAM 2006) Conference 25(4), 405–412 (2007b)
59. Koskinen, K.U.: Boundary brokering as a promoting factor in competence sharing in a project work context. International Journal of Project Organisation and Management 1(1), 119–132 (2008)
60. Eskerod, P., Skiver, H.J.: Organizational culture restraining in-house knowledge transfer between project managers – A case study. Project Management Journal 38(1), 110–122 (2007)

Competence Management in Knowledge-Based Organisation: Case Study Based on Higher Education Organisation

Przemysław Różewski and Bartłomiej Małachowski

West Pomeranian University of Technology in Szczecin,
Faculty of Computer Science and Information Systems,
ul. Żołnierska 49, 71-210 Szczecin, Poland
{prozewski,bmalachowski}@wi.ps.pl

Abstract. Authors present the method and tools for competency management in knowledge-base organisation. As an example, the higher education organisation working under Open and Distance Learning condition is examined. Competency management allows to management knowledge in the efficient and effective way on the stuff's and student's level. Based on the system analysis the management model for educational organisation was created. Next, method for quantitative competence assessment was presented. Moreover, the authors explore possibilities of integrating new competence description standards with existing mathematical methods of competence analysis into one common framework. The product of this idea is the concept of reusable Competence Object Library.

Keywords: competence, competence management, educational organisation, open and distance learning.

1 Introduction

The continuous development of Web systems allows to created complete Internet working environment on different organization level. This situation is special notable in knowledge management systems [23]. The knowledge management systems are a crucial component of modern organization [25]. In most cases the knowledge management systems decided about competitive advantage of organization [32]. New approach for knowledge management expends the organization dimension on competence concept [31]. Application of competence in companies allows to optimized task time, made more adequate human resources management and more efficient knowledge transfer process [29]. In addition competence concept plays important role in employ process. Every position can be described very precisely by the competence and related qualification [12].

The competence management system becomes important elements of the knowledge-base organization ecosystem. One of the good examples of the knowledge-base organization is a modern educational organization. Modern universities can be seen as continuously changing learning organizations [42]. An educational organization is

D. Karagiannis and Z. Jin (Eds.): KSEM 2009, LNAI 5914, pp. 358–369, 2009.
© Springer-Verlag Berlin Heidelberg 2009

considered as a complex system aimed at delivering knowledge to students in order to expand their competences to the expected level [35]. The idea of life-long learning (LLL) causes development of educational market. Moreover, the distance learning tools and system are very effective method of knowledge transfer [19]. In addition the concept of Open and Distance Learning (ODL) secures proper organization framework for LLL [33], [27].

In the paper authors discussed the concept of the competence management. In the first part different aspects of competence concept are analysed. In the second part the structure of the conceptual model of educational organization is presented. The proposed approach takes advantage of the theory of hierarchical, multilevel systems in order to define the decision-making process in educational organisation. The educational organisation is treated as a knowledge-based organisation. Then, on the basis of the fuzzy competence set model the cost estimation method is proposed. The method focuses on the cost of personal competence expansion caused by the knowledge development process. In the last part reusable Competence Object Library (COL) is proposed. The COL corresponding to the idea o competence management and can be used as a base for matching information system.

2 Research over the Competence Concept

2.1 Competence Concept

Within the space of the last decades many researches representing different fields of science and humanities (ex. sociology, philosophy, psychology, pedagogy and education etc.) attempted to give the definition of the term 'competence'. Shapes of these definitions were strongly conditioned by the context of their studies which resulted in many different meanings of competence that can be found in the literature. Romainville claims, that French word 'compétence' was originally used in context of vocational training to describe capabilities to perform a task [30]. In the later period the term competence settled in the general education, where it was mainly related to "ability" or "potential" to act effectively in a certain situation. According to Perrenoud competence is not only a knowledge how to do something, but it is a ability to apply it effectively in different situations [28]. The European Council proposed to define competence as general capabilities based on knowledge, experience, values that can be acquired during learning activities [3].

Many management researchers, for example [2], [26], [4] acknowledge the work of Boyatzis called "The competent manager" [1] as the first work that popularized and introduced the notion of competence into management literature. Boyatzis defined competence as description of a person characterized by motivation and skills expressed in his or her personal image, social role or actively applied knowledge. Such a broad definition could be use for any distinctive feature of a someone's personality, however Boyatzis narrowed it only to these individual features that are important in professional activities performed at workplace. Moreover, he distinguished competences from tasks and functions assigned to employees in their organizations claiming that competence was a kind of an added value brought into professional duties [1].

Woodruff defined competence as a set of behavior patterns required to successfully perform a task or function. According to him competence can comprise knowledge, skills and capabilities but also can exceed these traditional characteristics by taking into account some other personal factors like motivation and intentions [37]. Some researches consider even more so called "below-the-waterline" characteristics of a person like individual personality, temperament, outlook on life, etc. and claim that these factors can have equal or sometimes even more significant influence of professional tasks performance, than for example technical skills [10].

In the literature the notion of competence model is often used interchangeably with the term competence. Mansfield defined competence model as detailed, behavioral description of employee characteristic required to effectively perform a task [22]. The form of this description can be of any kind – from verbal description to formal, mathematical model. Competence model can be used to describe a set of competences connected with a task, job or role in organization.

The notion of competence and competence model is also researched by human resource management (HRM) specialists. In this field of management, competence models are considered as useful tool in solving main HR problems like recruitment and selection, training and development, performance evaluation, promotion and redundancy, payroll management, etc. [26]. Elaboration of different competence models provided efficient techniques for stuff and job description and analysis.

The above review of competence definitions shows, that its meaning is very broad and sometimes even ambiguous. Very brief and precise definition is provided by the International Standard Organization in its ISO 9000:2005 standard "Quality Management Systems" [17]. The definition that can be found in this document describes competence as "demonstrated ability to apply knowledge and skills".

The recent research initiatives on competence provide more and more elaborated formal models of competence. The good example of such an initiative is TENCompetence Project – a large research network founded by European Commission through the Sixth Framework Programme with fifteen partners throughout Europe [34]. The main aim of this project is to develop a technical and organizational infrastructure to support lifelong learning in Europe. Their research effort is focused mainly on the problem of managing personal competences. Within this effort the TENCompetence project group developed the TENCompetence Domain Model (TCDM) that covers many important issues related to the notion of competence [34]. TENCompetence project team proposed the definition of *competence*, which defines this term as "effective performance within a domain/context at different levels of proficiency". This definition of competence was contrasted with other definitions of competence and especially with definitions of *competency* from IEEE RDC [15] and HR-XML [12]. Comparison of these definitions shows, that competency is more general term for describing skills, knowledge or abilities (for example: *Java programming*), while competence is more precise by adding information about context and proficiency level (for example: *advanced Java web programming*). Furthermore, TCDM group proposed collecting competences into *competence profiles* and distinguished *acquired competence profile* (competences already acquired by a person) and *required competence profile* (competences to be acquired by a person). TCDM comprising notions like competency, competence, context, proficiency level and other related to personal competence management was built in as the UML class diagram.

2.2 Competence Applications

The competence analysis based mainly on the visible result. In similarity to knowledge the competence concept cannot be direct examined. The competence is a construct concept which is derived or inferred from existing instance [21]. Proper indicators of construct recognitions rely on many different factors: social, cultural and cognitive [5]. The concept of competence has become an important issue for researcher over the world [31], [32]. Tab. 1 discusses the competence issue following two main competence applications: job description and leaning outcomes description.

Table 1. Main competence application analysis

Competence application	Job description	Leaning outcomes description
Definition	Competence is shows by the actor who plays some role. a observable or measurable ability of an actor to perform necessary action(s) in given context(s) to achieve specific outcome(s) [16]	Competence is an outcome from learning process in Life-long Learning paradigm. The competence is defined as any form of knowledge, skill, attitude, ability or learning objective that can be described in a context of learning, education or training [14].
Motivation	The reason of research over the competence concept is believe of many managers that analysis of organization's structure and resources is not satisfactory. The managers focus on the output of competencies [7].	Open and Distance Learning [33] presumes that students are mobile across different university and educational systems in the frame of common learning framework i.e. European Higher Education Area [20]. Moreover, the Life-long Learning concept assumes that the student's knowledge can be supplemented and extended in other educational system over the life time. For both of these concepts the well established and transparent method for student's achievement recording is required.
Main characteristics	The competence in job context is a dynamic system and provides systemic, dynamic, cognitive and holistic framework for building management theory [32]. Moreover competence explains the formal and informal way in which human beings interact in the border of technology, human begins, organisation, culture [7], [6].	The student's competence can be certificated on every single step of training, learning, etc [11]. The competence achievement is mainly founded on the cognitive process [20]. However, the competence, likewise other human's characteristic, can be discussed from several different angels: pedagogical, philosophical and psychological [9].
Standards	HR-XML [12], ISO 24763 [16], TENCompetence [34]	IMS RDCEO [15], IEEE 1484.20.1 [14]
Projects		European Qualifications Framework (EQF) [8], ICOPER [13], TENCompetence [34]

2.3 Mathematical Model of Competence

The concept of competence still lacks formal models allowing to create quantitative methods for competence analysis. One of the most advanced idea of this type is the approach called *competence sets*. This approach was for the first time introduced by Yu and Zhang [38], [40] and is closely related to the concept of *habitual domain* [39].

These authors define competence set as the set containing skills, information and knowledge of a person (acquired competence set denoted *Sk*) or related to a given job or a task (required competence set *Tr*).

Most publications on competence sets deal with optimization and cost analysis of the competence set expansion process. This process is described as obtaining new skills and adding them to the actual acquired competence set of a person. The cost and pace of obtaining new skills depends on elements of actual competence set and how close these elements are related with the new skill. Methods of optimal competence set expansion consist of determining the order of obtaining successive competences that provides minimal cost. Competences that need to be obtained are defined by set Tr(E)\Sk(P). The optimisation problem is usually solved by finding the shortest path in an oriented graph, in which vertices represent competences and arcs represent the relations between them [38].

In the early stage of the research on competence sets, competence was presented as a classical set containing knowledge, skills and information necessary to solve a problem. However, assessing the presence of a competence in binary terms – one has a competence or not at all – turned out to be insufficient regarding the continuous nature of competence. On account of that, it was proposed to present human competence as a fuzzy set, defined as follows [36]:

$$A = \{(x, \mu_A(x)) | x \in X\}$$

where: $\mu_A(x)$ is the membership function assessing the membership of an element x in relation to set A by mapping X into membership space $[0; 1]$, $\mu_A : X \rightarrow [0; 1]$.

Regarding the definition of the fuzzy set it is possible to define the notion of *fuzzy competence strength*. For each competence g, its strength is a function of a person P or an event E in the context of which the competence is assessed: $\alpha : \{P \text{ or } E\} \rightarrow [0; 1]$

Expansion optimization methods of fuzzy competence sets are computationally more demanding but provide better accuracy and reproduction of nature of competence. Vast set of competence set expansion cost analysis methods can be found in literature [36],[38],[40].

The methods for competence set analysis with its quantitative models provides good background for development of competence management system. Example of a system build around these methods can be found in [18].

2.4 Quantitative Assessment of Competence

The methods of competence set expansion cost assessment described in Section 2.3 can be applied to perform quantitative analysis of personal competence, that allows to design and build different systems for competence management. Possible applications of this approach are vast, for example in HRM the cost of competence expansion can be used as criterion for candidate selection, in system for education competence set expansion cost can be a measure for individual learning path design.

The very significant issue in quantitative competence assessment is proper identification of the acquired (*Sk*) and required competence set (*Tr*). In case of required competence set identification it is useful to take advantage of universally accepted standards (ex. ISO), different best practice databases or taxonomies. Identification of acquired competence set comes down to detect competences from the set *Tr* in a person being tested. The process of quantitative competence assessment is presented in details in Figure 1.

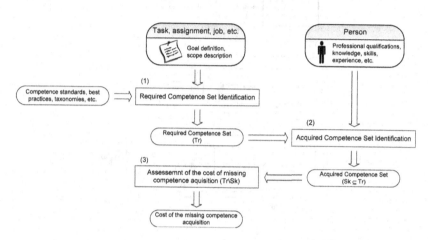

Fig. 1. Assessment of competence expansion cost

3 Educational Organization Analysis

Modern educational organisation is focused on the processing of information-knowledge-competence sequence. In this process, all members of educational organisation participate: students, university's stuff, and teachers. Based on the management model of educational organisation we analyzed the decision about competence on all management levels. Presented management model is created base on theory of hierarchical, multilevel systems [24]. There are two main reasons that educational organization is changing and becoming knowledge-base organization: (i) Open and Distance Learning concept [33], which creates open educational market based on standard and unification framework, (ii) Life-Long Learning [31]: which creates continuous requests for education and knowledge.

The analysis leads to identification of four embedded management cycles, which differ in their operating time (Tab. 2). As can be seen in fig. 2, each cycle includes a process that is being arranged by a certain decision maker [18]. Within the time limit of each cycle, the system decision-maker compares knowledge areas in order to make decisions by estimating their content and depth [42]. The management model described the process of sequential knowledge processing during: (i) syllabus preparation, (ii) providing education services, (iii) developing didactic materials, (iv) process of acquiring competences based on a specified knowledge model and (v) statistical evaluation of students' progress.

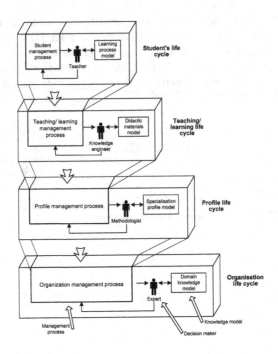

Fig. 2. Management model of educational organisation

Table 2. Educational organization's life cycles and competence processes

Name	Description	Competence management process	Tools
Organization life cycle	Subsystem of educational organization strategic management aims at maintaining a high position of organization's graduates at the job market.	Identification of commences which are demanded by the market.	Knowledge engineering
Profile life cycle	Subsystem of managing the process of adapting competence to the student's profile	Market's competences adaptation in student's profile	Curriculum design
Teaching/ learning life cycle	Subsystem meant to provide an intelligent and network space for the learning/teaching system, assuring effective use of network environment and developing or adapting knowledge repository to student's profile and students contingent.	Market's competences adaptation to the didactical material in every course. Ensure that each competence will be posses by the student in limited time.	Didactic material design
Student's life cycle	Subsystem that allows maintaining and monitoring administrative correctness of the learning process, and evaluating the competence-gaining process with the given knowledge model and teaching/learning system	Student's learning process, competence transfer based on the didactic material and cognitive methods.	Student support

4 Competence Object Library

Scientific research in the domain of competence and competence management has been developing dynamically recently. The biggest effort of the researchers is put on notions definition and competence structure modeling. Among several big initiatives researching competence some of them like IEEE Reusable Competency Definition [15] and HR-XML [12] has made a big step forward defining well common models for competency interoperability. The most complete model of this type was proposed by TENCompetence project [34]. TENCompetence Domain Model (TCDM) proposed by this initiative provides complex structure model supporting development of software tools for competence management purposes. However, TCDM still lacks mathematical models of competence, that would enable development of system performing quantitative analysis of competence.

One of the ideas of this article is to propose concept of reusable Competence Object Library (COL) that integrates into one framework objective data structures proposed in TCDM with the method for fuzzy competence set expansion cost analysis (FCSECA) proposed in [36] (Fig. 3). This library would allow rapid development of competence management systems for different purposes, for example: human resource management (employee assessment, staffing), e-learning (learning progress assessment, individual learning path planning, etc.)

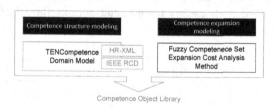

Fig. 3. Concept of the Competence Object Library

4.1 Conceptualization of the Competence Object Library

Two approaches TCDM and fuzzy competence sets provides completely different perspective of competence modeling. The first one is very precise in terms of notions definitions, structure modeling and link with competency standards, while the second one defines competence on very general level but provides accurate and verified methods for quantitative analysis. Development of software tools for advanced competence analysis requires qualities of both of these approaches: complete and well defined data structures and algorithms providing quantitative methods.

One of the main problem in integrating two approaches into one framework is to properly match notions taking into consideration their semantic meaning. For example fuzzy competence set analysis does not introduce the notion of *competency*. Semantically the closest notion to *competency* is *skill*. *Competence* in both approaches has different definitions and cannot be considered as semantically identical. Table 3 contains main notions from TENCompetence DM and Fuzzy Competence Set Analysis. A row from the table indicates the closest semantic equivalents. This allows to compare namespaces of two approaches and define namespace for COL.

Table 3. Comparison of main notions from COL with notions form its source approaches

TENCompetence Domain Model	Fuzzy Competence Set	Competence Object Library
Competency	Skill	Competency
Competence	Competence set	Competence
		Element of competence
		Competence set
Context	Habitual domain	Context
Competence profile	-	Competence profile
Proficiency Level	Competence strength	Proficiency level
		Competence strength

4.2 Data Structures for the Competence Object Library

Theoretical analysis of two source approaches for Competence Object Library provides background for its structure modelling. The COL structure was modelled with UML as UML Class Diagram. The choice of UML was determined by the fact that it is the most common and recognizable notation among software engineering professionals.

The UML Class Diagram depicted in Figure 4 is a synthesis of two background approaches: TENCompetence Domain Model and Fuzzy Competence Sets. The basis for class structure modelling comes from TENCompetence DM. The original structure was extended with classes representing notions form Fuzzy Competence

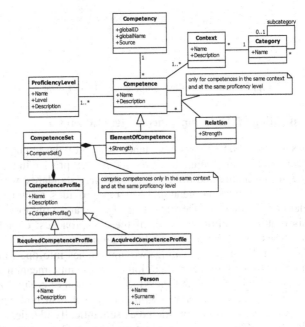

Fig. 4. UML Class Diagram for Competence Object Library

Sets in order to adapt COL to its analytical methods operating on sets, relations and graphs defined by this relations.

Implementation of competence expansion cost computing algorithm [36] can be called with methods CompetenceSet.CompareSet() and Competence-Profile.CompareProfile(). These methods return a real value proportional to the difference between two competence sets/profiles. Details of UML Class Diagram are presented in Figure 4, while Table 4 describes all classes defined in the library.

Table 4. Competence Object Library classes

Class name	Description
Competency	Any form of knowledge, skill, attitude, ability or learning objective that can be described in a context of learning, education, training or any specific business context.
Competence	Effective performance of a person within a context at a specific level of proficiency.
Context	Circumstances and conditions surrounding actions performed by a person.
Category	Indicates the relative level in a taxonomic hierarchy.
Proficiency Level	Indicates the level at which the activity of a person is considered.
Relation	Arbitrary association of competences within a context and at specific proficiency level.
Element of Competence	Entity derived from competence that can form a set.
Competence Set	Collection of elements of competence.
Competence Profile	Collection of competence sets.
Required Competence Profile	Requirements in terms of competence to be fulfilled by a person.
Acquired Competence Profile	Description of competences possessed by a person.
Vacancy	Activity, job, skill, attitude, ability or learning objective for which competence requirements can be specified.
Person	Competent actor performing activities.

5 Conclusion

The competence management issue is directly related to knowledge management. Both knowledge and competence management require development of new tools and methods which will be working on the cognitive level in quantitative way.

The concept of Competence Object Library proposed in the article allows rapid development of applications for competence management and analysis. The main idea of COL is to provide HR and e-Learning professionals with reusable library allowing developing applications for quantitative competence analysis. Additionally, the interface to open competence exchange standards like IEEE RCD and HR-XML creates new possibilities for web-centric HR and e-Learning systems.

References

1. Boyatzis, R.E.: The competent manager: a model for effective performance. Wiley, New York (1982)
2. Cardy, R.L., Selvarajan, T.T.: Competencies: Alternative frameworks for competitive advantage. Business Horizons 49, 235–245 (2006)

3. Council of Europe: Key competencies for Europe. Report of the Symposium in Berne March 27-30, 1996, Council of Europe, Strasbourg (1997)
4. Crawford, L.: Senior management perceptions of project management competence. International Journal of Project Management 23, 7–16 (2005)
5. Doninger, N.A., Kosson, D.S.: Interpersonal construct systems among psychopaths. Personality and Individual Differences 30(8), 1263–1281 (2000)
6. Drejer, A., Riis, J.O.: Competence development and technology: How learning and technology can be meaningfully integrated. Technovation 19(10), 631–644 (1999)
7. Drejer, A.: How can we define and understand competencies and their development? Technovation 21(3), 135–146 (2001)
8. EQF: European Qualifications Framework for Lifelong Learning, http://ec.europa.eu/education/lifelong-learning-policy/doc44_en.htm
9. Frixione, M.: Tractable Competence. Minds & Machines 11(3), 379–397 (2001)
10. Hofrichter, D.A., Spencer, L.M.: Competencies: The right foundation for effective human resource management. Compensation and Benefits Review 28(6), 21–24 (1996)
11. Holmes, G., Hooper, N.: Core competence and education. Higher Education 40(3), 247–258 (2000)
12. HR-XML: HR-XML Consortium (2006), http://www.hr-xml.org/
13. ICOPER - Interoperable Content for Performance in a Competency-driven Society. An eContentplus Best Practice Network (2008–2011), http://www.icoper.org/
14. IEEE 1484.20.1/draft - draft standard for Reusable Competency Definitions (RCD), http://ieeeltsc.wordpress.com/working-groups/competencies/
15. IMS RDCEO: Reusable Definition of Competency or Educational Objective (RDCEO), http://www.imsglobal.org/competencies
16. ISO 24763/draft: Conceptual Reference Model for Competencies and Related Objects (2009)
17. ISO 9000:2005, Quality Management Systems
18. Kusztina, E., Zaikin, O., Różewski, P., Małachowski, B.: Cost estimation algorithm and decision-making model for curriculum modification in educational organization. European Journal of Operational Research 197(2), 752–763 (2009)
19. Kusztina, E., Zaikin, O., Różewski, P.: On the knowledge repository design and management in E-Learning. In: Lu, J., Ruan, D., Zhang, G. (eds.) E-Service Intelligence: Methodologies, Technologies and applications. Studies in Computational Intelligence, vol. 37, pp. 497–517. Springer, Heidelberg (2007)
20. Kusztina, E., Zaikine, O., Różewski, P., Tadeusiewicz, R.: Competency framework in Open and Distance Learning. In: Proceedings of the 12th Conference of European University Information Systems EUNIS 2006. Tartu, Estonia, pp. 186–193 (2006)
21. Lahti, R.K.: Identifying and integrating individual level and organizational level core competencies. Journal of Business and Psychology 14(1), 59–75 (1999)
22. Mansfield, R.S.: Building competency models: Approaches for HR professionals. Human Resource Management 35(1), 7–18 (1996)
23. Marwick, A.D.: Knowledge management technology. IBM Systems Journal 40(4), 814–830 (2001)
24. Mesarovic, M.D., Macko, D., Takahara, Y.: Theory of Hierarchical. Multilevel Systems. Academic Press, New York (1970)
25. Nonaka, I.: A Dynamic Theory of Organisational Knowledge Creation. Organisation Science 5(1), 14–37 (1994)

26. Partington, D., Pellegrinelli, S., Young, M.: Attributes and levels of programme management competence: an interpretive study. International Journal of Project Management 23, 87–95 (2005)
27. Patru, M., Khvilon, E.: Open and distance learning: trends, policy and strategy considerations. UNESCO, kod: ED.2003/WS/50 (2002)
28. Perrenoud, P.: Construire des compétences dès l'école. Pratiques et enjeux pédagogiques. ESF éditeur. Paris (1997)
29. Petts, N.: Building Growth on Core Competences – a Practical Approach. Long Range Planning 30(4), 551–561 (1997)
30. Romainville, M.: L'irrésistible ascension du terme compétence en éducation, Enjeux, no. 37/38 (1996)
31. Sampson, D., Fytros, D.: Competence Models in Technology-Enhanced Competence-Based Learning. In: Adelsberger, H.H., Kinshuk, Pawlowski, J.M., Sampson, D. (eds.) Handbook on Information Technologies for Education and Training, 2nd edn., pp. 155–177. Springer, Heidelberg (2008)
32. Sanchez, R.: Understanding competence-based management: Identifying and managing five modes of competence. Journal of Business Research 57(5), 518–532 (2004)
33. Tait, A.: Open and Distance Learning Policy in the European Union 1985–1995. Higher Education Policy 9(3), 221–238 (1996)
34. TENCompetence - Building the European Network for Lifelong Competence Development, EU IST-TEL project (2005–2009), http://www.tencompetence.org/
35. Valadares, T.L.: On the development of educational policies. European Journal of Operational Research 82(3), 409–421 (1995)
36. Wang, H.-F., Wang, C.H.: Modeling of optimal expansion of a fuzzy competence set. International Transactions in Operational Research 5(5), 413–424 (1995)
37. Woodruffe, C.: What is meant by competency? In: Boam, R., Sparrow, P. (eds.) Designing and achieving competency. McGraw-Hill, New York (1992)
38. Yu, P.L., Zhang, D.: A foundation for competence set analysis. Mathematical Social Sciences 20, 251–299 (1990)
39. Yu, P.L.: Habitual Domains. Operations Research 39(6), 869–876 (1991)
40. Yu, P.L., Zhang, D.: Optimal expansion of competence sets and decision support. Information Systems and Operational Research 30(2), 68–85 (1992)
41. Zadeh, L.A.: Fuzzy sets. Information and Control 8, 338–353 (1965)
42. Zaikin, O., Kusztina, E., Różewski, P.: Model and algorithm of the conceptual scheme formation for knowledge domain in distance learning. European Journal of Operational Research 175(3), 1379–1399 (2006)

Knowledge Maturing Services: Supporting Knowledge Maturing in Organisational Environments

Karin Schoefegger[1], Nicolas Weber[1,2], Stefanie Lindstaedt[1,2], and Tobias Ley[2]

[1] Graz University of Technology
k.schoefegger@tugraz.at, nweber@know-center.at, slind@know-center.at
[2] Know-Center
tley@know-center.at

Abstract. The changes in the dynamics of the economy and the corresponding mobility and fluctuations of knowledge workers within organizations make continuous social learning an essential factor for an organization. Within the underlying organizational processes, Knowledge Maturing refers to the the corresponding evolutionary process in which knowledge objects are transformed from informal and highly contextualized artifacts into explicitly linked and formalized learning objects. In this work, we will introduce a definition of Knowledge (Maturing) Services and will present a collection of sample services that can be divided into service functionality classes supporting Knowledge Maturing in content networks. Furthermore, we developed an application of these sample services, a demonstrator which supports quality assurance within a highly content based organisational context.

1 Introduction

Since the world became more and more dynamic in recent years, the mobility and fluctuations of knowledge workers within organizations made continuous learning throughout one's career an essential factor. Within underlying organizational learning processes, knowledge is passed on, learned and taught. Corresponding knowledge objects are transformed from informal and highly contextualized artifacts into explicitly linked and formalized learning objects. This transformation of knowledge artifacts together with the corresponding organizational learning processes, is commonly known as Knowledge Maturing. The knowledge maturing process theory structures this process into five phases: expressing ideas, distribution in communities, formalization, ad-hoc learning and finally, formal training (see [Maier and Schmidt, 2007] and [Schmidt, 2005]). Knowledge Maturing Services refer to integrated support for this knowledge maturing process, bridging the separation along the dimensions of knowledge construction and knowledge sharing. They are needed not only to help knowledge workers to handle different knowledge assets, but also to entice them in sharing and negotiating among them.

D. Karagiannis and Z. Jin (Eds.): KSEM 2009, LNAI 5914, pp. 370–381, 2009.

In this work, we will present a definition of the concept of knowledge (maturing) services and in addition to that, five service functionality classes of knowledge maturing services will be described in detail where each of the described sample services can be classified into. These service functionality classes for classification of knowledge maturing services support negotiation,formalization, standardization and deployment of knowledge assets.

This paper is organized as follows: In section 2 we will present a definition of knowledge services in general and knowledge maturing services in particular, including related work. Section 3 introduces five service functionality classes and corresponding example services including their detailed description and impact on knowledge maturing and organizational learning. Finally, section 4 discusses our work and concludes the paper.

2   Knowledge Services

We define Knowledge Services as (composite) software services that are concerned with three knowledge entities people, content, and semantic structures and the relationships between them. The results of Knowledge Services improve or extend the knowledge available within these three entities and their relationships. This can be achieved by either enabling people to add or improve knowledge contained in the three entities or by providing automated services to discover knowledge based on the available knowledge entities and their relationships. In doing so Knowledge Services enable a knowledge (eco)system (comprised of the three knowledge entities) to learn. This learning can take place on very different levels. For example, people can learn by interacting with other people, content, and structures; here the knowledge entity improved is people. On the other hand, the (eco)system can learn more about its users by e.g. analyzing their interactions with people, content, and structures; here the knowledge entity improved might be a structure representing a user (e.g. user profile).

Knowledge workers need Knowledge Services to improve (learn) the knowledge (eco)system. Whereas the technical definition of services is supported by a set of standards (such as web services), it is the conceptual part (i.e. defining types of services that are useful) that is currently lacking. But exactly this conceptual part matters most when organizations attempt to profit from the promised benefits of service-oriented architectures.

Knowledge Maturing Services are a form of complex Knowledge Services, which in turn are composed of basic services. These may be either already offered in heterogeneous systems as part of an enterprise application landscape, implemented additionally to enrich the services offered in an organization or invoked over the Web from a provider of maturing services. They operate on several kinds of external knowledge representations, [Maier and Schmidt, 2007] suggests several of these knowledge representations that are worth considering: contents, semantics and processes. These are represented in external knowledge artifacts but also have their corresponding representation in individual or collective knowledge. Accordingly, we initially differentiate Maturing Services into two broad categories according to the type of external knowledge representation they operate on: **Structure Services** operate on a more or less formal knowledge structure. Semantic structures (corresponding to conceptual or declarative knowledge) and process structures (corresponding to procedural knowledge and drawing out temporal characteristics) can be distinguished. **Content Services** operate mainly on texts composed of natural language. **Usage Services** act as a third category of Maturing Services. Knowledge Maturing Services mostly work in the background to analyze several kinds of knowledge contents, processes, structures and their use within an organization to discover emergent patterns and support individuals, communities or organizations in dealing with the complexities of these underlying structures and their evolution over time.

Related Work

The concept of Knowledge Services has surfaced within research literature only recently. Due to the novelty of the concept it is not surprising that a widely agreed definition is not yet available. Nevertheless, it can be traced back to at least two schools of thought: Knowledge Management Systems (KMS) and Knowledge Market.

The Knowledge Management Service concept (and its subset Knowledge Services) were developed in response to monolithic knowledge management systems (KMS). Currently, one can observe a clear convergence of application development and service oriented application development. Important approaches in research are service oriented architectures (SOA) and (semantic) web service technologies. Knowledge Management Services or Knowledge Services are a subset of services, both basic and composed, whose functionality supports high-level knowledge management instruments as part of on-demand knowledge management initiatives, e.g., find expert, submit experience, publish skill profile, revisit learning resource or join community-of-interest ([Maier and Remus, 2008]). These services might cater to the special needs of one or a small number of organizational units, e.g., a process, work group, department, subsidiary, factory or outlet in order to provide solutions to defined business problems. Knowledge Management Services describe aspects of Knowledge Management instruments supported by heterogeneous application systems. [Dilz and Kalisch, 2004] and [Maier et al., 2009] both propose similar typologies of Knowledge Management Services. A related definition of Knowledge Services by [Mentzas et al., 2007] defines them as services for knowledge trading and managing electronic knowledge markets.

The knowledge services which will be described in the next section are specific services aiming at identifying and supporting Knowledge Maturing. Several kinds of Knowledge Services have already been developed for certain purposes: context-aware knowledge services ([Rath et al., 2008]), competence identification services ([Lindstaedt et al., 2006]), work-integrated learning services ([Lindstaedt et al., 2008], [Lindstaedt et al., 2007]) but in addition to that, other knowledge services can be envisioned, for example collaboration services, etc.

3   Service Classes

Knowledge maturing services can act on both, the organizational and personal perspective to support knowledge workers in their daily work and learning process. Whereas organizational learning (and maturing) services provide structural guidance that enables users to collaboratively develop knowledge assets of organizational interest and to align them with organizational goals, personal learning (and maturing) services involves guidance from the perspective of the guided person, i.e. it makes the results of the maturing activities visible and digestible for the individual.

This work is part of MATURE (http://mature-ip.eu/en), an ongoing EU-founded project, whose objective is to understand the maturing process and provide maturing support for knowledge workers in a collaborative environment. The sample Knowledge Maturing Services for each of the service classes described in the following, were gained from a maturing scenario to support quality assurance within the context of highly content based organisational environment, based on results of a previous design study ([Weber et al., 2009]). The underlying maturing scenario for this field is described in detail in [Attwell et al., 2008] where maturity stages necessary for a new knowledge worker in the field of career guidance are described. These maturing phases include access and search of information, aggregation of information, manipulation of documents, analyze and reflect previous work, present and represent created knowledge objects, share of information and finally network with other people.

The derived service functionality classes for classification of knowledge maturing services supporting the knowledge maturing phases for career organizations are the following: (I) Orientation Services, (II) Inspiration Services, (III) Creation and Refinement Services, (IV) Collaboration Services and (V) Reflection Services and will be described in detail in the following sections. Figure 1 depicts the interplay of these five knowledge maturing service classes. The service classes are arranged in a circle to emphasize that different services used by knowledge workers during their working process (to e.g. create a document of high quality for training) do not follow a strict consecutiveness of the use of services of one service classes and then from another, they rather enable a flexible process of knowledge object creation and refinement with iterative use of collaboration and reflection services to enable knowledge maturing in this setting. Though the service functionality classes for knowledge maturing support were derived from this setup of knowledge workers in the career guidance sector which is heavily content dependent, they can easily be adopted for other setting where knowledge is heavily content and structure based.

3.1   Development and Evaluation of the Service Demonstrator

As already mentioned, the description of the services (and corresponding software requirements), which we will describe in this section, were derived from a real-world setting to support quality assurance within the corresponding maturing scenario in career guidance sector. This scenario is heavily content dependent,

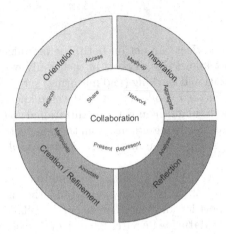

Fig. 1. Interactions between Service functionality classes: It should not be seen as a strict consecutiveness usage of service, they enable rather a flexible process of knowledge creation with iterative use of collaboration and refection services

such that we have chosen as a basis for the demonstrator the MediaWiki, which allows community driven knowledge creation. A first prototype has already been developed and evaluated within a design study ([Weber et al., 2009]) and the results of this evaluation including a more detailed requirement elicitation lead to the development of a demonstrator enhancing the existing services and providing additional ones. We used a bottom-up and top-down approach to meet both, the software requirements for implementing these knowledge maturing services within our project and the requirements of knowledge workers within career organizations by regular feedback from potential end-users throughout the software development cycle. The implementation of these services was done using rapid prototyping which involved iterative design phases using mock-ups and development phases combined with regular input to generate feedback of the viability of our approach on supporting knowledge maturing in these organizational settings. The sample services for each service functionality class described in the following are part of this demonstrator.

3.2   Orientation Services

The purpose of orientation services is to provide support for identifying already existing relevant knowledge objects, related topics or colleagues working in related fields. With the help of these services, the user gets an overview of what is important in relation to his topic of interest in order to ensure organizational consistency and avoid redundant actions. In accordance with the knowledge maturing for career guidance model, this set of services support the access and search phase. In relation to the knowledge maturing model, orientation services are supporting the first phase, namely, expressing ideas, since having an overview of the organizational knowledge is necessary to develop and express new ideas whilst trying to keep organizational coherency. This set of services supports a knowledge worker who is new to a certain area of interest, e.g. because (s)he is newly appointed within an organization or is supposed to write a report about a certain topic.

Visual Semantic Browsing Service. This service provides a visualization for the organizational and personal workspace. Resources in an organizational environment cover a broad range of information sources like web-pages, mail, wiki articles or local files and the visualization provides an overview over the organizational semantic model and the (semantic) relations between objects in this model. Each node in the graph represents an annotated resource, a user or a tag, directed edges represent a relation between two nodes: A user is for example related to artifacts which he has created/edited, a resource might have one or more assigned tags and a category might contain one or more artifacts. Clicking on a node in the graph leads to an update of the visualization.

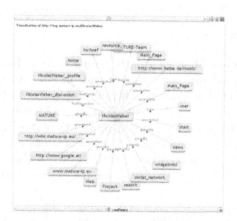

Fig. 2. Semantic Browsing Widget

This service supports the daily work of users by enabling visual browsing through the semantic content of a knowledge base from resources to resources via their relations, thus pulls together related information for a certain resource at a glance. During the browsing process, the user gains a detailed overview of the organizational knowledge base and how it is semantically structured. If a user takes a look at for example all resources and their tags which are related to a certain category, (s)he is able to identify missing topics within this category and is motivated to start creating a new resource with the missing information. Furthermore, depending on the choice of the user the service enables to show either the public, personal (or both) relations and resources, such that a user can analyze her/his own contributions within the organization in relation to certain topics. In addition to that, the visual browsing service allows for personal settings depending on individual preferences and learning styles. Thus, the features offered by the visual browsing service covers relevant requirements for getting an overview of the organizational knowledge base and its semantic model and operates on the structure of knowledge objects.

Summary Service. Knowledge workers might experience time pressure for certain tasks and the easier it is to get to know about relevant knowledge objects the faster the worker can execute what (s)he is supposed to. This service offers a short summary for each content based knowledge artifact such that the user dont need to go through the whole document to know what it is about.The summary of a knowledge artifact is based on its semantic markup and automatically generated by using information extraction (by using services of the KnowMiner framework, see [Klieber et al., 2009] for details) which can finally be extended and improved manually. In addition to that, it is updated (or users are asked to update the summary) if the content of a knowledge artifact remarkably changes. The summary service not only incorporates a summary regarding the content, but also giving an overview of document length, rating of the artifact or readability measures.

3.3   Inspiration Services

The purpose of this set of services is to support inspiration for probably interesting topics and offering new perspectives on already existing topics of a knowledge worker's field of interest. This class of services support brainstorming, organizing and analyzing ideas and encourages knowledge workers to gain new knowledge by presenting for example mash-ups of existing knowledge assets including videos or pictures or an aggregation of organizational knowledge assets related to a certain topic. In accordance with the knowledge maturing for career guidance model, this set of services supports the aggregate and scaffold phase.

Mashup Search Service. During the browsing of the organizational knowledge base with regard to a certain topic, a knowledge worker might have figured out that that information on a subtopic of a certain category is missing. In order to write a meaningful article about this topic, one of the first steps in this process is to collect related information in addition to the user's informal knowledge and context. This service provides a search interface which helps the user to aggregate information related to a certain topic without the need to use multiple search engines for different knowledge databases. Using different search facilities of various web resources (yahoo search, YouTube, wiki articles, Xing) and IBM OmniFind (Yahoo! edition) to enable including local information sources (which are for example only available within an organization or on one's own computer), the Search Support Service provides a combined interface. As users might have different background, e.g. one is already an expert on a certain topic, another one is a novice, different resources are ranked higher depending on the user's context. The wide range of information sources which are provided, varying between textual content, pictures, videos and persons stimulates the user's inspiration on a certain topic for extending and improving already existing knowledge artifacts or creating new ones.

Aggregation Service. This service is an extension of the Mash-up Search service. When the user is presented the search results of certain keywords, (s)he can choose from a variety of results and create one or more collections of the search results for later use. These collections can include complete knowledge artifacts (emails, pictures, documents, ...) but in addition to that, also parts of e.g.. a wiki article. A semantic description for the collections can be added (in form of semantic annotations or assigning categories) such that the service can automatically update collections with appropriate information when needed, for example by suggestion new knowledge objects with related semantic description that might be suitable. The user can assign knowledge objects to collections by using drag-and-drop of the search results into existing or new collections for easy use.

3.4   Creation and Refinement Services

The purpose of creation and refinement services is to generate knowledge artifacts that meet a certain form of (higher) quality standard (e.g. reports, presentations, learning material) by aggregating existing knowledge assets. Furthermore, the services help to identify available or newly created knowledge objects that need an improvement in their maturing levels or can contribute to a given maturing process, for example by offering access to meaningful maturing indicators. The aim is to provide knowledge workers at every moment with adequate information (e.g. when generating a report that satisfies a given quality standard) in order to guide the users towards a common organizational goal. In accordance with the knowledge maturing for career guidance model for, this set of services supports the manipulation, analyze and refection phases.

Maturing Indicator Service. The objective of analyzing content is to facilitate the assessment of the maturity of a content based knowledge artifact. This maturity level allows to decide whether the maturity of a certain document should be improved by supporting the user in creating or editing the knowledge artifact. The bottleneck in assessing the maturity of knowledge artifacts is the selection of qualified attributes reflecting the maturity of the artifact. Attributes describing an object which is an instance of content, semantic or processes can be used to determine the maturity of the artifact. Quantitative and qualitative parameters or a combination of them are the basis for the assessment of the maturity. In case of content we assume that the readability and the maturity have a strong correlation ([Braun and Schmidt, 2007]), two metrics for readability scores are offered in this service. Both of them analyze text samples: In the Flesch Reading Ease Score ([Si and Callan, 2001]), higher scores indicate material that is easier to read while lower numbers mark passages that are more difficult to read. The second score, theGunning fog index, is an indication of the number of years of formal education that a person requires in order to easily understand the text on first reading.

Since semantic mark-up is a crucial factor for identifying relevant information during search, a semantic indicator provides a quantitative measure for the semantic annotations of a document which enables the user to access the amount

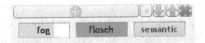

Fig. 3. Widget providing Maturing Indicator Services

of semantic mark-up of the current document. The value for this indicator is given by the relation of markup and length of a document. Figure 3 shows the graphical interface to these indicators.

Relation Service. Objective of the Relation Service is to interconnect knowledge artifacts with other resources. Referring to the functionality of Maturing Services, this service enables the improvement of emerging semantic structures. Relation Services aim at providing a tool enabling the user to create meaningful relations without having modeling skills or knowledge about description languages. So the service facilitates the informal expression (externalization) of relational knowledge and nevertheless the information is stored in a structured and formal representation. The formal relational knowledge provides the foundation for other maturing services like semantic search and visual browsing. Since the user contributes to the organizational knowledge and benefits only indirectly, creating annotations and relations between knowledge artifacts has to be easy and fast. Following this idea, this service offers the user to drag and drop knowledge objects a new relation should be added. When defining the new relation between these objects, the system suggests already existing relations to pre-consolidate the relations, though new relations can be easily added.

Accuracy Control Service. Accuracy control is necessary to make sure the data is accurate, up-to-date and relevant. Long articles are unlikely to be read and will be too time-consuming to search through for the information one is looking for. Additionally, the number of entries in a knowledge database could get too large. Within this service, accuracy control is divided into two relevant sections: Concerning the time, automatic date flags are used to remind authors and editors to update (parts of) a certain knowledge artifact. Though, depending on the live time of an article, one has to take into account that some articles are more general than others and could therefore have longer shelf-life. Furthermore, warnings are used to show knowledge workers how accurate information is. Concerning the content, a method of controlling the amount of data added is available to make sure information is concise. Therefore, authors are made aware of relatively similar articles and of the amount of content in documents and articles.

Mark-up Recommendation Service. Creating semantic mark-up conveys to the enrichment of wiki content. Meaningful semantic annotations offer an easy possibility for users to find useful and relevant information during search within a knowledge base. In addition, mark-up is used as a basis for recommendation of useful resources and visualization of emergent content structures. The mark-up recommendation services strive for two goals. First, lowering the barrier for creating mark-up which replaces the complex Semantic MediaWiki syntax and second, improving the quality of structure by recommendation of meaningful, pre-consolidated mark-up. Depending on the content of an article, the system analyses used words and their frequencies to recommend the most used keywords as tags for the article.In order to categorize articles, the system suggests already existing categories which corresponds best to the newly created content. Additionally, the user can add a certain category which seems to be appropriate and can train the service with this category such that the system can suggest this category in future for appropriate and related articles.

3.5 Collaboration Service

The aim of this set of services is to offer collaboration support in every phase of the maturing process. This includes for example the distribution of knowledge artifacts in communities. The goal is to create opportunities for discussing questions or problems with authors of related knowledge artifacts, share newly created knowledge artifacts with possibly interested colleagues, present/represent newly acquired knowledge to different target groups or simply start a collaboration for developing new ideas related to a topic. The initiation of collaboration and collaboration itself is necessary to reach a consensus and a common understanding of knowledge assets of organizational interest. In accordance with the knowledge maturing for career guidance model, this set of services supports the present/represent, share and networking with people phases.

Discussion Service. This offers the facility to initiate easy collaboration with authors of articles or interested persons by not having to switch to another tool since it is embedded into the framework and thus enables easier and faster use. The user can start a discussion concerning any wiki articles and documents in order to support negotiation and consolidation of knowledge artifacts. These discussions are bound to the knowledge artefacts by a relation in the knowledge data base such that other users can easily find them and read through them whenever she wants to improve an article or get more detailed background information. The maturing service in the background informs users who might be interested that a discussion about a certain topic has started and automatically invites them to join. Discussions can also be annotated with keywords to assign them to a certain topic or relations to other discussions, documents, people can be created.

3.6 Reflection Service

Reflection is necessary for organizational and individual development. Though it is mainly driven by actions of individuals, reflection can contribute towards improved organizational systems and practices. When people change their way of thinking and how they carry out their work tasks, this has organizational implications. Reflection services aims at guiding knowledge workers to improve organizational knowledge by comparing one's contributions to others with respect to the organization's goals.

Resource Activity Evaluation Service. The Activity Evaluation service is supposed to collect all activities related to resources of a user. Such activities are for example opening and reading documents, disseminating or modifying content, rating of a document or giving a feedback. A user is shown the list of activities in addition to who performed these actions (with regard to the user's privacy settings), related knowledge objects and (newly created) objects of possible interest. If for example a document is not read very often and the overall rating is not very good, the user might want to improve the quality of the resource by starting a discussion with the people who performed the rating.

4   Discussion and Outlook

In this paper, a definition for Knowledge Services in general and Knowledge Maturing Services in particular was presented, including a discussion of related definitions which includes Knowledge Management Systems and the Knowledge Market. We described in detail sample services of five service functionality classes, which have already been developed within a demonstrator to support knowledge workers in a highly content based organizational setting according to the knowledge maturing process of [Schmidt, 2005] with regard to quality assurance. These service classes are not thought to be used one after another, rather enable the user to use them iteratively during the overall process of content creation in oder to support knowledge maturing. Further evaluation of the work presented is still in progress in order to gain new ideas and deeper insight how knowledge maturing can be supported within MATURE. Thus, it has to be clear that the presented services do not provide an overall system supporting Knowledge Maturing, but give practical examples of the five service functionality classes that were presented and are thought to support knowledge maturing in heavily content based organizational environments.

Acknowledgments

This work has been partially funded by the European Commission as part of the MATURE IP (grant no. 216346) within the 7th Framework Programme of IST. The Know- Center is funded within the Austrian COMET Program - Competence Centers for Excellent Technologies - under the auspices of the Austrian Ministry of Transport, Innovation and Technology, the Austrian Ministry of Economics and Labor and by the State of Styria.

References

[Attwell et al., 2008] Attwell, G., Bimrose, J., Brown, A., Barnes, S.-A.: Maturing learning: Mash up personal learning environments. In: Wild, F., Kalz, M., Palmr, M. (eds.) Proceedings of the First International Workshop on Mashup Personal Learning Environments (MUPPLE 2008). Maastricht, The Netherlands (2008)

[Braun and Schmidt, 2007] Braun, S., Schmidt, A.: Wikis as a technology fostering knowledge maturing: What we can learn from wikipedia. In: 7th International Conference on Knowledge Management (IKNOW 2007), Special Track on Integrating Working and Learning in Business, IWL (2007)

[Dilz and Kalisch, 2004] Dilz, S., Kalisch, A.: Anwendungen und systeme für das wissensmanagement. In: Anwendungen und System für das Wissensmanagement. GITO-Verlag, Berlin (2004)

[Klieber et al., 2009] Klieber, W., Sabol, V., Muhr, M., Kern, R., Granitzer, M.: Knowledge discovery using the knowminer framework. In: Proceedings of the iadis international conference on informatin systems (2009)

[Lindstaedt et al., 2006] Lindstaedt, S., Scheir, P., Lokaiczyk, R., Kump, B., Beham, G., Pammer, V.: Integration von arbeiten und lernen kompetenzentwicklung in arbeitsprozessen, semantic web - wege zur vernetzten wissensgesellschaft. In: Proceedings of the European Conference on Technology Enhanced Learning (ECTEL) (September 16-19, 2008). Maastricht, The Netherlands (2006)

[Lindstaedt et al., 2007] Lindstaedt, S.N., Ley, T., Mayer, H.: Aposdle - new ways to work, learn and collaborate. In: Proc. WM 2007, pp. 381–382. GITO-Verlag, Berlin (2007)

[Lindstaedt et al., 2008] Lindstaedt, S.N., Scheir, P., Lokaiczyk, R., Kump, B., Beham, G., Pammer, V.: Knowledge services for work-integrated learning. In: Dillenbourg, P., Specht, M. (eds.) EC-TEL 2008. LNCS, vol. 5192, pp. 234–244. Springer, Heidelberg (2008)

[Maier et al., 2009] Maier, R., Hdrich, T., Peinl, R.: Enterprise Knowledge Infrastructures, 2nd edn. Springer, Heidelberg (2009)

[Maier and Remus, 2008] Maier, R., Remus, U.: Integrating knowledge management services: Strategy and infrastructure. In: Abou-Zeid, E.-S. (ed.) Knowledge Management and Business Strategies: Theoretical Frameworks and Empirical Research. Information Science Reference, Hershey, PA (2008)

[Maier and Schmidt, 2007] Maier, R., Schmidt, A.: Characterizing knowledge maturing: A conceptual process model for integrating e-learning and knowledge management. In: Gronau, N. (ed.) 4th Conference Professional Knowledge Management - Experiences and Visions (WM 2007), Potsdam, vol. 1, pp. 325–334. GITO, Berlin (2007)

[Mentzas et al., 2007] Mentzas, G., Kafentzis, K., Georgolios, P.: Knowledge services on the semantic web. Commun. ACM 50(10), 53–58 (2007)

[Rath et al., 2008] Rath, A.S., Weber, N., Krll, M., Granitzer, M., Dietzel, O., Lindstaedt, S.N.: Context-aware knowledge services. In: Workshop on Personal Information Management (PIM 2008) at the 26th Computer Human Interaction Conference (CHI 2008), Florence, Italy (2008)

[Schmidt, 2005] Schmidt, A.: Knowledge maturing and the continuity of context as a unifying concept for knowledge management and e-learning. In: Proceedings of I-KNOW 2005, Graz, Austria (2005)

[Si and Callan, 2001] Si, L., Callan, A.J.: Statistical model for scientific readability. In: Proc. of CIKM, pp. 574–576 (2001)

[Weber et al., 2009] Weber, N., Schoefegger, K., Ley, T., Lindstaedt, S.N., Bimrose, J., Brown, A., Barnes, S.: Knowledge maturing in the semantic mediawiki: A design study in career guidance. In: Proceedings of EC-TEL 2009, Nice, France (in press, 2009)

How Much Well Does Organizational Knowledge Transfer Work with Domain and Rule Ontologies?

Keido Kobayashi[1], Akiko Yoshioka[1], Masao Okabe[1,2], Masahiko Yanagisawa[2], Hiroshi Yamazaki[2], and Takahira Yamaguchi[1]

[1] Department of Administration Engineering, Faculty of Science and Technology,
Keio University, 3-14-1 Hiyoshi, Kohoku-ku, Yokohama-shi Kanagawa-ken 223-8522, Japan
`{k_koba,y_aki,yamaguti}@ae.keio.ac.jp`
[2] Tokyo Electric Power Co., Inc., 1-1-3 Uchisaiwai-cho, Chiyoda-ku Tokyo 100-8560, Japan
`{okabe.masao,yanagisawa.masahiko,yamazaki.hiroshi}@tepco.co.jp`

Abstract. Knowledge transfer to next-generation engineers is an urgent issue in Japan. In this paper, we propose a new approach without costly OJT (on-the-job training), that is, combinational usage of domain and rule ontologies and a rule-based system. A domain ontology helps novices understand the exact meaning of the engineering rules and a rule ontology helps them get the total picture of the knowledge. A rule-based system helps domain experts externalize their tacit knowledge to ontologies and also helps novices internalize them. As a case study, we applied our proposal to some actual job. We also did an evaluation experiment for this case study and have confirmed that our proposal is effective.

Keywords: domain ontology, rule ontology, rule-based system.

1 Introduction

A great number of skilled engineers are now retiring and Japanese industries are facing a problem to lose their skills. It is an urgent issue to transfer their skills to the next generation. However, the traditional transfer by OJT needs a lot time and money.

To solve this problem, there are several proposals using information technology, mainly, multimedia and virtual reality technology (see e.g. [1]). But, they focus on so-called "craft skill", which is a skill such as to create a complex mold with high precisions. Chuma, however, points out that in manufacturing, there is another crucial skill called "intelligence skill", which is a skill such as to detect expected flaws of products or production processes and to solve them [2]. In a recent automated and integrated manufacturing plant, to operate it efficiently, intelligence skill is becoming more and more important, while craft skill is being replaced by computerized numerical control. Intelligence skill requires integrated knowledge regarding various aspects of a whole plant but it can be decomposed into pieces of knowledge. Hence, the key issue is how to combine pieces of knowledge logically, depending on various situations. Therefore, for intelligence skill, ontologies and a rule-based system are more suggestive than multimedia and virtual reality technology.

For an expert system, knowledge acquisition is the bottleneck of its development and maintenance. To solve the problem, knowledge engineering treats building a

D. Karagiannis and Z. Jin (Eds.): KSEM 2009, LNAI 5914, pp. 382–393, 2009.
© Springer-Verlag Berlin Heidelberg 2009

rule-based system as a modeling activity. This knowledge modeling, such as expertise modeling of CommonKADS [3], improves the maintainability of the system, in addition to its cost-effective development, providing implementation-independent knowledge-level description of the system's problem solving processes. It consists of a reusable PSM (problem solving method) and domain ontologies. The former provides a template for describing knowledge-level problem solving processes and the latter describes the structure of the knowledge that is used to solve the problems.

In knowledge management, Nonaka proposed SECI model for organizational knowledge creation [4]. It shows how organizational knowledge is created by syntheses of tacit and explicit knowledge, Socialization, Externalization, Combination and Internalization. Nonaka also proposed "ba" where these syntheses are conducted [5]. Hijikata proposed a computerized "ba" where two domain experts externalize and combine their tacit knowledge efficiently with the help of a computer that points out inconsistency among tacit knowledge of two domain experts and knowledge created by inductive case learning [6].

SECI model focuses on organizational knowledge creation. On the other hand, our proposal mainly focuses on its transfer to next-generation engineers, and emphasizes that they can do jobs using the transferred knowledge at various situations. Therefore, internalization is important so that they can apply the knowledge to various situations. In this paper, we propose combinational usage of ontologies and a rule-based system for the organizational transfer of intelligence skill.

The next section outlines our proposal and also introduces an ontology repository called GEN. Section 3 presents a case study from Tokyo Electric Power Co., Inc. (hereafter, TEPCO). Section 4 is its evaluation and discussions. Finally, Section 5 summarizes our proposal and points out some future works.

2 Ontologies and a Rule-Based System for Organizational Knowledge Transfer

We propose ontologies and a rule-based system that help domain experts externalize their tacit intelligence skills and also help novices internalize them, for a specific domain that requires intelligence skill. Ontology is defined as "an explicit specification of a conceptualization" [7]. Usually, it consists of terms and semantic relations among them. But, in this paper, we take ontology in broader sense. It is not only about terms but also about pieces of knowledge. Therefore, what consists of pieces of knowledge and semantic relations among them is also an ontology. Well-structured explicit knowledge that is externalized from tacit intelligence skills and combined can be an ontology. Fig. 1 shows the overview of our proposal with a simplified example. Explanations follow in the subsequent sections.

2.1 Domain and Rule Ontologies

Both a domain ontology and a rule ontology are what tacit intelligence skills of domain experts are externalized and combined as. A domain ontology consists of the technical terms that are used in the domain and the semantic relations among them. This is used to standardize the description of intelligence skill externalized by domain experts and to eliminate dependency on each domain expert. It is also used for novices to understand the domain knowledge. So, each term in a domain ontology has a description of its meaning in a natural language.

We claim that in most cases, tacit intelligence skill can be decomposed into piece of simple knowledge, and that they can be externalized as engineering rules in a natural language. A domain ontology plays an important role for novices to understand

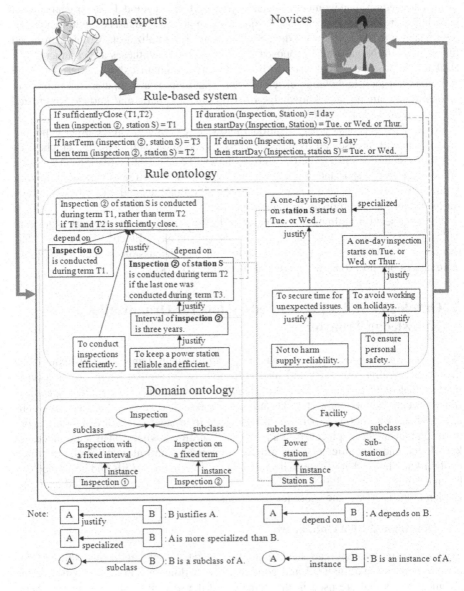

Fig. 1. Overview of the proposal

each engineering rule. But, it is not enough for novices to understand all the engineering rules as a whole and to be able to do the jobs using them. To make it possible, a rule ontology is introduced. A rule ontology consists of engineering rules as primitives and semantic relations among them. It gives novices a high-level picture of the intelligence skills. The semantic relations among rules are mainly as follows:

- relation "justify", which is a relation between a deep rule and its shallow rule
- relation "depend on", which is a relation between a rule and another rule whose application is prerequisite for its application
- relation "specialized", which is a relation between a specialized rule and its generalized rule
- relation "override", which is a relation between a specialized rule with some overrides and its generalized rule that it overrides

The deepest rules through relations "justify" mainly express the objectives of the jobs that the intelligence skill is used and have a similar role of the top-level task description by PSM. The reason why we introduce a rule ontology, rather than PSM, is that a rule ontology is more flexible to be applied to various situations and also easier for novices to understand.

A rule ontology is a kind of extension of explanation facilities based on deep knowledge in an expert system [8] and has some similarity to coarse-grain intelligent content [9]. A coarse-grain intelligent content consists of mostly single sentence is chunk for human to understand easily. But, a rule ontology consists of engineering rules, which are usually more than a single sentence to keep an engineering meaning.

2.2 Rule-Based System

These ontologies are expected to be developed and maintained by domain experts, but it is not easy for them because they themselves get the intelligence skills by OJT and have no experience of externalizing and combining them to ontologies. Especially, it is difficult for them to recognize all the rules explicitly and precisely since they do the job based on their tacit intelligence skills and not based on the ontologies.

One of the objectives of a rule-based system is to help them refine rules in a rule ontology. A rule-based system has executable rules that correspond to the shallowest rules that are directly used to do the job. Domain experts can refine rules in a rule ontology, checking the outputs from the rule-based system. A rule ontology, giving domain experts a high-level picture, reduces the difficulty in maintaining the rule-based system. Conversely, the rule-based system motivates domain experts to develop and maintain the ontologies, giving them the output of the job semi-automatically on behalf of them.

A rule-based system also helps novices internalize the intelligence skills of the domain experts. Even though the intelligence skills are externalized and combined as ontologies that novices can understand, it is still difficult for novices to internalize them. To internalize them, novices need experience to do the job using them. Novices can compare the outputs by themselves with the ones from the rule-based system as many times as necessary. Then, novices can sufficiently internalize them.

2.3 GEN (General knowlEdge Navigator)

To support and examine our proposal, we have also developed an experimental ontology repository, specialized for knowledge management, called GEN (General knowlEdge Navigator) [10]. GEN is similar to Protégé [11] but is more end-user oriented and suitable for structured textual information like a rule ontology. GEN provides a "ba" where domain experts externalize and combine their tacit intelligence skills to ontologies and also novices internalize them. Therefore, it supports multi-user concurrent use. GEN also has a rule-based system as its subsystem.

3 Case Study

Having described our proposal, let us turn to a case study and examine our proposal in details. In this case study, we have focused on some specific job on hydroelectric power stations at some remote control and maintenance office in TEPCO. It is so-called "inspection and maintenance work scheduling job" [10].

What we did first is similar to OJT. We, as novices, tried to do the job, referring to the previous outputs, under the direct supervision of the domain experts of this job. Second, we externalized what we learned, and the domain experts reviewed and re-fined them. Then, the domain experts, together with us, reorganized and combined them as ontologies in GEN. We also developed a rule-based system specific to this job, that is, a scheduling subsystem. It was first developed in prolog [12], and transferred to GEN. Using this rule-based system, the domain experts and we refined the ontologies repeatedly.

3.1 Description of Case Study

The system of hydroelectric power stations in TEPCO has a long history and is now highly automated and integrated. All the hydroelectric power stations are unmanned. A remote control and maintenance office is responsible to remote-operate and maintain all the hydroelectric power stations along a river system, which vary from very old small ones to state-of-the art large-scale pumped storage ones.

The "inspection and maintenance work scheduling job" is mainly to make out yearly inspection and maintenance work schedules of generators and other devices of all power stations controlled by a remote control and maintenance office. The schedule is made out so that it minimizes discharged water, which is water not used for power generation, under various constraints such as:

- statutory inspection interval of each device
- agreements with agricultural unions and other outside associations
- natural environment conditions
- operational conditions among interrelated facilities etc.

This is a typical job that needs intelligence skill since it requires a variety of integrated knowledge on the whole power stations. Moreover, this is not a well-defined optimization problem. Some of the constraints are not mandatory but desirable, and in most cases there is no feasible solution that satisfies all the constraints. In case that there is no strictly feasible solution, sophisticated intelligence skill is required to determine what constraints should be loosened, depending on a situation. Most of the knowledge has not been externalized and the skill for the job has been transferred by costly OJT.

3.2 Rule Ontology

What we learned from the domain experts was decomposed into 134 engineering rules. Among the 134 rules, 90 rules were the shallowest rules that can be used directly to make out the schedule, and the rest are deep rules that justify the shallowest rules. Since relation "justify" between a deep rule and a shallow rule is relative and hierarchal, we chose a set of the deepest rules, which are the objectives and basic constraints of the job. Fig. 2 shows a part of simplified fragment of the rule ontology. There are also rules that do not belong to the set of the deepest rules but that no other rules are deeper than, which are objective facts.

We also developed a class hierarchy of the rule ontology, where each rule is treated as an instance. There are two reasons why a rule is treated as an instance and a class hierarchy is introduced on it. One is to define semantic relations among rules uniformly in a schema and the other is to make it easy for novices to apply necessary rules to proper works. Hence, a subclass does not necessarily have additional structure to its superclass. Table 1 shows number of classes and instances of the rule ontology.

3.3 Domain Ontology

Domain ontologies were created from the technical terms in the rules. Table 1 shows number of classes and instances of them. Since the domain is scheduling of inspection and maintenance works of hydroelectric power station facilities, main domain ontologies are a facility ontology and a inspection and maintenance work ontology (hereafter, work ontology) [13].

The facility ontology should have information on facilities such as power stations, transmission lines etc. TEPCO already has a kind of class hierarchy of facilities with a long history although TEPCO does not call it a class hierarchy. Basically we adopted this class hierarchy for the facility ontology because the facility ontology should be commonly applicable to a broader domain. Based on this class hierarchy, information and semantic relations that are necessary to "inspection and maintenance work scheduling job" are introduced.

The work ontology should have information that is necessary for inspection and maintenance works. In real situations, it should have the information how works can be done and the information for scheduling them is a small part of it. But, since to develop the work ontology for real situations is far beyond the scope of the case

Table 1. Number of classes and instances

	Classes	Instances	Note
Rule ontology	20	134	90 instances among 134 instances were converted to a rule base.
Domain Ontology	55	292	

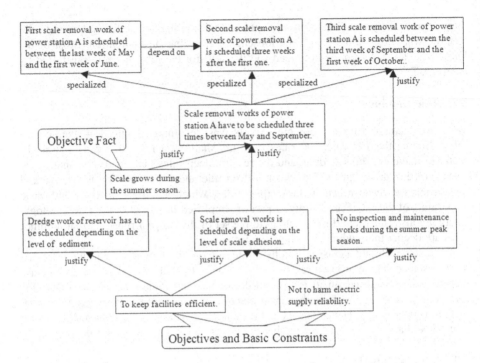

Fig. 2. Fragment of the rule ontology

study, the work ontology we developed for this case study had only a relation between a work and the rules that are applied to it. Each work here is not a particular work done at some particular date and time and is more natural to be treated as a class, but for the same reason as the rule ontology, it was treated as an instance and the class hierarchy is basically based on classification by power stations.

3.4 Scheduling Subsystem

The shallowest rules (90 rules among 134 rules, see Table 1.) were converted to the executable format on a reasoner in GEN. However, it was not easy to get a proper solution. As mentioned before, this is not a well-defined problem and most of the rules are not mandatory and may contradict. Here is a simplified example to illustrate how they contradict.

Rule 1. day(startDate(inspection1))=Wednesday
Rule 2. startDate(inspection1)= March 1st
       where inspection1 is an individual name for some specific inspection.

Then, a question is which of Rule 1 or Rule 2 should be excluded or loosened if March 1st is not Wednesday in the target year. This kind of situation occurs very often usually in much more complex manner. To treat these situations properly occupies an important part of the intelligence skill of this job. First, we tried to enumerate all such cases and give a specific solution for each case. But, there are a lot of cases and it is difficult to investigate all the cases. Moreover, even if all the cases are investigated at one time with a great effort, a lot of new cases may appear when a rule is added or updated and it is almost impossible to maintain them.

Hence, we changed the strategy. Giving up enumerating all the infeasible cases, we decided to give each rule a priority number. If there is no feasible solution, then rules that have the least priority are simply ignored. Now, the way to externalize the tacit knowledge to treat infeasible cases became dramatically simple. It now is to give priority numbers to all the rules by trial and error using the scheduling subsystem so that they can work well in any cases. Practically, we can assume that if they work well in many cases, they presumably work well in all other cases. By trial and error, we were successful in giving a suitable priority number to each rule [14], [15].

4 Evaluation and Discussion

In this section, we evaluate the case study in the previous section, from the point of externalization and combination by domain experts and also from the point of internalization by novices. For the former, the evaluation was done by domain experts who joined this case study with us, and for the latter, we have done an experiment for the evaluation [16].

4.1 For Externalization and Combination by Domain Experts

In the case study, the domain experts and we successfully developed the rules and the ontologies. Their evaluation was that at least together with us, it was not difficult to develop the rules and the ontologies using the scheduling subsystem.

The domain experts, however, had little idea what extent of detail and carefulness was necessary in externalization so that the novices could understand it. For example, they tended to externalize only the deepest and shallowest rules because they did not recognize that the novices had difficulty in relating them without in-between rules. The ontologies we developed had a major role in solving the problem. First, the description in a natural language of each rule in the rule ontology was a good guide to show what extent of detail and carefulness was necessary. Also the semantic relations among the rule ontology showed what were necessary to be externalized.

Nevertheless, it was also true that the ontologies we developed, especially the rule ontology, were incomplete before the trial and error refinements using the outputs of the scheduling subsystem. The scheduling subsystem was highly evaluated by the domain experts and was successful to motivate them to maintain the ontologies. They highly evaluated it because it could do easy but painful routines that may cause human errors, such as simple calendar checking.

All the results are based on the ontologies that the domain experts developed together with us. It is still not clear what kind of support is necessary for domain experts to develop ontologies and a rule-based system by themselves without us.

4.2 For Internalization by Novices

To evaluate the internalization by the novices, we have done an experiment, where three groups internalized the knowledge by three different ways, including by our proposal, and each group solved scheduling problems for evaluation. The first group is a group that learned the knowledge by our proposal (hereafter, the proposed group). For this group, a person who had neither experience nor background of TEPCO's work was selected as an examinee (hereafter, examinee A). The second group is one which consists of persons who are almost equivalent to the ones who get the knowledge by OJT (hereafter, the OJT group). The examinees of this group were selected from the members of our project, who developed the rules and the ontologies together with the domain experts, interviewing them and under the direct supervision of them (hereafter, examinee B and C). The third group is one between the proposed group and the OJT group (hereafter, the intermediate group). The examinee of this group was a person who joined our project after the rules and the ontologies were almost developed but were not stored in GEN, and who were mainly responsible for developing the rule-based system (hereafter, examinee D).

Table 2 shows the experiment result for each examinee, with his or her profile. The effectiveness of the proposal was confirmed since the accuracy rate of the proposed group was almost the same as that of the OJT group (88% vs. 91%), although the learning time of the proposed group was less than a quarter of that of the OJT group (55 hours vs. 221 hours). Although not by self-learning by GEN, the intermediate group got even better accuracy rate as that of the OJT group (96% vs. 91%), with the learning time less than two thirds of that of the OJT group (138 hours vs. 221 hours). This shows the effectiveness of systematic learning using ontologies, even without GEN. The reason why the intermediate group got such high accuracy rate is analyzed.

Erroneous answers were classified into three categories by cause. The first one is the erroneous answers caused by "lack of description", which mean the ones caused by lack of description of necessary knowledge in GEN. All 3 erroneous answers in this category were made by the proposed group (see column "Proposed" and row "Lack of description" at Table 2).

There were two points that lack description. One is about inclusive relation among inspection works. Inspection works are classified into several types, from the one that is simple but whose interval is short, to the one that is minute but whose interval is long. A minuter one is also counted as a simpler one. But, this rule was not described in GEN explicitly for two reasons. One is that the maintenance and inspection work ontology where this rule should have been stated was simplified in this case study as stated in Sect. 3.3. The other is that this rule is almost obvious for the person who has the least background in this job. But, examinee A has no background in any jobs of TEPCO and made 2 erroneous answers where a simple inspection is allocated just after minute one. This shows that practically an ontology cannot be independent of how much background its user has since a fully comprehensive ontology can be hardly developed because of its development cost.

The other point is lack of the procedural information how scheduling should be done efficiently. As stated before, the problems do not necessarily have a strictly feasible solution that satisfies all the rules, and some of the rules may have to be loosened. A domain expert can easily and intuitively find a schedule that satisfies most of the rules and that needs only a few adjustments. But, a novice could hardly find such an initial schedule, the initial schedule tended to need a large amount of adjustments, and consequently he or she loosened the rules that should have not been loosened. The last erroneous answer that examinee A made was such a kind of one. As stated in

Table 2. Result and profile of each examinee

Group		Proposed	OJT		Intermediate
Examinee		A	B	C	D
Method of learning	For knowledge acquisition	Self-learning by GEN	By interview domain experts		Self-learning by ontologies
	For internalization	Self-learning by GEN	Through developing ontologies	Through hand simulations	Through developing a rule-based system
Learning time (in hours)		55	220	222	138
Material for solving		GEN	GEN	Private documents	Private documents
Solving time (in minutes)		320	330	255	255
Accuracy rate (%)		88	90	92	96
Erroneous answers (among 130 questions)		15	13	10	5
Break-down by cause	Lack of description	3	0	0	0
	Ambiguity of description	5	5	5	1
	Careless mistake	7	8	5	4

Sect. 3.4, each rule in the scheduling subsystem has its own priority number. In the case that a complete feasible solution cannot be found, rules are loosened in order of increasing priority number until a solution can be found. Similarly, examinee A should have applied the rules in order of decreasing priority number to reduce the amount of necessary adjustments. But, in GEN, the priority numbers were only for reasoner and did not have a suitable interface for human.

The second one is the erroneous answers caused by "ambiguity of description", which mean the ones caused by ambiguity of natural language description of the rules in GEN. For the erroneous answers in this category, both the proposed group and the OJT group made 5 erroneous answers on an average, but the intermediate group made only 1 (see row "Ambiguity of description" at Table 2). This is the main reason why the intermediate group attained the higher accuracy rate than the OJT group. The reason why the intermediate group made only 1 erroneous answer is that examinee D in the intermediate group was responsible for developing the rule-based system and had to disambiguate the rules. This is collateral evidence that a rule-based system can contribute higher accuracy of externalization by domain experts and internalization by novices.

The last one is careless mistakes. The proposed made 7 careless mistakes on an average and the OJT group made 6.5 (see row "Careless mistake" at Table 2), and there was not a big difference between them. But, the group who used GEN for solving the problems (examinee A and B) made 7.5 careless mistakes and the group who used private documents (examinee C and D) made 4.5, and there was a significant difference between them. In addition, the average solving time of the former group was 325 minutes, although the one of the latter was 255 minutes. The reasons are:

- GEN is an experimental system and is not equipped with a good user interface, especially is not good at browsing information at a glance.

- A paper document is better than a computer display for browsing this size of information (134 rules) at a glance.
- For each individual, his or her private documents that reflect his or her background has better understandability than GEN that does not have a facility to personalize its information.

The fact that even with these kinds of weaknesses of GEN, the proposed group gained almost the same accuracy rate as the one of the OJT group, only with about quarter of the learning time of the OJT group shows the effectiveness of our proposal.

5 Conclusion and Future Work

In this paper, we have proposed ontologies and a rule-based system that enable organization transfer of intelligence skill without OJT. Intelligence skill is decomposed into pieces of knowledge and they are externalized as engineering rules and technical terms that constitute them. They are combined as a rule ontology and domain ontologies respectively. A rule ontology consists of engineering rules as primitives and semantic relations among them such as "justify", "depend on" etc. The shallowest rules are translated into a rule base so that they can be executed on a rule-based system. The rule-based system motivates domain experts to externalize and combine their tacit intelligence skill to the ontologies and also help domain experts refine them and novices internalize them.

Accompanied with GEN, this proposal was experimentally applied to the actual job at a remote control and maintenance job in TEPCO and the effectiveness of our proposal has been confirmed by the evaluation of this case study. The lessons learned from the case study include that novices, in addition to domain experts, are encouraged to participate in developing ontologies since ontologies are practically depend on the backgrounds of its users.

In the next step, we will do a more comprehensive and long-term case study using enhanced GEN, to examine what kind of support is necessary for domain experts:

- to develop ontologies and a rule-based system from scratch by themselves
- to maintain them for longer term

In addition, since we have found that novices, as well as domain experts, have an important role to develop and improve ontologies, we will enhance GEN to provide a "ba" where both domain experts and novices collaboratively externalize their tacit intelligence skills and refine them by communicating each other.

Acknowledgments. The authors wish to thank Masahiro Ohira, Miho Kato, and Yurie Takeda for their contribution.

References

1. Watanuki, K., Kojima, K.: New Approach to Handing Down of Implicit Knowledge by Analytic Simulation. Journal of Advanced Mechanical design, Systems, and Manufacturing 1(3), 48–57 (2007)

2. Chuma, H.: Problem Finding & Solving Skill in Manufacturing Factories. The Japanese Journal of Labor Studies 510 (2002) (in Japanese)
3. Schreiber, G., Akkermans, H., Anjewierden, A., Hoog, R., Shadbolt, N., Velde, W.V., Wielinga, B.: Knowledge Engineering and Management The CommonKADS Methodology. MIT Press, Cambridge (1999)
4. Nonaka, I., Takeuchi, H.: The Knowledge-Creating Company: How Japanese Companies Create the Dynamics of Innovation. Oxford University Press, Oxford (1995)
5. Nonaka, I., Konnno, N.: The Concept of 'ba': Building a foundation for knowledge creation. California Management Review 40(3), 40–54 (1998)
6. Hijikata, Y., Takenaka, T., Kusumura, Y., Nishida, S.: Interactive knowledge externalization and combination for SECI model. In: Proceedings of the 4th international conference on Knowledge capture, Whistler BC Canada, pp. 151–158 (2007)
7. Gruber, T.R.: A Translation Approach to Portable Ontology Specifications. Knowledge Acquisition 5(2), 199–220 (1993)
8. Yamaguchi, T., Mizoguichi, R., Taoka, N., Kodaka, H., Nomura, Y., Kakusho, O.: Explanation Facilities for Novice Users Based on Deep Knowledge. The Transactions of the Institute of Electronics, Information and Communication Engineers J70-D(11), 2083–2088 (1987) (in Japanese)
9. Hashida, K.: Semantics Platform Based on Ontologies and Constraints. Journal of the Japanese Society for Artificial Intelligence 21(6), 712–717 (2006)
10. Iwama, T., Tachibana, H., Yamazaki, H., Okabe, M., Kurokawa, T., Kobayashi, K., Kato, M., Yoshioka, A., Yamaguchi, T.: A consideration of ontology that supports organized business knowledge accumulation and retrieval. In: 3rd Annual Conference of Information Systems Society of Japan, Niigata Japan (2007) (in Japanese)
11. Protégé: http://protege.stanford.edu/
12. Kobayashi, K., Kato, M., Yoshioka, A., Yamaguchi, T.: Constructing Business Rules and Domain Ontologies to Support Externalization of Business Knowledge in Knowledge Transfer. In: The 22nd Annual Conference of the Japanese Society for Artificial Intelligence, Asahikawa Japan (2008) (in Japanese)
13. Yoshioka, A., Ohira, M., Iijima, T., Yamaguchi, T., Yamazaki, H., Yanagisawa, M., Okabe, M.: Supporting Knowledge Transfer and Scheduling task with Ontologies. In: The 21st Annual Conference of the Japanese Society for Artificial Intelligence, Miyazaki Japan (2007) (in Japanese)
14. Okabe, M., Yanagisawa, M., Yamazaki, H., Kobayashi, K., Yoshioka, A., Yamaguchi, T.: Organizational Knowledge Transfer of Intelligence Skill Using Ontologies and a Rule-Based System. In: The 7th Practical Aspects of Knowledge Management, Yokohama Japan (2008)
15. Okabe, M., Yanagisawa, M., Yamazaki, H., Kobayashi, K., Yoshioka, A., Yamaguchi, T.: Ontologies that Support Organizational Knowledge Transfer of Intelligence Skill, Interdisciplinary Ontology. In: Proceedings of the Second Interdisciplinary Ontology Meeting, Tokyo Japan, vol. 2, pp. 147–159 (2009)
16. Kobayashi, K., Takeda, Y., Yoshioka, A., Okabe, M., Yanagisawa, M., Yamazaki, H., Yamaguchi, T.: Organizational Knowledge Transfer with Ontologies and a Rule-Based System. In: The 23rd Annual Conference of the Japanese Society for Artificial Intelligence, Kagawa Japan (2009) (in Japanese)

Ontology Evaluation through Assessment of Inferred Statements: Study of a Prototypical Implementation of an Ontology Questionnaire for OWL DL Ontologies

Viktoria Pammer[1,*] and Stefanie Lindstaedt[2,**]

[1] Knowledge Management Institute, TU Graz
viktoria.pammer@tugraz.at
[2] Know-Center
slind@know-center.at

Abstract. Evaluation and consequent revision of ontologies is a critical task in the process of ontology engineering. We argue the necessity and merits of reviewing implicitly entailed knowledge as part of a methodical evaluation. We study this process by means of a prototypical implementation for OWL DL ontologies, called ontology questionnaire, and discuss support for dealing with unwanted inferences based on experiences with applying this questionnaire in some real-world modelling scenarios.

1 Motivation

Creating and maintaining formal ontologies requires a significant amount of human effort. Automatic creation of ontologies, also called ontology learning, provides only partial support since state-of-the art ontology learning techniques is at the level of term and relation extraction when learning from natural language text [2]. Additionally, ontology engineering can not always be seen as merely "re-writing" already known knowledge in a formal language. Sometimes it is precisely this act of formal specification in which new knowledge is generated. This is the reason, we argue, that some manual interference with ontologies will always be necessary where a certain quality is expected from these formal models. However, also for humans ontology engineering is a challenging task. Naively, it could be assumed that domain experts and knowledge engineers only state correct things anyway, and that the only point of argument is whether what ends up being in an ontology really corresponds to a consensus. But modelling mistakes do occur and a variety of reasons is at their root, apart from the simple possibility that errors of negligence may slip in. The used formalism may be too little understood, the ontology in question may simply be large and complex, or heterogeneous if multiple people were involved in modelling. Sometimes such mistakes manifest themselves as

---

* Viktoria Pammer is a PhD Student at Graz University of Technology.
** The Know-Center is funded within the Austrian COMET Program - Competence Centers for Excellent Technologies - under the auspices of the Austrian Ministry of Transport, Innovation and Technology, the Austrian Ministry of Economics and Labor and by the State of Styria. COMET is managed by the Austrian Research Promotion Agency FFG.

D. Karagiannis and Z. Jin (Eds.): KSEM 2009, LNAI 5914, pp. 394–405, 2009.

inconsistencies[1], at other times as undesired inferences or less obviously via the implemented ontology's models. The latter means that either there are unintended models which satisfy the given ontology or there are intended models which do not satisfy the modelled ontology[2]. One fundamental difficulty for humans is to anticipate the implications single formal statements may have if reasoning is applied to the whole ontology (see also [18] specifically for OWL DL). At the same time, reasoning is exactly what machines can do very well, and thus this difficulty lends itself naturally to being supported in an automated way.

With this work we contribute to the topic of ontology evaluation by studying the review of entailed statements as part of an ontology evaluation procedure, with the purpose to verify whether what people modelled is what they meant. The trick hereby is that entailed statements as well as simple solutions to get rid of unwanted entailments can be generated automatically. Ontology engineers benefit from reviewing inferred statements in two ways: First, they get an insight into knowledge implicit in the ontology, and second they can review the ontology without getting bored by having to go through what they have explicitly modelled yet another time.

2 Related Work

Technical support for ontology engineering is available through ontology editors which take care e.g. of writing syntactically correct files, e.g. Swoop [23] or Protégé [14] for OWL[3] or of addressing a reasoner to check for satisfiability. Some higher-level support is available through ontology engineering methodologies as e.g. Noy's Ontology 101 [11], more complex methodologies such as Methontology [3] or On-To-Knowledge [22] or more general knowledge engineering approaches such as CommonKADS [20]. Even though ontology evaluation is a crucial task in ontology engineering however, there are relatively few systematic approaches and none that can be called standard. In general, different authors have come up with different categorisations for relevant dimensions which to consider when evaluating an ontology, but mostly they can be thought of as belonging to either category of usability-related, functionality-related or structure-related [4]. In this paper we are concerned with the fundamental requirement that an ontology be correct, i.e. that the relevant people (domain experts, participants of a community etc.) agree to the knowledge expressed in the ontology. Using the categories above, this approach falls into the category of functionality. Among existing ontology evaluation methodologies, OntoClean [6] stands out as the most formal methodology, based on metaproperties like essence, identity and unity of concepts, inspired by philosophical considerations about the nature of concepts. After having identified the relevant metaproperties of each concept, the ontology's taxonomic relations can be evaluated with regard to their conceptual correctness. The main drawback

---

[1] Note that not all inconsistencies are necessarily modelling mistakes.

[2] The latter two modelling mistakes are inspired by the discussion by Guarino in [5] of conceptualizations, ontologies and intended models, where mismatches between intended models and implemented ontologies are discussed as problem in the context of ontology integration.

[3] A longer list of available ontology editors can be found e.g. in Wikipedia:
http://en.wikipedia.org/wiki/Ontology_editor

of this method is that it is difficult to apply in practice, and thus will probably be applied only for high-quality or top-level ontologies. A more practical approach to evaluation is the usage of so called competency questions, first mentioned in [24]. Competency questions are questions the ontology should be able to answer and can be posed either informally e.g. as natural language questions, or formally as queries or logical axioms. This approach serves very well for using competency questions as requirements on the ontology against which it can at different stages be tested. There is also a proposal to express expected inferences and unexpected inferences as logic formulae [25] similar to unit tests in software engineering. Unfortunately, this approach has never been applied in practice to the best of our knowledge. Modelling mistakes which result from lack of knowledge of OWL DL as underlying formalism could also be addressed through a tutorial in the line of [18] for instance. There, the authors address a variety of modelling mistakes which they have observed while teaching OWL DL. Summarizing, the review of entailed statements during an ontology evaluation procedure can be seen as complementary in particular to the usage of competency questions. While the latter validates the ontology's conformity to requirements, the review of entailed statements validates the agreement of domain experts and ontology engineers with what they modelled.

A variety of tools could be used to review entailed statements, since functionalities for inference explanation and ontology repair are available in most modern ontology editors, for instance in Swoop [23], in Protégé 4 using the explanation plugin [15,7] and in the NeOn toolkit through the RadOn plugin [16]. Because of a different focus, the access to the inferred statements and to their explanations is often not very direct, and the approach to repair is more technical than conceptual. For instance, Protégé 4 includes a list of inferred statements[4]. Otherwise, inferred concept inclusions are often displayed implicitly through a tree visualisation of the subsumption hierarchy in ontology editors.

3   The Ontology Questionnaire

We study the following procedure: The ontology engineer reviews statements which logically follow from the ontology but are not explicitly stated. In case of agreement, the statement "passes" the review. In case of disagreement, the ontology engineer asks for an explanation for the statement in question and can act on it in different ways.

In the ontology questionnaire, a tool to support exactly this procedure, partial support is given to the decision of how to act. Although the functionality necessary to review entailed statements exists in a number of ontology editors (see Section 2 above), we developed our own tool in the course of this work in order to study in a focused manner first, how the process of reviewing entailed statements helps ontology engineers and domain experts validate that the ontology expresses what they actually mean and second, how ontology engineers and domain experts creatively deal with undesired inferences. The name "ontology questionnaire" expresses that this process can be seen as going through an automatically generated questionnaire for assessment of the ontology under review. Starting point for using the ontology questionnaire therefore is a content-wise

---

[4] In the "Active Ontology" tab, as "Inferred Axioms".

Entailed Statements:

⊙ Imaginäres_Brainstorming is a(n) Intuitiv_-_kreative_Methoden
○ Imaginäres_Brainstorming is a(n) Kreativitätstechniken
○ Attribute_Listing is a(n) Kreativitätstechniken
○ Kernkompetenzanalyse is a(n) Tool
○ Benchmarking is a(n) Analyse
○ Benchmarking is a(n) Tool
○ Funktionenanalyse is a(n) Analyse
○ Funktionenanalyse is a(n) Tool
○ Ideen-Delphi is a(n) Intuitiv_-_kreative_Methoden
○ Ideen-Delphi is a(n) Kreativitätstechniken
○ Methode_635 is a(n) Intuitiv_-_kreative_Methoden
○ Methode_635 is a(n) Kreativitätstechniken
○ Wertanalyse is a(n) Analyse
○ Wertanalyse is a(n) Tool
○ Morphologischer_Kasten is a(n) Kreativitätstechniken
○ Wettbewerbsanalyse is a(n) Analyse
○ Wettbewerbsanalyse is a(n) Tool
○ Markt-_und_Wettbewerbsanalysen is a(n) Tool

[Justify!]

Explicit Statements:

In this box you can see the axioms the specified ontology exists of. By checking one of the checkboxes and then ch

⊙ Markt-_und_Wettbewerbsanalysen is a(n) Analyse
○ APOSDLE is a(n) EU_Projekt
○ Lehrveranstaltung is a(n) Aktivität
○ ReifTechnologie-Roadmapping is a specific instance of a(n) ReifiedConcepts
○ Moderation is a(n) Tool
○ ReifProzessinnovation is a specific instance of a(n) ReifiedConcepts
○ ReifKernkompetenzanalyse is a specific instance of a(n) ReifiedConcepts
○ ReifThemenspezifisches_Prasentationsmaterial is a specific instance of a(n) ReifiedConcepts
○ ReifRecherche is a specific instance of a(n) ReifiedConcepts
○ Unternehmen is a(n) Projektpartner
○ ReifEndbericht Is_part_of ReifProjekt
○ Neuigkeitsgrad is a(n) Bewertungskriterien
○ ReifWorkflow is a specific instance of a(n) ReifiedConcepts

Fig. 1. The ontology questionnaire displays inferred subsumption axioms (upper box) as well as explicitly stated axioms (lower box)

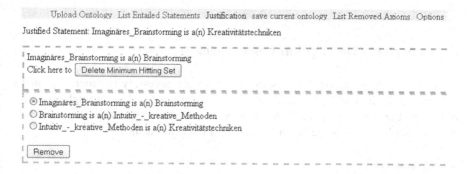

Upload Ontology   List Entailed Statements   Justification   save current ontology   List Removed Axioms   Options

Justified Statement: Imaginäres_Brainstorming is a(n) Kreativitätstechniken

Imaginäres_Brainstorming is a(n) Brainstorming
Click here to [Delete Minimum Hitting Set]

⊙ Imaginäres_Brainstorming is a(n) Brainstorming
○ Brainstorming is a(n) Intuitiv_-_kreative_Methoden
○ Intuitiv_-_kreative_Methoden is a(n) Kreativitätstechniken

[Remove]

Fig. 2. Why is imaginary brainstorming a creativity technique? The ontology questionnaire retrieves an explanation for the inference. In case of disagreement, the user can choose to remove the suggested statement in the blue box, either of the three statements in the pink box, or deal more indirectly with it by changing the ontology's structure.

arbitrary ontology. The ontology questionnaire displays inferred statements, as for instance "Imaginary brainstorming is a creativity technique" for an ontology about the domain of innovation management (see Figure 1). For each inferred statement, the ontology questionnaire offers a *justification*, i.e. a reason why the statement is entailed by the ontology. While reviewing the inferred statements, the user of the questionnaire (domain expert or ontology engineer) might disagree with an inferred statement and wish to remove it. This is now the interesting part: Removing an inferred statement is not directly possible precisely because it is inferred and not explicitly stated. The ontology questionnaire finds one or more explanations for an inferred statement, and the user can choose to remove the reason for the inference. The statement "Imaginary brainstorming is a creativity technique" for instance is inferred because of the explicit statements "Imaginary brainstorming is a brainstorming technique", "Brainstorming is an intuitive-creative creativity technique" and "An intuitive-creative creativity technique is a creativity technique"[5]. Usually there is more than one option (see Figure 2), and sometimes it is also necessary to remove more than one statement. There is of course always the option to react more indirectly and change the structure and design of the ontology instead of simply removing some statements. Returning to the example from above, if the user disagreed with the inferred statement that "Imaginary brainstorming is a creativity technique" she can either remove the statement "Imaginary brainstorming is a brainstorming technique" or remove the statement "Brainstorming is a creativity technique" in the ontology in order to get rid of the unwanted inference.

The ontology questionnaire accepts ontologies formulated in OWL DL, which is the decidable version fragment of the web ontology language OWL [12]. OWL DL in turn is based on a family of logics called description logics. A prototypical implementation is online[6] at the time of writing but is not under development anymore. Instead, current efforts are spent on a version which will be integrated into MoKi[7] [19], a wiki-based modelling environment for business processes and ontologies. For reasoning, the ontology questionnaire accesses the open-source OWL DL reasoner Pellet [17,21]. Justifications are computed according to the built-in explanation functionalities in the Swoop [10,23] libraries.

4   Underlying Description Logic Theory

Some notions from the field of description logics are necessary in order to formally express the fundamental questions which need to be answered in an implementation of the ontology questionnaire: Which statements are entailed by an ontology, and how must an ontology be modified in order no longer entail a statement?

4.1   Terminology

Terminological axioms describe the relation of concepts (unary predicates, usually denoted by C, D) and roles (binary predicates, usually denoted by R) to each other. An *inclusion axiom* is an expression of the form $C \sqsubseteq D$ and is often expressed as "D subsumes C" or C is a subclass of D. An *equality axiom* is an expression of the form

---

[5] In this ontology there are two kinds of creativity techniques, one of which are called intuitive-creative and the other are called systematic-analytic.

[6] http://services.know-center.tugraz.at:8080/
InteractiveOntologyQuestionnaire

[7] http://moki.fbk.eu

$C \doteq D$ and can be seen as abbreviation for writing the two inclusion axioms $C \sqsubseteq D$ and $D \sqsubseteq C$. A set of terminological axioms constitutes a *terminological box*, the so called TBox. Terminological axioms express general truths (within the domain of discourse) about whole sets of things or about abstract concepts. *Assertional axioms* describe knowledge about individuals. A *concept (resp. role) assertion* is a statement of the form $C(x)$ resp. $R(x, y)$. These are often expressed verbally as "x is of type C" and "x is related to y via R". Concept and role assertions constitute an *assertional box*, the so called ABox. Assertional axioms express truths about specific individual entities. An *interpretation* $\mathcal{I}$ consists of a non-empty set $\Delta^\mathcal{I}$, the domain of interpretation, and an interpretation function that assigns to every primitive concept $A \in \mathcal{P}$ a set $A^\mathcal{I} \subseteq \Delta^\mathcal{I}$, to every primitive role $R \in \mathcal{R}$ a binary relation $R^\mathcal{I} \subseteq \Delta^\mathcal{I} \times \Delta^\mathcal{I}$ and to any instance $x \in \mathcal{X}$ an element $x^\mathcal{I} \in \Delta^\mathcal{I}$. An interpretation satisfies an inclusion axiom $C \sqsubseteq D$ iff $C^\mathcal{I} \subseteq D^\mathcal{I}$, a concept assertion $C(x)$ iff $x^\mathcal{I} \in C^\mathcal{I}$ and a role assertion $R(x, y)$ iff $(x^\mathcal{I}, y^\mathcal{I}) \in R^\mathcal{I})$. By *ontology* a TBox $\mathcal{T}$ is understood, while by *knowledge base* a TBox and an ABox $\mathbf{KB} = (\mathcal{T}, \mathcal{A})$ is meant. An ontology (resp. knowledge base) is satisfiable iff there is an interpretation that satisfies all its axioms. An ontology (resp. knowledge base) entails an axiom α if and only if all its interpretations satisfy α. This is written as $\mathcal{T} \models \alpha$ resp. $\mathbf{KB} \models \alpha$. Throughout the paper the term *statement* is used interchangeably with *axiom*, where the first is used rather when talking about ontology engineering issues and the latter is used rather when talking about logic/theoretical issues.

4.2 Formulation of Relevant Problems

Given an ontology $\mathcal{O} = \{C_1 \sqsubseteq D_1 \ldots C_n \sqsubseteq D_n\}$, explicitly stated axioms can be removed from $\mathcal{O}$ by simply deleting them. For any inferred axiom α, i.e. $\mathcal{O} \models \alpha$ and $\alpha \notin \{C_i \sqsubseteq D_i | i = 0 \ldots n\}$, a more complex solution is obviously necessary. The task is formalised as follows: Modify $\mathcal{O}$ such that it results in a modified ontology $\mathcal{O}'$, for which $\mathcal{O}' \not\models \alpha$ is true. This modification is called "removing an inferred axiom from $\mathcal{O}$" throughout this paper, although clearly not the inferred axom itself can be removed but only its causes. Since OWL DL is based on a monotonic logic, it can be seen at this point already that $\mathcal{O}' \subset \mathcal{O}$, i.e. some statements need to be removed from $\mathcal{O}$ in order to remove an inferred axiom α.

Two questions are fundamental to realising the ontology questionnaire as sketched conceptually above. Given an ontology $\mathcal{O} = \{C_1 \sqsubseteq D_1, \ldots C_n \sqsubseteq D_n\}$:

1. Apart from the explicitly modelled general inclusion axioms $C_i \sqsubseteq D_i, i = 1 \ldots n$, which further general inclusion axioms are entailed by $\mathcal{O}$?
2. Given a general inclusion axiom α such that $\mathcal{O} \models \alpha$, and $\alpha \notin \mathcal{O}$, how must $\mathcal{O}$ be modified into $\mathcal{O}'$ such that $\mathcal{O} \not\models \alpha$?

It is strictly necessary to answer the first question in order to construct the list of inferred statements which the ontology engineer shall review at all. It not necessary in the technical sense but clearly desirable to be able to answer the second question, in order to give hints to the ontology engineer on how to revise the ontology. As will be seen in the discussion later on however, the kind of support given is limited under a certain aspect.

Limitation to Explicitly Mentioned Concepts. In general, OWL DL ontologies entail an infinity of subsumption axioms. Limiting the entailments considered by the ontology questionnaire to concepts that are explicitly mentioned, i.e. which occur in an explicit statement in the ontology, allows dealing with a finite number of entailments however. Then, listing entailed but not explicitly stated general inclusion axioms simply amounts to comparing concepts pairwise and checking which of the two is subsumed by the other, or if no subsumption is entailed either way. Any subsumption axiom which is not explicitly stated is listed as entailed statement.

Within the ontology questionnaire, an additional heuristic is used for display, namely subsumption axioms including `owl:Thing` as superclass, i.e. subsumption axioms stating that a concept is satisfiable, and subsumption axioms including `owl:Nothing` as subclass are hidden from the user.

To illustrate the limitation to known concepts, consider axiom $Human \sqsubseteq Mammal$, which entails an infinity of axioms like $\forall parent.Human \sqsubseteq \forall parent.Mammal$[8], $\forall parent.(\forall parent.Human) \sqsubseteq \forall parent.(\forall parent.Mammal)$ etc. in OWL DL, given that $Human, Mammal$ are concepts and *parent* is a role within the ontology. If however, $\forall parent.Human$ and $\forall parent.Mammal$ are not mentioned explicitly in an axiom within the ontology, the entailment $\forall parent.Human \sqsubseteq \forall parent.Mammal$ is not displayed in the questionnaire.

Justifications in OWL. The second question has been treated with slightly varying focus under the names of ontology debugging, ontology repair and ontology explanations. First we define justifications of entailments, which correspond to what has also been called "the reason for an entailed axiom" in this paper.

Definition 1 (Justification). *Let $\mathcal{O} \models \alpha$ where α is an axiom and $\mathcal{O}$ an ontology. A set of axioms $\mathcal{O}' \subseteq \mathcal{O}$ is a justification for α in $\mathcal{O}$ if $\mathcal{O}' \models \alpha$ and $\mathcal{O}'' \not\models \alpha$ for every $\mathcal{O}'' \subset \mathcal{O}'$. (e.g. [9, p269])*

In other words, a justification is a minimal set of axioms from which the statement α follows. There may be more than one justification for any particular α. Clearly, in order to remove the entailment α, at least one axiom from *each* justification must be removed. Such a set of axioms (at least one from each justification) HS can be formally defined $\forall s \in S : HS \cap s \neq \emptyset$ where S is the set of all justifications. The technical term for HS is *hitting set*. This explains why sometimes multiple statements need to be deleted from an ontology in order to remove an inferred statement. Several algorithms to find all justifications for OWL entailments are described in [9], and some are implemented as part of the Swoop [23] libraries.

The ontology questionnaire leaves the choice of which axiom(s) of a justification to remove to the user. However, a minimal hitting set, i.e. the smallest possible set of axioms which make the undesired inference disappear if they are deleted, is suggested by the ontology questionnaire.

---

[8] In accordance with the conventional notation and semantics used in description logics, a concept $\forall R.C$ is interpreted as $\{x | R(x,y) \rightarrow C(y)\}$. Verbalizing the axiom $\forall parent.Human \sqsubseteq \forall parent.Mammal$ would give something like "For everyone for whom it is true that all his/her parents are Humans, it is also true that all his/her parents are Mammals".

5 Experimental Study

5.1 Application Setting

The ontology questionnaire was applied in the scope of APOSDLE[9] to model six different learning domains. In this setting, for each learning domain not only a domain model, but also a task and a competency model was created. Furthermore, models were created following a modelling methodology, the Integrated Modelling Methodology (IMM), which is described in detail in [1]. It has been designed to offer revision support at all its steps [13]. Therefore all ontologies which "reached" the point of being reviewed in the ontology questionnaire had already gone through several different reviews, as is sketched in Figure 3.

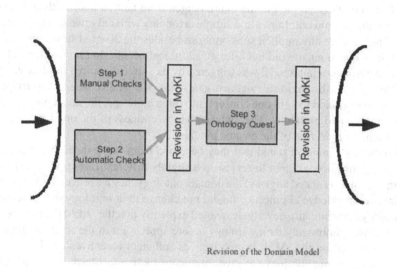

Fig. 3. A zoom into the revision process of the domain model in the IMM. All revision steps imply, where the necessity of changes in the models is identified, an iteration loop going back to revising the domain model in MoKi and another entry into the revision process.

All application partners used the ontology questionnaire to review the domain models created for their own APOSDLE installation. The domain ontologies concerned the topics of consulting on industrial property rights (ontology with 95 concepts), the RES-CUE requirements engineering methodology (ontology with 79 concepts), the Information Technology Infrastructure Library (ontology with 101 concepts), electromagnetic simulation (ontology with 116 concepts), innovation management (ontology with 134 concepts) and statistical data analysis (ontology with 71 concepts).

[9] APOSDLE (www.aposdle.org) has been partially funded under grant 027023 in the IST work programme of the European Community.

5.2   Evaluation Procedure and Results

After the ontology questionnaire was completed by an ontology engineer, a telephone interview with him or her was conducted to elicit information about the invested effort, the perceived benefit and consequences of reviewing the ontology via the ontology questionnaire. In order to guarantee some confidentiality to the involved application partners, the following analysis does not refer to the domain ontologies by their domain.

The questions were:

Invested effort. How long did it take to go through the questionnaire? Was this a difficult task?
Benefit. Did the ontology engineers benefit from going through the questionnaire?
Effect. Did going through the questionnaire trigger any changes in the model? If yes, which?

The effort to go through the questionnaire ranged from half an hour to three hours. In two cases, the domain ontology was a simple taxonomy with only two levels. Therefore they did not contain any implicit subsumptions besides the asserted subsumptions, and the ontology questionnaire did not offer an additional benefit. In two cases, no change in the underlying domain model was triggered by its application. In a third case, which took three hours, a discussion about the organisation of domain concepts was triggered. Finally it was decided to split one concept into three more specific concepts in order to obtain a more valid hierarchy. This indicates the usefulness of the ontology questionnaire to support *reflection* of the created model.

All ontology engineers stated that they benefited from going through the questionnaire, since this allowed them to review in a slightly different form (i.e. minus everything they explicitly stated anyway) the domain ontology they had created. Interestingly, also the two knowledge engineers who did not change their ontology due to review in the ontology questionnaire nevertheless noted explicitly that they liked the tool, but had already revised thoroughly their ontology before applying it in the scope of the IMM. This reflects well on the IMM, in that it gives sufficient incentives and ideas for revision, but unfortunately does not give more constructive feedback on the reviewing process itself.

Concerning the usability of the ontology questionnaire, the ontology engineers were satisfied in general with the rather straight-forward presentation, but dissatisfied with the presentation of statements as lists without prioritisation. Both inferred statements and the originally explicated ontology were presented as such lists.

5.3   Discussion

Although it is difficult to derive general conclusions from just five use cases, such an experience nevertheless gives useful indications. One, for normal-sized ontologies, the effort to review entailed statements is manageable. Two, it is desirable in a tool to develop a prioritised list of statements to review. Such a prioritisation will become more important if even larger ontologies are to be reviewed. Three, all involved persons judged such a review to make sense. This is highly relevant to keep up the motivation of people in carrying out the ontology evaluation. Together, this feedback indicates that further research into the proposed direction of ontology evaluation is worthwhile. One point which definitely influenced the given feedback however was that the ontology questionnaire was applied in situations where the creation of the models was embedded in a thorough modelling methodology. From some statements given by the ontology engineers it can be suspected that the small number of changes triggered is due to the rigour of revisions which happened *before* the ontology questionnaire was even applied.

Also some more informal experiments than the above described application study indicate that removing existing statements may not always be the method of choice to deal with an undesired inference. Frequently, more subtle and creative actions were taken, which ranged from simple renaming to removal/addition of concepts and relations, and to a restructuring of the ontology.

5.4 Future Evaluation

We plan to carry out a more rigorous study in which the question whether and to what degree such a review procedure supports the identification of modelling mistakes shall be researched. For this study we will create several small test ontologies with modelling mistakes taken from the categories described in [18] and from our own experiences. Since the focus is not on ontology repair however, no logically inconsistent but merely conceptually wrong ontologies will be provided. Test persons for this planned experiment will have a background in knowledge engineering or computer science, but not necessarily be experts specifically in description logic modelling. The test persons will be divided into three groups. The first group will get the assignment to review the inferred statements, with a list of explicitly stated axioms as supplement. This group will work with an improved version of the current questionnaire's implementation. A second group will get the same assignment, but will work with an ontology editor like Swoop or Protégé in conjunction with a reasoner. The third group serves as control group that will get the assignment to review only the explicitly stated axioms. This group will work with an ontology editor with the reasoner turned off. A comparison between the first and second group on the one hand and the third group on the other hand will give indications on the validity of the assumption that reviewing inferred statements supports the creation of conceptually correct ontologies. A comparison between the first and second group will serve to receive input on the influence of the user interface. Throughout this study, special attention will be paid to complex solutions to modelling mistakes that exceed simple removal or addition of axioms. This may give input to further research about supporting the creative parts inherent in ontology engineering.

6 Outlook and Conclusion

A first improvement of the ontology questionnaire will be to to include assertional axioms in the ontology questionnaire, i.e. assertional axioms such as for instance "Alice is a project leader" and "Alice works together with Bob" will also show up as inferred statements. This does not present technical problems, and such functionality is already integrated for instance in Swoop. In the context of the experiment carried out so far however, this did not represent a limitation since the ontologies in question only contained concepts. Some directions for improvement can be found in the representation of the ontology questionnaire. Prioritisation of inferred statements will definitely be taken up in the upcoming new version. A simple heuristic may be to prioritise all subsumption axioms which do *not* stem from the simple transitivity of subsumption. Another issue is the representation of side-effects of removing axioms. When removing asserted axioms, it is possible that not only the undesired entailment but also some actually desired inferences are also "removed". Such information is currently not presented in the ontology questionnaire. On the other hand, this touches on the issue of verbalisation. A more closely natural language representation of the formal axioms would push the ontology questionnaire more into the direction of the knowledge experts, i.e. more towards the source of knowledge. A critical issue with this respect is that although work on English verbalisation for OWL exists (e.g. [8]), this seems to be lacking for most other languages. Concerning the hints given on how to modify the ontology in order to improve it, it is desirable to give hints not only on which axioms to remove but also on which (kinds of) axioms then to add. The rationale behind this goal is of course that if many undesired inferences are removed by removing explicit statements, in the end only the empty ontology with no undesired inferences remains. A first step in this direction could be based on the catalogue of modelling mistakes in [18] and take the form of simple modelling guidelines or specific examples. In general however it is not obvious how such complex ontology revision behaviour can be supported automatically.

To conclude, we think that reviewing inferred statements is a valuable part of an ontology evaluation procedure which contributes to evaluating an ontology's correctness. This is also indicated by the experiments carried out so far, and will be evaluated more rigorously in the future. Reviewing inferred statements can be tightly integrated with modelling, and lends itself naturally to being integrated in a modelling environment.

References

1. APOSDLE Deliverable 1.6. Integrated modelling methodology version 2 (April 2009)
2. Buitelaar, P., Cimiano, P., Magnini, B.: Ontology Learning From Text: Methods, Evaluation and Applications. In: Ontology Learning from Text: An Overview. Frontiers in Artificial Intelligence and Applications, vol. 123, pp. 1–10. IOS Press, Amsterdam (2005)
3. Fernández-López, M., Gómez-Pérez, A., Sierra, J.P., Sierra, A.P.: Building a chemical ontology using methontology and the ontology design environment. IEEE Intelligent Systems 14(1), 37–46 (1999)
4. Gangemi, A., Catenacci, C., Ciaramita, M., Lehmann, J.: Modelling ontology evaluation and validation. In: Sure, Y., Domingue, J. (eds.) ESWC 2006. LNCS, vol. 4011, pp. 140–154. Springer, Heidelberg (2006)
5. Guarino, N.: Formal ontology and information systems. In: Formal Ontology and Information Systems, Proceedings of FOIS 1998, pp. 3–15. IOS Press, Amsterdam (1998)
6. Guarino, N., Welty, C.: Handbook on Ontologies. In: An Overview of OntoClean. International Handbooks on Information Systems, pp. 151–172. Springer, Heidelberg (2004)
7. Horridge, M., Parsia, B., Sattler, U.: Laconic and precise justifications in owl. In: Sheth, A.P., Staab, S., Dean, M., Paolucci, M., Maynard, D., Finin, T., Thirunarayan, K. (eds.) ISWC 2008. LNCS, vol. 5318, pp. 323–338. Springer, Heidelberg (2008)
8. Kaljurand, K., Fuchs, N.E.: Verbalizing OWL in Attempto Controlled English. In: Proceedings of Third International Workshop on OWL: Experiences and Directions, Innsbruck, Austria, June 6-7, vol. 258 (2007)

9. Kalyanpur, A., Parsia, B., Horridge, M., Sirin, E.: Finding all justifications of OWL DL entailments. In: Aberer, K., Choi, K.-S., Noy, N., Allemang, D., Lee, K.-I., Nixon, L.J.B., Golbeck, J., Mika, P., Maynard, D., Mizoguchi, R., Schreiber, G., Cudré-Mauroux, P. (eds.) ASWC 2007 and ISWC 2007. LNCS, vol. 4825, pp. 267–280. Springer, Heidelberg (2007)
10. Kalyanpur, A., Parsia, B., Sirin, E., Grau, B.C., Hendler, J.: Swoop: A web ontology editing browser. Elsevier Journal of Web Semantics 4(2), 144–153 (2006)
11. Noy, N.F., McGuinness, D.L.: Ontology development 101: A guide to creating your first ontology. Technical report, Stanford University (March 2001)
12. OWL Web Ontology Language Reference, http://www.w3.org/tr/owl-ref/ (last visited 07-13-2009)
13. Pammer, V., Kump, B., Ghidini, C., Rospocher, M., Serafini, L., Lindstaedt, S.: Revision support for modeling tasks, topics and skills. In: Proceedings of the i-Semantics 2009, Graz, Austria (to appear, 2009)
14. Protégé, http://protege.stanford.edu/ (last visited 07-07-2009)
15. Protégé - Explanation Workbench, http://owl.cs.manchester.ac.uk/explanation/ (last visited 09-24-2009)
16. RaDON - Repair and Diagnosis in Ontology Networks, http://radon.ontoware.org (last visited: 09-24-2009)
17. Pellet The Open Source OWL DL Reasoner, http://clarkparsia.com/pellet/ (last visited: 03-01-2009)
18. Rector, A., Drummond, N., Horridge, M., Rogers, J., Knublauch, H., Stevens, R., Wang, H., Wroe, C.: Owl pizzas: Practical experience of teaching owl-dl: Common errors & common patterns. In: Motta, E., Shadbolt, N.R., Stutt, A., Gibbins, N. (eds.) EKAW 2004. LNCS (LNAI), vol. 3257, pp. 63–81. Springer, Heidelberg (2004)
19. Rospocher, M., Ghidini, C., Pammer, V., Serafini, L., Lindstaedt, S.: MoKi: the modelling wiki. In: Proceedings of the Forth Semantic Wiki Workshop (SemWiki 2009), co-located with 6th European Semantic Web Conference (ESWC 2009). CEUR Workshop Proceedings, vol. 464 (2009)
20. Schreiber, G., Akkermans, H., Anjewierden, A., de Hoog, R., Shadbolt, N., Van de Velde, W., Wielinga, B.: Knowledge Engineering and Management The CommonKADS Methodology. MIT Press, Cambridge (1999)
21. Sirin, E., Parsia, B., Grau, B.C., Kalyanpur, A., Katz, Y.: Pellet: A practical OWL-DL reasoner. Journal of Web Semantics 5(2), 51–53 (2007)
22. Sure, Y., Studer, R.: On-to-knowledge methodology. In: On-To-Knowledge: Semantic Web enabled Knowledge Management, pp. 33–46. Wiley, Chichester (2002)
23. Swoop, http://code.google.com/p/swoop/ (last visited 07-07-2009)
24. Uschold, M., Gruninger, M.: Ontologies: Principles, methods, applications. Knowledge Engineering Review 11, 93–155 (1996)
25. Vrandecic, D., Gangemi, A.: Unit tests for ontologies. In: Proceedings of the 1st International Workshop on Ontology content and evaluation in Enterprise. LNCS. Springer, Heidelberg (2006)

Knowledge-Based Process Modelling for Nuclear Inspection

Florin Abazi[1] and Alexander Bergmayr[2]

[1] University of Vienna
(Doctoral Candidate)
a0309798@unet.univie.ac.at
[2] University of Vienna
Department Knowledge Engineering, Bruenner Strasse. 72,
1210 Vienna, Austria
ab@dke.univie.ac.at

Abstract. The nuclear safeguards system is a very specialized domain which demands dedicated modelling languages to capacitate the preservation of nuclear knowledge while improving the efficiency and effectiveness of safeguards related activities. Inspection modelling method represents a process oriented and a knowledge based approach created through the integration of modelling languages which cover one or more of three perspectives of the underlying conceptual framework. First, the operational perspective which covers all safeguards objectives and the related inspection processes. As second is the knowledge management perspective which models the enterprise nuclear knowledge. Finally, the technology perspective with modelling languages enabling the use of IT services in support of inspection processes. This study represents a conceptual proposal for merging the three perspectives into a knowledge based and process oriented modelling framework. The approach is expected to improve operational aspects of inspection and contribute to the overall nuclear knowledge management.

Keywords: Nuclear Inspection, Knowledge-based Inspection Modelling, Knowledge Engineering, Nuclear Knowledge Management.

1 Introduction

By modelling we refer to the act of representing something [1]. In Business Process Management (BPM) modelling as an activity, referred to as Business Process Modelling, seeks representing business processes as "an ordering of work activities in time and place, with a beginning, and end, and clearly identified inputs and outputs" [2].In contrast to graphical models used in BPM, in mathematics models may also be used for representing a theory since "models are instantiations or application of the theory" [3]. If we focus on graphical and textual models as described in [4] how can models represent knowledge that can be operationalized? Knowledge based view of modelling can be twofold; the first view is that modelling is an activity which entails engineering techniques to extract knowledge from experts and transferring it to machines.

D. Karagiannis and Z. Jin (Eds.): KSEM 2009, LNAI 5914, pp. 406–417, 2009.

This constitutes the Knowledge Engineering (KE) models that enable interpretation by machines and therefore creating such models necessitates a certain level of formalism (e.g. syntactic and semantic correctness) for the computers to interpret them. The second view of the knowledge based approach is to use models to represent a reality or vision [5] for documentation, specifications or training purposes. Such models can facilitate understanding of humans who interpret them. Commonly supporting Knowledge Management (KM) tasks such models are referred to in [4] as KM models.

If knowledge is hidden in organizational processes and embodied in individuals the use of models to externalize knowledge seems as a suitable approach. How do process and knowledge models complement one another and help achieve this objective? Through process models (e.g. Unified Modeling Language UML – activity diagrams) and still vaguely defined "knowledge" models, critical knowledge hidden in processes and individuals may be identified, synthesized, preserved and processed to directly or indirectly improve business performance. As if such models would require knowledge absorptive capacity that can enable its preservation and eventually transformation into performance. It is suggested that knowledge engineering through a model based approach supports BPM as a domain, its application as a management method and its execution or deployment (based on unpublished work from Karagiannis et al) For the purpose of the paper the term equivalent to BPM is nuclear Inspection Process Management (nIPM) which will be used since there is an intended departure from the for-profit business terminology.

Knowledge used to build and operate nuclear facilities is essential for IAEA inspectors to continuously provide assurance that nuclear material and technology is used exclusively for peaceful purposes. The problem with nuclear knowledge is that great amount of facts about knowledge processes are hidden in the heads of individuals. Externalizing implicit knowledge from nuclear workers that can be used by inspectors and vice versa can facilitate nuclear knowledge management - by nuclear knowledge management we understand all efforts made towards implementing knowledge management in nuclear facilities as well as internal undergoing KM efforts at the IAEA [6]. It is not suggested that such knowledge is externalized only for use by inspectors instead having a "lingua franca" for modelling nuclear processes in the industry, governmental institutions and the nuclear community as a whole also represents an efficient way of sharing critical nuclear knowledge. From the inspection point of view there is a knowledge gap between the nuclear processes taking place in facilities and the inspection processes which are framed under the auspices of legal agreements. This leads to our research which is to create a framework for a knowledge based approach for modelling inspections. Inspection modelling is expected to improve nuclear knowledge management and inspection process management. At least two main objectives need to be achieved. The first is nuclear knowledge preservation and the second is inspection effectiveness and efficiency. This paper will discuss the initial research by detailing efforts to integrated process models that can be used for inspection modelling. Inspection knowledge models, technology models as well and their integration with inspection process models will be subject to future research. In the next section of the paper the conceptual framework is discussed. Thereafter components of the inspection modelling method are described along with a use case scenario.

2 nIPM Modelling Framework

In order to support the above mentioned nuclear Inspection Process Management or nIPM a conceptual modelling framework is presented. Such a framework represents a high level structure that serves as common foundation for creating the inspection modelling method that is based on the concepts introduced by [4][7]. Modelling methods provide a theoretical basis for enabling modelling activities by introducing major components such as modelling languages, modelling procedures, mechanisms and algorithms that can be applied onto models which correspond to meta-models defining the syntactical elements of the used modelling languages. Introducing a proper set of interconnected modelling languages to support the three perspectives – operational, knowledge management, and technology - of the conceptual framework is core to the work at hand. Before deeper insights to the inspection modelling method are given the above mentioned perspectives are introduced in more detail.

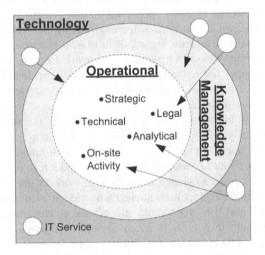

Fig. 1. Conceptual framework for Inspection Modelling. It consists of three perspectives *operational, knowledge management* and *technology*. The operational perspective consists of 5 components.

Figure 1 above depicts the three perspectives of the framework. The operational perspective consists of five components: *technical, strategic, legal, analytical and on-site activity*. It covers aspects specifically related to the nuclear safeguards system. Most of the components do not need further clarification whereas the word "technical" may be ambiguous since it is the term used to refer to equipment and instruments placed at facilities. Such equipment is used for performing measurements which is an important part of inspection. Another component that may require explanation is the "analytical" component that stipulates analytical services for samples taken at nuclear facilities. To offer a complete overview of all operational aspects all these components have to be "en suite" in a framework by using process oriented modelling languages. The figure also shows the knowledge management perspective which covers aspects not included in the operational perspective focused on facilitating knowledge management and providing context. Technology perspective depicted as the environment around the operational and knowledge management perspective is a layer that makes services accessible in support of data acquisition, processing and execution of models.

2.1 Operational Perspective

This is represented by modelling languages that document the operational elements of nuclear inspections. Inspection procedures, strategic and technical objectives, legal requirements and analytical processes used to reach conclusions are incorporated here. Nuclear processes and the nuclear fuel cycle are also represented in this view since they serve as specifications for conceiving inspection processes. Reference models of facilities within a state and the related safeguards objectives for each state/site/facility are modelled in order to contextualize actual inspection processes.

2.2 Knowledge Management Perspective

Based on process-oriented knowledge management [8], here the so called knowledge management processes are modelled describing the tasks for creating, storing and disseminating inspection knowledge. The same way inspection processes are derived from the nuclear processes, knowledge perspective includes languages that help derive knowledge processes relevant to knowledge at nuclear facilities. Additionally, languages represented under this perspective are used to map knowledge resources available within the organization so to identify and close the gap between "what is known" and "what needs to be known" to support inspection processes. Skill Models, Knowledge Map Model and Knowledge Management Process Models are used to account for the knowledge necessary and available for performing effectively inspections.

2.3 Technology Perspective

It includes modelling languages to represent the technology layer. Languages used for the IT Architecture or orchestration of services provisioned by SOA and similar architectures. Languages such as OWL-S, OWL-WS and UML can be used to represent this perspective. Another interesting outlook of this perspective is the growing trend to perform inspections through remote monitoring. Containment and surveillance in addition to unattended safeguards systems can be integrated using a model based approach. Such an approach would require an enterprise architecture where remote monitoring services are also modelled part of the IT architecture.

3 Inspection Modelling Method

In order to support the three different views introduced in the previous section with models a set of modelling languages relevant to the nuclear inspection domain are discussed. Although each of the modelling languages presented are from a conceptual viewpoint independently defined, connections between these languages are supported through inter-model relationships to enable highlighting the coherence on the model level. Thereby, the goal of modelling is to externalize knowledge (1) about the processes applied at the facilities/states throughout all phases of the nuclear fuel cycle and

(2) processes required by the IAEA to perform the nuclear inspections at facilities/states. Process-oriented models such as the nuclear fuel cycle process model or the inspection process model can be used to document and create a "protocol" of the approach applied by the IAEA in order to ensure that nuclear activities in states are accordingly inspected and comply with the undersigned safeguards agreement. The Agency's authority to apply safeguards stems from its Statute - "pursuant to this authority, the Agency concludes agreements with States, and with regional inspectorates, for the application of safeguards" [9].

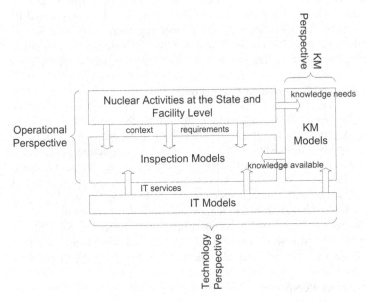

Fig. 2. Inspection Modelling Framework high-level requirements

As depicted in Fig. 2 the operational perspective is viewed as the layer of "business" of the facilities/states in terms of the nuclear domain. In a sense an attempt is made to "reconstruct" the activities performed by the facilities/states. A common reference model or models for each type of facility and safeguards approach is available and can be adjusted for each facility/site/state (a kind of a virtual state file). A top-down approach is represented by which requirements can be derived from the top layer (nuclear activities) for the definition of the inspection processes. Because the nuclear processes and inspection processes are represented at a different level of detail, inspection objectives are used to link the two. The requirements derived also define the context for the inspection process. Inspection models layer actually contains the models related to the nuclear inspection processes and the facilities inspected. Whereas the KM Models layer describes and identifies available knowledge roles, skills and resources that are needed to actually perform the inspections. KM models are also derived from the top layer context and act in support the inspection process modelling (e.g. making inferences on automatically derived ontologies used to validate inspection processes). The layer of IT models provides accessibility of IT services in order to support the knowledge based inspection modelling. Due to a large

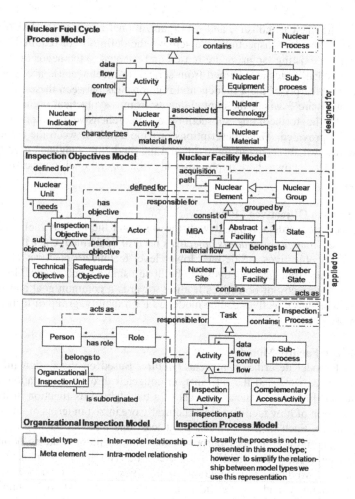

Fig. 3. Syntactical elements of the inspection model types relevant for the operational perspective

and heterogeneous composition of technologies available for supporting inspections this layer can also support the integration of technology at the inspection process level [10]. This layer uses the knowledge management layer to reduce the semantic gap between the inspection process requirements and the services available.

3.1 Operational Meta-models

Based on the requirements outlined for the inspection modelling method, modelling languages are derived that support the operational perspective by capturing (a) strategic issues in terms of objectives of the participating parties (e.g. IAEA inspectors, member states), (b) description of nuclear facility/site organizations and relationships between these facilities/sites, (c) inspection processes applied for nuclear inspection activities, (d) common nuclear knowledge within the IAEA and knowledge about processes in the nuclear industry, and (e) organizational issues of the IAEA relevant for execution of nuclear inspections. To achieve the definition of required modelling languages either existing language are reused and adapted to the needs of the nuclear domain or new languages are defined from scratch to meet specific needs of the nuclear domain. Excerpts of meta-models and relationships between these meta-models are shown in Figure 3 where the core elements from a syntactical point of view are introduced. In the forthcoming subsections a comprehensive description for each meta-model is provided. These descriptions provide a link between the aspect of nuclear inspection we are trying to represent and the modelling language used to do so. The meta-models shown here represent only the operational perspective of the framework.

3.1.1 Inspection Objectives Model

Based on the safeguards agreement with the state and the nuclear fuel cycle in the state high level strategic *safeguards objectives* need to be defined. Once the state level or facility level objectives are defined lower level technical objectives are also depicted. Modelling strategic objectives can be done using a goals model. Goals model will show the hierarchy of objectives leading towards a common goal and how they relate to each other. Objectives can be modelled for a facility, group of facilities, state or a group of states.

3.1.2 Nuclear Facility Model

This type of model describes a nuclear *Facility*. Based on the design information received from the state and the information collected through verification activities Facility Model (FM) represents all aspects of a facility and its function. It includes a conceptualization of how facilities are structurally organized in terms of nuclear material accountancy. Each facility consists of at least one material balance area (MBA) for nuclear material accountancy purposes; selection of those strategic points which are key measurement points (KMPs[1]) for use in determining nuclear material flows and inventories. Modelled representation of facilities provides reference for nuclear material accountancy and the verification activities that need to be implemented to ensure a permanent continuity of knowledge for the material received, processed or shipped to other facilities. The same model can be used to represent one facility or a group of facilities based on a common location referred to as site. For complementary access purposes this model can be used also to describe any other type of location.

Material flow between modelled facilities can be used to represent their relationship. Such relationships can exist between facilities within the same state or internationally.

3.1.3 Nuclear Fuel Cycle Process Model

The Nuclear Fuel Cycle Process Model (NFCM) is meant to describe the planned or on-going *nuclear activities* in a state. It identifies all known and declared nuclear processes which serve as basis for deriving the corresponding *inspection activities*. In addition to representing the fuel cycle in a state it can be used to identify, describe and

---

[1] KMP is usually modelled as a property and properties are not depicted in our model types.

characterize every known process for carrying out each step necessary for the acquisition of weapons-usable material. It can be used to represent all plausible acquisition paths for highly enriched uranium (HEU) and separated plutonium (Pu) as well as the associated *equipment, material, technologies* that characterize such processes. Specific items known as *Indicators* associated to nuclear processes are evaluated in the context of the activities declared by the state (cf. *designedfor* relationship). NFCM models and associated ontology models of indicators are used to provide context for generating *inspection processes* and enable knowledge based modelling.

3.1.4 Inspection Process Model

Inspection Process Model represents a business process model for *inspection activities* with an extension for inspection pathways. Here random inspection activities as well as complementary access activities are incorporated. *Complementary Access* (CA) can be performed with a 2 hours or 24 hours advanced notice. It is provisioned by the additional protocol to allow the agency to request access for any of the following reasons:

(a) to assure the absence of undeclared nuclear material and activities at sites of facilities or LOFs, or at mines, concentration plants or other locations declared under as containing nuclear material;
(b) to resolve a question relating to the correctness and completeness of the information provided or to resolve an inconsistency relating to that information; and
(c) to confirm, for safeguards purposes, the State's declaration of the decommissioned status of a facility or of a LOF [9].

Inspection implementation activities based on the criteria can be graphed using inspection process models. They can be used also to externalize expert know how and serve as an entry point to knowledge management models. It is the intention to automatically or semi-automatically generate inspection process models based on the reference models generated by domain experts. Guided creation of inspection pathways tailored for the specific state and nuclear activity will help implement integrated safeguards approach in states under traditional safeguards. Such an approach is founded on the principles of effective but also efficient *inspection processes*.

3.1.5 Organizational Inspection Model

On the one hand actors are represented through states that pursue different objectives in the context of nuclear material acquisition or inspections through the IAEA. The organizational foundations for actually executing inspection processes, defining objectives for such inspections and evolving the models that support inspectors performing their activities is expressed using the organizational inspection model. Some relevant organizational roles within the IAEA for nuclear inspections are process designer that are in charge designing optimal inspection processes in accordance with safeguards agreements and other legal aspects. While senior inspectors are responsible for devising safeguards approaches other inspectors are the users of inspection process models by practicing the defined activities on a mission. The IAEA management also benefits from the process-oriented approach as the documentation of inspection processes derived from nuclear processes in a state provisions process optimization while maintaining the effectiveness of inspections. Furthermore, it supports the continuing effort of implementing integrated safeguards universally by designing and maintaining state specific approaches tailored to the state's nuclear activities and profile.

4 Use Case: Integrated Safeguards

In order to clarify this use case of the inspection modelling framework the nuclear safeguards system needs to be briefly described. Each member states signs a safeguards agreement with the IAEA in which the obligations of both parties are agreed to. Safeguards activities are based on the non-proliferation treaty and international safeguards agreements. Historically, the so called comprehensive (now referred to as traditional) safeguards agreement was implemented in the respective states. This agreement is characterized by the obligation of the IAEA to verify declared nuclear material. The Safeguards systems has evolved since and so called strengthened measures were introduced. Moreover, a protocol additional was created which gives more access to the IAEA inspectors in order to verify that declared and undeclared material is not diverted for non-peaceful purposes [9]. Not all countries are signatories to the protocol additional however it is expected in the future that it will be universally accepted. A progress from the traditional safeguarding is the so called Integrated Safeguards (IS) which provisions for greater access and more inspection effectiveness while the total number of inspections is considerably reduced at the same detection probability of undeclared activities. This does not only represent a benefit for the IAEA but also for the member states who find interest in allowing inspections but having to deal with less intrusiveness and interruption to their nuclear operations.

Modelling inspection processes provided by the modelling framework documents procedures and gives the ability to optimize processes while increasing efficiency and effectiveness of inspections. Due to a number of routine like activities there may be a gap between the inspection process knowledge and the nuclear process verified.

Through the use of the inspection modelling method all inspection activities and the related nuclear activities in a state can be modelled. Furthermore, knowledge models can be used to model all other aspects (i.e. qualitative) not included in the processes models however relevant to providing the context for generating semantic structures that support the design and execution of inspection processes. This provides the basis for implementing the idea of inspection pathways discussed in [11]. By processing machine and human interpretable models of inspection activities, pathways that include most strategic facilities to be inspected can be generated automatically or semi-automatically. Under integrated safeguards already existing algorithms can be applied to increase the unpredictability of inspections while maintain the efficiency in comparison to the traditional inspections. Inspection modelling and related knowledge management and engineering methods can provide the support for the inspection process management in three ways:

1. Analyzing the similarities of traditional inspection processes (as is) and integrated safeguards processes (should be)
2. Generating and simulating inspection processes along the potential nuclear weapons acquisition paths

3. Serve as requirements for statisticians and software engineers for creating algorithms and mechanisms in order to calculate the minimum number of inspections which have an unpredictability component and the detection probability of undeclared activities remains.

5 State of the Art/Related Work

Management of the knowledge in the area of nuclear technology has been there for a while. Due to security concerns KM efforts in this area was intensified as depicted by ISCRAM in [12] In order to fulfill the needs and mandatory requirements related to nuclear technologies, it involved knowledge workers to face immense amount of important information in the process of decision-making in contrast to different approaches suggested today in order to support such tasks – e.g.. [13], [14], [15].

This task on one side provided boost to involvement of different technologies to support the decision-makers, but on the other, most of the systems were created on a isolated ad-hoc basis working against the system as whole (seen as a worldwide KM in nuclear technology area) – facts about the knowledge processes were still hidden in the heads of persons responsible for the functioning of the KM system.

Example for such system was the task toolkit developed by the LAIR in early 90's [16] that was designed to support mission critical decisions regarding operations in the nuclear plant. Support was provided in three layer system: Monitoring, Diagnosing (failure) and taking Procedural action in order to minimize chance of complete malfunction.

It can be said that "procedural management" was a long time synonym for successful knowledge management in this area (as discussed in [17]), that is, predecessors of today's knowledge management processes (KMP's) were managed as procedures that were invoked and carried out, e.g. in case of emergency.

KM initiatives in this area acknowledged that one of the most important issues to be tackled is the goal to enhance usage of the Organizational Learning (OL) in order to improve the effectives and efficiency. Example for such initiative – to profit from OL - is the DOE lessons learned program [18] established by the US Department of Energy (DOE) that aimed to transform the OL into best/recommended practices available for all interested organizations (a framework for transfer of best practices in the NPP area was proposed by MIT in 1993 – see [19] for more details). Another study carried out by Jennex and Olfman [20] investigated productivity increase in a nuclear plant over period of five years that pointedly utilized OL came to the findings that usage of OL does enhance overall productivity.

A similar research to the one studied in this paper was initiated by the European Commission Joint Research Centre as described by Poucet et al [21]. Also, with the objective of supporting IAEA and the nuclear inspections an approach to integrating numerous information sources is suggested. Under the Additional Protocol of the IAEA, the authorized use of open source information and satellite imagery and numerous other sources introduced the need to integrate this information. The approach suggests a Geographical Information System (GIS) to structuring all inspection related information for use by inspectors. It proposes the use of geographic models for representation of data. Although, there are some common objectives, our research is focused on nuclear knowledge management through a concise model based approach achieved via process oriented models.

6 Outlook

The research presented is in the stage of conceptualization. A framework and a method for modelling inspections for the nuclear inspection domain were introduced. It is expected that benefits of the suggested approach will extend beyond the possibility of externalizing knowledge through transformation of implicit-to-explicit knowledge. The objective is to create a modelling framework that will provision for the creation of models across the nuclear industry. It can be used to promote nuclear knowledge and help the development of nuclear technology and safety related standards. A model-based approach will:

- improve the preservation and exchange of nuclear knowledge,
- improve the performance of inspection processes
- standardize a modelling languages used in inspection (e.g. training, performing inspections, inspection reports)
- increase transparency, quality and monitoring capabilities of inspection processes
- provide basis for a model based platform supported by KE and artificial intelligence applications

Based on the meta-modeling framework introduced in [4] syntactical elements of the inspection modelling languages will have to be further defined in terms of the semantics and notation. Also, the modelling procedures will have to be developed enabling the creation of mechanism and algorithms that can be applied to models representing different perspectives of the framework (e.g. inspection pathways)

In addition to the operational perspective described in this paper the KM and IT perspectives are to be defined and integrated with the operational perspective into a common meta-model. The research that lies ahead requires selection of modelling languages that best fit the nuclear knowledge management and complement the operational and IT perspectives of the framework.

References

1. Davis, R.: Business Process Modelling with ARIS: A practical Guide. Springer, Heidelberg (2001)
2. Davenport, H.T.: Process Innovation: Reengineering Work through Information Technology. Harvard Business School Press, Boston (1993)
3. Hommes, B.: The Evaluation of Business Process Modelling Techniques. Ph.D. Thesis. Delft University of Technology, Netherlands (2004)
4. Karagiannis, D., Grossmann, W., Hoefferer, P.: Open Model Initiative: A Feasibility Study, http://www.openmodel.at
5. Whitten, L.J., Bentley, D.L.: Systems Anlaysis and Design Methods, 7th edn. McGraw-Hill, New York (2005)

6. About Nuclear Knowledge Management. International Atomic Energy Agency (2008), http://www.iaea.org/inisnkm/nkm/aboutNKM.html
7. Karagiannis, D., Kühn, H.: Metamodelling Platforms. In: Bauknecht, K., Tjoa, A M., Quirchmayr, G. (eds.) EC-Web 2002. LNCS, vol. 2455, p. 182. Springer, Heidelberg (2002)
8. Karagiannis, D., Telesko, R.: The EU-Project PROMOTE: A Process-oriented Approach for Knowledge Management. In: Proceedings of the 3rd International Conference on Practical Aspects of Knowledge Management, PAKM (2000)
9. The Safeguards System of the International Atomic Energy Agency, IAEA, http://www.iaea.org/OurWork/SV/Safeguards/safeg_system.pdf
10. Karagiannis, D., Abazi, F.: Knowledge Management Approach to Emergency Procedures in Nuclear Facilities. In: IAEA. International Conference on Knowledge Management in Nuclear Facilities (2007)
11. Abazi, F., Woitsch, R., Hrgovcic, V.: "Inspection Pathways" to Nuclear Knowledge Management. In: Proceedings of IM 2009 in Gdańsk, Poland (2009)
12. Nieuwenhuis, K.: Information Systems for Crisis Response and Management. In: Löffler, J., Klann, M. (eds.) Mobile Response 2007. LNCS, vol. 4458, pp. 1–8. Springer, Heidelberg (2007)
13. Augusto, J.C., Liu, J., Chen, L.: Using Ambient Intelligence for Disaster Management. In: Gabrys, B., Howlett, R.J., Jain, L.C. (eds.) KES 2006. LNCS (LNAI), vol. 4252, pp. 171–178. Springer, Heidelberg (2006)
14. Crichton, M.T., Flin, R., McGeorge, P.: Decision making by on-scene incident commanders in nuclear emergencies. In: Cognition, Technology & Work, vol. 7(3). Springer, Heidelberg (2005)
15. Menal, J., Moyes, A., McArthur, S., Steele, J.A., McDonald, J.: Gas Circulator Design Advisory System: A Web Based Decision Support System for the Nuclear Industry. In: Logananthara, R., Palm, G., Ali, M. (eds.) IEA/AIE 2000. LNCS (LNAI), vol. 1821, pp. 160–167. Springer, Heidelberg (2000)
16. Hajek, E.K., Miller, D.W., Bhatnagar, R., Stasenko, J.E., Punch III, W.F., Yamada, N.: A generic task approach to a real time nuclear power plant fault diagnosis and advisory system. In: IEEE, International Workshop on Artificial Intelligence for Industrial Applications (1988)
17. Sturdivant, M., Carnes, W.E.: Procedure System Management. In: IEEE, 7th Human Factor Meetings (2002)
18. Carnes, W.E., Breslau, B.: Lessons learned - improving performance through organizational learning. In: IEEE, 7th Human Factor Meetings (2002)
19. Ruecker, L., Lal, B.: A Framework for Incorporating Best Practices at Nuclear Power Plants. In: Nuclear Science Symposium and Medical Imaging Conference, Conference Record of the 1992 IEEE, pp. 781–783 (1992)
20. Jennex, M.E., Olfman, L.: Organizational Memory/Knowledge Effects on Productivity, a Longitudinal Study. In: IEEE Proceedings of the 35th Annual Hawaii International Conference on System Sciences (2002)
21. Poucet, A., Contini, S., Bellezza: A GIS-Based Support System for Declaration and Verification. In: Symposium on International Safeguards (2001)

Two Dependency Modeling Approaches for Business Process Adaptation

Christian Sell[1], Matthias Winkler[1], Thomas Springer[2], and Alexander Schill[2]

[1] SAP Research CEC Dresden, SAP AG
01187 Dresden, Germany
{Christian.Sell,Matthias.Winkler}@sap.com
[2] TU Dresden, Faculty of Computer Science, Institute for System Architecture
01069 Dresden, Germany
{Thomas.Springer,Alexander.Schill}@tu-dresden.de

Abstract. Complex business processes in the form of workflows or service compositions are built from individual building blocks, namely activities or services. These building blocks cooperate to achieve the overall goal of the process. In many cases dependencies exist between the individual activities, i.e. the execution of one activity depends on another. Knowledge about dependencies is especially important for the management of the process at runtime in cases where problems occur and the process needs to be adapted. In this paper we present and compare two approaches for modeling dependencies as a base for managing adaptations of complex business processes. Based on two use cases from the domain of workflow management and service engineering we illustrate the need for capturing dependencies and derive the requirements for dependency modeling. For dependency modeling we discuss two alternative solutions. One is based on an OWL-DL ontology and the other is based on a meta-model approach. Although many of the requirements of the use cases are similar, we show that there is no single best solution for a dependency model.

Keywords: Dependency model, ontology, meta-model, process adaptation.

1 Introduction

Complex business processes are usually modeled and implemented by workflows or composite services consisting of a set of building blocks (we call them entities) which contribute to a common goal. To achieve that, the different entities need to collaborate in the sense that the execution of one entity may have certain dependencies on other entities. A dependency describes a relationship between entities, whereas an entity can be for instance a web service in a service composition or an activity in a workflow. Each dependency has specific features (e.g., uni- or bidirectional, inverse, disjoint) which describe the characteristics of the relationship between the entities or the relation to other dependencies. Within processes different types of dependencies like time or data can occur.

Knowledge about dependencies between entities of workflows or service compositions is often not explicitly available, but is rather implicitly contained in the descriptions

D. Karagiannis and Z. Jin (Eds.): KSEM 2009, LNAI 5914, pp. 418–429, 2009.

of entities, workflows, and compositions. It may also be available as domain knowledge of an expert or in process related artifacts like a service level agreement (SLA). To handle dependencies, however, it is necessary to have this knowledge available in explicit form, i.e. represented by a dependency model (DM).

Knowledge about dependencies between the entities of business processes is important for the management of these processes. It is used for instance for activity scheduling [11, 12] or to analyze and optimize large software systems with regard to dependencies between their building blocks [4]. In our work we are interested in the handling of changing conditions during the execution of business processes. This may be the occurrence of a problem during execution of a workflow which requires its adaptation or the renegotiation of a service level agreement by a participant of a service composition which can affect the composition as a whole or its individual building blocks [13]. In such situations it is necessary to actively manage the occurring event to assure successful process execution. A DM captures important knowledge about dependencies between the entities of a business process. This knowledge needs to be considered when adapting processes since changes with regard to one entity may affect multiple other entities. In order to adopt knowledge about dependencies for adaptation processes at runtime, dependencies have to be captured and explicitly modeled before. Moreover, dependent on the usage, different types and characteristics of dependencies might be of interest.

In this paper we present and compare two approaches for modeling dependencies as a base for managing adaptations of complex business processes. As a first step we discuss related work in the field (section 2). Following that we explore two use cases (section 3) where processes need to be adapted at runtime and derive requirements for a DM (section 4). Based on these requirements we present two approaches for capturing dependency knowledge (section 5) and discuss their advantages and disadvantages (section 6). We show that there is no single best solution for a DM. Finally, we conclude our paper with summary and outlook (section 7).

2 Related Work

The description of dependencies has been a focus in different research areas. Next to services [11] and activities, dependencies between classes or software modules [4], features (mostly in area of product line management) [9], and requirements [6] are captured and evaluated for different purposes such as dependency based service composition or optimizing the design of large software systems. The approaches for representing dependencies vary. In [4] the authors present the idea to use a *Dependency Structure Matrix (DSM)* to model dependencies where dependencies between entities are represented by a mark in a matrix. The automatically generated matrix is used to optimize product development processes. It highlights e.g. cyclic dependencies between entities, which thereupon can be removed. However, a DSM is not suitable to create a DM such as required for the use cases described in section 3. It does not support the explicit modeling of dependency features such as *symmetric* and *asymmetric* dependencies. Also, within a DSM one entity cannot have more than one dependency to exactly one other entity, since only one value for each entry of the matrix can be specified.

Wu et al. use the DAG Synchronization Constraint Language (DSCL) to model different types of dependencies [12]. Data and control dependencies are expressed by synchronization constraints such as *happenBefore* or *happenTogether*. However, due to the nature of expressing dependencies as synchronization constraints, this limits its applicability to certain use cases. For the handling of dependencies between services and their SLAs (first use case) this is not sufficient. For our second use case the modeling of dependency types such as *parallel* or *alternative* would be important. In DSCL those can only be expressed in terms of the constraint *happenTogether*. Thus the modeling of those types is hindered and not intuitive. Furthermore, dependency features such as inverse or bidirectional are not supported.

A common approach to capturing dependencies are *dependency graphs*, where nodes represent the entities which are dependent on each other and edges represent the dependencies between these entities. One example is the work by Zhou et al. [11], who developed an approach for the automatic discovery of dependencies based on semantic web service and business process descriptions. They use a dependency graph to capture control and data dependencies and create a minimal dependency graph presenting all dependencies. Dependency type information and specific properties of the dependency are missing. This DM is limited and does not allow the specification of typed dependencies or properties such as bi-directional, inverse, and disjoint dependencies.

3   Use Cases

In this section we present two use cases from service engineering and workflow management domain, namely a composite logistics service and a workflow-based plan for emergency management. Both use cases illustrate the need for considering dependencies between the entities of business processes while performing process adaptation at runtime.

3.1   Composite Logistics Services

In our scenario logistics services are offered on a service marketplace. They are consumed, composed to business processes, and resold. The 3PL (third party logistics provider) Trans Germany Logistics (TGL) acts as organizer of a logistics process offered as composite service Transport Germany. The process is created from services offered by different providers. The service is bought by Dresden Spare Parts (DSP), which needs to ship its products from Dresden to Hamburg. During the pre-carriage phase goods are picked up by Truck Dresden (pallet P1 and P2) and Express DD (pallet P3 and P4) and are transported to a warehouse where they are loaded to a large truck. TGL uses its own service TGL Truck to realize the main carriage. Three different logistics providers are responsible for picking up goods from the TGL Hamburg warehouse and delivering goods to different customers in the Hamburg region. The process is illustrated in Fig. 1. During the provisioning of this composite service the 3PL manages the underlying process.

In this use case dependencies occur regarding the goods being handled (e.g. TGL Truck is dependent on resources P1 and P2 by Truck DD and resources P3 and P4 by

Express DD), pickup and delivery times (e.g. late delivery of Truck DD and Express DD affect TGL Truck), price (price of the composite service depends on its building blocks), and quality of goods transport (quality of composite service depends on its building blocks). These aspects are regulated by SLAs. It is the responsibility of composite service providers to assure that the SLAs negotiated with the different providers enable the smooth execution of the composition. Requests for changes (re-negotiation) or problems (delivery time cannot be met) during provisioning of one service may affect the composition as well as different services of the composition, for instance if the failure of a service requires the adaptation of other process parts.

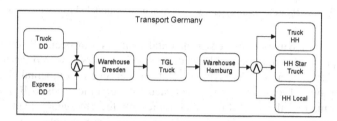

Fig. 1. Composite logistics process

Dependencies occur between the composite service and its atomic services (e.g. quality of goods transport and price) as well as between the different atomic services (e.g. handled goods and delivery times). Knowledge about dependencies is necessary to support the handling of problems. Because these dependencies are not available explicitly but only implicitly in the process description and SLAs, we need to capture this knowledge in a suitable model for further usage. If, for example, DSP changes its order and wants to ship fewer goods, TGL needs to adapt the SLAs with its subservices (e.g. cancel the contracts with Express DD and HH Local). To automate this process, knowledge regarding the different dependencies needs to be captured. Thus, different types of dependencies are needed.

The different dependencies may occur as *unidirectional* or *bidirectional*. For example, the price of a composition depends on the price of its building blocks (unidirectional) while the delivery and pickup times of two interacting services have a bidirectional dependency. Furthermore, it is valuable to capture information which describes the dependency in more detail. For example, a formula describing composite price calculation enables the automation of this calculation upon changes.

3.2 Emergency Management

In Germany when disaster strikes, usually a so called executive staff is convened to manage the disaster in the best possible way. In support of their work the staff refers to predefined emergency plans for dedicated events, e.g. an evacuation. As we have described in [2], emergency plans are similar to business processes and hence can be modeled as workflows, where one measure of the emergency plan is represented by one activity within the workflow. Fig. 2 illustrates a part of the emergency plan *Evacuation* modeled as a workflow.

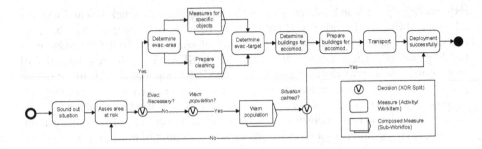

Fig. 2. Part of the emergency plan Evacuation modeled as a workflow

During an evacuation many environmental changes can occur. They require the staff to adapt the plan and therewith the corresponding workflow in order to guarantee consistency with the actual deployment [2]. Consider an environmental change where the concentration of toxic gases over the evacuation area increases to 100 ppm. Based on this information, the workflow which represents the emergency plan *Evacuation*, has to be extended by an activity for the decontamination of people.

Approaches as described in [3] deal with the automation of such workflow adaptations based on environmental changes. To be able to perform the insertion of the activity "Decontamination", the correct position within the workflow has to be calculated by such an approach. The position of the activity is, to a large extent, determined by its interdependencies to other activities. Therefore, when inserting the activity "Decontamination", it has to be taken into consideration that this activity always has to be performed before the activity "Transport" and after the activity "Determine evacuation area". To specify this knowledge, approaches for the automatic adaptation of workflows require an adequate DM. Without such a model, the activity dependencies have to be integrated directly into the workflow logic in the form of alternative paths [3]. This would lead to a more complex management of the dependencies and redundant workflow artifacts.

Within the domain of emergency management the following types of dependencies have to be considered: *before, after, parallel, alternative* and *require*. A dependency of the type *parallel/alternative* between two activities a_1 and a_2 describes that a_1 has to be performed in parallel/alternatively to a_2. The *before/after* dependency states that a_1 has to be performed before/after a_2. *Require* means that if a_1 is performed then a_2 also has to be performed. The *require* dependency is orthogonal to the *before* and *after* dependencies. Thus, there can be two dependencies between an activity a_1 and a_2, one of the type *require* and the other either of the type *before* or *after*. For example, the activity "Decontamination" is *required* and *after* dependant on the activity "Determine evacuation area". Meaning, if this activity has to be inserted, the activity "Determine evacuation area" must be a part of the workflow and located before it (with regard to the execution path of the workflow). Next to the orthogonal feature, dependencies, which describe the relationship between activities in the area of emergency management, can be *unidirectional or bidirectional as well as disjoint or inverse* to each other. *Unidirectional* and *bidirectional* have the same meaning as in the use case described before. Two dependency types t_1 and t_2 are *inverse* to each other if after the creation of a dependency of the type t_1 from the activity a_1 to a_2

automatically a_2 has a dependency of the type t_2 to a_1. If t_1 and t_2 were *disjoint*, a dependency from a_1 to a_2 could be either of type t_1 or t_2 (not both).

4 Requirements

From our use cases we derived a number of requirements for modeling dependencies. Depending on the use case different types and characteristics of dependencies as well as different methods and aspects of modeling are important. Therefore, it is important to analyze the suitability of different approaches. The following requirements have been derived to guide the comparison and assessment of different approaches:

(R1) Support of different dependency types: The DM must support different types of dependencies. For the use case Emergency Management, for instance the dependency types *parallel, alternative, before, after* and *require* must be considered.

(R2) Support of multiple dependencies: In both use cases described above, one entity can have dependencies to several other entities, whereas one entity can also have several dependencies of a different type to exactly one other entity.

(R3.1) Support for unidirectional and bidirectional dependencies: In both use cases there are *unidirectional* and *bidirectional* dependencies. The DM must support the specification of *unidirectional* and *bidirectional* dependencies. This means, that after the creation of a *bidirectional* dependency of type t_1 from the activity a_1 to a_2, the DM automatically creates a dependency of type t_1 from a_2 to a_1. The aim is to reduce the modeling effort. Furthermore, this requirement must be fulfilled to be able to meet requirement R4. In detail, this requirement differs for each described use case. So, while in the logistics use case it has to be specified individually at instance level for each dependency whether it is *unidirectional* or *bidirectional*, in the emergency management use case it is defined at type level, whether a dependency of a certain type is *unidirectional* or *bidirectional*.

(R3.2) Support of inverse and disjoint dependencies: To support *inverse* dependencies adequately, after the creation of one dependency, the DM should automatically create its inverse dependency. This reduces the adaptation time and the number of possible inconsistencies which arise if one inverse dependency has been forgotten to model. The support of the *disjoint* dependencies is required to detect inconsistencies within the model and therewith precondition to meet requirement R4.

(R4) Validation: It should be possible to semantically and syntactically validate the DM to avoid inconsistencies and therewith errors within the adaptation process. The syntactic validation has to detect whether the model is consistent with the used meta-model. The semantic validation must recognize whether the features of the entities (e.g. *bidirectional* or *disjoint*) do not lead to inconsistencies.

(R5) Use of a standardized approach: To simplify the specification of the DM and to increase its global understanding, it should be modeled using a standardized modeling language.

(R6) Tool support: For the creation of the DM, it should be possible to access existing tools. These tools should be suitable for domain experts. An adequate tool support simplifies the specification of the DM and therewith saves time and costs.

(R7) Dependency description: Dependencies may require a more detailed description which helps to automate the handling of related events at runtime.

5 Dependency Model

In this section we will describe two approaches to explicitly represent dependency information: an ontological approach and a meta-model based approach. Both approaches allow the modeling of dependencies at design time and their evaluation at runtime. We decided for the explicit modeling of dependencies because that facilitates the evaluation at runtime.

5.1 Ontology Based Approach

In this section we present the idea of using an ontology to create a DM. An ontology is defined as an explicit specification of a conceptualization [1] and can be used to formally describe knowledge about terms of a domain (e.g., activity or service names) and their interrelations. In this paper we refer to the Web Ontology Language (OWL) [7] in the Version OWL-DL to model ontologies. OWL is standardized by the W3C and includes among other the following parts:

- *Classes*: combine a group of individuals, which have certain common features. They can be embedded into a hierarchy by using *sub-classes*.
- *Individuals*: represent instances of classes.
- *Properties*: can be used to specify relationships between exactly two individuals (*object property*) or an individual and a data value (*data property*). By defining the *domain* of a property the number of individuals which can have this property can be reduced. The *range* of a property p can be used to limit the number of individuals to which the individuals of the domain of p can be related to via p. The relationship of properties, individuals and classes is illustrated in Fig. 3.

OWL supports the *inverse, symmetric, asymmetric* and *disjoint* feature for its object properties. If an individual a_1 is related to an individual a_2 via a *symmetric* property p, a_2 is also related to a_1 via p. A property p is *asymmetric* if either a_1 can be related to a_2 or a_2 to a_1 via p. Two properties are *disjoint* if either the one or the other can be used to describe the relationship between the same activities. Let p_1 be the *inverse* property of p_2. If p_1 is created between individual a_1 and a_2, a_2 is autom. related to a_1 via p_2.

The concept of ontologies can be used to specify a DM. To do so, the individuals are used to represent entities. All individuals belong to one class. If not defined differently, this is the class *Thing* by default. The dependency between two entities can be modeled via object properties, whereas the range as well as the domain of each property is of the class *Thing*. Furthermore, based on the dependency type it represents, a property can be extended to the features described above.

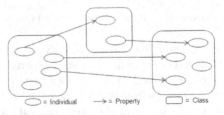

Fig. 3. Relationship of properties, individuals and classes within OWL

To specify a DM based on an ontology for the emergency management use case, properties for the *before, after, parallel, alternative* and *require* dependency have to be modeled. The properties representing the *before, after, parallel* and *alternative* dependency have to be *disjoint*. Furthermore, the properties for the *before* and *after* dependency must be *asymmetric* and *inverse* to each other. The properties representing the *parallel* and *alternative* dependencies must be *symmetric*. Since a *require* dependency can be combined with a *before* and *after* dependency but not with a *parallel* and *alternative* dependency, its representing property has to be *disjoint* from those representing the *parallel* and *alternative* dependency. Fig. 4 shows an example of a DM, which can be used within the emergency management use case to support the automatic adaptation of workflows based on an environmental change.

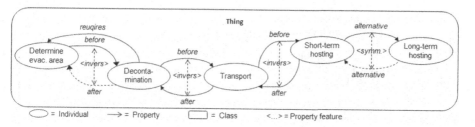

Fig. 4. Activity dependency model specified by an OWL ontology

If, for instance the activity "Short-term hosting" is supposed to be integrated into an emergency plan this has to be done before the activity "Transport" and alternatively to the activity "Long-term hosting".

5.2 Meta-model Based Approach

An alternative approach to modeling a dependency in form of an ontology is the development of a meta-model which allows the creation of DMs. An XML Schema allows the specification of a XML data structure and is thus a meta-model for these data structures. DMs are model instances conforming to the schema. In this section we present our schema which has been developed with a focus on capturing dependencies between services in compositions.

The DM consists of two core elements: *serviceEntity* and *dependency*. A service entity is any service within a composition as well as the composition itself. In a DM each entity occurs only once. Service entities are described by their name and unique identifier. Furthermore they reference a SLA through which the provisioning of the service is regulated and which contains the relevant contractual information from which the dependencies originate.

The abstract *dependency* element represents a dependency between service entities. It is either uni- or bidirectional. Services can be dependent on each other with regard to different aspects such as pickup and delivery time, location, resource, price, and different quality of service aspects. Different dependencies (e.g. *timeDependency*) extend *dependency* and model dependency type specific information. The dependency references the service entities as well as the service level objectives regarding which

Table 1. Example dependencies for logistics use case

```
<dep:serviceEntity serviceName="Transport Germany" serviceKey="tg-c1" slaID="sla-cust" />
<dep:serviceEntity serviceName="Truck DD" serviceKey="tdd-a1" slaID="sla-dl1" />
<dep:serviceEntity serviceName="TGL Truck" serviceKey="tgl-a2" slaID="sla-tgl2" />
<dep:resourceDependency id="dep_v-1" isBidirectional="false">
    <dependant>tdd-a1</dependant>
    < antecedent >tgl-c1</antecedent>
    <dep:resourceIDList>
        <dep:resourceID>P1</dep:resourceID>
        <dep:resourceID>P2</dep:resourceID>
    </dep:resourceIDList >
</dep: resourceDependency>
<dep:timeDependency id="dep_h-1" isBidirectional="false">
    <dependant> tgl-a2 </dependant>
    < antecedent > tdd-a1</ antecedent >
    <dep:timeOperator>finish-to-start</ dep:timeOperator>
</dep:timeDependency>
<dep:locationDependency id="dep_v-2" isBidirectional="true">
    <dependant>tdd-a1</dependant>
    < antecedent >tg-c1</ antecedent >
    <dependantSLO>//guaranteeTerm[name=pickupLocation_guarantee]</dependantSLO>
    <antecedentSLO>//guaranteeTerm[name=pickupLocation_guarantee]</antecedentSLO>
    <dep:locationOperator>equals</dep:locationOperator>
</dep:locationDependency>
```

the dependency exists. Finally, it contains a *dependencyDescription* which describes the dependency in more detail. The description is specific for each dependency type. It is realized as a formula for the calculation of a composite value for the dependent entity (e.g. price and quality calculation of a composite service). A resource dependency is described by a list of resources regarding which the dependency exists. Time dependencies are described by time relations as defined by Allen [10] and in the field of project management [15]. In Table 1 we present a brief excerpt from a DM based on the logistics use case presented earlier in this paper. The composite service Transport Germany and the atomic services Truck DD and TGL Truck are represented as service entities. Furthermore, three (shortened) dependency examples are presented: a resource dependency, a time dependency, and a location dependency.

6 Discussion

In this paper we have discussed a set of requirements which have to be met by a DM to suitably specify knowledge about dependencies for our use cases. Our solutions address these as follows: The meta-model based approach allows the realization of almost all requirements specified in section 4. Dependencies can be marked as bidirectional (R3.1). However, no dependency type has fixed uni- or bi-directional features. Inverse and disjoint relationships are currently not supported (R3.2), since the logistics use case does not require such features. Different types of dependencies can be modeled (R1) as well as the fact that one entity depends on multiple entities with regard to different aspects (R2). The validation of the DM is not automatically supported (R4). It would require the development of specific validation support.

Compared to the ontological approach this is a disadvantage. Since the meta-model approach is based on XML Schema and XML it is based on standard web technologies (R5). A major advantage of the meta-model based approach is the good tool support (R6). For the creation of DM instances and the extraction of dependency information a Java API can be automatically generated. Advanced tools for graphical or tree based modeling can be generated based on Eclipse GMF [14]. This is an important aspect as instances of the DM are created for each instance of a composite service as opposed to having a DM for a class of composite services. This poses the requirement that the developers of composite services must be able to create the DM instances. No specific technological knowledge should be required, but instead the complexity should be hidden. This can be achieved by integrating the required dependency modeling components with the tools for composite service development. Finally, the meta-model allows the specialized descriptions of properties (R7).

For the logistics use case the usage of specific dependency features such as being inverse or disjoint to other dependencies are of no relevance. On the other hand good tool support is important to allow end users to modify the DM. For these reasons the meta-model approach is more reasonable.

In OWL it is possible to specify different properties between different individuals (entities). Hence, the ontology approach introduced in section 4.1 can meet the requirements R1 and R2. For OWL-DL, there is a set of reasoners [8] which can be used to query information from the ontology and to syntactically and semantically validate the ontology. The validation checks whether the syntax and the constraints corresponding to the object property features are violated. Therewith, the ontology approach also meets the requirements R3 and R4. A further advantage of OWL is that it supports object property features such as *symmetric, asymmetric, inverse* and *disjoint*. Thus, creating a symmetric property p from the individual a_1 to a_2 implicitly means also the creation of that property from a_2 to a_1. In contrast to XML/XSD, there are only few tools such as Protégé [5] for the management of OWL ontologies. Moreover, these tools are not suitable for application domain users but require expert knowledge regarding ontologies. Thus, requirement R6 is only partly addressed. For the modeling of the dependency graph required in the emergency management use case, the OWL ontology approach is more adequate in contrast to the XML/XSD approach. Especially the automatic support of *symmetric, asymmetric, disjoint* and *inverse* dependencies by the OWL object property features is of great advantage. Using XML/XSD these features would have to be modeled manually. This increases the required modeling time and opens a new source for potential mistakes. Moreover, the manual creation of these features hinders the semantic validation of the DM. Table 2 presents an overview about the different requirements presented in section 3 and in how far these requirements are met by the two approaches.

Table 2. Overview fulfilled requirements

Heading level	R1	R2	R3.1	R3.2	R4	R5	R6	R7
XML/XSD	x	x	x	o	o	x	x	x
Ontology (OWL)	x	x	x	x	x	x	o	x

In this section we have shown that none of the introduced dependency modeling approaches is optimal. The selection of the right modeling approach is dependent on the use case it is applied to and the weighting of its requirements. Hence, for each use case it has to be discussed individually which modeling approach is more suitable. The requirements described in section 4 can be used to assess different modeling approaches and to guide the selection of the one most appropriate for a use case.

7 Summary and Outlook

The consideration of dependencies between entities of business processes is necessary to consistently adapt business processes at runtime. Knowledge about such dependencies is often not explicitly available but has to be extracted from process and entity description or SLA definitions. Moreover, the extracted knowledge has to be explicitly represented in a DM to be accessible during adaptation at runtime. In this paper we discussed the problem of efficiently modeling knowledge about dependencies in explicit DMs. Using two use cases from the logistics and emergency management domains we analyzed the requirements for DMs. As a first contribution, we presented a general set of requirements which can be used to assess and compare different approaches for dependency modeling. Based on the requirements we compared two modeling approaches using OWL-DL ontologies and a meta-model based on XML/XSD. As our discussion has shown, both of the introduced approaches for the modeling of dependencies are not optimal. The ontology approach is based on the standard language OWL-DL. It supports properties which can be adopted for modeling dependencies and their characteristics. The DM can easily be extended and of-the-shelf reasoning tools can be used for validating and processing the model. A major drawback is the lack of specific tools for dependency modeling. The meta-model approach uses open standards which allow to easily create specific tools for dependency modeling. With the creation of a domain specific language, this approach is very flexible and specific DMs can be tailored to the needs of particular use cases. However, there are no off-the-shelf tools available for reasoning about the created models.

In summary, the selection of the right modeling approach is rather dependent on the use case it is applied to and the weighting of its requirements. Although there are many common requirements for both use cases, there are also some specific requirements which lead to different DM solutions. As next steps we plan to implement different runtime adaptation approaches for workflows and service compositions which use the DMs as base for their work.

Acknowledgements

The project was funded by means of the German Federal Ministry of Economy and Technology under the promotional reference "01MQ07012". The authors take the responsibility for the contents.

References

1. Gruber, T.R.: Toward principles for the design of ontologies used for knowledge sharing, KSL-93-04, Knowledge System Laboratory. Stanford University (1993)
2. Sell, C., Braun, I.: Using a Workflow Management System to Manage Emergency Plans. In: Proceedings of the 6th International ISCRAM Conference, Gothenburg (2009)
3. Sell, C., Springer, T.: Context-sensitive Adaptation of Workflows. In: Proceedings of the 7th joint meeting of the European Software Engineering Conference and the ACM SIGSOFT Symposium on the Foundations of Software Engineering, Amsterdam (2009)
4. Sangal, N., Jordan, E., Sinha, V., Jackson, D.: Using dependency models to manage complex software architecture. In: Proceedings of the 20th annual ACM SIGPLAN conference on Object oriented programming systems languages and applications (2005)
5. Noy, N., Sintek, M., Decker, S., Crubezy, M., Fergerson, R., Musen, M.: Creating semantic web contents with Protégé-2000. IEEE Intelligent Systems (2001)
6. Zhang, W., Mei, H., Zhao, H.: A Feature-Oriented Approach to Modeling Requirements Dependencies. In: Proceedings of the 13th IEEE International Conference on Requirements Engineering, Washington, DC, USA (2005)
7. Dean, M., Schreiber, G.: OWL Web Ontology Language Reference. W3C Recommendation 10 (2004)
8. Sirin, E., Parsia, B., Grau, B.C., Kalyanpur, A., Katz, Y.: Pellet: A practical OWL-DL reasoner, Web Semantics: Science, Services and Agents on the World Wide Web (2007)
9. Lee, K., Kang, K.C.: Feature Dependency Analysis for Product Line Component Design. In: Proceedings of the 8th International Conference, ICSR, Madrid (2004)
10. Allen, J.F.: Time and time again: The many ways to represent time. Journal of Intelligent Systems 6(4) (1991)
11. Zhou, Z., Bhiri, S., Hauswirth, M.: Control and Data Dependencies in Business Processes Based on Semantic Business Activities. In: Proceedings of iiWAS 2008. ACM, New York (2008)
12. Wu, Q., Pu, C., Sahai, A., Barga, R.: Categorization and Optimization of Synchronization Dependencies in Business Processes. In: Proceedings of IEEE 23rd International Conference on Data Engineering, ICDE 2007 (2007)
13. Winkler, M., Schill, A.: Towards Dependency Management in Service Compositions. In: Proceedings of the International Conference on e-Business, Milan (2009)
14. The Eclipse Foundation: Graphical Modeling Framework. Project page (2009), http://www.eclipse.org/modeling/gmf/
15. PMI, A Guide to the Project Management Body of Knowledge (PMBOK Guide), Project Management Institute (2008)

The IT-Socket:
Model-Based Business and IT Alignment

Robert Woitsch and Wilfrid Utz

BOC Asset Management, Bäckerstraße 5, 1010 Wien, Austria
{Robert.Woitsch,Wilfrid.Utz}@boc-eu.com

Abstract. As a result of the ongoing evolution in Information Technology (IT) from an enabler towards an industrial sector at its own right, the necessity to align business requirements with IT resources and services is constantly increasing. Within the introduced EU project plugIT we are assuming that businesses in different sectors require IT for various reasons in order to reach a level of strategic differentiation therefore envisioning the idea of an "IT-Socket" to be developed. The "IT-Socket" uses the analogy to use IT for business similar to the way electrical power is consumed, setting up means of "plugging-in" business into IT. The results of the project are demonstrated within 3 demonstration scenarios, namely (1) "Certification of IT-Infrastructure" in the area of compliance management of IT infrastructures, (2) "Virtual Organization" taking the service orientation to a higher level of abstraction for business usage, as well as (3) "Governance of IT infrastructures" supporting and introducing the business context in distributed and complex systems.

The "IT-Socket" is realized by applying a model-based approach targeted by the research challenge of plugIT to tightly link human interpretable graphical models with machine interpretable formalisms to allow an involvement of domain experts in formalizing knowledge, providing modelling languages/tools/services that are fitting to the respective context and establish mechanisms for domain specific notations for semantic concepts.

Keywords: Next Generation Modelling Framework, Knowledge Management, Semantics, IT-Socket, Virtual Organisation, IT-Governance, Certification.

1 Introduction

In recent years a evolution of Information Technology (IT) and change of the role could be observed, from an enabler of business transaction towards an industrial sectors in its own right [13]. This evolution resulted in the necessity to continuously and comprehensively align Business and IT [11], [13]. The triggers for this are manifold: legal aspects, regulations contexts, business requirements, economic factors from a business viewpoint but also technological trends related to Service-Oriented Architectures (SOA), Software as a Service (SaaS) and Virtualisation are influencing the way IT services are offered and consumed nowadays.

D. Karagiannis and Z. Jin (Eds.): KSEM 2009, LNAI 5914, pp. 430–441, 2009.
© Springer-Verlag Berlin Heidelberg 2009

Various approaches from the area of IT-Governance and IT Service/Architecture Management are well-known candidates to bridge the gap between evolving business contexts and IT, enabling adaptive provisioning of IT for business needs. From the perspective of plugIT semantic technologies are regarded as an important factor as means for successfully including domain experts from both domains (business and IT) and enable a consequent and continuous alignment between them.

Assuming that businesses require custom IT solutions and services to reach a level of strategic differentiation, the EU project plugIT - FP7-3ICT-231430 [37] aspires to develop an IT-Socket that will realize the vision of businesses "plugging-in" into IT. Information Technology has continuously evolved from back-office support, to the support of the core business process and has now reached the level of strategic differentiation [6] establishing nowadays the decisive factor within various industries and branches such as Airline Transport, Finance, Automotive Industry or Health Care [14].

The results achieved in plugIT are demonstrated within 3 demonstration scenarios that have been selected on different abstraction levels to show the applicability of results:

(1) "Certification Use Case" showing how the alignment between business area and IT domain during certification processes for regulations such as SOX, EuroSOX, ITIL®, CoBIT®, ISO20000 or BASEL II is supported.
(2) "Virtual Organization Use Case" demonstrating how virtual organizations can be supported using business driven requirements and semantically described SLA's for intelligent interpretation.
(3) "Governance Use Case" providing a scenario on how intelligent agents are applied to identify IT infrastructure elements of data centers. In this scenario graphical models are regarded as means of mediation between system administration and intelligent discovery systems.

The paper provides first insights on the analysis of the IT-Socket performed in cooperation with demonstration partners and research partners by establishing a framework for the IT-Socket analysis and identifying challenges within the execution of the demonstration scenarios as valuable input for the research challenges targeted.

2 IT-Socket Idea – Industrialization of Alignment

Industrialisation, as a well-known concept from various other domains, is a phenomenon that is observable in today's IT[1] providers work and can be compared with the industrialisation of electricity, where electric power is provided and consumed via power sockets [16] , [26]. plugIT envisions and targets the development of an IT-Socket, where IT can be consumed by businesses in a similar way, as electric power is consumed by electronic devices when plugged into a power socket [37].

In an initial phase of the project, different alignment approaches have been analyzed and investigated starting from formal methods, to unstructured and ending at intuitive mechanisms. This analysis builds up the basis for defining the model-based alignment framework of the IT-Socket.

---

[1] The term "information technology" (IT) is used as a synonym to "information and communication technology" (ICT), because of the acronym of the project.

The hypothesis of the IT-Socket is that knowledge of experts involved in the process of alignment can be externalized by using graphical semi-formal models, which build up the foundation for further formalization, hence supporting business and IT alignment via semantic technologies.

2.1 Relevant Work

Business and IT alignment has been and still is the focus of various research and industrial initiatives and projects, focusing on technical transformation ([9], [41]) on one hand or involving business views ([4], [43]) on the other.

Despite the ongoing research in the domain, initiatives by vendors and practice guidelines [45], the business level of SOA approaches is still underrepresented. A literature survey of 175 research papers about SOA from 2000 to 2008 outlines this statement [48].

plugIT's goal of integrating the business aspects in the alignment process is established through externalization current participants' knowledge within the alignment interaction. The model-based approach envisioned by plugIT provides a way to conceptual integrate business requirements and IT [22], [49].

The alignment between the two domains is described in detail by graphical models on different levels of abstraction regarded as the initial step of formalization. The key challenge identified in that area is the integration of different modelling languages on different levels of abstraction from different domains.

2.2 Scoping the Analysis of the IT-Socket

During the initial tasks of the project the analysis of the IT-Socket at the use-case partners' sites were in focus. The identification tasks of the relevant aspects from the business domain as well as from the IT domain for the alignment processes have been targeted.

The starting point for the identification of elements on business level was the business processes currently implemented by demonstration partners. For the IT level a detailed investigation on the IT products, regarded in the context of plugIT as bundles of IT-services that allow a commercial exploitation, has been executed.

The analysis executed used three investigation channels:

(1) Expert interviews carried out with iTG[2], HLRS[3] and CINECA[4], involving experts from the business as well as IT domain,
(2) Complementing experiments at universities establishing the visionary challenges for the IT-Socket,
(3) Literature survey in [3], [7], [10], [15], [24], [30], [32], [50] and [42].

As a result a common framework for the analysis of the IT-Socket could be established, consisting of six core elements separated in two horizontal perspectives and three vertical aspects. Fig. 1 provides a graphical representation of the perspectives and aspects identified, detailed in the following as:

---

[2] ITG, Innovation Technology Group SA, http://www.itg.pl
[3] HLRS, High Performance Computing Center Stuttgart, http://www.hlrs.de
[4] CINECA, Consorzio Interuniversitario, http://www.cineca.it

Business Perspective (derived from the business processes environment):

- Competence aspect: analyses the competences needed to correctly specify the IT products needed for the execution of a business process.
- Technical aspect: technical view on the requirements of the IT products.
- Organisational aspect: organisational frame to correctly specify the IT products.

IT Perspective (derived from IT products as a composition of IT services):

- Competence aspect: analyses of IT services provided as competences and skills to the consumer such as helpdesk, training, consulting, etc.
- Technical aspect: technical view on IT services on different abstraction levels such as applications, middleware or housing and combination of the services
- Organisational aspect: setup of the organisational aspect to handle provision of IT services such as maintenance processes, user administration or monitoring of services.

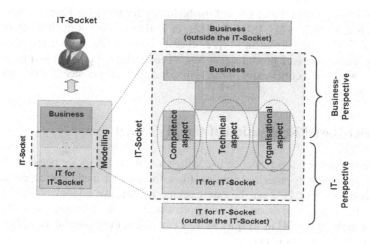

Fig. 1. IT Socket Framework (high level view)

The analyzed environment of CINECA, ITG and HLRS resulted in IT services on different abstraction layers. For example, a specific IT service such as "Housing of Infrastructure", a service offered by data centres to provide physical environment for IT infrastructure is considered as being on a lower abstraction than an "ERP-Application" service, which is considered on a high abstraction layer eventually combining services from lower levels. A services assignment to an abstraction layer indicates the conceptual "distance" from the consuming IT transparent[5] business

---

[5] IT-transparent means that the higher-levels such as the business process level do not need to be aware of the specifics and details of the lower levels such as IT infrastructure. IT is transparent from the perspective of the business process.

process. At the current stage a concrete definition of the abstraction layers applied and a generalization is not possible, as different abstraction layer frameworks such as [19], [46], as well as demonstration partner specific adaptations like in the use case of CINECA and HLRS are available.

Applying the paradigm of the electric power socket, the elements on the business perspective are regarded as the "plug" depending on the abstraction level can result in different representations. The IT-Socket has to provide means to integrate plugs on any abstraction level (that are made available by the information integration capabilities of semantic technologies and formalization) therefore handling any business plug that is plugged into the IT-Socket.

Consequently the framework introduced earlier must be flexible enough to be adaptable to the context of the applying party resulting in custom and specific IT-Sockets.

The alignment methodology itself between the businesses requirement identification and the actual provision of the IT products is performed through formalisation of the business perspective to an extent that: the level of abstraction is identifiable, IT product parameters are explicitly described to allow a matching of service demand and service supply and the appropriate IT product components in terms of the vertical aspects can be derived.

The identification of the alignment methodology for plugIT is based on approaches found in bodies such as [1], [8], [17], [18], or [33]. 17 different methods to reach the alignment goal are discussed in [47]. A classification of the approaches from formal procedures over heuristic procedures towards semi-structure or intuitive procedures is needed to handle the complexity of the IT product, the competence, the organisational culture and the like.

3 Model-Based Realisation of the IT-Socket

From a methodological perspective the realization of the IT-Socket needs to deal with different alignment approaches ranging from formal and mathematic definitions of requirements [12] or [51] on the one extreme to intuitive alignments on the other extreme including in between hybrid approaches such as questionnaires and model based methods [23], [44] or [35] as well as communication based approaches [29] or [40] adaptive alignment in [1].

The model-based approach handles the externalisation of knowledge related to the methodology selected in a semi-formal and formal way using graphical models. This approach is regarded as a simplification of the externalization process since as domain experts in both fields can use modelling languages they are familiar with in order to produce semi-formal or formal knowledge representations.

3.1 IT-Socket's Modelling Framework

Based on the IT-Socket analysis framework introduced in the previous chapter, an overview on the modelling elements for each of the perspectives and aspects has been established. Within the initial working phase of the project the consortium did analyze the use-case partners in accordance with the framework introduced and deriving concrete realisation scenarios for demonstration. A detailed description of the current situation at use-case partners is available in deliverable 2.1 [36], the challenges per scenario are available in deliverable 5.1 [38]. As an initial step, the PROMOTE® modelling language has been used to depict the 6 elements of the IT-Socket. In the realisation phase of the demonstration scenarios a detailed analysis of modelling languages for each element and scenario will be performed resulting in more specific and targeted modelling languages for the elements. A classification schema on modelling languages is available in D5.1 as an initial step. The mapping for the IT-Socket elements to the PROMOTE® approach is shown in Fig. 2.

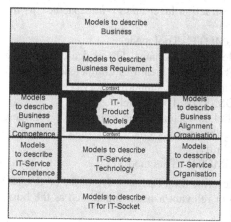 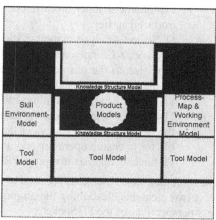

Fig. 2. IT-Socket Framework (model-based view and initial PROMOTE® mapping)

A graphical overview on the elements is provided in Fig. 2, a definition of each element is provided in the following chapter identifying the content to be described within the elements.

i.   Business: The business is described using models for different aspects such as business models, business processes, data and knowledge, business rules and the like. The IT-Socket modelling framework distinguishes between aspects relevant and within the IT-Socket (e.g. business processes using IT services) and aspects that are considered outside of the IT-Socket (business strategy definition).

ii. Business Requirements: The business requirements formulate the needed IT product in terms of technical, organizational and competence aspects. The representation on this level is strongly interlinked with the selected alignment approach as introduced earlier. A formal alignment approach would establish mathematical formulas to express business requirements whereas in a heuristic alignment approach, the requirement specification may be a collection of answers to question and interview results. In an informal alignment approach unstructured text is analyzed in contrast to an intuitive alignment approach where vague expressions build up the business requirements.

These two elements are interpreted as the "plug". The counterpart is the IT-Socket itself that is described using the following six elements.

i. IT for IT-Socket: The IT for IT-Socket level includes software, hardware and infrastructure required to provide the IT services. The exact border to IT not relevant to the IT-Socket and hence classified as outside the IT-Socket depends on the level of detail in which initiatives are applied by the IT service providing organization. Commonly this level understood as "IT infrastructure" meaning infrastructure elements, network elements, and the like.
IT-Socket Analysis mapping: not in focus of analysis

ii. IT Service Technology: services are identified on top of infrastructure level, such as software, hardware and IT-infrastructure services in a commercially exploitable manner on different abstraction layer and IT-transparency as introduced earlier.
IT-Socket Analysis mapping: PROMOTE® Tool model

iii. IT-Service Competence: in parallel to the technical IT service competence provision to the business level. The element identifies competence required by the provider to handle requests and provide supporting services.
IT-Socket Analysis mapping: PROMOTE® Tool model

iv. IT-Service Organization: in parallel to technical and competence services, the organizational IT services describe processes the IT provider is responsible for to ensure operation and usability.
IT-Socket Analysis mapping: PROMOTE® Tool model

The aforementioned elements are used to describe the IT perspective of the IT-Socket. Next two elements, describing the alignment relevant aspects considered as the business perspectives of the IT-Socket.

v. Business Alignment Competence: explicitly defined skills and competences needed to apply an alignment method considering different approaches describing those skills in detail.
IT-Socket Analysis mapping: PROMOTE® Company map and business process model

vi. Business Alignment Organization: processes performed during the alignment approach selected. Different methods require different description when distinguishing between mathematical models on the one extreme and intuitive actions on the other extreme. The aforementioned business requirements that are seen as the business plug are correlated with the processes for the alignment as well as the skills that are necessary to perform this alignment.
IT-Socket Analysis mapping: PROMOTE® Skill Environment model

By applying a model-based approach, each of the elements is described using a graphical, semi-formal models resulting in a collections of modelling languages of different formal expressiveness for each IT-Socket implementation, leading to a heterogeneous collection of different models expressed in different modelling languages in different formal levels for each user of the system. Semantic technologies are regarded as mediator within this heterogeneous environment.

3.2 The Semantic within the IT-Socket

When applying the model-based approach to realize the IT-Socket, the user faces a set of different modelling languages each classified in different perspectives, aspects, and formalization levels depending on different modelling language families.

Therefore the main objective for applying semantic technologies is to act as mediator between these modelling languages and models, thus ensuring that requirements expressed in the business perspective in one modeling language are considered in the IT-perspective using a different modelling language with a different formal expressiveness.

As mentioned earlier information integration on different levels is regarded as crucial. The following levels of information integration can be distinguished:

- Schema integration: Design a global unified schema for the knowledge space
- Data Integration: Take into account both schema and actual data
- Semantic integration: Take into account ontologies, schemata and data which can be structured, unstructured or semi-structured

Semantic integration uses ontologies to exemplify the semantics of data and schema. Normally they are used to identify and resolve heterogeneity problems at schema and data level, as a means for establishing explicit formal vocabulary to be shared between applications.

In the IT-Socket plugIT discusses three dimensions of semantic integration techniques:

- Mapping discovery: to find similarities within two given ontologies/schemata, determine which concepts and properties represent similar notions, and the like.
- Declarative formal representations of mappings: to represent the mappings between two given ontologies enabling reasoning and mappings
- Reasoning with mappings: reasoning within the knowledge space once the mappings are defined

With those semantic integration techniques at hand, plugIT addresses the research challenges in the IT-Socket, namely:

- The modelling language integration, which tries to integrate modelling languages on different levels, aspects and pillars through a common ontological representation of the modelling language and meta-model. The so-called Modelling Language Ontology (MLO) allows mapping, comparison, transformation and translation of modelling languages on syntactical and semantical layers.
- On model level means will be establish to generate machine interpretable models out of graphical models (making models meaningful similar as above with modelling languages) and allow translation between these models based upon a Model Ontology (MO).
- The automatic domain ontology generation from those models as a basis for information integration is necessary to allow establishing cross-domain exchange of information between experts between experts from both domains. The tight integration of the MO, MLO and Conceptual Reference Ontology (CRO) will provide the bridge to enable exchange of knowledge.

The research challenges identified above are targeted by the research partners in the consortium and resulting in the implementation of the so-called Semantic Kernel for the Next Generation Modelling Framework.

4 Conclusion

This article introduced the EU project plugIT and the envisioned IT-Socket for business and IT alignment utilizing a model-based approach. The project is co-funded by the European Commission and started in March 2009; this article therefore introduced the vision and the approaches of the project and provides initial results of the IT-Socket analysis within the demonstration scenarios.

The role of semantic has been initially drafted based on the research challenges for the project. An in-depth investigation on the use of semantic and meta modelling matching patterns as the conceptual integration is scheduled during the upcoming months. From a technical perspective, Next Generation Modelling Framework to be developed in the course of the project is regarded as the platform for technical integration made available publicly to assess reference models of different realisation of IT Sockets as well as the design of own IT-Sockets.

Following the Open Model paradigm[6] the "Next Generation Modelling Framework" targets to establish a community as a means for dissemination and exploitation of results but also as possibility to include experts from the domain in the development process.

Acknowledgments. The authors would like to thank the members of Elsag Datamat, Consorzio Interuniversitario CINECA, Innovation Technology Group SA and University of Stuttgart, High Performance Computer Centre as members of the plugIT consortium for the cooperation of this publication.

The authors especially acknowledge the contribution from Prof. Dr. Dimitris Karagiannis (University of Vienna), Prof. Dr. Plexousakis (Foundation for Research and Technology-Hellas) and Prof. Dr. Hinkelmann (Fachhochschule Nordwestschweiz) for their contribution to this publication.

References

1. ACS, Governance of Information and Communication Technology Committee (2009), http://www.acs.org.au/governance/ (accessed May 29, 2009)
2. Andrea, J.: An Agile Request For Proposal (RFP) Process. In: ADC 2003: Proceedings of the Conference on Agile Development. IEEE Computer Society, Washington (2003)
3. Bloomberg, J.: SOA Governance – IT Governance in the context of service orientation (2004), http://www.zapthink.com/ report.html?id=ZAPFLASH-10272004 (accessed May 29, 2004)
4. BREIN, Business objective driven REliable and Intelligent grids for real busiNess (2009), http://www.eu-brein.com/ (accessed May 29, 2009)
5. Calvanese, D., Giacomo, G., Lenzerini, M., Nardi, D., Rosati, R.: Description Logic Framework for Information Integration. In: International Conference on Knowledge Representation and Reasoning (KR), pp. 2–13 (1998)
6. Carr, N.G.: IT doesn't matter, HBR, BCG Analysis (May 2003)
7. Charlsworth, I., Davis, D.: Getting to grips with SOA governance. Ovum Report #040077, 2006-23 (2006)

---

[6] www.openmodels.at

8. CISR, Center for Information Systems Research (2009),
   `http://mitsloan.mit.edu/cisr/research.php` (accessed May 29, 2009)
9. COMPAS, Compliance-driven Models, Languages, and Architectures for Services (2009),
   `http://www.compas-ict.eu/` (accessed May 29, 2009)
10. Dostal, W., Jeckle, M., Melzer, I., Zengler, B.: Service-orientierte Architekturen mit Web
    Services: Konzepte - Standards - Praxis, 2nd edn. Spekturm Akademischer Verlag,
    München (2007)
11. EITO, in cooperation with IDC, European Information Technology Observation (2006),
    `http://www.eito.com` (access: 04.04.2008)
12. Esswein, W., Weller, J., Stark, J., Juhrisch, M.: Identifikation von Sevices aus
    Geschäftsprozessmodellen durch automatisierte Modellanalyse. In: Proceedings of WI
    2009, pp. 513–522 (2009), `http://www.dke.univie.ac.at/wi2009/`
    `Tagungsband_8f9643f/Band2.pdf` (accessed May 15, 2009)
13. Forrester Research, European IT Services (2007), `http://www.forrester.com/`
    `Research/Document/Excerpt/0,7211,38932,00.html` (accessed: April 4,
    2009)
14. Guide Share Europe, "IT Value Delivery", Working Group "IT Governance" (2007)
15. Hedin, M.: SOA-Driven Organizational Change Management: A Market Trends and Ven-
    dor Landscape Analysis of Major Service Players to Address this Emerging Opportunity.
    IDC Competitive Analysis #204727 (2006)
16. Hochstein, A., Ebert, N., Uebernickel, F., Brenner, W.: IT-Industrialisierung: Was ist das?
    Computerwoche, 15 (2007)
17. ISACA, Information Systems Audit and Control Association (2009),
    `http://www.isaca.org/` (accessed May 29, 2009)
18. ITGI, IT Governance Institute (2009), `http://www.itgi.org` (accessed May 29,
    2009)
19. ITIL, IT Infrastructure Library (2009), `http://www.itil-officialsite.com/`
    `home/home.asp` (accessed May 29, 2009)
20. Kappel, G., Kapsammer, E., Kargl, H., Kramler, G., Reiter, T., Retschitzegger, W.,
    Schwinger, W., Wimmer, M.: On Models and Ontologies – A Layered Approach for
    Model-based Tool Integration. In: Mayr, H.C., Breu, R. (eds.) Modellierung 2006, pp. 11–
    27 (2006)
21. Karagiannis, D., Höfferer, P.: Metamodels in Action: An overview. In: Filipe, J., Shishkov,
    B., Helfert, M. (eds.) ICSOFT 2006 - First International Conference on Software and Data
    Technologies: IS27-36. Insticc Press, Setúbal (2006)
22. Karagiannis, D., Utz, W., Woitsch, R., Eichner, H.: BPM4SOA Business Process Models
    for Semantic Service-Oriented Infrastructures. In: eChallenges e-2008, IOS Press. Stock-
    holm (2008)
23. Kätker, S., Patig, S.: Model-Driven Development of Service-Oriented Busienss Applica-
    tion Systems. In: Proceeding of Wirtschaftsinformatik 2009, pp. 171–180 (2009),
    `http://www.dke.univie.ac.at/wi2009/Tagungsband_8f9643f/`
    `Band1.pdf` (accessed May 29, 2009)
24. Kohnke, O., Scheffler, T., Hock, C.: SOA-Governance – Ein Ansatz zum Management
    serviceorientierter Architekturen. Wirtschaftsinfor 50(5), 408–412 (2008)
25. Kühn, H.: Methodenintegration im Business Engineering, PhD Thesis (2004)
26. Lamberti, H.-J.: Banken Technologie als Schlüssel für die Bank der Zukunft. Leistung aus
    Leidenschaft Deutsche Bank (2009), `http://wi2009.at/fileadmin/`
    `templates/downloads/2009_0227_Lamberti_WI2009_Wien_`
    `Versand.pdf` (accessed May 29, 2009)

27. Lembo, D., Lenzerini, M., Rosati, R.: Review on models and systems for information integration, Technical Report, Universita di Roma "La Sapienza" (2002)
28. Manakanatas, D., Plexousakis, D.: A Tool for Semi-Automated Semantic Schema Mapping: Design and Implementation. In: Proceedings of the Int. Workshop on Data Integration and the Semantic Web (DISWeb 2006), pp. 290–306 (2006)
29. Mhay, S.: Request for. Procurement Processes, RFT RFQ RFP RFI (2009), http://www.negotiations.com/articles/procurement-terms/ (accessed May 29, 2009)
30. Mitra, P., Noy, N., Jaiswal, A.R.: OMEN: A Probabilistic Ontology Mapping Tool. In: Gil, Y., Motta, E., Benjamins, V.R., Musen, M.A. (eds.) ISWC 2005. LNCS, vol. 3729, pp. 537–547. Springer, Heidelberg (2005)
31. Murzek, M.: The Model Morphing Approach - Horizontal Transformation of Business Process Models, PhD Thesis (2008)
32. OASIS: OASIS (2009), http://www.oasis-open.org (accessed May 29, 2009)
33. OCG, AK IT-Governance – Aufgaben und Ziele (2009), http://www.ocg.at/ak/governance/index.html (accessed May 29, 2009)
34. OCG, AK IT-Governance – Aufgaben und Ziele (2009), http://www.ocg.at/ak/governance/index.html (accessed May 29, 2009)
35. Offmann, P.: SOAM – Eine Methode zur Konzeption betrieblicher Software mit einer Serviceorientierten Architektur. Wirtschaftsinformatik 50(6), 461–471 (2008)
36. plugIT D2.1, Use Case Analysis and Evaluation Criteria Specification (2009), http://plug-it-project.eu/CMS/ADOwebCMS/upload/ plugIT_D2.1_Use_Case_Description_Evaluation.pdf (accessed May 29, 2009)
37. plugIT, EU-Project FP7-3ICT-231430, plugIT HomePage (2009), http://www.plug-it.org (accessed May 29, 2009)
38. plugIT D5.1, Use Case Modelling And Demonstration (2009), http://plug-it-project.eu/CMS/ADOwebCMS/upload/ plugIT_D5.1_Use_Case_Modelling_Demonstration.pdf (accessed August 31, 2009)
39. Rahm, E., Bernstein, P.A.: A Survey of Approaches to Automatic Schema Matching. The VLDB Journal 10(4), 334–350 (2001)
40. Shawn, J.: Beyond the Template: Writing a RFP that Works (2009), http://www.sourcingmag.com/content/c070228a.asp (accessed May 29, 2009)
41. SOA4ALL, Service Oriented Architectures for All (2009), http://www.soa4all.eu (accessed May 29, 2009)
42. Strohm, O., Ulich, E.: Unternehmen arbeitspsychologisch bewerten: ein Mehr-Ebenen-Ansatz unter besonderer Berücksichtigung von Mensch, Technik und Organisation. Hochschulverlag AG an der ETH, Zürich (1997)
43. SUPER, Semantics Utilized for Process Management within and between Enterprises (2009), http://www.ip-super.org/ (accessed May 29, 2009)
44. TEC, Technology Evaluation Centers (2009), http://www.technologyevaluation.com (accessed May 29, 2009)
45. Teubner, A., Feller, T.: Governance und Compliance. Wirtschaftsinformatik 50(5), 400–407 (2008)
46. The Open Group TOGAF Version 9 (2009), http://www.opengroup.org/togaf/ (accessed May 29, 2009)

47. Thomas, O., Leyking, K., Scheid, M.: Vorgehensmodelle zur Entwicklung Serviceorientierter Softwaresysteme. In: Proceedings of Wirtschaftsinformatik 2009, pp. 181–190 (2009), http://www.dke.univie.ac.at/wi2009/Tagungsband_8f9643f/Band1.pdf (accessed May 29, 2009)
48. Viering, G., Legner, C., Ahlemann, F.: The (Lacking) Business Perspective on SOA - Critical Themes in SOA Research. In: Proceedings of Wirtschaftsinformatik 2009, pp. 45–54 (2009), http://www.dke.univie.ac.at/wi2009/Tagungsband_8f9643f/Band1.pdf (accessed May 29, 2009)
49. Willcocks, L.P., Lacity, M.C.: Global Sourcing of Business & IT Services. Palgrave Macmillan, New York (2006)
50. Windley, P.J.: Teaming up for SOA (2007), http://www.infoworld.com/t/architecture/teaming-soa-620 (accessed May 29, 2009)
51. Zimmermann, S.: Governance im IT-Portfoliomanagement – Ein Ansatz zur Berücksichtigung von Strategic Alignment bei der Bewertung von IT. Wirtschaftsinformatik 5, 357–365 (2008)

Identifying and Supporting Collaborative Architectures

I.T. Hawryszkiewycz

School of Systems, Management and Leadership
University of Technology, Sydney

Abstract. Knowledge management is becoming more integrated into business processes rather than as a standalone activity. Knowledge workers increasingly collaborate and share knowledge as the process proceeds to create new products and services. The paper proposes the development of a collaborative architecture to define the collaboration and a collaborative infrastructure to support the collaboration. It defines the collaborative architecture in terms of an enterprise social network. It then shows how to convert the collaborative architecture to a collaborative infrastructure based on Web 2.0 technologies.

Keywords: Knowledge management, collaboration, social networks.

1 Introduction

Knowledge management can no longer be considered as a separate activity in any organization. It must be part of any business process. It is also often a social process where knowledge workers share knowledge and work together on shared goals. For example, Pralahad and Krishnan [9] see social networking and its influence on knowledge sharing as essential in supporting innovation in processes. They also suggest that social networking be considered early in design and not just as an afterthought. This is particularly the case given the growth of Web 2.0 technologies and their impact on organizations. McAffee [7] for example suggests the trend to what is called Enterprise 2.0, which emphasizes the growth and emphasis on business networking and the consequent collaborative activities that characterize such networking.

Much of the emphasis on social networking comes from what are known as knowledge workers [1, 2]. These workers must quickly assess complex situations and respond to them. Efforts to reengineer the work of knowledge workers into prescribed forms have proven unworkable [1]. Knowledge work is characterized by greater emphasis on social connectivity and interactivity, autonomy and quickly changing practices that lead to changes in social connectivity and interactivity. Support systems for such workers cannot be predefined but organizations must provide knowledge workers with abilities to change processes and the ability to adopt any new technology, and assimilate it in their work [10]. Such systems are increasingly known as lightweight systems.

Lightweight systems can enhance the dynamic capabilities and enable knowledge workers to be more effective in such environments. The most common implementations of lightweight workspaces are workspace management systems or groupware. Most of these, however, are stand alone systems and do not easily support integration

D. Karagiannis and Z. Jin (Eds.): KSEM 2009, LNAI 5914, pp. 442–449, 2009.

with corporate ERP systems as often needed in supply chains. Another evolving support system is middleware that provides the interface between corporate systems and the interface provided by users. However, most such middleware still requires interfaces to be set up by information technology professionals thus making it difficult for users to drive the change. The other requirement is the ability to capture knowledge, which often emerges through interactions between users. Social software is suggested for this purpose. For example, blogs are increasingly used in communication and are increasingly used in practice. They can become a repository of knowledge but still need to be integrated and shared between workspaces.

This paper proposes ways to select the collaborative support to be provided for a particular network. The questions addressed are how to choose the collaborative architecture for a particular business network and how to support it with technology. The paper distinguishes between business architecture, collaborative architecture and collaboration infrastructure. It focuses on defining a collaborative architecture that provides value as suggested by in [8] rather than simply providing communications tools that sometimes do not generate value as outlined in [6]. In this terminology:

- Business architecture defines the way business activities are combined to work towards a particular objective,
- The collaboration architecture are the ways that business entities are to collaborate within the business architecture to achieve business goals, proposing the enterprise social network (ESNs) as the modeling tool to define the collaborative architecture. The collaborative architecture defines the team structures and other relationships in business activities.
- The collaboration infrastructure is the support the collaboration infrastructure including technology support through social software. This is setup by information technology professionals but adapted as a process emerges by knowledge workers.

The paper focuses on the modeling methods for designing the collaborative infrastructure. It sees the emergent nature of knowledge based processes as difficult to modeling methods designed for structured processes. What is thus needed are more lightweight methods that can both be understood and even used by knowledge workers and that can be easily converted to collaborative infrastructure. These should cater for the different perspectives as these are always important in emerging environments. At the same time they should provide a framework for flexibility and include sufficient structure that addresses organizational requirements but at the same time provide the needed flexibility for change. The model describes collaboration from a number of perspectives to address the social knowledge intensive nature of collaborative processes.

2 Combining the Perspectives

The perspectives used in modeling are shown in Figure 1. It shows the following perspectives:

- The business activities and processes that describe what people do in the business and the processes they follow. These are modeled by the business activity model.

- The social perspective that considers the roles and responsibilities as well as the social interactions between the people in the business. This perspective describes the collaborative architecture.
- The knowledge requirements of the different roles and ways to create, capture and reuse knowledge articulated during the interactions.
- The technology and how it supports the business activities, processes, interactions and knowledge.

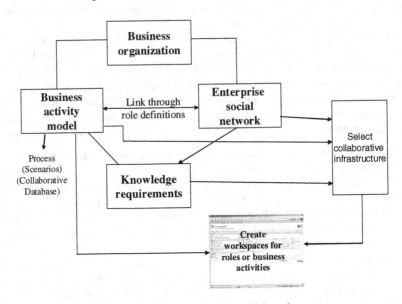

Fig. 1. Objects for modeling complex collaborative systems

The model of collaborative systems is thus made up of three parts – the business activity model (BAM), and the enterprise social network (ESN) and rich picture to model knowledge flows. These two parts concerns knowledge about changing the system as a situation evolves. Other components describe the domain knowledge captured during the business activities and are not covered here. This model is then converted to the collaborative infrastructure.

3 The Modeling Methods

Modeling requirements corresponds to the perspectives. The paper outlines in broad form a business activity model for business activities, an enterprise social networking (ESN) for the collaborative architecture and uses rich pictures to describe knowledge requirements. More detailed descriptions can be found in [5].

3.1 Modeling the Business Activities

Figure 2 illustrates a business activity diagram. It is based on a set of collaborative concepts [3]. The main concepts used here are activities, roles and artefacts. Here the

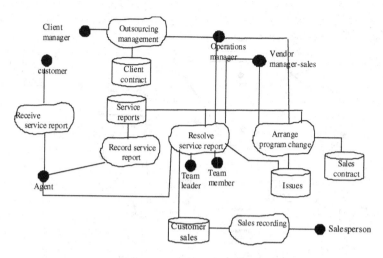

Fig. 2. A business Activity Model

clouded shapes represent activities, black dots represent roles, and disk shapes represent artefacts. The clouded shape is used instead of a regular shape to indicate the emergent nature of the activity.

Figure 2 is a model where a global client outsources the provision of sales application to a technology solutions provider. The provider is required to support a customer service center and respond to a customer reports. The outsourcing is managed in the 'Outsourcing management activities' which includes the development of contracts. The customer reports are received by an agent and recorded in the service reports database. The technology solutions business in turn obtains the applications from the software vendor and customizes them to the needs of the client. Resolution of customer problems requires extensive communication between the various roles in the system to provide the necessary solutions. It often requires negotiations between the solutions provider and the vendor manger to determine ways to respond to a customer report.

The model includes a variety of commands that can be used up by knowledge workers to setup and change systems specified in terms of the model. These include creating new groups, activities and work-items and their associated views. They also include creating workflow events and issuing notifications.

3.2 Modeling the Collaborative Architecture from the Social Perspective

The ESN is an extension of social network diagrams. The ESN is derived from the business activity diagram. It shows the interactions that are required as part of the business activity. The social relationships diagram in Figure 3 the roles in activities and their interactions.

In the ESN in Figure 3 the roles are shown as black dots. The thick lines between the roles indicate work connections, which define the essential communication paths for the participants. The dotted lines show informal connections. The labels attached to each role show the role responsibilities and the labels attached to the lines joining the

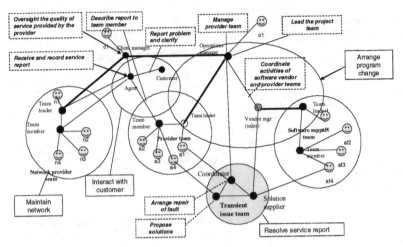

Fig. 3. The collaborative architecture

roles show the interactions between the roles. It is these interactions that often capture much of the needed knowledge to be used in future decisions. The ESN shows:

- The responsibilities of the roles and their interactions and the knowledge they require, and
- The interactions between the roles.

For a larger problem the responsibilities and interactions can be recorded on a list.

3.3 Modeling the Knowledge Requirements

The exchange of knowledge is in most cases difficult to define in terms of a well specified process. The proposed approach here is to use the rich picture, which comes from soft systems methodologies. A rich picture can indicate in an informal way the kinds of knowledge needed by different roles and what they expect from others.

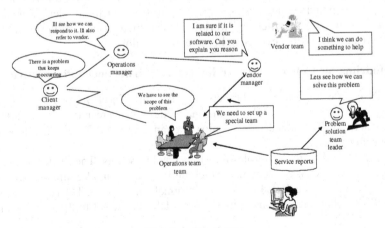

Fig. 4. A rich picture

It also illustrates the flow and sharing of knowledge between the different roles. Figure 4 shows an example of a rich picture. It is by no means a complete picture of all the flows but simply indicates the modeling method.

4   Conversion to the Collaborative Infrastructure

Conversion to the collaborative infrastructure simply looks at each of the interactions and responsibilities in the ESN and selects a service most appropriate to that interaction or responsibility. An example of such mappings is shown in Figure 5.

In complex adaptive systems the roles themselves must be able to reorganize their connections and activities. It also show the knowledge needed by the roles to carry out the activities. For example:

- The agent needs knowledge of how to deal with a trouble report. This knowledge is partly their tacit knowledge and partly the knowledge captured in a fault log,
- The operations manager needs knowledge about when to initiate a special transient team to make changes to prevent the recurrence of a fault. This again includes their tacit knowledge as well as knowledge about the system as a whole.

Figure 5 also identifies the social software that can be used to share knowledge between activities and individuals. For example, creating a fault log and allowing team leaders and members to access this log to identify similar faults can expedite solutions, as well as providing invaluable information to people new to the work. Similarly an issues board can provide the rationale used to make earlier decisions and provide guidelines for addressing new problems and setting up transient teams.

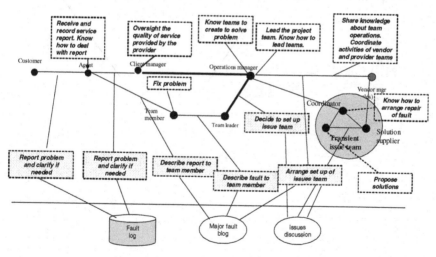

Fig. 5. Converting to the collaborative Infrastructure

5.1   Lightweight Technologies for Supporting the Collaborative Architecture

Lightweight platforms [4] that support adaptive systems are needed. Lightweight here means easy to learn, web based and easy to use. Lightweight systems should be built on the concepts defined for the different perspectives. They should include the concepts defined for the collaborative model while providing commands to easily create and change the structures of workspaces. Our experimental system, LiveNet, demonstrates the kind of support needed by workspace systems. Figure 6 shows the LiveNet interface and its typical commands.

It provides a menu that can be used to create new collaborative objects, including activities, roles, and artefacts. It also enables people to be assigned to the roles. Apart from these elementary operations the system includes ways to implement governance features as for example allowing roles limited abilities to documents. The system includes support for sharing artifacts across workspaces and a permissions structure to control such sharing. Social software such as blogs or discussion systems are supported and can be shared across workspaces.

Commercial systems in this area focus on middleware software that provides the commands that allows users to use the middleware functionality to create workspaces. Furthermore, it should allow users to change the workspaces as work practices change. Many manufacturers are now providing ways to integrate the kind of software with enterprise applications. A typical example here is Websphere provided by IBM. The challenge in many such systems is to provide ways to share knowledge across activities. They provide access to corporate databases but often do not support the sharing of knowledge collected in the course of knowledge work in identifying and solving problems, and making decisions.

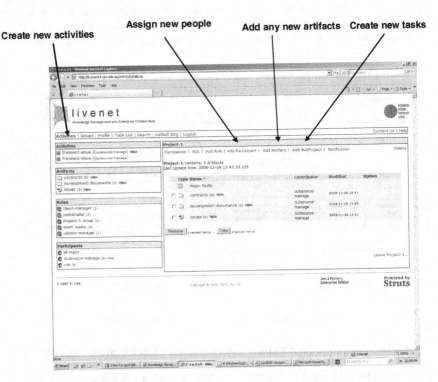

Fig. 6. A Demonstration Workspace

6 Summary

The paper defined the need to integrate knowledge management into everyday business processes and emphasized the need to create a collaborative architecture to support the collaboration needed for sharing knowledge within business processes. It proposed a number of perspectives to model the emergent social nature of knowledge based processes. It proposed a business activity model for the business activity perspective, an enterprise social network (ESN) to model the social perspective and rich pictures to model the knowledge perspective. The paper identified ways to convert the model to a collaborative infrastructure based on Web 2.0 technologies to support the business process.

References

1. Chen, A., Edgington, T.: Assessing Value in Organizational Knowledge Creation: Considerations for Knowledge Workers. MIS Quarterly 29(2), 279–309 (2005)
2. Davenport, T.: Thinking for a Living. Harvard Business School Press, Boston (2005)
3. Hawryszkiewycz, I.T.: A Metamodel for Modeling Collaborative Systems. Journal of Computer Information Systems XLV(3), 63–72 (spring 2005)
4. Hawryszkiewycz, I.T.: Lightweight Technologies for Knowledge Based Collaborative Applications. In: Proceedings of the IEEE CEC/EEE 2007 Conference on E-Commerce Technology, Tokyo, July 2007, pp. 255–264 (2007)
5. Hawryszkiewycz, I.T.: Knowledge Management: Organizing the Knowledge Based Enterprise. Palgrave-Macmillan, Oxford (2009)
6. Hansen, M.T.: When Internal Collaboration is Bad for Your Company. Harvard Business Review 84(3), 83–88 (2009)
7. McAfee, A.P.: Enterprise 2.0: The Dawn of Emergent Collaboration. MIT Sloan Management Review, 21–28 (spring 2006)
8. Pisano, G.P., Verganti, R.: What Kind of Collaboration is Right for You. Harvard Business Review 83(8), 80–86 (2008)
9. Prahalad, C.K., Krishnan, M.S.: The New Age of Innovation. McGraw-Hill, New York (2008)
10. Swanson, E., Ramiller, N.: Innovating Mindfully with Information Technology. MIS Quarterly 28(4), 563–583 (2004)

An Examination of Talent Management Concepts in Knowledge Intensive Settings

Eoin Whelan[1], David Collings[2], and Brian Donnellan[3]

[1] Dept. of Management and Marketing, Kemmy Business School,
University of Limerick, Ireland
eoin.whelan@ul.ie
[2] J.E Cairnes School of Business and Economics, NUI Galway, Ireland
david.collings@nuigalway.ie
[3] Brian Donnellan, Innovation Value Institute, NUI Maynooth, Ireland
brian.donnellan@nuim.ie

Abstract. Despite more than a decade of hype around the concept of talent management, we still have a relatively limited knowledge regarding its application in practice. In this paper we examine how the concepts of talent management apply in knowledge intensive settings. A case study of a high-technology R&D group is conducted. Extending the technological gatekeeper theory, we apply social network analysis (SNA) techniques to identify those employees critical to the knowledge flow network. The specific talents exhibited by these individuals are then explored and we point to some organisational level interventions which can facilitate knowledge intensive organisations in fully exploiting their resources to maximise innovative capabilities.

Keywords: Talent management, technological gatekeeper, knowledge diffusion, social network analysis, R&D.

1 Introduction

The global economic climate has altered significantly since a group of McKinsey consultants coined the phrase "The War for Talent" which brought the topic of talent management to the fore for practitioners and academics alike. While the economic context may have altered radically since the 1990s when the concept emerged, the underlying premise remains valid. It is a firm's human resources that provide a key source of sustainable competitive advantage [1, 2]. However, as we know from the resource based theory, possessing resources is insufficient to create competitive advantage. Firms must be appropriately organised to fully exploit their resources to attain a competitive advantage [3]. Talent management plays a key role in assisting the organisation to ensure its human resources are utilised to the fullest extent. Talent management is regarded by many as one of the most critical HR challenges that organisations will face over the next decade [4]. Yet, despite a decade of debate on the importance of talent management, the concept itself remains somewhat under-developed and under-explored. In spite of the rhetoric of strategic integrated talent management systems in

D. Karagiannis and Z. Jin (Eds.): KSEM 2009, LNAI 5914, pp. 450–457, 2009.

the practitioner literature, paradoxically the evidence suggests that relatively few organisations manage talent on a coordinated or effective basis [5, 6].

In light of the above challenges, the purpose of this paper is to advance our understanding of talent management by identifying and examining talented individuals in the context of Research & Development (R&D). We specifically choose this setting given the strategic importance of R&D in driving an organisation's innovation capabilities [7]. In this effort, we revisit the highly influential technological gatekeeper theory and argue that the talented individuals who will contribute most to organisational success in R&D settings are the small number of individuals who occupy pivotal positions in the knowledge flow network. Consequently we seek to answer two key questions: 1) Is the technological gatekeeper still a pivotal position in the modern R&D group, and 2) What are the specific competencies required by those individuals who occupy pivotal positions in the R&D knowledge flow network? Drawing on social network analysis (SNA) and interview evidence from a single case study, we find that the gatekeeper role is indeed pivotal. However, the role has evolved and undergone a division of labour. It is now rare for a single individual to possess all the talents necessary to effectively acquire and disseminate external knowledge.

2   What Is Strategic Talent Management?

Despite the widespread use of the terminology and its perceived importance, there is a degree of debate, and indeed confusion around the conceptual and intellectual boundaries of talent management. Broadly there are four key streams of thought on what talent management is [8]. Some authors merely substitute the label talent management for HR management. Studies in this tradition often limit their focus to particular HR practices such as recruitment, leadership development, succession planning and the like. A second strand of authors emphasises the development of talent pools focusing on "projecting employee/staffing needs and managing the progression of employees through positions" [8, p.140]. The third stream focuses on the management of talented people. Finally, there is an emerging body of literature which emphasises the identification of key positions which have the potential to differentially impact the competitive advantage of the firm [9, 10].

We adopt Collings et al (in press) definition: as activities and processes that involve the systematic identification of key positions which differentially contribute to the organisation's sustainable competitive advantage, the development of a talent pool of high potential and high performing incumbents to fill these roles, and the development of a differentiated human resource architecture to facilitate filling these positions with competent incumbents and to ensure their continued commitment to the organisation. They argue that the first step in any talent management system should be the identification of the pivotal talent positions which have the greatest potential to impact on the organisation's overall strategic intent. This perspective calls for a greater degree of differentiation of roles within organisations and an emphasis on strategic over non-strategic jobs [9] or organisational roles which have the potential for only marginal impact vis-à-vis those which can provide above-average impact [10]. However, the extent to which a variation in performance between employees in strategic roles is also a significant consideration [11]. This contrasts with the status-quo in many firms where over-investment in non-strategic roles is commonplace [10, 11].

3 The Technological Gatekeeper: A Pivotal Position in R&D?

Throughout the 1970s and 1980s, a rich stream of research examined the processes through which knowledge of the latest technological advances enters the R&D group. This particular stream was headed by MIT's Thomas Allen and his seminal book *Managing the Flow of Technology* [12] documents over a decade's worth of studies with some of the largest American R&D corporations. Allen discovered that knowledge of the latest scientific and technological developments entered the R&D group through a two-step process. Not every R&D professional was directly connected with external sources of knowledge. Instead, a small minority had rather extensive external contacts and served as sources of knowledge for their colleagues. These individuals were termed 'technological gatekeepers' [12, 13, 14] as they served as the conduit through which knowledge of external technology flows into the R&D group. Essentially, a gatekeeper is an individual who acquires technological knowledge from the outside world (step 1) and disseminates this to his or her R&D colleagues (step 2). A more formal definition explains that technological gatekeepers are those key individual technologists who are strongly connected to both internal colleagues *and* external sources of knowledge, and who possess the ability to translate between the two systems [12, 13, 14].

Gatekeepers make a significant contribution to the innovation process by virtue of their pivotal position in the knowledge flow network. Not only do they act as the firm's antennae tuned to a variety of external broadcasting sources, they also exploit their familiarity of the internal knowledge network to internalise emerging technologies. Allen & Cohen [13] noted when studying gatekeepers in the R&D division of a large aerospace firm that "...if one were to sit down and attempt to design an optimal system for bringing in new technological information and disseminating it within the organisation, it would be difficult to produce a better one than that which exists." Indeed, subsequent studies have provided the empirical evidence to support this claim. Development focused R&D projects containing gatekeepers have been found to be significantly higher performing than those without [15, 16].

While we argue that the gatekeeper theory provides a useful lens to examine talented individuals in R&D, we acknowledge that the theory is a little outdated. It has been over 20 years since any significant investigation into the gatekeeper concept has been conducted. In the time since, there have been huge advances in information and communication technologies. The gatekeeper existed in a time when it was a difficult and time consuming process for the average R&D professional to acquire knowledge from beyond the company's boundaries. Thus, the gatekeeper mediated with the outside world on their behalf. What technologies such as the World Wide Web have changed is the ease and speed with which employees at all organisational levels can access and disseminate information. As a result, recent studies suggest that the modern gatekeeper may have morphed into another role providing an altogether different range of services [17]. While we have a good understanding of the role and characteristics of the traditional gatekeeper, scant attention has been paid to how the gatekeeping function is performed in the modern R&D group. From the talent management perspective, this study seeks to explore whether the technological gatekeeper remains a pivotal position in the modern R&D setting and further highlight how organisations can identify and define those performing the gatekeeping function in the modern

R&D group. Thus, our first research question asks: Is the technological gatekeeper still a pivotal position in the modern R&D group?

Once the pivotal positions are identified, the strategic talent management system advocates the development of a talent pool of high potential and high performing incumbents to fill these roles. In order to groom potential incumbents, management needs to know the specific talents of those occupying key positions in the R&D knowledge flow network. Thus, our second research question asks: What are the specific competencies required by those individuals who occupy pivotal positions in the R&D knowledge flow network?

4 The Case Study Site

Utilising a case study approach, we studied the R&D group of a medical device manufacturing firm operating in Ireland, MediA[1]. The R&D group, referred to in the rest of the paper as Group A, consisted of 42 engineers who specialised in the designed and development of catheter-based minimally invasive devices.

4.1 Identifying Pivotal R&D Positions through Social Network Analysis

Identifying pivotal positions is something organisations find difficult. We propose a novel approach in this regard. Given the well established centrality of knowledge flows in the R&D innovation process, we use SNA techniques to identify pivotal talent positions. SNA or sociometry is an established social science approach of studying human relations and social structures by "disclosing the affinities, attractions and repulsions between people and objects" [18]. In simple terms, SNA is the mapping and measuring of relationships and flows between people, groups, organisations, computers or other information/knowledge processing entities [19].

The goal of this study is to demonstrate how SNA supports the identification of talented individuals in knowledge intensive settings. In our argument, these are the handful of individuals performing the gatekeeping role. We adopt the classic definition of a gatekeeper as an individual who is *both* an internal communication star (i.e. in the top 20% of internal communication measures) and an external communication star (i.e. in the top 20% of external communication measures). While it can be argued that this is an arbitrary measure, it serves our purpose of identifying the key individuals in the R&D knowledge flow network.

Figure 1 presents the SNA of Group A. To collect these data, all group members were asked to complete a short online questionnaire on their internal and external communications. The SNA software package UCINET [20] was used to produce this diagram. The nodes in the diagram are the individual members of Group A and the lines represent the flow of technical knowledge between them. The more connected nodes tend to gravitate towards the centre of the network while those nodes with fewer connections are found on the periphery. Nodes 4, 16, 35 and 40 did not complete the questionnaire hence the reason they are isolated on the left. Nodes 2, 11, 38 and 42 are also isolates because they have no reciprocated interactions with another group member. The external communication stars of the group are represented as

---

[1] Company names are fictitious to preserve anonymity.

triangles. The size of the triangle is reflective of how well connected that individual is to external knowledge sources. For example, node 9 is the biggest triangle as this individual is the most frequent user of external knowledge sources.

Figure 1 reveals a number of key people in Group A's knowledge flow network. Firstly, there are nodes 7 and 37. Using the classic definition, only these two members (or 5%) of Group A can be classified as technological gatekeepers. While external knowledge is imported and disseminated around the group by these two gatekeepers, the SNA evidence indicates that separate communication specialists also combine to perform the gatekeeping role. One set of boundary spanning individuals acquire external knowledge, and a largely different set of individuals distribute this knowledge around the group. The relationship between node 5 and node 25 can be used to demonstrate this process (the relationship between nodes 17 and 28, nodes 9 and 6, or nodes 15 and 6 could also have been used). Node 5 is an external communication star. This individual is well connected to external knowledge sources but is not very well connected internally. Node 5 acquires external knowledge and communicates this to node 25. Node 25, on the other hand, is well connected internally and can distribute this knowledge around the group through his or her many connections.

It must be noted however that the SNA evidence, and our interpretation of that evidence, only suggests that such a sequence of knowledge flow is evident. Semi-structured interviews with selected group members were also conducted to validate this interpretation, and to explore the specific talents exhibited by these key individuals. Based upon the interview findings, table 1 summarises the specific competencies exhibited by those individuals occupying key positions in the knowledge flow network.

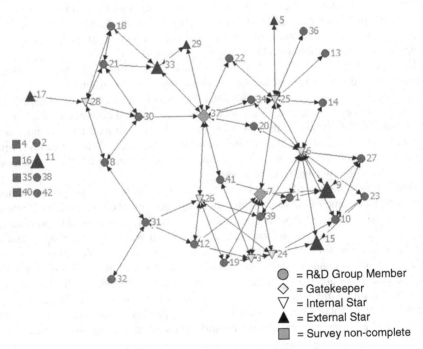

Fig. 1. A Social Network Analysis of Group A

Table 1. Summary Table of those Performing the Gatekeeping Role

	Key Skills	Motivation/ Attitudes	Preferred Media
External Communication Stars	• Ability to acquire relevant knowledge of external developments • Narrow and deep technology domain knowledge • Strong analytical skills	• Genuine interest in keeping abreast of emerging trends in their specialty • Primarily acquire knowledge for own use but lack the skills to disseminate effectively	• Predominately Web-based e.g. Google search, online communities, materials websites
Internal Communication Stars	• Ability to translate complex external knowledge into a form understandable and relevant to internal colleagues • Wider knowledge base which facilitates understanding the context of new knowledge and how it fits with extant knowledge	• Enjoy helping others • Develop their own knowledge from these interactions • Expect reciprocation	• Email and oral
Gatekeepers	• Display both depth of knowledge of external communication star and breadth of knowledge of internal communication star • Highly sociable with very good networking skills enabling them to develop extensive internal and external networks	• May acquire knowledge for their own use but also transmit it others • Enjoy helping others	• External – both Web-based and oral • Internal – Email and oral

5 Discussion and Conclusions

This paper provides a very clear example of how pivotal positions can be identified in organisations. Recognising the importance of knowledge flows in the context of innovation in R&D settings, we argue that the technological gatekeeper role continues to represent a pivotal position. While we find that the gatekeeping tasks of acquiring and disseminating knowledge are integral to the R&D operation, we also find that these tasks no longer need to be performed by a single individual. Indeed, it is more likely that the gatekeeping role will be performed by external and internal communication specialists combining their unique talents together. Gatekeepers do exist, but they are rare. When Allen [12] first formulated the theory, the gatekeeping role could only be performed by a single individual because technical communications were predominately oral based. Among other skills, the traditional gatekeeper needed excellent social networking abilities in order to effectively acquire and disseminate knowledge orally. While other R&D engineers may have wanted to perform the gatekeeping role, the lack of these social networking skills possibly impeded them. From the R&D group we have studied, we find that Web technologies now enable the individuals that are interested in external developments to easily access that knowledge. Rather than having social networking skills, these external communication stars possess analytical and Internet search skills. However, the lack of excellent social networking skills inhibits the ability of the external stars to distribute that knowledge around the R&D network themselves. This is the domain of a different set if individuals, the internal communication stars, who possess those excellent social networking abilities. A key contribution of the current paper is to identify the competencies evident in both internal and external communication stars (see table 2). This will provide organisations with the information required to identify these competencies in the individuals within their talent pools, to focus development interventions for the talent pool in developing these competencies and to facilitate the placing of internal and external communication stars in each R&D group. Managers would also be interested to know what they can do to facilitate the external and internal communication star positions. While we would not advocate that management formally appoint individuals to these positions, we do advocate that the handful of key individuals who exhibit the competencies of the communication star be given the opportunity to display their talents. External stars could be freed any mundane administrative duties and allocated the time they need to scan the external environment for emerging technologies and trends. In terms of resources, all they need is a PC with an internet connection. However, it would more beneficial if external stars are given priority for external networking events such as conferences or tradeshows. Internal stars have a natural flair for getting to know others. If management fails to recognize the valuable role performed by these individuals, there is a danger that their knowledge dissemination efforts could be stifled. Internal stars need the opportunity to network. Involving these individuals in multiple projects throughout the firm will enable them to build their network more rapidly, allowing them to become more effective disseminators of knowledge.

Finally, our approach demonstrates the usefulness of SNA in identifying such positions. This is a concrete tool that can be utilized in practice to confirm the technological gatekeeper role as a pivotal position in R&D settings. Additionally it is a tool that practitioners can use in the process of identifying high potential and high performing individuals for the organisation's talent pool.

References

1. Lowe, K.B., Milliman, J., De Cieri, H., Dowling, P.J.: International compensation practices: a ten-country comparative analysis. Human Resource Management 41(1), 45–66 (2002)
2. Caligiuri, P.M., Lazarova, M., Tarique, I.: Training, learning and development in multinational organizations. In: Scullion, H., Linehan, M. (eds.) International Human Resource Management: A critical text. Palgrave Macmillan, Hampshire (2005)

3. Barney, J.: Gaining and sustaining competitive advantage. Addison-Wesley, Reading (1997)
4. Boston Consulting Group. The future of HR: Key challenges through 2015. Dusseldorf, Boston Consulting Group (2007)
5. Cappelli, P.: Talent management for the twenty-first century. Harvard Business Review, 74–81 (March 2008)
6. Cheese, P., Thomas, R.J., Tapscott, D.: The talent powered organization: Strategies for globalization, talent management and performance. London and Philadelphia: Keogan Page (2008)
7. Cohen, W., Levinthal, D.: Absorptive capacity: A new perspective on learning and innovation. Administration Science Quarterly 35, 128–152 (1990)
8. Lewis, R.E., Heckman, R.J.: Talent management: a critical review. Human Resource Management Review 16(2), 139–154 (2006)
9. Becker, B.E., Huselid, M.E., Beatty, R.W.: The differentiated workforce: Transforming talent into strategic impact. Harvard Business School Press, Boston (2009)
10. Boudreau, J.W., Ramstad, P.M.: Talentship, talent segmentation, and sustainability: A new HR decision science paradigm for a new strategy definition. Human Resource Management 42, 129–136 (2005)
11. Huselid, M.A., Beatty, R.W., Becker, B.E.: 'A players' or 'A positions'? The strategic logic of workforce management. Harvard Business Review, 110–117 (December 2005)
12. Allen, T.J.: Managing the flow of technology. MIT Press, Cambridge (1977)
13. Allen, T.J., Cohen, S.I.: Information flow in research and development laboratories. Administrative Science Quarterly 14(1), 12–19 (1969)
14. Allen, T.J., Tushman, M.L., Lee, D.M.S.: Technology transfer as a function of position in the spectrum from research through development to technical services. Academy of Management Journal 22(4), 694–708 (1979)
15. Tushman, M., Katz, R.: External communication and project performance: an investigation into the role of gatekeepers. Management Science 26(11), 1071–1085 (1980)
16. Katz, R., Tushman, M.: An investigation into the managerial roles and career paths of gatekeepers and project supervisors in a major R&D facility. R&D Management 11, 103–110 (1981)
17. Assimakopoulos, D., Yan, J.: Sources of knowledge acquisition for Chinese software engineers. R&D Management 36(1), 97–106 (2006)
18. Moreno, J.L.: Sociometry in relation to other social sciences. Sociometry 1(1/2), 206–219 (1937)
19. Scott, J.: Social network analysis; a handbook. Sage Publications, London (2000)
20. Borgatti, S.P., Everett, M.G., Freeman, L.C.: Ucinet for Windows: Software for social network analysis. Analytic Technologies, Harvard (2002)

A 'Soft' Approach to TLM Requirements Capture to Support Through-Life Management

Huseyin Dogan, Michael Henshaw, and Esmond Urwin

Loughborough University, Systems Engineering, Garendon Wing, Holywell Park,
Leicestershire, LE11 3TU, United Kingdom
{H.Dogan,M.J.d.Henshaw,E.N.Urwin}@lboro.ac.uk

Abstract. Loughborough University and BAE Systems are sponsoring a research programme to develop an enterprise Knowledge Management system for Through Life Management (TLM) in support of Through Life Capability Management (TLCM). This paper summarises the finding of a requirements analysis case study which captured, analysed and synthesised the key stakeholder requirements for this Knowledge Management research within the aerospace and defence industry. This study consists of two approaches; (1) an Interactive Management (IM) workshop and (2) semi-structured interviews with the Subject Matter Experts (SMEs). Three of the group methodologies used in the Interactive Management workshop were Idea-Writing, Nominal Group Technique and Interpretive Structural Modelling. Soft systems rich pictures were also constructed by the SMEs to provide a diagrammatic representation of the systematic but non-judgmental understanding of the problem situation. The difficulties and benefits of adopting this 'soft' approach and future research plans are also discussed here.

Keywords: Knowledge Management, Through-Life Management, Interactive Management, Requirements Analysis, Soft Systems.

1 Introduction

The UK aircraft and aerospace industry (civil air transport, defence and space) which employs over 124,000 people directly and over 350,000 indirectly is one of the UK's top exporting industries and is the largest in the world outside the USA [1]. The aerospace and defence enterprises can be considered as large complex and adaptive systems that require Knowledge Management (KM) models and architectures to capture and analyse behaviour, structure and knowledge. Such enterprises need to know what their knowledge assets are in addition to managing and making use of these assets to maximise return. They conduct business as a service-product mix, in which they simultaneously manage the product and service lifecycles of many projects. Employees need to have sufficient knowledge or competency (both tacit and formal) to carry out the tasks allocated to them [2]. Employees also spend a very consistent 20 to 25 percent of their time seeking information. Line business managers and administrators spend as much of their time seeking information as do research scientists [3]. Knowledge Management, which has been defined as *"the effective learning processes*

D. Karagiannis and Z. Jin (Eds.): KSEM 2009, LNAI 5914, pp. 458–469, 2009.

associated with exploration, exploitation and sharing of human knowledge (tacit and explicit) that use appropriate technology and cultural environment to enhance an organisation's intellectual capital and performance [4]" can become a key enabler in this context due to the multiple instances of collaboration, co-ordination and cooperation between the stakeholders.

Through Life Management (TLM) is the philosophy that brings together the behaviours, systems, processes and tools to deliver and manage projects through the acquisition lifecycle [5]. The implementation of TLM within aerospace companies was analysed and it was discovered that data and knowledge sharing, whether within an organisation or with customers and suppliers, could be improved [6]. The same research also identified that current information systems do not appear to align with the requirements of the advanced services being developed for TLM.

The overall aim of this research is to develop an enterprise Knowledge Management (KM) system for TLM in support of Through Life Capability Management (TLCM) where TLCM is defined as *"an approach to the acquisition and in-service management of military capability in which every aspect of new and existing military capability is planned and managed coherently across all Defence Lines of Development from cradle to grave* [7]". The results of the overall research will be expressed as a practical methodology leading to the development of a tool for potential incorporation within an integrated business framework that relates the engineering and commercial activities of a project to each other.

The aim of this particular study was to interact with the domain experts and stakeholders within the aerospace and defence industry to capture, analyse and synthesise the through-life enterprise Knowledge Management requirements to determine the stakeholders' needs in addition to contextualising the problem situation. This paper therefore introduces TLM and KM followed by a requirements analysis case study which consists of two approaches; (1) an Interactive Management (IM) workshop and (2) semi-structured interviews with the Subject Matter Experts (SMEs). The results from this case study are illustrated through a set of IM diagrams and rich pictures. The difficulties and benefits of adopting this 'soft' approach and future research plans are also discussed.

2 Through-Life Knowledge Management

In the emerging defence acquisition environment, which is characterised as Through-Life Capability Management (TLCM), commercial success will be significantly determined by the ability of companies to manage the value of knowledge within the various enterprises in which they participate [8]. Knowledge of the systems, for which a company has through-life management responsibility, may be distributed around an enterprise that comprises several commercial organisations and the customer. The provision of seamless through-life customer solutions *"depends heavily on collaboration, co-ordination and co-operation between different parts of an enterprise, different companies within a group, other manufacturers, support contractors, service providers and all their respective supply chains"* [6]. This entails the importance of an approach to manage and value knowledge. It is also important to understand the differences between data, information, knowledge and wisdom before KM is discussed in detail. Data is known facts or things used as a basis of inference and hence data depends on context. Information is systematically organised data. Knowledge is considered as actionable information and Wisdom is the ability to act critically or practically in a given situation [4].

Definitions of KM are presented from selected and classified sources to provide an overview of the most important and the most promising approaches of defining KM in terms of proportions such as strategy, technology and organisation [9]. The three stages of KM are also identified as (1) intellectual capital; (2) the human and cultural dimensions; and (3) the content and retrievability stage [3]. The successive key phases of these stages encompass lessons learned; communities of practice; and content management and taxonomies. Knowledge mapping is considered as part of a KM methodology to develop conceptual maps as hierarchies to support knowledge scripting, profiling and analysis [10] [11] [12]. Knowledge audit is also used for evaluating knowledge management in organisations [13].

The value is very difficult to measure and extract when knowledge is used [17]. Approaches to value knowledge are still deficient in the current literature. In contrast, the cost of considering or implementing KM through practical tactics such as Return on Investment (ROI) and Key Performance Indicators (KPIs) are very common [9] [14]. The 'content value chain', which involves embedding value in the form of a message or signal contained within all elements of the value chain, is considered as a way of measurement [16]. A set of information characteristics with associated metrics to assist the measurement of information quality or value is also established [17]. The Knowledge and Information Management (KIM) project [18] has developed an approach to build on an information evaluation assessment systems based on the information characteristics establishment [17], Bayesian Network (BN) theory, and conditional probability statistical data.

This requirements analysis study through interactions with Subject Matter Experts (SMEs) started guiding the development of knowledge categories and taxonomy relevant to knowledge value.

3   Requirement Analysis Case Study

The aim of this case study was to interact with the domain experts and stakeholders to capture, analyse and synthesise the through-life enterprise KM requirements to determine the stakeholders' needs and how they can benefit from this research. Two approaches have been identified; (1) an Interactive Management (IM) workshop and (2) semi-structured interviews with the Subject Matter Experts.

3.1   Method

Two structured IM workshops were conducted to capture requirements. The first workshop predominantly consisted of academics (n=8) from Loughborough University whereas the second workshop exclusively involved industrial participants (n=13) from BAE Systems. Four semi-structured interviews were also conducted with senior managers to capture and represent KM centric problems within the defence and aerospace industry. A technique called Interactive Management (IM) was used during the workshop. Four of the group methodologies typically used with IM are Idea-Writing

(IW), Nominal Group Technique (NGT), Interpretive Structural Modelling (ISM) and Field and Profile Representations [19] [20] [21]. In addition to these Interactive Management techniques, Soft systems rich pictures were also constructed to have a diagrammatic representation of the thorough but non-judgmental understanding of the problem situation [22] [23].

The authors adopted a 'soft' systems approach by using the Interactive Management and SSM rich pictures rather than a more engineered approach (e.g. using Quality Function Deployment and Functional Modelling) to define the problem space. The 'hard' systems approach, and hence the traditional systems analysis concepts, will be exploited as the research progresses to develop a Knowledge Management (KM) framework including a toolset. In addition, a step-by-step guide to implement Systems Engineering processes as described in the INCOSE Systems Engineering Handbook [24] will be utilised to adopt a systems thinking approach in order to gain insights and understanding to this situation [25]. Therefore, the Interactive Management technique was identified as the most appropriate technique as a result of potential user involvement and classification of objectives to this 'fuzzy', multidimensional and complex problem situation.

3.2 Procedure

An IM session was conducted during the two workshops with a total of four groups; Academics A, Academics B, Industrials A and Industrials B. The participants were invited as they were either researching or managing TLM or TLCM. There was no one from the government i.e. Ministry of Defence (MoD). The participants were also briefed about the workshop through a series of slides consisting of the following four sessions.

- *Session I: Idea Writing (IW)*. An Idea-Writing session was conducted to partly produce the issues related to a given KM trigger question and partly as an enabling process to aid consensus decision-making (e.g. categorising ideas) amongst the SMEs during the workshop. The trigger question presented and agreed with the members of the group was *"what are the issues for through-life enterprise knowledge management?"*. The IW session therefore included generating ideas individually; exchanging lists of ideas; identifying headings to categorise the ideas; and editing the ideas generated.
- *Session II: Nominal Group Technique (NGT)*. The NGT was used to generate, clarify, edit and obtain a preliminary ranking of a set of objectives. The trigger question used was, *"what are the objectives for through-life enterprise knowledge management?"*. This process was similar to IW and consisted of generating a set of objectives in writing; round-robin recording of those objectives; serial discussion of objectives for clarification; and also voting on items of importance.
- *Session III: Interpretive Structural Modelling (ISM)*. ISM is a method that helps members of a group examine the inter-relationships between elements gained using the NGT process and provides a structure for tackling its complexity. In this session, an Intent Structure was achieved by using the relation, *'help to achieve'*. The process comprised generating the element set; completing the matrix of element interactions; and also displaying, discussing and, if necessary, amending the ISM.
- *Session IV: Soft Systems Rich Pictures*. This session consisted of constructing rich pictures for the expression of the problem situation. There are no rules in

constructing rich pictures. Rich pictures are usually free form diagrams or cartoons. Participants made up their own icons as they went along. Also, an example of a rich picture expressing the problem situation of university student accommodation [23] was used to illustrate creation of a rich picture from each group.

Semi-structured interviews which consisted of identification and contextualisation of through-life KM requirements were also conducted with four senior participants in addition to the workshop sessions described above. The interviewees were not the same as workshop participants. The semi-structured interviews focused on a series of key KM topics including communication, collaboration, searching, storing, sharing, tools, corporate knowledge and security.

4 Results

The results of the Interactive Management workshop were transcribed and analysed. The products of the sessions consisted of four IW categories, four different lists of finalised NGT objectives, four ISM models and also four sets of SSM rich pictures. These were from the two groups of academics (n=8) in addition to the two groups of industrials (n=13). The details are stated below.

Table 1. IW session categories

KM Strategy	Organisational Culture
- Knowledge capability - Institutionalise KM - Governance/ownership - Funding/investment - Management structure - Leadership/champion - Succession planning and career paths - Transform tacit knowledge into explicit	- Trust - Competencies - Retention of key skills/expertise - Social and physical environment - Learning culture/experience - Change - Sharing for advantage - Language – meaning/usage - Demographics - Education/training - Creativity
Knowledge Assessment	**Knowledge Configuration**
- Knowledge about knowledge - Information distillation - Cumulative property - Time (freshness/staleness) - Risks - Value proposition - Maturity - Scale - Rigour - Quality - Accessibility - Capacity - Lifecycle of knowledge	- How and what to capture? - How to interpret knowledge? - How to communicate? - How to store and what to store? - How to access? - How to present or represent? - How to choose what to keep)? - Security or privacy - Ontology - Support mechanism

4.1 Idea Writing (IW) Results

The total number of IW statements from the four groups (n=21) was 268. This gives an average of approximately 13 statements per person. Academics A used a more traditional Systems Engineering approach and categorised the statements under the 'people', 'process' and 'technology' headings, which was similar to the results of Industrials B. The only difference was that Industrials B used the heading 'business' rather than 'process' and introduced a new heading called 'knowledge creation'. The authors summarised the categories from the four groups under the four main headings shown in Table 1.

4.2 Nominal Group Technique (NGT) Results

The total number of NGT objectives from the four groups (n=19 due to 2 dropouts) was 199. This gives an average of approximately 10 objectives per person. Each group member was asked to indicate their five most important objectives and rank them accordingly ('5' being the most important and '1' the least important objective). These voting scores provide an initial assessment of the participants' judgements regarding relative importance of objectives. A ranking system using a form of Single Transferable Vote (STV) was adopted to minimise 'wasted' votes. Some of the important implications of the NGT findings are:

- the academics focused more on creating and adapting a KM vision and strategy within and outside the enterprise whereas the industrials didn't;
- both of the groups (academics and industrial) stated the importance of developing a culture of co-operation and trust across all enterprise communities that considers KM as a core discipline;
- the groups also touched on points such as enabling to become a learning organisation; adding to the accumulated store of knowledge; and ensuring that lessons are learned;
- all four groups emphasised the importance of exploiting, measuring and managing the corporate knowledge and information for the business benefit.

Table 2. NGT results from Academics A

		Rankings		
Rank	**Finalised Objectives**	**P1**	**P2**	**P3**
1	Create/adapt a vision and strategy	5	4	5
2	Develop organisational culture and structure	3	5	4
3	Be able to measure success/failure (value success)	2	3	3
4	Identify current status & implement continuous improvement activities	4	2	2
5	Develop/update/implement KM processes, procedures and practices	1	~	1
6	Build a trust environment within enterprise	~	1	~
7	Develop knowledge management system	~	~	~

The participants only voted against the items finalised by their own group. The NGT approach can be improved by merging the items derived from the consecutive groups from disperse locations and time. An example of a finalised NGT list and the corresponding votes are shown in Table 2.

4.3 Interpretive Structural Modelling (ISM) Results

The product of the ISM process is an 'intent structure'; this shows how the objectives captured through NGT inter-relate with one another. An example of an ISM from the academic and industrial group is shown in Figure 1 and 2, with the boxes representing the objectives and the arrows indicating a 'help to achieve' intent structure. This relationship is termed 'transitive', which means that elements at the foot of the figure help to achieve all other objectives above it to which they can be linked via one or more arrows. The process of completing the matrix of element interactions were found to be complicated due to not having a distinct transitive set of objectives hence some groups decided not to follow this process. Consequently, those groups ended up with arrows pointing downwards as shown in Figure 2.

The key points captured in ISM from the four groups reflect the NGT results which highlighted the need to build a well developed KM vision and strategy; to embed a culture and trust within the enterprise; and enable efficient sharing and application of knowledge across all appropriate stakeholders.

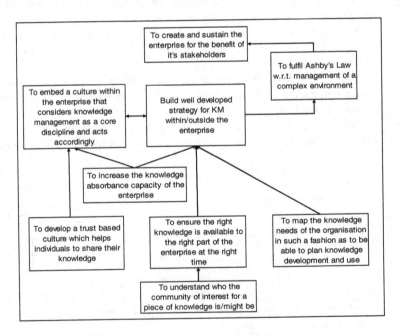

Fig. 1. ISM results from Academics B

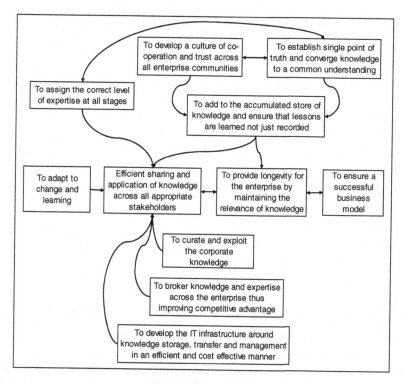

Fig. 2. ISM results from Industrials B

4.4 Soft Systems Rich Pictures

Participants constructed SSM rich pictures for the expression of the problem situation during this session. Academics A focused on external actors such as competitors, government, Ministry of Defence (MoD), Non-Governmental Organisations (NGOs), suppliers as well as trying to define a boundary of the organisation. The organisational implications included the development of a vision, culture and structure within different business units. Social and economic circumstances were also considered to be part of the external environment. Academics B focused more on knowledge flow within the enterprise. This included business requirements captured by the Management Board or CEO being fed back to the Chief Information Officer (CIO), Functional Delivery Managers (FDM) and Human Resources (HR). CIO then forwards these business requirements to the KM group which may start the Research and Development (R&D) process or even seek external expertise. The personal needs including the training and skill needs are handled by HR who has access to the whole organisation hence employees. The authors merged these academic rich pictures by incorporating the external environment, actors, internal knowledge flow and processes derived from the two academic groups as shown in Figure 3.

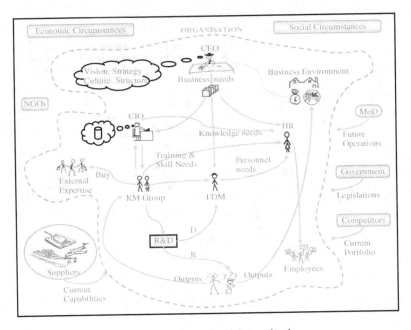

Fig. 3. Rich picture from the academics

Industrials A as opposed to the other three groups illustrated the through life elements of enterprise KM. They compared the timeline of a human being (e.g. from cradle to grave) to the timeline of a product or system (e.g. from the concept all the way to the disposal phase). Towards the start of the timeline there is a low output in contrast to the latter stages where there is a high output. This timeline of a human life also expresses the importance of training and learning from experience which also entails a similar approach as being a learning organisation. Industrials B presented how a defence company operates in the existing environment. This comprise having an information repository which needs to be analysed to encapsulate and store the corporate knowledge in addition to considering the stakeholders and other interested parties (e.g. competitors) in this process. They have also drawn boundaries around areas such as core engineering and information services domains.

4.5 Semi-structured Interview Results

Semi-structured interviews which consisted of identification and contextualisation of through-life KM requirement were also conducted with only four participants due to time constraints. Further interviews will be scheduled. The results from the current semi-structured interviews revealed the following key topics:

- *Communication*: communicating and collaborating within the enterprise and blockages between individuals.
- *Searching and storing*: knowledge hiding and waiting to be unleashed; tapping into existing, potential and most critical knowledge; creating and managing knowledge in addition to its succession into the business.

- *Sharing*: making best practice ad examples of excellence available from across the company; changing people's attitudes and willingness to share best practise to overcome the barriers of any "not-invented-here" syndromes.
- *Knowledge support and tools*: reducing knowledge deficiencies; empowering people; documenting one's own knowledge for repeated use; addressing re-inventing the wheel and duplication of the effort; analysis of existing KM tools, processes and systems; and sourcing outside knowledge.
- *Corporate knowledge and security*: analysis of knowledge flow and maps; security of corporate knowledge and its policy; managing the threat of the loss of key people and their know-how; sharing knowledge across bounda-ries; standardisation; and becoming a better learning organisation.

5 Discussion

5.1 Interactive Management Workshop

As a result of analysing the data especially the rich pictures, it can be concluded that for these samples the academics are more organisational oriented e.g. less constrained by time whereas industrials were more product oriented e.g. more consideration of product development and lifecycle. The two Idea Writing (IW) groups focused on traditional Systems Engineering approaches and came up with categories such as people, process and technology whereas the other groups followed a more knowl-edge-centric approach and synthesised categories more specific to organisational knowledge management. The authors followed a subjective approach and categorised the finalised IW statements from the four groups under the headings of (1) KM strat-egy, (2) organisational culture, (3) knowledge assessment and (4) knowledge configu-ration. The sub-headings are also to be studied further when developing a through life enterprise KM framework.

The ISM process is usually continued until the relationships between all necessary pairs of ideas have been explored. Often, ISM software is used to display a structural map showing the result of the group's judgments. The length of time required to com-plete discussion of all necessary pairs of ideas depends on the total number of ideas in the set, but generally the process requires between five to eight hours of group delib-eration. Therefore, it was decided to stick to a more flexible approach by allowing the participants to develop a model similar to ISM through discussions. The ISM software was not used due to time constraints. The relationship within ISM is termed 'transi-tive', which means that elements at the foot of the figure help to achieve all other objectives above it to which they can be linked via one or more arrows. Some partici-pants struggled to put the objectives in this transitive hierarchical format as the final-ised objectives were identified to be relatively high level. As the processes of IW, NGT, ISM, SSM rich pictures and semi-structured interviews progressed, it was clear that a better understanding emerged as to what this through-life enterprise KM research should be trying to achieve.

5.2 Traditional versus Soft Systems Requirements Capture

Soft systems approaches are a set of techniques that are informal and generally include multiple stakeholders. They have been developed in contrast to the driven formalistic approaches that are commonly used for requirement engineering. Galliers & Swan [26] argue that much more is needed than just standard hard methods when trying to generate a comprehensive set of representative requirements. It is put forward that to be able to arrive at such a set, information is generally informal by nature and that perceptions of the real world are *"cognitively and socially constructed by actors"*. The aim of SSM is to allow a set of stakeholders to make a bigger contribution to requirements engineering activities. One way of characterising a soft systems requirement capture approach against a traditional approach is that it is much more naturalistic. Traditional requirement capture is much more systematic and formal in approach using such tools, techniques and methods as QFDs, formal modeling, document analysis and linguistics analysis.

6 Conclusion

An objective identified within the strategic framework of a leading aerospace and defence company [27] revealed the importance of sharing of expertise, technology and best practice between the company and its global business. This research illustrated the result of applying a 'soft' systems approach to define the problem space and represent the objectives to capture the requirements centred on through life enterprise Knowledge Management. A 'hard' systems approach hence the traditional Systems Engineering concepts, will be exploited as the research progresses to develop a KM framework including a toolset. A future study will direct the focus on creating a use case diagram, domain model and package dependency diagram to guide the development of the architecture of the system derived from this 'soft' systems study.

References

1. UK Trade & Investment - Department for Business, Enterprise and Regulatory Reform (DERR): An overview of Aerospace and Defence sector (2008)
2. Kennedy, G.A.L., Siemieniuch, C.E., Sinclair, M.A.: Towards an Integrated Model of Enterprise Systems. In: INCOSE International Symposium (2007)
3. Koenig, M.E.D., Srikantaiah, T.K.: The evolution of knowledge management. In: Srikantaiah, T.K., Koenig, M.E.D. (eds.) American Society for IS, Medford, NJ (2000)
4. Jashapara, A.: Knowledge Management: An Integrated Approach. Prentice Hall, Englewood Cliffs (2004)
5. UK Ministry of Defence (MoD) Acquisition Operating Framework (AOF) (2008)
6. Ward, Y., Graves, A.: Through-life management: The provision of total customer solutions in the aerospace industry. Int. Journal Services Technology & Management, 455–477 (2007)
7. Taylor, B.: Presentation: TLCM Overview. Defence Science and Technology Laboratory (Dstl). Ministry of Defence (MoD). United Kingdom (2006)
8. Defence Industrial Strategy: Defence White Paper. Ministry of Defence (MoD) (2005)
9. Maier, R.: Knowledge Management Systems: Information and Communication Technologies for Knowledge Management. Springer, Berlin (2002)
10. Wiig, K.: Knowledge Management Methods. Schema Press, Arlington (1993)

11. Newbern, D., Dansereau, D.: Knowledge Maps for Knowledge Management. In: Wiig, K. (ed.) Knowledge Management Methods, Schema Press, Arlington (1993)
12. Liebowitz, J.: Knowledge Mapping: An Essential Part of Knowledge Management. In: White, D. (ed.) Knowledge Mapping and Management. IRM Press (2002)
13. Hylton, A.: A KM initiative is Unlikely to Succeed without a Knowledge Audit (2002), http://www.annhylton.com/siteContents/writings/writings-home.htm
14. Goodman, J.: Measuring the value of knowledge management. Ark Group, London (2006)
15. Stewart, T.A.: IC: The New Wealth of Organizations. Doubleday, New York (1997)
16. Simard, A.: Knowledge Services: Technology Transfer in the 21st Century Pres. to: Federal Partners in Technology Transfer. Halifax, NS (2007)
17. Zhao, Y.Y., Tang, L.C.M., Darlington, M.J., Austin, S.A., Culley, S.J.: Establishing information valuing characteristics for engineering design information. ICED (2007)
18. Tang, L.C.M., Yuyang, Z., Austin, S.A., Darlington, M.J., Culley, S.J.: A characteristics based information evaluation model. CIKM - WICOW (2008)
19. Delbeq, A.L., Van De Ven, A.H., Gustafson, D.H.: Group techniques for program planning: A guide to Nominal Group and DELPHI processes. Scott, Foresman, Glenview (1975)
20. Broome, B., Keever, D.: Facilitating Group Communication: The Interactive Management Approach. In: Paper presented at the 77th Annual Meeting of the Eastern Communication Association, Atlantic City, NJ (1986)
21. Warfield, J.N., Cardenas, A.R.: A Handbook of Interactive Management. Iowa State University Press, Ames (1995)
22. Checkland, P., Jim, S.: SSM in Action. John Wiley and Sons, Toronto (1990)
23. Lewis, P.J.: Rich picture building. EU Journal of Information Systems 1(5) (1992)
24. International Council on Systems Engineering (INCOSE), Systems Engineering Handbook, V2.0 (2000)
25. Landtsheer, B., Aartman, L., Jamar, J., Liefde, J., Malotaux, N., Reinhoudt, H., Schreinemakers, P.: Implementing Systems Engineering: A Step-By-Step Guide. In: Fifth European Systems Engineering Conference. Published by INCOSE (2006)
26. Galliers, R.D., Swan, J.A.: There's more to Information Systems Development than Structured Approaches: Information Requirements Analysis as a Socially Mediated Process. Requirements Engineering 5, 74–82 (2000)
27. Group Strategic Framework (GSF), BAE Systems (2009), http://www.annualreport08.baesystems.com/en/strategy/strategic-overview.aspx

Author Index